INTEGRATED PEST MANAGEMENT IN DIVERSE CROPPING SYSTEMS

INTEGRATED PEST MANAGEMENT IN DIVERSE CROPPING SYSTEMS

Edited by
Devendra Pal Singh, PhD

APPLE ACADEMIC PRESS

First edition published 2023

Apple Academic Press Inc.
1265 Goldenrod Circle, NE,
Palm Bay, FL 32905 USA

760 Laurentian Drive, Unit 19,
Burlington, ON L7N 0A4, CANADA

CRC Press
6000 Broken Sound Parkway NW,
Suite 300, Boca Raton, FL 33487-2742 USA

4 Park Square, Milton Park,
Abingdon, Oxon, OX14 4RN UK

Library and Archives Canada Cataloguing in Publication

Title: Integrated pest management in diverse cropping systems / edited by Devandra Pal Singh, PhD.

Names: Singh, D. Pal (Agriculturist), editor.

Description: First edition. | Includes bibliographical references and index.

Identifiers: Canadiana (print) 20220407959 | Canadiana (ebook) 20220407983 | ISBN 9781774911167 (hardcover) | ISBN 9781774911174 (softcover) | ISBN 9781003304524 (ebook)

Subjects: LCSH: Agricultural pests—Control. | LCSH: Phytopathogenic microorganisms—Control.

Classification: LCC SB601 .I58 2023 | DDC 632—dc23

Library of Congress Cataloging-in-Publication Data

Names: Singh, Devendra Pal, editor.

Title: Integrated pest management in diverse cropping systems / edited by Devandra Pal Singh.

Description: First edition. | Palm Bay, FL, USA : Apple Academic Press, 2023. | Includes bibliographical references and index. | Summary: "Climate change has intensified in recent decades, which has affected crop production as well as facilitated the emergence of new diseases and insect pests, causing serious threats to agriculture. Farmers have mostly taken a crop-based approach to insect pest management (IPM); the authors of this new volume, Integrated Pest Management in Different Cropping Systems, however, take the unique approach that IPM based on specific cropping systems is more efficient, resulting in reduced cultivation costs, increased yield and profitability, and decreased residue from crop produce and products. This volume presents the results of research done by crop protection scientists on integrated pest management in diverse cropping systems based on rice, wheat, maize, pulses, food legumes, oilseeds, groundnut, potato, and other horticulture crops. With chapters written by well-known and experienced scientists in their fields, this volume provides in-depth knowledge on integrated pest management in conjunction with an array of specific cropping systems, taking into consideration all the elements, including the crops, crop sequences, spatial and temporal aspects of managing an agricultural system, and other aspects. This volume will be valuable for entomologists, plant pathologists, agronomists, as well as for farmers-both small and industrial sized, agricultural extension centers, faculty and students, and many others involved with crop cultivation"-- Provided by publisher.

Identifiers: LCCN 2022038773 (print) | LCCN 2022038774 (ebook) | ISBN 9781774911167 (hardback) | ISBN 9781774911174 (paperback) | ISBN 9781003304524 (ebook)

Subjects: LCSH: Insect pests--Integrated control. | Pests--Integrated control. | Plants, Protection of.

Classification: LCC SB950.A3 I58 2023 (print) | LCC SB950.A3 (ebook) | DDC 632/.7--dc23/eng/20220829

LC record available at https://lccn.loc.gov/2022038773

LC ebook record available at https://lccn.loc.gov/2022038774

ISBN: 978-1-77491-116-7 (hbk)
ISBN: 978-1-77491-117-4 (pbk)
ISBN: 978-1-00330-452-4 (ebk)

About the Editor

Devendra Pal Singh, PhD

Devendra Pal Singh, PhD, was formerly Principal Scientist (Plant Pathology) and Principal Investigator (Crop Protection Programme) at the ICAR-Indian Institute of Wheat and Barley Research, Karnal, India (1991–2019), dealing with integrated pest management. In addition, he also worked on bast fiber crops, food legumes, rice, maize, millets, fruits, vegetable crops, and aromatic plants at different institutes. He is currently working as Senior Consultant at Weather Risks Management Services, Gurugram, India. Dr. Singh served as a United Nations (V) Specialist in Zambia on a United Nations Development Programme/Food and Agriculture Organization project; as a visiting scientist at Université Catholique de Louvain, Louvain-la-Neuve, Belgium; as an Indian National Science Academy fellow working at the Nepal Agricultural Research Council; as an Indian Technical and Economic Cooperation expert in Guyana at the National Agricultural Research and Extension Institute, Mon Repos; and as Coordinator at the Rice Research Station, Guyana Rice Development Board, Guyana. He has published over 118 research papers and edited three books. He is fellow of the Indian Phytopathological Society, Society of Mycology and Plant Pathology, and Indian Society of Plant Pathologists as well as on the editorial boards and panels of reviewers for several international journals in plant pathology, including those published by the American Phytopathological Society. Dr. Singh earned his PhD in Plant Pathology from G. B. Pant University of Agriculture & Technology, also known as Pantnagar University, Pantnagar, India.

Contents

Contributors

T. Aravind
Department of Plant Pathology, College of Agriculture, Govind Ballabh Pant University of Agriculture and Technology, Pantnagar, India

K. Srinivasa Babu
ICAR-Indian Institute of Millets Research, Rajendranagar, Hyderabad 500030, India

Sanjay M. Bandi
ICAR-Indian Institute of Pulses Research, Kanpur, India

Amitava Basu
Department of Plant Pathology, Bidhan Chandra Krishi Viswavidyalaya, Mohanpur, Nadia, West Bengal, India; E-mail: basuamitava1961@gmail.com

Biswaranjan Behera
ICAR-Indian Institute of Water Management, Bhubaneswar 751023, Odisha, India

Pooja Bhatt
Department of Plant Pathology, College of Agriculture, Govind Ballabh Pant University of Agriculture and Technology, Pantnagar, India

Ashok Bhattacharyya
Former Vice Chancellor of Assam Agricultural University (AAU), Presently Vice Chancellor, Assam Women University cum Director of Research (Agric.), AAU, Jorhat 785013, Assam, India

Parul Dobhal
Department of Entomology, College of Agriculture, Govind Ballabh Pant University of Agriculture and Technology, Pantnagar, India

P. Duraimurugan
ICAR-Indian Institute of Oilseeds Research, Rajendranagar, Hyderabad 500030, Telangana State, India; E-mail: duraimuruganp@rediffmail.com

Pranab Dutta
School of Crop Protection, College of Post Graduate Studies in Agricultural Sciences, Central Agricultural University, Umiam 793103, Meghalaya, India; E-mail: pranabdutta74@gmail.com

B. Gangaiah
ICAR-Indian Institute of Millets Research, Hyderabad, Telangana, India

Harish Gangaiah
ICAR-Directorate of Groundnut Research, Junagadh, Gujarat 362001, India

H. S. Gaur
School of Agricultural Sciences, Sharda University, Greater Noida 201306, India; E-mail: hs.gaur@sharda.ac.in

Himani Jeena
Department of Plant Pathology, College of Agriculture, Govind Ballabh Pant University of Agriculture and Technology, Pantnagar, India

Rupak Jena
ICAR-Directorate of Groundnut Research, Junagadh, Gujarat 362001, India

Gururaj Katti
ICAR-Indian Institute of Rice Research, Rajendranagar, Hyderabad, India;
E-mail: gururajkatti59@gmail.com

A. Kalaisekar
ICAR-Indian Institute of Millets Research, Hyderabad, Telangana, India

U. K. Khare
Department of Plant Pathology, College of Agriculture, Jawaharlal Nehru Krishi Vishwa Vidyalaya, Jabalpur, M.P., India

M. K. Khokhar
ICAR-National Research Centre for Integrated Pest Management, Pusa Campus, New Delhi 110012, India

Amitava Konar
Department of Plant Pathology, Bidhan Chandra Krishi Viswavidyalaya, Mohanpur, Nadia, West Bengal, India

R. Mahender Kumar
ICAR-Indian Institute of Rice Research, Rajendranagar, Hyderabad, India

Narendra Kumar
ICAR-Indian Institute of Pulses Research, Kanpur 208024, Uttar Pradesh, India;
E-mail: nkumar.icar@gmail.com

Sanjeev Kumar
Department of Plant Pathology, College of Agriculture, Jawaharlal Nehru Krishi Vishwa Vidyalaya-Jabalpur, M.P.; India; E-mail: sanjeevcoa@gmail.com

Arti Kumari
Department of Plant Pathology, Assam Agricultural University, Jorhat 785013, Assam, India

Ananth Kurella
ICAR-Directorate of Groundnut Research, Junagadh, Gujarat 362001, India

Nataraja V. Maheshala
ICAR-Directorate of Groundnut Research, Junagadh, Gujarat, 362001, India

P. D. Meena
ICAR-Directorate of Rapeseed-Mustard Research, Bharatpur 321303, Rajasthan, India;
E-mail: director.drmr@gmail.com

P. N. Meena
ICAR-National Research Centre for Integrated Pest Management, Pusa Campus, New Delhi 110012, India

R. P. Meena
ICAR-Directorate of Medicinal and Aromatic Plants Research, Boriavi, Anand, Gujarat 387310, India

R. K. Mishra
ICAR-Indian Institute of Pulses Research, Kanpur, India; E-mail: rajpathologist@yahoo.com

Kancheti Mrunalini
1ICAR-Indian Institute of Pulses Research, Kanpur-208024, Uttar Pradesh, India;
E-mail: mrunalinikancheti@gmail.com

S. Nakkeeran
Department of Plant Pathology, Tamil Nadu Agricultural University, Coimbatore 641003, India;
E-mail: nakkeeranayya@tnau.ac.in

Ch. Padmavathi
ICAR-Indian Institute of Rice Research, Rajendranagar, Hyderabad, India

Neetu Pandey
Mewar University, NH-79, Gangarar, Chittorgarh, Rajasthan 312901, India

Girija Prasad Patnaik
Odisha University of Agriculture & Technology, Bhubaneswar 751003, Odisha, India

Durga Prasad
Department of Plant Pathology, Banda University of Agriculture and Technology (BUAT), Banda, Uttar Pradesh, India

G. Shyam Prasad
ICAR-Indian Institute of Millets Research, Rajendranagar, Hyderabad 500030, India;
E-mail: shyam@millets.res.in

D. Raghavendra
ICAR-National Research Centre for Integrated Pest Management, Pusa Campus, New Delhi 110012, India

P. K. Rai
ICAR-Directorate of Rapeseed-Mustard Research, Bharatpur 321303, Rajasthan, India

P. Renukadevi
Department of Plant Pathology, Tamil Nadu Agricultural University, Coimbatore 641003, India

Satyajit Roy
ICAR-Directorate of Medicinal and Aromatic Plants Research, Boriavi, Anand, Gujarat 387310, India;
E-mail: sroycob1959@gmail.com

R. Saravanan
Department of Plant Pathology, Tamil Nadu Agricultural University, Coimbatore 641003, India

J. C. Sekhar
Winter Nursery Centre, ICAR-Indian Institute of Maize Research, Rajendranagar, Hyderabad 500030, India; E-mail: jcswnc@rediffmail.com

U. P. Shahi
Department of Soil Science and Agricultural Chemistry, Sardar Vallabhbhai Patel University of Agriculture and Technology (SVPUAT)-Meerut, Uttar Pradesh, India

O. P. Sharma
ICAR-NCIPM, LBS Building, Pusa Campus, New Delhi 110012, India; E-mail: opsharmadelhi@gmail.com

K. T. Shivakumara
ICAR-Directorate of Medicinal and Aromatic Plants Research, Boriavi, Anand, Gujarat 387310, India

Devendra Pal Singh
ICAR-Indian Institute of Wheat and Barley Research, Karnal 132001, India; E-mail: dpkarnal@gmail.com

K. P. Singh
Department of Plant Pathology, College of Agriculture, Govind Ballabh Pant University of Agriculture and Technology, Pantnagar, India

Monika Singh
Mewar University, NH-79, Gangarar, Chittorgarh, Rajasthan 312901, India

Ramji Singh
Department of Plant Pathology, Sardar Vallabhbhai Patel University of Agriculture and Technology (SVPUAT)-Meerut, India; E-mail: singh.ramji@gmail.com

N. P. Singh
ICAR-Indian Institute of Pulses Research, Kanpur, India

P. Lakshmi Soujanya
Winter Nursery Centre, ICAR-Indian Institute of Maize Research, Rajendranagar, Hyderabad 500030, India

B. Sreenayana
Department of Plant Pathology, Tamil Nadu Agricultural University, Coimbatore 641003, India

J. Stanley
ICAR-Indian Institute of Millets Research, Hyderabad, Telangana, India

B. Subbarayudu
ICAR-Indian Institute of Millets Research, Hyderabad, Telangana, India

S. B. Suby
ICAR-Indian Institute of Maize Research, Pusa Campus, New Delhi 110012, India

M. Sujatha
ICAR-Indian Institute of Oilseeds Research, Rajendranagar, Hyderabad 500030, Telangana State, India

Kumari Surbhi
Department of Plant Pathology, College of Agriculture, Govind Ballabh Pant University of Agriculture and Technology, Pantnagar, India; E-mail: surbhi.gbpuat2018@gmail.com

Radhakrishnan Thankappan
ICAR-Directorate of Groundnut Research, Junagadh, Gujarat 362001, India;
E-mail: radhakrishnan.nrcg@gmail.com

Ajay Tomer
Institute of Plant and Microbial Biology (IPMB), Academia Sinica, Taipei, Taiwan

S. Vinodkumar
Department of Plant Pathology, Tamil Nadu Agricultural University, Coimbatore 641003, India

H. S. Viswanath
Department of Plant Pathology, Sardar Vallabhbhai Patel University of Agriculture and Technology (SVPUAT)-Meerut, India

K. R. Yathish
Winter Nursery Centre, ICAR-Indian Institute of Maize Research, Rajendranagar, Hyderabad 500030, India

Abbreviations

AB	Alternaria blight
ABA	abscisic acid
AESA	Agro-Ecosystem Analysis
BCAs	biocontrol agents
BGM	Botrytis Grey Mold
BPH	brown plant hopper
CA	conservation agriculture
CGD	citrus greening disease
CLCV	cotton leaf curl virus
DAP	di-ammonium phosphate
DHA	dehydrogenase activity
DRR	dry root rot
EDB	ethylene dibromide
EIL	economic injury level
ELS	early leaf spot
EPNs	entomopathogenic nematodes
ETL	economic threshold level
FAC_S	fatty acid amino conjugates
FAW	fall armyworm
FFS	Farmer Field School
FYM	farm yard manure
GAP	good agricultural practices
GIS	Geographical Information Systems
GOX	glucose oxidase
HIPVs	herbivore-induced plant volatiles
HLB	Huanglongbing
HPR	host plant resistance
ICM	integrated crop management
IDM	integrated disease management
IFS	integrated farming systems
IGP	Indo-Gangetic Plains
IJs	infective juveniles
INM	integrated nutrient management
IPM	integrated pest management

JA	jasmonic acid
LAMP	loop-mediated isothermal amplification
LER	land equivalent ratio
LFDs	lateral flow devices
LLS	late leaf spot
LMO	live modified organisms
MABC	marker assisted backcrossing
MYMV	mungbean yellow mosaic virus
NE	North Eastern
NGS	next generation sequencing
NOP	National Organic Program
NPV	nuclear polyhedrosis virus
NSKE	neem seed kernel extract
PAL	phenylalanine ammonia lyase
PGPR	plant growth-promoting rhizobacteria
PIs	protease inhibitors
PLRV	potato leaf roll virus
PM	powdery mildew
PMTV	potato mop-top virus
PSB	pink stem borer
PSB	Phytophthora stem blight
PTM	potato tuber moth
RAD	restriction-site associated DNA
RKN	root knot nematode
RW	rice-wheat
RWCS	rice-wheat cropping system
SA	salicylic acid
SAR	systemic acquired resistance
SC	suspension concentrate
SR	Sclerotinia rot
SMBC	soil microbial biomass carbon
SOC	soil organic carbon
SOM	soil organic matter
SRI	system of rice intensification
SSB	spotted stem borer
ToT	transfer of technology
TPM	total pest management
WBPH	white backed plant hopper
YSB	yellow stem borer

Foreword 1

There are ever-increasing food demands to feed the burgeoning population, with shrinking natural resources like land, water, and other resources, which necessitates multiple cropping for crop intensification practices in both dimensions, vertical as well as horizontal. Crop health is an important and prime requirement to have full benefit of multiple cropping. Thus, it is pertinent to study pests over the entire cropping period, and the pests of one crop may be influenced by the previous crop and also by companion crops under intercropping system. A part of pest management research in multiple cropping systems may be predictive in nature. Therefore, development of pest management technology for multiple cropping systems must be the in the lines of resources available to a farmer. Many variables in crops in the cropping system and their effects on pests need to be determined in the situations of introduction of new crop species, replacing old varieties with new high yielding varieties, and arranging crops in time and space. There is, therefore, a regular need for research on pest management in multiple cropping systems that should take into consideration crop-crop interactions, under intercrop, relay, or sequential, and the performance of individual crops. The global movement of serious pests like fall army worm, locust, rust pathotypes in wheat, wheat blast, etc. along with changes in climate and cropping systems has further demanded coordinated research on survey and surveillance as well as research on integrated management of pests.

I take this opportunity to congratulate Dr. Devendra Pal Singh, editor of the book, and all the contributors of chapters for bringing out new knowledge and technologies for integrated pest management in different cropping systems based on cereals, food legumes, oilseeds, and horticultural crops. The book is unique in the sense that it covers the entire cropping systems and will save resources and environment while achieving sound crop health and finally higher yields and better produce quality. The publication is the need of the hour, and IPM technologies given need to be quite relevant to field situations and act as stimulus in achieving food and nutritional security using an agroecological approach to growing crops with the most economic use of natural resources.

—**Prof. J. S. Sandhu**
Vice Chancellor
Sri Karan Narendra Agriculture University,
Jobner, District - Jaipur, Rajasthan 303325, India

Foreword 2

Integrated pest management (IPM) is used as a tool in the production of healthy food. It is a strategic approach to manage crop pests through minimizing pest populations by combining environment-friendly pest control methods and economically viable farming practices. However, the approach should be to keep these interventions to a minimum, without sacrificing efficacy to make them more farmer-friendly and more cost effective. Also IPM is more relevant if applied in contiguous areas in a cropping system. Therefore, it is highly desirable to evolve IPM technologies under different cropping systems and in different agroecological conditions. The ease of global movement of plant pathogens and insect pests through planting material, human activities, and wind is posing new global problems of pests and threats to food and nutritional security in many countries.

The compilation of different chapters written by eminent experts in the field of IPM in the present form of this book is, therefore, highly welcoming and will serve different stakeholders in the food value chain. System-based IPM is a comparatively new area, and information is still evolving. The IPM technologies discussed will be quite useful in minimizing pest invasion, losses, and spread. They will also reduce the pesticide residue in crop produce and save resources as well as provide a safer environment.

I congratulate the contributors and the editor in doing a commendable job. I am quite confident that the book will be highly beneficial to those involved in crop and seed production, pest management, extension workers, and students.

—Prof. Dr. Uma Shankar Singh
South Asia Advisor for Research & Partnerships
International Rice Research Institute (IRRI)
IRRI-India Office
NASC Complex, DPS Marg
Pusa Campus, New Delhi 110012
India

Preface and acknowledgments

Healthy crops are the prime requirement for getting the best quality produce and enough food to feed the ever-increasing human population. Over time food habits have changed and still continue to change, thus posing new challenges in the field of agriculture. This has resulted in changes in kind and intensity of cropping systems. In addition, climate change has intensified in recent decades, affecting crop production and new diseases and insect pests that cause serious threats to agriculture.

The information on pests and their integrated management is mostly available for different crops but not in the context of cropping systems. Hence, there is a need for compiling the available literature and results of research done by crop protection scientists on integrated pest management (IPM) in different cropping systems dealing with (1) cropping systems and IPM, and crops that are; (2) cereals based, (3) food legumes based, (4) oilseeds based, and (5) horticulture crops based. The chapters in this volume have been written by well-known and experienced scientists in their fields and provide in-depth knowledge on the theme. It had been a pleasant opportunity to bring entomologists, plant pathologists, and plant namatologists to one platform and bring the integrated pest management modules keeping in view of insect pests and diseases of each cropping system. The information thus generated is unique and will be quite applicable for achieving healthy crops and managing any outbreak of diseases and insect pests effectively.

This book was written to serve those who deal with integrated disease and insect pest management in different cropping systems and will serve as a reference book for students, teachers, and researchers in the fields crop protection, seed production, and plant quarantine.

My grateful thanks to all eminent authors for doing the good job and agreeing to contribute or the theme of book inspite of short notice. My special thanks to Professor J. S. Sandhu and Dr. U. S. Singh for writing forward. My appreciations to Mr. Rakesh Kumar and Mr. Ashish Kumar for arranging the publication of book. I convey my heartful thanks to my family (Sadhna, Anubhav, and Surbhi) for help and encouragements and late Shri Sonuji, MD, WRMS for support.

—Devendra Pal Singh
Editor

CHAPTER 1

Efficient Diversified Cropping Systems of Field and Horticultural Crops for Livelihood Security

NARENDRA KUMAR[1*], KANCHETI MRUNALINI[1*],
GIRIJA PRASAD PATNAIK[2], and BISWARANJAN BEHERA[3]

[1]*ICAR-Indian Institute of Pulses Research, Kanpur 208024, Uttar Pradesh, India*

[2]*Odisha University of Agriculture & Technology, Bhubaneswar 751003, Odisha, India*

[3]*ICAR-Indian Institute of Water Management, Bhubaneswar 751023, Odisha, India*

[*]*Corresponding author. E-mail: nkumar.icar@gmail.com; mrunalinikancheti@gmail.com*

ABSTRACT

The cropping system refers to the yearly sequence, spatial arrangement of crops, their interactions with farm resources, other farm enterprises. There are several factors that affect determine the cropping system of a region such as prevailing climate, soil type, irrigation facilities, market demand, farmers socioeconomic condition, government policy, etc. But in general, farmers prefer to accommodate large numbers of crops on their lands to meet out family food requirements, also to reduce the risk of aberrant weather during crop season, fluctuating market prices. In India, there are about 250 different types of cropping systems, in which 30 are being the most common. Large variability in cropping systems exists in rainfed dryland regions, with

Integrated Pest Management in Diverse Cropping Systems. Devendra Pal Singh, PhD (Ed.)

intercropping being more common. Furthermore, cropping systems have a significant impact on the sustainability of conservation agriculture.

1.1 INTRODUCTION

The ever-increasing global population strives for a holistic approach to ensure food security. Besides, climate change further adds fuel to the fire by adversely affecting crop, fish, and livestock productivity.[1] The climatic vagaries, associated risks in crop harvesting, instable market prices of farm produce urging for a cropping system approach, which is potentially holistic in nature. The cropping system approach can be a breakthrough for getting sustainability in production, ensuring environmental security. The concepts underpinning the cropping system have the potential to revolutionize food production and socioeconomics of farm family.[2] Cropping system is one of three agricultural system components that depicts the cropping pattern (sequence of crops on a piece of land), as well as its relationship with the other two components (livestock, soil, water, and household resources).[3] A recent projection shows that the population in India is expected to increase by 25%, as compared to 2011, to 1.52 billion by 2036. As per the vision 2030 documents of ICAR, the foodgrain demand of the country will increase to 345 Mt by 2030, which will have to be met from a stagnated net cultivated area. This could be achieved by increasing crop productivity/or cropping intensity. Cropping intensity must be enhanced by increasing the quantity of crops grown per unit land per unit time, leaving no resources unutilized or underutilized. There have been many cropping systems developed for utilizing the available resources more efficiently that can serve the purpose of ensuring food, income security of farm family. This chapter encompasses the basic concept of cropping system, various types of cropping systems, advantages of the cropping system, major cropping systems (both field crop-based, horticulture crops based) that have been evolved, followed in India under different soil type, climatic conditions of that locale.

A crop system is a component of a cropping system in which the plant population is organized in such a way that solar energy, nutrients, water, and other inputs are efficiently transformed into useable biomass. Soils, crops, weeds, pathogens, and insects are all part of the crop system. The crop can be of various species and varieties, but if managed as a unified unit, it only constitutes one-crop system. The cropping system is made up of these crop systems and their interactions with other farm resources. Cropping systems in agriculture, such as crop diversification, crop rotation, intercropping, and other agronomic practices, have a variety of spatial and temporal effects on soil health and quality. Cropping systems are basically designed to maximize agro-system productivity; however, modern agriculture is greatly concerned with cropping systems' environmental sustainability. Soil health maintenance aims to ensure long-term high productivity and environmental sustainability of cropping systems by evaluating five fundamental functions: nutrient cycling, water interactions, biodiversity and habitat, filtering, buffering, physical stability, and support.

1.2 BENEFITS OF CROPPING SYSTEM OVER CROP SYSTEM

The cultivation of several crops (together or in a sequence) on a piece of land in a year has direct (due to mutual interaction) or indirect (due to interaction with farm resources) effects on subsystems of the crop systems.

1.2.1 IMPROVED SOIL PROPERTIES

In a cropping system, some crops like cereals may exhaust soil nutrients, whereas some crops like legumes may restore soil fertility. The cultivation of diverse group of crops helps in maintaining a higher production level through balanced utilization of soil nutrients. Some cover crops like groundnut, cowpea, green gram, etc., add a huge quantity of biomass/organic matter in soil that improve soil physico-chemical and biological properties.[4,5] The deep-rooted crops like pigeonpea, cotton breaks hard pan, facilitates nutrient, water movement to upper soil layer that helps in the establishment of shallow-rooted crops in the system.[6]

1.2.2 ENHANCED CROP GROWTH

The crop growth is enhanced in a cropping system due to the derived mutual benefits. The residues of previous crop in a cropping system may improve the winter survival of crops like wheat in subzero temperature.[7] Some crops can act as windbreak and can reduce the lodging of the companion crops.

1.2.3 MINIMIZED SPREAD OF DISEASES

Unlike monoculture, crop rotation in the cropping system helps in reducing the incidence of pest, diseases. Lesser the duration of crop, lesser would be the chance for disease susceptibility. Brooks et al.[8] found 3-year rotations of rice, soybean, and maize were more effective in suppressing false smut and kernel smut in rice.

1.2.4 REDUCED WEED GROWTH

Rotating crops in a cropping system help to prevent the building up of weeds. For example, in rice–wheat (RW) cropping system and growing berseem in place of wheat in every 2–3 years help in suppressing little canary grass (*Phalaris minor*) and chicory (*Cichorium intybus*). Growing trap crops/catch crops in a cropping system help in suppressing parasitic weeds like *Striga* and *Orobanchae*.

1.2.5 REDUCED INSECT PEST ATTACK

The insect pest attack can be reduced by adopting push–pull strategy in the cropping system. This strategy was initiated in Nairobi and Kenya in which insect repellent crops (push crops) like *Desmodium* are grown with main crop like maize or sorghum. Napier grass is grown on the border of the field as insect attractant crop (pull crop) to manage stalk borer. In addition, crop rotation reduces insect pest attacks in the cropping system. Francis et al.[9] coined the term "biological structuring" to describe the use of crop rotation and management alternatives and biological phenomena and environmental conditions and interactions of these factors to manage crop pests, such as diseases, weeds, and insects. Reynolds et al.[10] found that rice crop rotation with barley, wheat, mungbean, and black gram reduced the incidence of *Pratylenchus indicus* in rice.

1.2.6 SOIL MOISTURE CONSERVATION

Diversified crops help in covering the ground with more foliage and capture the solar radiation more efficiently. Some early growing pulse crop such as cowpea grows quickly, cover the ground, and conserve soil and water.

1.2.7 EFFICIENT UTILIZATION OF RESOURCES

Available resources can be utilized more efficiently as there are diverse groups of crops in a cropping system. To reduce the possibility of nutritional deficits or drought situations, above-ground resources like sunlight and underground resources such as water and nutrients are being utilized evenly in the system. Other resources like farm machinery, labor, and animal draft are properly utilized as the efforts spent on planting and harvesting crops are more spread out over the harvesting period. An increase in productivity and energy use efficiency and nutrient uptake due to increase in cropping intensity was also observed.

1.2.8 REDUCED RISK OF CROP FAILURE

Crop weather is responsible for about 70% of the variability in crop productivity. During aberrant weather conditions, growing diverse groups of crops in a cropping system can reduce the risk of total crop failure. If one crop is failed, then there is chance to get monetary return from other crops which are least affected. Particularly, under mixed cropping and the system is more resilient toward weather abnormalities.

1.2.9 IMPROVED FOOD AND FINANCIAL SECURITY

Choosing the more diverse number of crops will lead to a more regular food production around the year. There would be a greater reliability on food production and income generation with a lower risk of crop failure. Multiple cropping has a potential to generate employment for unskilled rural people throughout the year.

1.3 MAJOR FIELD CROP-BASED CROPPING SYSTEMS IN INDIA

The cropping system of a region is generally decided by many factors, such as agro-climatic condition, available resources, crop productivity, profitability, consumer preference, marketing facility, the interaction between ecologically practicable crops, present methods of pest-and-disease control, etc. Because of its diverse agro-climatic zones, India has a place for a variety of cropping systems.

Sequential Cropping System

Sequential multiple cropping, using short duration crop cultivars, intensive input management, is a common way of increasing land-use efficiency and especially under irrigated ecosystems. Food crops are grown in the most amenable seasons and thus a particular crop is the basis of a cropping system. Sequential cropping is instead called nonoverlapping cropping systems. For describing a *kharif* season cropping system, rice, sorghum, pearl millet, maize, groundnut, and cotton may be considered as a base crop, whereas in *rabi* season wheat, chickpea, and mustard lentil may be considered.

1.3.1 RICE-BASED CROPPING SYSTEM

Rice is cultivated in areas of high rainfall or in areas where supplemental irrigation ensures a higher yield. However, only 9% of India's rice area receives more than 30 cm/month of rainfall during the entire growing period. Nearly half of India's total rice acreage receives 30 cm of rain per month for at least 2 months (July and August) during the south-west monsoon and lesser during other months. Conventionally and planting of long duration single crop of rice is practiced. On the other hand, due to photo-insensitive, short-duration varieties, multiple cropping is feasible in the rice-based cropping system. The eastern and southern states, including Odisha, Assam, West Bengal, coastal, most of Karnataka, Tamil Nadu, and Kerala, receive a fair amount of rainfall amenable for rice production. Around 30 rice-based cropping systems have been identified in India. The major rice-based cropping systems in different regions of India have been shown in Figure 1.1.

Monoculture of Rice

In the monoculture of rice and the crop is repeatedly grown twice to thrice a year. In a triple rice system and the land is cropped at least 10 months in a year with three crops of rice. The rice can be grown all-round the year in areas having assured irrigation facilities. Water stress-related issues generally do not exist in these areas. Dry season crops are more productive due to better solar radiation utilization, mere natural calamities like cyclones and heavy rainfall and strong winds. The double rice system, having two rice crops in a year and is mostly, found in many conventionally rice-growing areas. The first rice crop is short duration and grown during *kharif* season and followed by a *rabi* rice crop.

Similarly, single rice crop with medium or long duration varieties is most prevalent in canal command areas. The cultivation in these areas is restricted to the water availability in the canals. Rice monoculture is practiced in a few sections of Odisha, Andhra Pradesh, Karnataka, and Tamil Nadu river command areas.

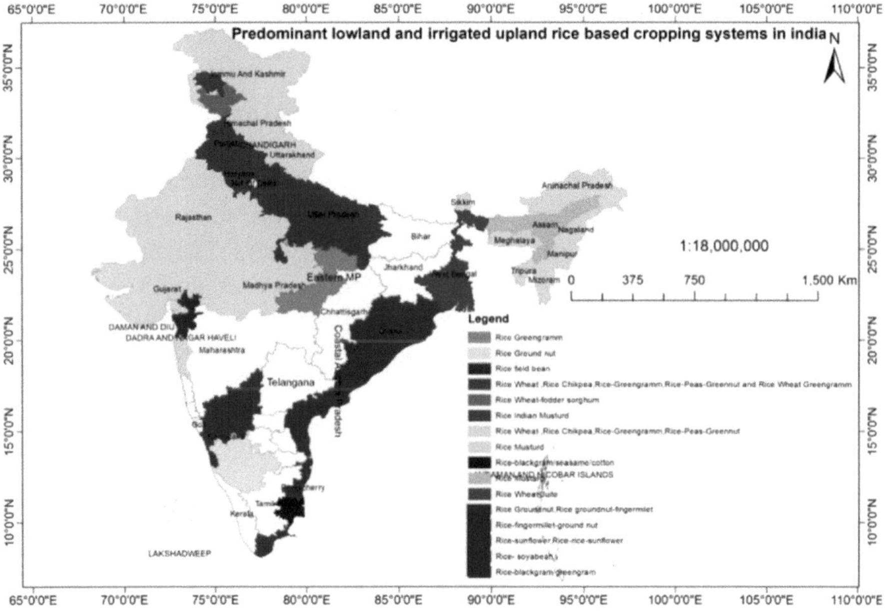

FIGURE 1.1 Major rice-based cropping systems in India.

Rice-Upland Crop Sequence

In the places with low rainfall, only a single rice crop is cultivated during the peak rainy season. A fast-maturing annual crop is cultivated between the first monsoon rain and the transplanting of rice in certain regions. Vegetables including green corn, cowpea, and green soybean, as well as green manure crops, are especially well-suited to these conditions. It is usually possible to grow another crop soon after the primary rice crop has been harvested. Rice is harvested at the end of the rainy season when the soil is still wet and there is plenty of time to use the remaining moisture. Green gram, black gram, cowpea, soybean, and other crops are good candidates for using residual soil moisture in a rice-upland crop sequence.

Kharif Cereals Other Than Rice

Maize, sorghum, and pearl millet are the main *kharif* cereals, whereas ragi and small millets are grown in limited patches. Maize is commonly grown in larger pockets where rainfall is high, and soil has higher water retaining capacity. Under medium rainfall region, sorghum is grown, whereas ragi is grown under low rainfall areas with light-textured soils. Though these crops are grown all over the western, northern, southern India, the regions of these cropping patterns are demarcated well to the west of 80° latitudes, except maize. Finger millet is mainly grown in Karnataka, Tamil Nadu, and Andhra Pradesh which shares about 60% of the area under finger millet.

Maize is cultivated in sequence with other crops under various agro-ecologies because of its greater adaptability and compatibility under a variety of soil and climatic conditions. As a result, maize is seen as one of the agricultural diversification's potential drivers. Maize–wheat ranks first among diverse maize-based cropping systems, with 1.8 million hectares cultivated, mostly in rainfed ecologies. After RW and rice–rice, maize–wheat is the third most significant cropping system, accounting for around 3% of the national food basket. Uttar Pradesh has the most Kharif maize planted, followed by Rajasthan, Madhya Pradesh, and Bihar. A total of 12 maize-based cropping systems have been identified in India. The major maize-based cropping systems in different regions of India have been shown in Figure 1.2.

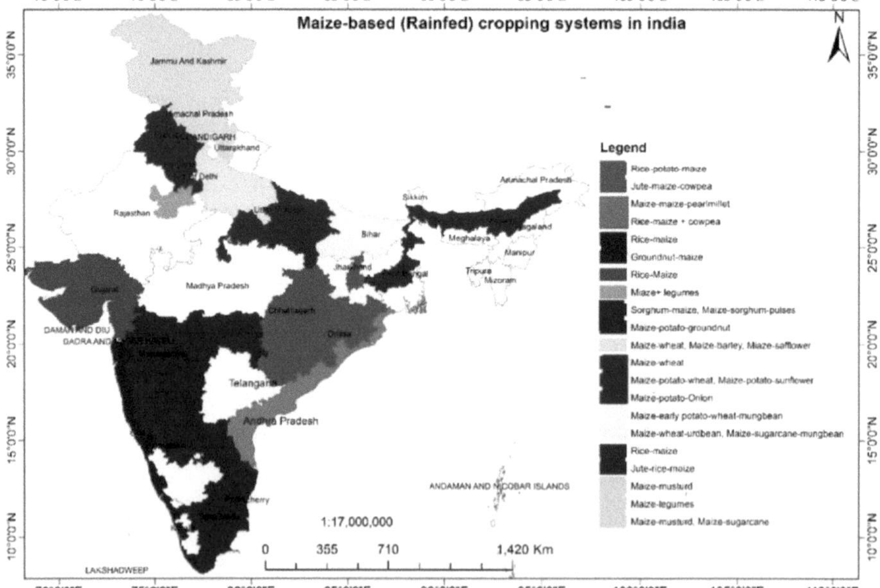

FIGURE 1.2a Major irrigated maize-based cropping system in India.

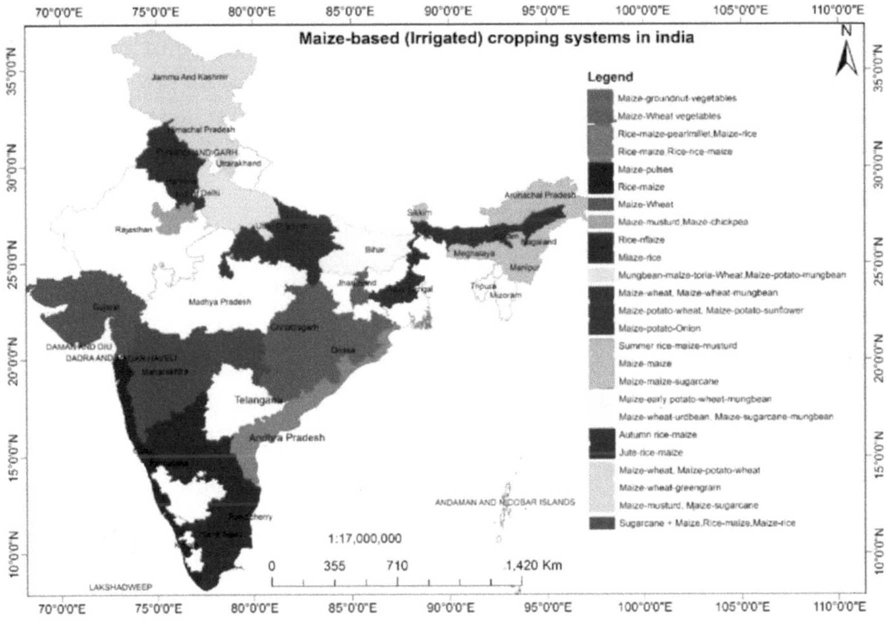

FIGURE 1.2b Major rainfed maize-based cropping system in India.

1.3.2 WHEAT-BASED CROPPING SYSTEMS

These areas are mainly confined to regions with low temperatures during winter. The RW sequential system is a dominant system throughout the wheat belt of India. The core of the wheat region and responsible for 70% of the area and 76% of production, comprises Punjab, Haryana, Delhi, Uttar Pradesh, Madhya Pradesh, Rajasthan, Gujarat, and Bihar. These areas have assured irrigation facilities. More than 19 wheat-based cropping patterns have been identified in India. The productivity of the wheat-based cropping system is higher than that of other systems (Figure 1.3).

1.3.3 OILSEED-BASED CROPPING SEQUENCES

India's average oilseeds crop productivity is 1168.9 kg/ha.[11] Due to its different agro-ecological circumstances, India has a large variety of annual oilseed crops. In India, nine annual oilseed crops are cultivated, including seven edible oilseeds (groundnut, rapeseed-mustard, soybean, sunflower, sesame, safflower, and Niger) as well as two nonedible crops (castor and linseed). Except for groundnut and mustard, the most of other oilseeds are grown as rainfed crops. The predominant oilseed-based cropping systems of India have been marked in Figure 1.4.

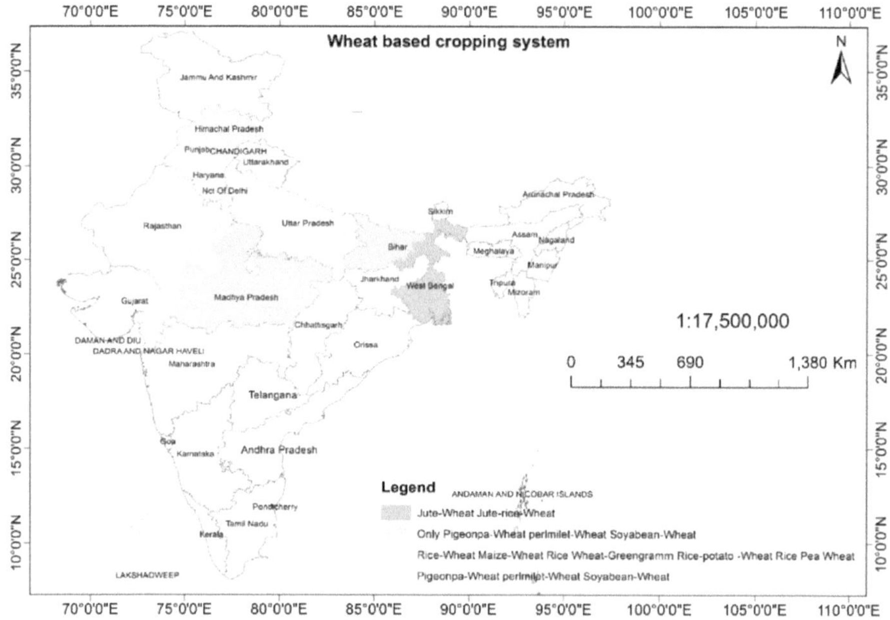

FIGURE 1.3 Major wheat-based cropping systems in India.

1.3.4 COTTON-BASED SEQUENTIAL CROPPING

Cotton is grown under 13 M ha and Maharashtra shares the highest area, followed by Gujarat, Andhra Pradesh, and Punjab. Altogether these four states contribute 70% of the area under cotton grown in the country. The maximum area under the cotton crop falls in the medium to high rainfall zone. Rainfed cotton is grown in the states like Madhya Pradesh, Maharashtra, and Karnataka, whereas partly irrigated cotton areas are available in Tamil Nadu and Gujarat. This crop can be well suited for intercropping with various short-duration crops as it is a long-duration crop.

1.4 INTERCROPPING SYSTEMS

In a region with a dry, humid, semiarid tropical climate, multiple cropping in the form of intercropping is suitable. Intercropping is when two or more crops are grown side by side on the same piece of land with specific row spacing. Intercropping major goals are to protect against total crop failure due to unusual weather or insect epidemics, boost total production per unit of land area, and make efficient use of resources including land, labor, and inputs. Intercropping is commonly used in places with rainfall ranging from 600 to 800 mm. Even if portion of the season is not ideal for other crops, at least one component of crops succeeds in providing economic yield in such a condition. Row ratios are optimized to provide minimal competition in order to achieve maximum biological production. Two crops are chosen to be cultivated together in such a way that their growth durations differ. Willey[12] defined temporal complementarity as the principle that the two crop species' peak growth periods should not coincide. More difference in crop maturity and growth demands leads to more resource use efficiency and productivity. Food crops are combined with cash crops to ensure the subsistence of farmers along with cash income.

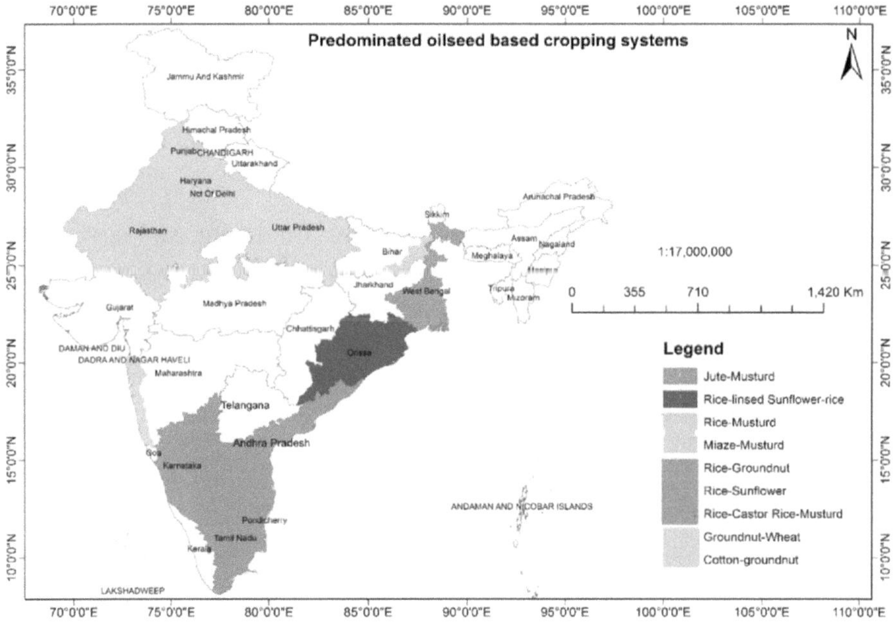

FIGURE 1.4 Predominant oilseed-based cropping systems in India.

Planting geometry plays an essential role in optimizing crop yield levels in intercropping systems. In a semiarid ecosystem and planting of intercrop at a normal spacing between two rows of a usual spaced main crop gives a higher yield than paired and skip row planting. Under a humid ecosystem, paired row planting may provide better results than usual and skip row planting.

1.4.1 CEREAL-BASED INTERCROPPING

The intercropping with cereals results in higher resource use efficiency. The cereal utilizes the rainy season resources while late-maturing crop exploits the postrainy season resources, that is, residual moisture. Sorghum can be included as component crop in an intercropping system with pigeonpea in varieties of soils. The sorghum is harvested at 105–135 DAS, whereas pigeonpea requires about 6–9 months to mature. Similarly, groundnut, grain legumes as cowpea, green gram, and black gram are cultivated with sorghum. Pearlmillet is also a good option as it is a quick tillering, fast-growing crop that develops a full canopy within 20–30 days. Sorghum can also be intercropped with black gram, groundnut, or castor. Fingermillet is also widely grown with field beans in Karnataka where 4–12 rows of millet are alternated with one row of beans.

1.4.2 PULSE-BASED INTERCROPPING SYSTEM

In the peninsular region of India, black gram, groundnut, and soybean are the best companion crops can be intercropped with pigeonpea. Sorghum is commonly grown as intercrop with pigeonpea. Some of the common intercropping systems involving pulses are pigeonpea + sorghum/black gram/green gram/pearl millet, chickpea + mustard/wheat, sorghum/pearl millet + cowpea/black gram/green gram and lentil + wheat, etc.

1.4.3 SUGARCANE-BASED INTERCROPPING SYSTEM

Sugarcane is slow-growing up to 80–90 days. Since sugarcane is planted in wide row spacing, that is, 80–100 cm and a wide range of short duration intercrops can be fitted into the space available. Black gram and soybean are found to be most suitable for sugarcane. Some of the major sugarcane-based intercropping systems of India are listed below (Table 1.1).

TABLE 1.1 Sugarcane-Based Intercropping Systems in India.

Region	Systems	Row ratio
Uttar Pradesh, Haryana, Punjab, Madhya Pradesh, Maharashtra	Sugarcane + maize	1 : 1
	Sugarcane + rajmash	1 : 1
	Sugarcane + lentil	1 : 1
	Sugarcane + mustard	1 : 1
Andhra Pradesh, Tamil Nadu, Karnataka	Sugarcane + green gram	1 : 2
	Sugarcane + black gram	1 : 2
	Sugarcane + cowpea	1 : 1
	Sugarcane + soybean	1 : 2
	Sugarcane + onion	1 : 2
Uttar Pradesh, Punjab, Haryana	Sugarcane—rice—potato	

The efficient intercropping systems along with row ratio for different rainfed areas are as follows:

A. Vertisols

Akola: Sorghum + pigeonpea (2:1), Cotton + green gram/cowpea (2:1), cotton + pigeonpea (2:1), Pigeonpea + green gram (1:1)

Indore: Maize + soybean (2:2), soybean + pigeonpea (2:2)

Arija: Maize + black gram (1:1, 2:1), groundnut + sesame (6:2)

Rajkot: Groundnut + castor (3:1), groundnut + pigeonpea (4:1), pearlmillet+ castor (1:1)

Jhansi: Sorghum + pigeonpea (2:1), pearlmillet + fodder legume (1:1)

Rewa: Sorghum + pigeonpea (2:1), soybean + pigeonpea (4:2)

Bijapur: Pearlmillet + pigeonpea (1:1), sunflower + pigeonpea (2:1), chickpea + safflower (3:1)

Solapur: Pearlmillet + pigeonpea (2:1), sunflower + pigeonpea (2:1, 2:2), chickpea + safflower (3:1), pearlmillet + moth bean (2:1, 3:1)

B. Alfisoils and red soils

Kovilpatti: Sorghum + black gram (2:1), sorghum + cowpea (2:1), cotton + black gram (2:1)

Anantapur: Groundnut + pigeonpea (7:1, 15:1), groundnut + cowpea/green gram (7:1)

Hyderabad: Sorghum + pigeonpea (3:1), castor + cluster been (2:2)

Bangalore: Fingermillet + pigeonpea (8:2), groundnut + pigeonpea (8:2)

Phulbani: Rice + pigeonpea (3:1, 4:1), castor + tuber crops (1:1), maize + pigeonpea (2:1), pigeonpea + radish (2:1)

Ranchi: Maize + pigeonpea (1:1), rice + pigeonpea (3:1), groundnut + pigeonpea (3:1).[13]

1.5 DOUBLE CROPPING

Double cropping has a clear possibility in places getting more than 800 mm of annual rainfall and a depth of stored soil moisture of around 200 mm/m. For rainfed regions of Odisha, Bihar, Madhya Pradesh, and Eastern Uttar Pradesh, double cropping is generally suggested.

1.6 HORTICULTURE-BASED CROPPING SYSTEM

1.6.1 VEGETABLE BASED

The importance of vegetables in nutrition security and increased farm income is very well recognized. The income from vegetable cultivation in per unit area and time is much higher than any of the field crops. In India, some parts of land are utilized by most of the farmers for vegetables cultivation either for their own consumption of for income through sale. Some of the regions are more suitable for the cultivation of vegetables like hills. The agriculture in periurban areas is also dominated by vegetable crops. Some of the locations specific vegetables-based crop rotations are described in Table 1.2.

1.6.2 FRUIT-BASED CROPPING SYSTEMS

Fruit-based cropping system is considered as an ideal strategy to provide food and nutrition and income security.[14] In this cropping system, natural resources (land, solar radiation, water, soil), socioeconomic inputs (labor, credit, power, market infrastructure) are efficiently utilized. Crop productivity under a tree canopy is enhanced due to improved soil fertility,[15] ameliorative effect on temperature, and evapotranspiration due to tree shade.[16] Some of the location-specific fruit-based cropping systems are described in Table 1.3.

TABLE 1.2 Some Location-Specific Vegetable-Based Crop Rotations in India.

States/region	Major vegetable-based cropping systems
Tarai Region, Uttarakhand, UP	(a) Colocasia–radish–French bean
	(b) Brinjal–vegetable pea–chili
	(c) Sponge gourd–palak–brinjal
	(d) Okra–tomato–amaranth
	(e) Cowpea–cauliflower–okra
	(f) Paddy–potato–tomato
	(g) Okra–carrot–bitter gourd
	(h) Round melon–radish–onion
Karnataka	(a) Okra–tomato–French bean
	(b) Tomato–French bean–okra
	(c) Brinjal–cauliflower–chili
	(d) Cowpea–cauliflower–okra
	(e) Paddy–tomato–bitter gourd
	(f) Okra–vegetable pea–onion
Eastern Uttar Pradesh	(a) Maize–potato–okra
	(b) Brinjal–vegetable pea–chili
	(c) Okra–tomato–bitter gourd
	(d) Bottle gourd–vegetable pea–tomato
	(e) Cowpea–cauliflower–onion
	(f) Paddy–radish–onion
Western UP	(a) Maize–potato–onion
	(b) Paddy–potato–onion
	(c) Maize–potato–okra
	(d) Okra–tomato–okra
MP	(a) Maize–potato–onion
	(b) Maize–potato–okra
	(c) Okra–tomato–cowpea
	(d) Cowpea–cauliflower–okra
	(e) Brinjal–vegetable pea–chili
	(f) Round melon–vegetable pea–onion
	(g) Potato–green gram–maize
	(h) Potato–cluster bean–green gram
	(i) Potato–okra–soybean
Vellanikkara, Kerala	(a) Paddy–tomato–okra
	(b) Paddy–tomato–chili

TABLE 1.2 *(Continued)*

States/region	Major vegetable-based cropping systems
	(c) Brinjal–watermelon–chili (d) Cowpea–chili–okra
	(e) Colocasia–tomato–bitter gourd
Hisar, Haryana	(a) Potato–green gram–maize
	(b) Potato–cowpea–okra
	(c) Potato–tomato–okra
	(d) Onion–bitter gourd–okra–potato
Tamil Nadu	(a) Cereals–tapioca
	(b) Tapioca–groundnut
	(c) Tapioca–black gram/green gram
	(d) Cereal–sweet potato–tapioca

TABLE 1.3 Fruit Crop-Based Intercropping Systems in Some States and Regions.

S. No.	States and regions	Cropping systems	References
1.	Andhra Pradesh	Mango + tomato	Dhulipalla[17]
		Mango + brinjal	
		Coconut + brinjal	Padma et al.[18]
		Coconut + cocoa	Rahim et al.[19]
		Coconot + banana, elephant foot yam, ginger, turmeric, cassava, vegetable cowpea	Devi et al.[20]
2.	Arunachal Pradesh	Apple + fodder (*Trifolium repens*)	–
3.	Assam	Coconut + patachauli (*Pogostemon cablin*)	Nath et al.[21]
		Cococnut + long pepper (*Piper longum*)	
		Coconut + cucumber	Nair et al.[22]
		Coconut–colocassia–betelvine–Assam lemon	Chowdhury and Deka[23]
		Coconut + turmeric + black pepper	AAU, Jorhat
		Potato—cucumber—ridge gourd	
		Potato—bitter gourd—okra	
		Tomato—late cauliflower—okra	
4.	Bihar	Litchi–banana	NRC on Litchi
		Potato (*Solanum tuberosum* L.)—green gram (*Vigna radiata* L)—rice (*Oryza sativa* L.)	Singh and Lal[24]
		Potato–maize (*Zea mays* L)—rice	
		Potato–onion (*Allium cepa* L)–rice	

TABLE 1.3 *(Continued)*

S. No.	States and regions	Cropping systems	References
5.	West Bengal	Guava + banana	Ghosh et al.[25]
		Guava + brinjal	
6.	Odisha	Guava + cowpea	Swain[26]
		Guava + Frenchbeans	
		Mango + guava + cowpea	Swain et al.[27]
		Mango + guava + mango ginger	
7	Kerala	Coconut (*Cocos nucifera*) + jamun (*Syzygium cumini*) coconut + mango (*Mangifera indica*)	Bhagya et al.[28]
8	Rajasthan	Ber + cluster bean	Gupta et al.[29]
		Ber + green gram	Saroj et al.[30]
		Ber + groundnut	
		Ber + aloe	Awasthi et al.[31]
		Aonla + moth bean–chickpea	Arya et al.[32]
		Aonla–ber–karonda–clusterbean–brinjal	Krishan et al.[33]
		Aonla + ber + cluster bean + fennel	
9	Haryana	Khejri (*Prosopis cineraria*) +	Puri et al.[34]
		Barley/jowar/mothbean/clusterbean	Kaushik and Kumar[35]
		Khejri + pearlmillet	
10	Hisar	Tomato + spinach, onion, radish	Singh[36]
11	Tamil Nadu	Turmeric + fenugreek (1:3)	Singh[36]
		Turmeric + onion	Nimbolkar et al.[37]
		Turmeric–bean/beetroot	Khan and Krishnakumar[38]
		Coconut + cocoa + banana + moringa + pineapple	Thiruvarssan and Maheswarappa[39]
		Coconut + blackpepper + cocoa + pineapple	
		Coconut + black pepper + banana + elephant foot yam	
		Turmeric + fenugreek (1:3)	Singh[36]
		Turmeric + onion	
		Turmeric–bean/beetroot	Nimbolkar et al.[37]
		Coconut + cocoa + banana +	Khan and Krishnakumar[38]

TABLE 1.3 *(Continued)*

S. No.	States and regions	Cropping systems	References
		Coconut + blackpepper + cocoa + pineapple	Thiruvarssan and Maheswarappa[39]
		Coconut + black pepper + banana + elephant foot yam	
		Moringa + pineapple	
12	Karnataka	Okra + beetroot, knol–khol, pea	Singh[36]
		Capsicum+ beetroot, knol–khol, pea	
		Chili + French bean, onion	Roy et al.[40]
		Coconut + cocoa + lime + banana + drumstick	
13	New Delhi	Mango + cowpea + Indian mustard	Mirjha and Rana[41]
14	Madhya Pradesh	Mango + cowpea (Kharif) + Bengal gram (Rabi)/mango + pigeonpea + tomato	Baghel et al.[42]

1.7 CONSERVATION AGRICULTURE

RW cropping system became one of the most important systems in India after "Green Revolution" which played a great role in achieving self-sufficiency in foodgrains production in the 1970s. The area, production of rice, and wheat increased tremendously in the country due to assured yield and less risk involved in its production. These crops have replaced the area under pulses, oilseeds, and coarse grains in Indo-Gangetic Plains (IGP). The maximum gain in area, production is recorded in wheat which increased from 12.93 M ha, 11.0 Mt during 1960–1961 to 24.17 M ha and 55.14 Mt during 1990. However, the area, production of pulses, and oilseeds remained almost stagnant. The continuous following of exhaustive system of RW cropping over the decades has leads to the degradation of natural resources to the extent that restrict their use efficiency to produce the level matching meet the future demand of foodgrains of the country. The nonsustainability of this cereal-based system is evident from the declining factor productivity, yield due to an undesirable decline in soil physical environment, and excessive mining of essential plant nutrients from soil.[43–45] The overexploitation of groundwater for irrigation to RW system has crossed the limit of natural rechargeability in regions like Punjab, Haryana, and western Uttar Pradesh. Because of these reasons, conservation agriculture (CA), crop diversification came into existence in the research, policy agenda in the recent past. CA is a type of agriculture that not only conserves natural resources but also ensures sustainable crop yields. It is based on three principles: minimal soil disturbance (i.e., no-till farming), crop residue retention as a permanent soil cover, and crop diversification (i.e., diversified cropping systems). The success of CA is heavily reliant on the deployment of diversified cropping systems. Crop diversification has been practiced as a vital component of profitable sustainable agriculture. The major driving forces for crop diversification in the Indian context are increasing farm income, withstand price fluctuations, aberrant weather, fulfilling diverse food demand, conserves natural resources, decreased incidence of insect-pests, diseases, less dependent on external inputs, and increased availability of fodder for livestock. The rotation of crops not only offers diverse diet to soil microbes but also minimizing the leaching loss of nutrients by recycling of from different depths. Further, crop rotation supports diverse groups of soil flora and fauna.

Pulses by virtues of short duration, low inputs (nutrients and water) requirement, biological N2 fixation, deep root system, reversing the adverse effects of cereal-based systems, and many more advantages are considered as ideal crops for crop diversification. Following of diverse crop rotations with the inclusion of pulses is an age-old practice in the country and it is an important source of protein in the diets of vegetarian population. Pulses can be used as an intercrop, catch crop, as replacement of low-yielding cereal/millet, green manure crop, or alley crop in existing cereal-based systems.

Benefits of Crop Diversification in CA

1.7.1 SOIL HEALTH IMPROVEMENT

The major concern for sustainability in cereal-based cropping system is deterioration in soil physico-chemical and biological properties. The inclusion of pulses can reverse the process and promote overall soil health improvement. Pulses-based crop rotations in general improve soil aggregate stability and formation of a favorable soil structure.[46] The fungi present in rhizosphere of pulse cops produce a glycoprotein called "glomalin." The stickiness of glomalin is mainly responsible for development of stable soil macroaggregates. Therefore, the microbiology of rhizospheric regions of pulse crops is directly involved in the improvement of soil structure in the pulse-based crop rotations. In pulse-based crop rotation, Meena et al.[47] found a greater percentage of soil aggregates exceeding 0.25 mm. Because pulse crop residues have a low C/N ratio, they decompose more quickly, improving soil organic carbon (SOC), and lowering soil bulk density. According to Ganeshamurthy et al.,[48] adding mungbean to an RW combination reduced bulk density and increased hydraulic conductivity. Two sets of long-term studies in sandy loam (TypicUstochrept) soil of the IGP[49] show an improvement in overall soil physical parameters under grain legume inclusive cropping systems.

In a long-term study at IIPR Kanpur recorded significantly higher SOC in rice–chickpea and RW–mungbean over RW. Similarly in another long-term trial and higher SOC was recorded under pigeonpea–wheat and maize–chickpea, and maize–wheat–mungbean. Further, it is also evident that the inclusion of summer mungbean in RW or maize–wheat rotation improved the total organic carbon content. In comparison to RW, Ghosh et al.[50] found a 6% rise in SOC and an 85% increase in soil microbial carbon in the RW–mungbean system. From 7 years of field trial on CA in IGP, Nath et al.[51] reported higher values of SOC, soil microbial biomass carbon (SMBC), SMBN, SMBC/SOC ratio and the activities of fluorescein diaceate, dehydrogenase, protease, urease, phosphatases, and β-glucosidase in the soil. Diversification of maize–wheat cropping system to maize–wheat–mungbean; MCS, maize–chickpea–*Sesbania* resulted in significant improvement in SOC content, labile SOC fractions, SMBC, and dehydrogenase activity.[52] The improvement in overall soil chemical properties was also recorded in pulses inclusive of diversified crop rotations under a 7 years of field study at Kanpur (Table 1.4).

TABLE 1.4 Soil Chemical Properties After 7 Years of Field Study (0–20 cm Soil Depth).

Crop rotation	P (kg/ha)	K (kg/ha)	S (kg/ha)	Zn (ppm)	Fe (ppm)
Rice–wheat	21.0	119.0	14.5	0.781	0.249
Rice–chickpea	22.4	139.9	14.8	0.796	0.257
Rice–wheat–mungbean	24.2	149.8	14.1	0.989	0.272
CD (p = 0.05)	0.89	4.2	NS	0.049	0.017

Source: unpublished data of the first author.

1.7.2 NUTRIENT RECYCLING

The deep root system of pulses enables them to recycle the nutrients from the deeper layer and thus helps to prevent loss of applied nutrients especially nitrate from below root zone of cereal crops in rotation. Further, the association of VAM with pulses roots helps in increasing the availability of nutrients and water to crop plants. Meisinger et al.[53] reported that grain legumes have reduced the concentration of NO_3 by 20–80% in leachate over nonleguminous crops (control). Due to the low C/N ratio, faster decomposition of leaf litter and roots of pulses adds to SOC. The nitrogen-fixing capacity of the pulses enables them to fulfill the large quantity of their requirement, and also helps to economize the N in succeeding crops. In general, pulses can leave 18–70 kg N/ha to soil and thus considerable amount of nitrogenous fertilizers can be saved in succeeding crops. A long-term study at Kanpur revealed that summer mungbean in RW rotation contributes up to 40–60 kg N/ha in the succeeding rice crop. Similarly in another study and N economy due to preceding pigeonpea over sorghum was found to be 51 kg N equivalent/ha. The effect of *rabi* (winter) grain legumes on yield, N economy in the following rice revealed that chickpea, rajmash, and lentil exhibited a most favorable effect in economizing N to the extent of 40 kg/ha.[54] In addition, pulses also release organic acids in soil and thereby mobilizing unavailable soil nutrients. Similarly, the root exudates released by pulses and the organic matter added to the soil make unavailable soil nutrients in plant-available forms. It is evident that acidification of the rhizosphere can solubilize a few low solvent macronutrients and micronutrients.

1.7.3 WATER ECONOMY

In general, pulses require less water than cereal crops. As per an estimate, cereals consume around 60% of global water as against 4% in pulses. The morphological and physiological and rooting patterns of pulses make them more efficient in the utilization of water. Pulses are able to draw moisture from deeper soil layer due to their deep tap root systems and thus thrive well under drought conditions. To produce 1 kg of rice requires 2500–4000 L of water and for 1 kg wheat 1500–2000 L, and for 1 kg lentil or peas only 50–100 L. The water requirement of rice crop is 900–2500 mm, wheat 400–450 mm, sugarcane 1400–2500 mm; however, pulses need only 250–300 mm of water. In general, pulses in India are grown under rainfed conditions. About one to two irrigations are, however, given to *rabi* pulses like chickpea and lentil in northern and central India.[55]

1.7.4 SMOTHERING WEEDS

Weeds remain the major constraints in the agriculture. The inclusion of pulses in cropping systems affects the relative dominance of various weeds. Continuous following of a similar kind of crop rotation leads to dominance of certain weed species. For example, *P. minor* became menace in the RW cropping regions in IGP. Likewise, weedy rice is dominant in rice monoculture regions in India. Therefore, crop rotation should have diverse group of crops with different growth habits. The following period can also be utilized to grow pulses to smother the weeds. From long-term experiments at Kanpur[56,57] reported that the inclusion of pulses in RW or maize wheat system can reduce the population of *P. minor* and *Avena fatua* (wild oat) in winter crops.

In comparison to sole crops, intercropping maize with legumes resulted in a higher soil cover and reduced light accessibility for weeds, resulting in a decrease in weed populations and dry matter.[58,59] Similarly, when compared to a solo crop, a pulse crop integrated intercropping system reduced relative weed cover by 41%, reduced *Senecio vulgaris* L. population by 58%, and increased crop yield by 10%.[60] Weed population is reduced by 30–40% in pigeonpea-based intercropping and fast-growing early maturing grain legumes (cowpea and mungbean), compared to 22% in sorghum.[61] Weeds reduced pigeonpea grain yield by 79.93% if they were allowed to grow until harvest; however, grain yield losses were just 38.19% in the pigeonpea + soybean intercropping system.[62] In addition, chickpea + wheat[63] and chickpea + mustard[64] intercropping systems reported similar effects.

Liebman and Davis[65] observed an allelopathic reaction of crop residue decomposition in small weed seed germination. Many weeds have been found to be stifled and reduced by growing various pulses (lentil, cowpeas, and lupins) due to crop–weed competition or allelopathic properties. Soybean crop residues smother weeds and improve the performance of summer squash (*Cucurbita pepo* L.) and tomato (*Solanum lycopersicum* L.).[66] Furthermore, water-soluble extracts from lentil crop residues are lethal to stinkweed (*Thlaspi arvense* L.), downy brome (*Bromus tectorum* L.), and flixweed (*Descurainia sophia* [L.] Webb.) rather than wheat, indicating that residues of these crops could be used for specific weed control in wheat, reducing reliance on herbicides.[67]

1.7.5 HIGHER PRODUCTIVITY AND SUSTAINABILITY

Declining factor productivity and yield plateauing forced the researchers, farmers, and policymakers to diversify the existing cereal-based cropping systems. Pulses because of N-fixing ability and short duration (some pulses), stand well under adverse situations of biotic and abiotic stresses are suitable candidate for diversification of cereal-based systems. In long-term studies at IIPR, Kanpur revealed that the inclusion of pulses in cereal-based cropping systems increased the system productivity as well as yield of component crops. In the rice-based system and highest system productivity was recorded in RW–mungbean followed by RW–rice–chickpea (2 years rotation) and least in RW. Similarly, in another set of study and system productivity was more in the case of maize–wheat–mungbean than maize–wheat. A similar trend was observed under CA, with RW–mungbean having the highest productivity and RW having the lowest. Singh et al.[68] in a study in mid-Himalaya at Nainital reported higher system productivity (rice equivalent yield), land-use efficiency, and profitability under diversified cropping systems over RW.

Among pulses, pigeonpea is planted in wider rows; its initial growth is low which provides an opportunity for intercropping with crops like mungbean, urdbean, sorghum, etc. Pigeonpea intercropped with short-duration pulses (mungbean and urdbean) is the most popular combination in north India. The unique feature of this system is that while the productivity of the base crop pigeonpea, is unaffected, an additional 0.4–0.5 t/ha of mungbean or urdbean or 0.6–0.8 t/ha of sorghum can be obtained with no additional inputs. Weeds are suppressed by intercropping mungbean and urdbean in pigeonpea, sorghum, or maize. In rainfed areas, winter pulses such as chickpeas and lentils are commonly intercropped with oilseeds. According to the literature, chickpea + mustard, lentil + linseed, and wheat + lentil intercropping systems can yield greater productivity and monetary returns.[69,70] Similarly, in the Himalayan mid-hills, horsegram can be intercropped with early pigeonpea, finger millets, and maize.[71] In rainfed conditions, lentil, and linseed make an excellent intercropping combination when compared to other rabi crops. Several other intercropping systems have also been reported.[71–74] Growing one row of mungbean provided about half a tone/h additional mungbean yield without affecting sugarcane yield.[75,76] Studies conducted at IISR, Lucknow, established that urdbean, mungbean had synergistic effects on cane yield in spring planted crop and provided 0.4–0.5 t/ha additional yield of pulses grains.[77] Similarly, lentil can be intercropped with autumn-planted sugarcane.

1.7.6 ECOSYSTEM SERVICES

To improve the crop resilience to biotic, abiotic stresses under the climate change and there is a need to increase the heterogeneity (both in temporal and spatial) in cropping system. The diversification of cereal-based crop rotations with pulses curtails the need of N-fertilizers in the system and thus reduces the GHGs emission per unit cropping area. Due to a modest input demand in pulses, both fossil and N loss use can be minimized. Drinkwater et al.[78] reported increased retention of soil C and N, improved environmental quality due to low C:N ratio of pulses residues. However, pulses sometimes favor higher N_2O emission. The groundwater pollution due to nitrate leaching is reported in RW growing regions of IGP. The proper crop selection and management practices can minimize nitrate leaching and improves the N-use efficiency. Grain legume intercropping in cereals has been shown to reduce nitrate leaching.[79] In comparison to sole cropping of sugarcane and maize, sugarcane + urdbean and pigeonpea + maize resulted in low nitrate N leaching.[80] Furthermore, including short-duration summer grain legumes reduces the fallow period between two crops (RW), reducing C loss during the hot summer and increases C sequestration. Pulses do not require fine tilth and can be grown in both conventional and CA. The addition of organic residue through roots, leaf litters, and other plant parts improves soil C-pools. The cultivation of summer pulses reduced the fallow period between two crops and thus reduces the soil carbon loss during hot summer.

1.8 CONCLUSION

The cropping system approach is a holistic approach to tackle the growing food demand and utilizing the available resources more efficiently. The goal of any cropping system-based research is to evolve a resource-efficient and location-specific cropping system that can make use of all the below and above ground resources. For sustainable management of our ecosystem, environment, cropping system approach can be taken as key, which can ensure nations food security and help in the continuous income of farmer community. Further, the declining factor productivity and the soil health can be reverse with the following of CA. Cropping systems are also important part in the success of CA. Thus, for overall development, sustainability, profitability, and nutritional, livelihood security of the region or country and agriculture production system is largely depends on the prevailing cropping systems.

KEYWORDS

- **conservation agriculture**
- **intercropping**
- **mono-cropping**
- **productivity**
- **sequential cropping**
- **sustainability**

REFERENCES

1. Bilali, H. E.; Bassole, I. H. N.; Dambo, L.; Berjan, S. Climate Change and Food Security. *Agric Forest.* **2020,** *66* (3), 197–210.
2. Teklewold, H.; Kassie, M.; Shiferaw, B.; Köhlin, G. Cropping System Diversification, Conservation Tillage, Modern Seed Adoption in Ethiopia: Impacts on Household Income, Agrochemical Use, Demand for Labor. *Ecol. Econ.* **2013,** *93*, 85–93.
3. Pearson, C. J.; Ison, R. L. *Agronomy of Grassland Systems*; Cambridge University Press: Cambridge, 1987; 169 p.
4. Daryanto, S.; Fu, B.; Wang, L.; Jacinthe, P. A.; Zhao, W. Quantitative Synthesis on the Ecosystem Services of Cover Crops. *Earth Sci Rev.* **2018,** *185*, 357–373.
5. Gabriel, J. L.; Garrido, A.; Quemada, M. Cover Crops Effect on Farm Benefits, Nitrate Leaching: Linking Economic, Environmental Analysis. *Agric. Syst.* **2013,** *121*, 23–32.
6. Gowda, C. L.; Saxena, K. B.; Srivastava, R. K.; Upadhyaya, H. D.; Silim, S. N. Pigeonpea: From an Orphan to a Leader in Food Legumes. In *Biodiversity in Agriculture: Domestication, Evolution and Sustainability*; Cambridge University Press: New York, 2012; pp. 362–373.
7. Snapp, S. S.; Swinton, S. M.; Labarta, R.; Mutch, D.; Black, J. R.; Leep, R.; Nyiraneza, J.; O'Neil, K. Evaluating Cover Crops for Benefits, Costs, Performance Within Cropping System Niches. *Agron. J.* **2005,** *97* (1), 322–332.
8. Brooks, S. A.; Anders, M. M.; Yeater, K. M. Influences from Long-Term Crop Rotation, Soil Tillage, and Fertility on the Severity of Rice Grain Smuts. *Plant Dis.* **2011,** *95* (8), 990–996.
9. Francis, C. A.; Harwood, R. R.; Parr, J. F. The Potential for Regenerative Agriculture in the Developing World. *Am. J. Altern. Agric.* **1986,** *1*, 65–74.
10. Reynolds, L. B.; Potter, J. W.; Ball-Coelho, B. R. Crop Rotation with *Tagetes* sp. an Alternative to Chemical Fumigation for Control of Root- Lesion Hematodes. *Agron. J.* **2000,** *92* (5), 957–966.

11. DAC. Directorate of Economics, Statistics, Department of Agriculture & Corporation, Ministry of Agriculture, Government of India, 2014. http://eands.dacnet.nic.in/APY_96_To_06.htm
12. Willey, R. W. Intercropping Its Importance, Research Needs 1. Competition, Yield Advantage, 2. Agronomy and Research Approaches. *Field Crop Abstracts* **1979,** *32*, 73–85.
13. Singh, K. K.; Ali, M.; Venkatesh, M. S. Pulses in Cropping Systems. Technical Bulletin, IIPR, Kanpur, 2009; 39 pages.
14. Awasthi, O. P.; Pareek, O. P. Horticulture-Based Cropping System for Arid Regions-a Review. *Range Manage, Agroforest.* **2008,** *29* (2), 67–74.
15. Young, A. Agroforestry for Soil Conservation; International Council for Research in Agroforestry: Narobi, 1989.
16. Bunderson, W. T.; Wakeel, A. El.; Saad, Z.; Hashim, I. Agroforestry Practices in Western Sudan. In *Planning for Agroforestry*; Budd, W. et al., Eds.; Elsevier Science Publisher: New York, 1990.
17. Dhulipalla, A. Evaluation of Performance of Certain Mango Based Intercropping Systems in Krishna District of Andhra Pradesh. *Int J. Commun. Syst.* **2017,** *5* (5), 409–412.
18. Padma, E.; Ramanandam, G.; Kumar, K. R.; Chalapathi Rao, N. B. V.; Maheswarappa, H. P. Performance of Coconut-Based High Density Multispecies Cropping System Under East Godavari Dist. of Andhra Pradesh. *Green Farm.* **2016,** *7*, 1431–1435.
19. Rahim, M. A.; Bari, M. S.; Islam, F.; Naher, N. Multistoried Cropping System—A New Room for Indigenous Vegetables, Medicinal Plants. *ActaHorticulturae* **2007,** *752*, 493–497.
20. Devi, L. G.; Gladis, R.; Biju, J. Performance Studies of Various Crops in Coconut Based Cropping System Compared to Sole Cropping. *J. Progress. Agric.* **2011,** *2* (1), 47–50.
21. Nath, J. C.; Deka, K. K.; Saud, B.; Maheswarappa, H. P. Intercropping of Medicinal, Aromatic Crops in Adult Coconut Garden Under Brahmaputra Valley Region of Assam. *J. Plant Crops* **2015,** *43*, 17–22.
22. Nair, A. K.; Nair, S. A.; Dubey, R. R.; Ravisankar, N.; Pramanik, S. C. Evaluation of Cucurbitaceous Vegetables as Intercrops in Coconut (*Cocus nucifera*) Plantations of Andamans. *Indian J. Agric. Sci.* **2000,** *70* (4), 241–242.
23. Chowdhury, D.; Deka, K. K. Coconut Based Inter Mixed Cropping Under Assam Conditions. *J. Plant. Crops* **1997,** *25* (1), 106–108.
24. Singh, S. K.; Lal, S. S. Integrated Nutrient Management in Potato Based Cropping Systems in South Bihar Alluvial Plains. *Potato J.* **2011,** *38* (2), 162–169.
25. Ghosh, S.; Sarkar, S.; Sau, S.; Karmakar, S.; Brahmachari, K. Influence of Guava (*Psidium guajava* L.) Based Intercropping Systems on Soil Health, Productivity in Alluvial Soil of West Bengal, India. *Int. J. Curr. Microbiol. Appl. Sci.* **2017,** *6* (11), 241–251.
26. Swain, S. C. Influence of Intercropping Systems on Soil Health, Productivity and Quality of Guava (*Psidium guajava* L.) in Eastern India. *J. Plant Nutr.* **2016.** DOI: 10.1080/01904167.2016.1187751
27. Swain, S. C.; Dora, D. K.; Sahoo, S. C.; Padhi, S. K.; Sanyal, D. Influence of Mango-Based Intercropping Systems on Improvement of Soil Health under Rainfed Situation. *Commun. Soil Sci. Plant Analys.* **2012,** *43* (15), 2018–2026.
28. Bhagya, H. P.; Maheswarappa, H. P.; Surekha, Bhat, R. Carbon Sequestration Potential in Coconut-Based Cropping Systems. *Indian J. Hort.* **2017,** *74* (1), 1–5.
29. Gupta, J. P.; Joshi, D. C.; Singh, G. B. Management of Arid Agrosystem. In *National Resource Management for Agriculture Production in India*; Yadava, J. S. P., Singh,

G. B. International Conference on Managing Natural Resource for Sustainable Agriculture Production in the 21st Century, 2000.
30. Saroj, P. L.; Dhandar, D. G.; Sharma, B. D. Bhargava, R.; Purohit, C. K. Ber (*Zizizphus mauritiana*L.) Based Agri-Horti System: A Sustainable Land Use for Arid Ecosystem. *Indian J. Agroforest.* **2003,** *5* (1&2), 30–35.
31. Awasthi, O. P.; Singh, I. S.; More, T. A. Performance of Intercrops During Establishment Phase of Aonla (*Emblica officinalis*) Orchard. *Indian J. Agric. Sci.* **2009,** *79* (8), 587–591.
32. Arya, R.; Awasthi, O. P.; Singh, J.; Arya, C. K. Comparison of Fruit Based Multi-Species Cropping System Under Arid Region of Rajasthan. *Indian J. Agric. Sci.* **2010,** *80* (5), 423–426.
33. Krishan, H.; Singh, I. S.; Bhargava, R.; Sharma, S. K. Fruitbased Cropping Systems for Sustainable Production. *ICAR News* **2013,** *9* (2), 9.
34. Puri, S.; Kumar, A.; Singh, S. Productivity of *Cicer arietinum* (chickpea) Under a *Prosopis cineraria* Agroforestry System in the Arid Regions of India. In *Spl.Issue: Acacia, Prosopis. J. Arid Environ.* **1994,** *27* (1), 85–98.
35. Kaushik, N.; Kumar, V. Khejri (*Prosopis cineraria*) Based Agroforestry System for Arid Haryana, India. *J. Arid Environ.* **2003,** *55*, 433–450.
36. Singh, S. P. *Cropping Systems in Vegetable Crops, Principles of Vegetable Production*; Agrotech Publishing Academy, 1997.
37. Nimbolkar, P. K.; Awachare, C.; Chander, S.; Husain, F. Multi Storied Cropping System in Horticulture-A Sustainable Land Use Approach. *Int. J. Agric. Sci.* **2016,** *8* (55), 3016–3019.
38. Khan, H. H.; Krishnakumar, V. Spices in Coconut Based Cropping System. In *Proceedings of the National Seminal on Strategies for Increasing Production and Export of Spices*; Calicut, 2002; pp 1–17.
39. Thiruvarasan, S.; Maheswarappa, H. P. 2014. Performance of Medicinal and Aromatic Plants as Intercrops in Coconut Garden. *J. Plant. Crops* **2014,** *42* (2), 238–40.
40. Roy, S.; Raj, S.; Choudhury, M.; Dey, S. K.; Nazeer, M. A. Intercropping of Banana, Pineapple in Rubber Plantations in Tripura. *Indian J. Nat. Rubber Res.* **2001,** *14* (2), 152–158.
41. Mirjha, P. R.; Rana, D. S. Yield, Yield Attributes, System Productivity, Economics of Mango *Mangiferaindica* Based Intercropping Systems as Influenced by Mango Cultivars and Nutrient Levels. *Indian J. Agron.* **2016,** *61* (3), 307–314.
42. Baghel, B. S.; Tiwari, R.; Gupta, N. Productivity and Profitability of Mango Based Intercropping System Under Rainfed Agro-Climatic Conditions of Madhya Pradesh. *South Indian Hortic.* **2003,** *52* (1–6), 1–4.
43. Yadav, R. L.; Dwivedi, B. S.; Gangwar, K. S.; Prasad, K. Overview, Prospects for Enhancing Residual Benefits of Legumes in Rice, Wheat Cropping Systems in India. In *Residual Effect of Legumes in Rice and Wheat Cropping Systems of the Indo-Gangetic Plains*; Kumar Rao, J. V. D. K., Johansen, C., Rego, T. J., Eds.; International Crops Research Institute for Semi Arid Tropics, Patancheru, India. Oxford & IBH Publishing Co. Pvt. Ltd.: New Delhi, 1998; pp 207–225.
44. Timsina, J.; Connor, D. J. Productivity Management of Rice-Wheat Cropping Systems: Issues and Challenges. *Field Crops Res.* 2001. *69*, 93–132.
45. Nadarajan, N.; Kumar, N. Role of Pulses in Conservation Agriculture. In *System Based Conservation Agriculture*; Singh, V. K., Gangwar, B. Westville Publishing House: New Delhi, 2018; pp 134–154.

46. Kumar, N.; Nath, C. P.; Hazra, K. K.; Das, K.; Venkatesh, M. S.; Singh, M. K.; Singh, S. S.; Praharaj, C. S.; Singh, N. P. Impact of Zero-Till Residue Management, Crop Diversification with Legumes on Soil Aggregation, Carbon Sequestration. *Soil Tillage Res.* **2019,** *189* (6), 158–167.
47. Meena, V. S.; Maurya, B. R.; Meena, R. S.; Meena, S. K.; Singh, N. P.; Malik, V. K. Microbial Dynamics as Influenced by Concentrate Manure, and Inorganic Fertilizer in Alluvium Soil of Varanasi, India. *Afr. J. Microb. Res.* **2014,** *8* (1), 257–263.
48. Ganeshamurthy, A. N.; Ali, M.; Srinivasarao, C. Role of Pulses in Sustaining Soil Health, Crop Production. *Indian J. Fert.* **2006,** *1* (12), 29–40.
49. Kumar, N.; Singh, M. K.; Ghosh, P. K.; Hazra, K. K.; Venkatesh, M. S.; Nadarajan, N. Resource Conservation Technology in Pulse Based Cropping Systems. Technical Bulletin 1/2012. Indian Institute of Pulses Research, Kanpur, 2012; 35 p.
50. Ghosh, P. K.; Venkatesh, M. S.; Hazra, K. K.; Kumar, N. Long-Term Effect of Pulses, Nutrient Management on Soil Organic Carbon Dynamics, and Sustainability on an Inceptisol of Indo-Gangetic Plains of India. *Exp.Agric.* **2012,** *48* (4), 473–487.
51. Nath, C. P.; Kumar, N.; Das, K.; Hazra, K. K, Praharaj, C. S.; Singh, N. P. Impact of Variable Tillage Based Residue Management, Legume Based Cropping for Seven Years on Enzymes Activity, Soil Quality Index, and Crop Productivity in Rice Ecology. *Environ. Sustain. Indic.* **2021,** 10, 100–107.
52. Parihar, C. M.; Jat, S. L.; Singh, A. K.; Datta, A.; Parihar, M. D.; Varghese, E.; Bandyopadhyay, K. K.; Nayak, H. S.; Kuri, B. R.; Jat, M. L. Changes in Carbon Pools, Biological Activities of a Sandy Loam Soil Under Medium-Term Conservation Agriculture and Diversified Cropping Systems. *Eur. J. Soil Sci.* **2018,** *69*, 902–912.
53. Meisinger, J. J.; Hargrove, W. L.; Mikkelsen, R. L.; Williams, J. R.; Benson, V. W. Effects of Cover Crops on Ground Water Utility. In: *Cover Crop for Clean Water*; Hargrove, W. L.; Soil, Water Conservation Society: Ankeny, 1998; pp 9–11.
54. IIPR. *25 Years of Pulses Research at IIPR*;. Indian Institute of Pulses Research: Kanpur, 2009.
55. Kumar N.; Hazra K. K.; Nath C. P.; Praharaj C. S.; Singh U. Grain Legumes for Resource Conservation and Agricultural Sustainability in South Asia. In *Legumes for Soil Health, Sustainable Management*; Meena, R., Das, A., Yadav, G., Lal, R. Springer: Singapore, 2018; pp 77–107.
56. Kumar, N.; Singh, K. K. Weed Management in Pulses: Why, When and How. *Indian Farm.* **2009,** *60* (4), 9–12.
57. Hazra, K. K.; Kumar, N.; Venkatesh, M. S.; Ghosh, P. K. Inclusion of Pulses in Rice-Wheat System Can Reduce *Phalaris minor* Population. *Pulses Newslett.* **2012,** *23* (3), 6.
58. Kumar, N.; Prakash, V.; Mina, B. L.; Gopinath, K. A.; Srivastva, A. K. Effect of Sowing Methods, Growth Retardant, and Intercropping on Horse Gram (*Microtyloma uniflorum*) Productivity. *Indian J. Agric. Sci.* **2010,** *80* (4), 335–337.
59. Meena, R. S. Response to Different Nutrient Sources on Green Gram (*Vigna radiate* L.) Productivity. *Indian J. Ecol.* **2013,** *40* (2), 353–355.
60. Baumann, D. T.; Kropff, M. J.; Bastiaans, L. Intercropping Leeks to Suppress Weeds. *Weed Res.* **2000,** *40*, 359–374.
61. Ali, M. Weed Suppressing Ability, Productivity of Short Duration Legumes Intercropped with Pigeonpea under Rainfed Conditions. *Trop. Pest Manage.* **1988,** *34*, 384–387.
62. Ali, M.; Singh K. K. Management of Pulse Crops Under Intercropping Systems. In *Recent Advances in Pulses Research*; Asthana, A. N., Ali, M., Eds.; Indian Society of Pulses Research, Development, IIPR: Kanpur, 1997; pp 489–508.

63. Banik, P.; Midya, A.; Sarkar, B. K.; Ghose, S. S. Wheat, Chickpea Intercropping Systems in an Additive Series Experiment: Advantages, and Weed Smothering. *Eur. J. Agron.* **2006,** *24*, 325–332.
64. Kaur, R.; Kumar, A.; Sharma, B. C.; Brijnandan, K. P.; Sharma, N. Weed Indices in Chickpea + Mustard Intercropping System. *Indian J. Weed Sci.* **2014,** *46*, 333–335.
65. Liebman, M.; Davis, A. S. Integration of Soil, Crop and Weed Management in Low-Eternal Input Farming Systems. *Weed Res.* **2000,** *40*, 27–47.
66. Barker, A. V.; Bhowmik, P. C. Weed Control with Crop Residues in Vegetable Cropping Systems. *J. Crop Prod.* **2001,** *4*, 163–184.
67. Moyer, J. R.; Huang, H. C. Effect of Aqueous Extracts of Crop Residues on Germination and Seedling Growth of Ten Weed Species. *Bot. Bull. Acad. Sinica.* **1997,** *38*, 131–139.
68. Singh, D. K.; Singh, R.; Singh, G. D.; Singh, A. P.; Chaturvedi, S.; Singh, J. P.; Rathi, A.; Singh, M. Diversification of rice (*Oryza sativa*)–wheat (*Triticum aestivum*) System, Its Influence on Productivity, Profitability and Energetics Under On-Farm Situation. *Indian J. Agron.* **2017,** *62* (3), 255–259.
69. Ali, M.; Mishra, J. P. Genotypic Compatibility with an Interplay of Row Orientation in Gram (*Cicer arietinum*) + Indian Mustard (*Brassica juncea*) Under Irrigated Conditions of Indo-Gangetic Alluviums. *Indian J. Agric. Sci.* **1992,** *72* (2), 97–100.
70. Singh, K. K.; Rathi, K. S. Dry Matter Production, Productivity as Influenced by Staggered Sowing of Mustard Intercropped at Different Row Ratios of Chickpea. *J. Agron. Crop Sci.* **2003,** *189*, 169–175.
71. Kumar, R.; Gopal, R.; Jat, M. L.; Gupta, R. K. Conservation Agriculture Based Strategies for Sustainable Weed Management in Maize (*Zea mays*); Training Manual, Maize for Freshers. Directorate of Maize Research: New Delhi, India, 2010.
72. Ahlawat, I. P. S.; Gangaiah, B.; Singh, O. Production Potential of Chickpea (*Cicer arietinum*)-Based Intercropping Systems Under Irrigated Conditions. *Indian J. Agron.* **2005,** *50* (1), 27–30.
73. Kumar, R.; Ali, M.; Arya, R. L.; Mishra, J. P. Enhancing Productivity, Profitability of Chickpea *(Cicer arietinum)* + Indian Mustard *(Brassica juncea)* Intercropping System. *Indian J.Agron.* **2006,** *51* (1), 27–30.
74. Kumar, N.; Prakash, V.; Meena, B. L.; Gopinath, K. A.; Srivastva, A. K. Evaluation of *toria* (*Brassica compestris*), Lentil (*Lens culinaris*) Varieties in Intercropping System with Wheat (*Triticum aestivum*) Under Rainfed Conditions. *Indian J. Agron.* **2008,** *53* (1), 47–50.
75. Yadav, R. L.; Prasad, S. R.; Singh, K. Fertilizer Requirement, Row Arrangements of Pulses in Sugarcane Based Intercropping Systems. *Indian J. Agron.* **1987,** *32*, 80–84.
76. Panwar, B. S.; Singh, B. V.; Sharma, J. C. Feasibility of Intercropping in Autumn Planted Sugarcane. *Indian Sugar* **1990,** *39*, 755–756.
77. Lal, M.; Singh, A. K.; Ali, M.; Kumar, R. Agronomic Evaluation of Intercropped Mungbean/Urdbean Genotypes in Spring Planted Sugarcane; Annual Report, IIPR: Kanpur, 1999.
78. Drinkwater, L. E.; Wagoner, P.; Sarrantonio, M. Legume-Based Cropping Systems Have Reduced Carbon, Nitrogen Losses. *Nature* **1998,** *396*, 262–265.
79. Yadav, R. L. Intercropping Pigeonpea to Conserve Fertilizer Nitrogen in Maize and Produce Residual Effect on Sugarcane. *Exp. Agric.* **1981,** *17*, 311–315.
80. Yadav, R. L. Minimising Nitrate Nitrogen Leaching by Parallel Multiple Cropping in Long Duration Row Crops. *Exp. Agric.* **1982,** *18*, 37–42.

CHAPTER 2

IPM Concept and Strategies for Sustainable Agriculture

MONIKA SINGH[1], NEETU PANDEY[1], and O. P. SHARMA[2]

[1]*Mewar University, NH-79, Gangarar, Chittorgarh, Rajasthan-312901, India*

[2]*ICAR-NCIPM, LBS Building, Pusa Campus, New Delhi-110012, India*

[*]*Corresponding author. E-mail: opsharmadelhi@gmail.com*

ABSTRACT

The ever-increasing world's population requires farmers to produce more crops from existing land with limited resources. For increasing crop yields, regular up-gradation of agricultural technologies is being applied to minimize crop losses without altering the environment. It is being increasingly adopted in developing as well as developed countries and a mandatory for world-bank-supported programs. Integrated pest management (IPM) is required for the sustainable agricultural system for achieving safe and eco-friendly and pesticide-free food production. It will improve farmers' livelihood and will conserve nonrenewable resources. IPM requires a timely and coordinated approach involving prevention, monitoring, and intervention. Monitoring of crop health may be taken for the management of pest and disease at the primary level. Agro-eco system analysis (AESA) approach is based on crop health at different stages of growth, pest, and natural enemies population dynamics, soil condition, climatic factors, and farmer's past experiences, field scouting, use of sticky traps, pheromone traps, soil sample analysis for soil-borne pathogens and nematodes, diagnostic techniques, economic threshold level (ETL) and pest forecasting models to give assistance in proper scheduling of interventions. Intervention by cultural and physical

Integrated Pest Management in Diverse Cropping Systems. Devendra Pal Singh, PhD (Ed.)

control of weeds control manually or with tractor-mounted appliances; physical removal of aggregated egg masses or larvae; disease control by removing infected plant debris or older leaves. The use of straw mulch or yellow polythene mulch repels certain insects, which are potential vectors of some viral diseases, and thus helps in reducing population. Biological control by augmenting with predators or parasitoids or conservation or spray with entomophagous fungi or bacteria or virus to control pest and harmful insect. Botanical sources, that is, secondary metabolites or allelochemicals provide for the fundamental or physiological or biochemical process of the plant.

2.1 INTRODUCTION

Agriculture was the second great technological revolution of mankind, discovery, and control of fire being the first and industry the third. The quest to feed ever-increasing world population has led to agricultural and industrial revolutions has led to injudicious way necessities like food, fiber, and medicines are extracted from nature. Perhaps the most significant transformation in agriculture was brought about by chemistry, especially with invent of DDT and urea rapidly took over traditional agriculture, especially after World War II. One key problem on which the post-World War II agricultural revolution focused was that of pests as its epidemic affecting cereals and horticultural crops resulting in famine. The chemical poisons that arose from war research seemed to make the total annihilation of pests achievable. It was not realized until the publication of Silent Spring in 1962 by Carson[1] highlighted the issue to the common man. However, general ecological and evolutionary mechanisms behind the problems were formulated.[2] The "pesticide treadmill" in which pest resistance to pesticides, pest resurgence, and the formation of secondary pest outbreaks were recognized, as the three mechanisms leading farmers into the need to spray ever greater quantities of pesticides. It is generally known that often only 1% of the active ingredients reach the target site,[3] rest pollute the environment.[4] The chemical pesticides not only affect the pest and their natural enemies but also contribute to the environmental pollution (soil, as well water). It interrupts parasitoid and predator populations and thus causing outbursts of secondary pests ultimately leading to pest resistance. Continuous injudicious use of pesticides put pressure on altering the genetic composition of a pest population enabling them to survive and the survivors pass on the resistance trait to the next generation. The proportion of the resistant pest population increases. In India, thirty-three (33) key insect pests such as *Helicoverpa*, whitefly, diamondback moth, tobacco caterpillar, and mustard aphids have developed resistance to almost all the pesticides being currently used to control them.[5–7]

2.1.1 ORIGIN OF INTEGRATED PEST MANAGEMENT CONCEPT

The integrated pest management (IPM) has been employed under an assortment of different associations.[8] Formerly, the term was used as "integrated control" by Barlett[9] and was expanded on by Stern and co-workers.[10] There is no universal definition of IPM[11] except one advocated by FAO. During the late 19th and initial 20th centuries, farmers relied on pest biology, cultural practices, and multitactical approaches as part of the present IPM system. The integrated control concept was first conceived and stated "biological and chemical control is considered as complementing to one another."[12] Nature's own equilibrium provides the major protection hence, insecticides should be used only in an emergency to prevent interference with nature pest control mechanisms. Michelbacher and Bacon[13] coined the word IPM. It gives permanent, satisfactory, and economical insect control.[14] The integrated control was first defined as "applied pest control."[10] The "protective population management" was also coined and later modified to "pest management."[15,16] The insect numbers are crucial for the integrated management of pest population.[17] The agreement on the following issues has reached:[19]

- "Integration" means the coordinated use of multiple methods to control individual pests or multiple pests.
- "Pests" are any organism causing economic losses.
- "Management" consists of decision rules based on economic, social, and environmental. The backbone for pest management in any agricultural system is the principle of economic injury level (EIL) for insects and "action level" for diseases.
- "IPM" is a multidisciplinary approach based on unstructured defense options.

2.1.2 DEFINITIONS OF IPM

Currently, there are above 72 definitions of IPM,[20] issued by the government, research organizations, NGOs, and universities. Some thought that IPM will completely remove the use of chemical pesticides. Extreme views equated IPM with "pest free" farming, which will become increasingly marginalized, and more balanced views will prevail. The IPM was backed by the FAO International Code of Conduct on the Distribution and Use of Pesticides (Article 2).[21]

The IPM is an integration of all suitable management techniques.[22] The experiences in southern Maharashtra[23] demonstrated that farmers are capable to use IPM strategy.

IPM is defined as an interdisciplinary ecological approach to the pest population management, which utilizes a range of control strategies compatibly in a single coordinated pest management system.[24] The IPM is effective, economical, and safe for farm workers and consumers health and the environment.[25]

The IPM keeps the pest population under thresh hold level.[26] The IPM concept[27] includes:

- Correct pest control methods.
- The gain to both growers and consumers.
- Monitoring and selection of the control action.
- To assess the combined effects of multiple pests.

The IPM is a rule-based decision support system.[27]

2.1.3 WHAT IS IPM?

The IPM is a systematic strategy of crop protection that uses information and dynamic decision-making models to reduce the use of inputs. It also improves economic, social, and environmental conditions.[28] It is a comprehensive approach to reduce the pest's status to bearable losses while keeping a quality environment.[29] The IPM is the intellectual selection of strategies warranting economic, ecological, and sociological consequences.[30] The IPM is an active and dynamic approach in which all the suitable management tactics utilizing available real-time surveillance and forecasting information to develop a comprehensive plan.[31]

The IPM is the crop protection system that meets the necessities of sustainable agriculture. IPM is essential for integrated crop management (ICM). ICM has been established as a farming system to meet long-term sustainability. It safeguards the farm's natural properties apart from shunning waste, improves energy efficiency, and reduce pollution.

2.2 PEST MANAGEMENT SCENARIO IN PREVAILING MODERN FARMING SYSTEMS AND ITS IMPACT ON ECOSYSTEM

Due to climatic changes and cultivation practices, there is a gradual change in the pest situation where in secondary pests having negligible status have now become well-established key pests in many regions. The following scenario has emerged from current plant protection practices[32]:

1. The cost-efficiency of chemical pesticides is losing fast, especially when pesticides are solely relied as the principal mode of pest management and being applied on a calendar base.
2. Because of adverse side effects of pesticides residues, environmental awareness and increasing demand of international markets for pesticide-free foods, the demand for stricter pesticides regulation is rising.[33]
3. Confidence is growing in preventive, biologically based IPM systems, which appear slow but are sustainable and steady in action.

The convergence of these trends is strongly advocating interest in IPM in the agricultural community. A broad-based agenda is unfolding to accelerate the transition from chemical-based pest management systems to multiple components as well as biologically based interventions.[34]

Indian agriculture has undergone unique and spectacular technological transformations that gave rise to "Green Revolution" (Table 2.1). The technology of intensive agriculture, no doubt, generated the Green Revolution producing more food from the available natural resources; however, its misuse has led to unexpected problems, especially for smaller farmers:

i) The misuse of soil and water resources.
ii) Replacement of a rich diversity of traditional crop varieties with a fewer high-yielding varieties.
iii) Injudicious and indiscriminate use of chemical fertilizers and pesticides[35] to get higher yields.

The intensification of cropping systems has led to an increase in biotic stresses on account of:

a. Introduction of new pest problems, for example, cotton leaf curl virus (CLCV), B Biotype of whitefly (*Bemisia tabaci* = *B. argentifolii*), spiral whitefly (*Aleurodicus dispersus*), wooly aphid (*Ceratovacuna lanigera*), Subabul psyllid, (*Heteropsylla cubana*), coffee berry borer (*Hypothenemus hampei*).

b. Increased intensity of the existing pests, for example, white rust of mustard (*Albugo candida*), leaf blight of wheat caused by *Helminthosporium* and *Alternaria* spp, sheath blight of rice (*Rhizoctonia solani*), downy mildew of pearl millet (*Sclerospora graminicola*) whiteflies (*B. tabaci*), rhinocerous beetle of coconut and oil palm (*Oryctes rhinoceros*), eriophyid mite of coconut (*Aceria guerreronis*), coreid bug (*Paradasynus rostratus*), mosquito bug of cashewnut (*Helopeltis antonii*), cotton mealy bug (*Plannococcus* sp.).

c. Development of resistance in pests against various pesticides, for example, American bollworm, whiteflies, and storage pests, for example, singhara beetle (*Galerucella birmanica*) (Jacoby), tobacco caterpillar (*Spodoptera litura*) (Fab.), Gram pod-borer (*Helicoverpa armigera*) (Hubner), White flies (*B. tabaci* (Gennadius), diamond back moth (*Plutella xylostella*) (linnaeus), Jassid (*Empoasca kerri* Pruthi), Aphid (*Aphis craccivora* Koch, *Lipaphis erysimi*), red floor beetle (*Tribolium castaneum*), rice weevil (*Sitophilus oryzae*), Gram beetle (*Oryzaephilus surinamensis*), cigarette beetle (*Lasioderma serricerne*), leather beetle (*Dermestes maculatus*), lesser grain borer (*Rhyzopertha dominica*), and khapra beetle (*Trogoderma granarium*).[37]

TABLE 2.1 Estimates of Crop Loss During Pre and Post Green Revolution Due to Insect Pests (%).

Crop	Green revolution		
	Pre	Post	
		1983	1993–1994
Cotton	18	50	50
Rice	10	20	25
Brassica veg	NA	37	35
Pulses	5	10	30
Sugarcane	NA	NA	20

Sugarcane woolly aphid has become a great concern to cane growers of Maharashtra and Karnataka. Chemical control has failed to provide a permanent solution as the pest reappears 20–30 after insecticide sprays. Biointensive IPM approach proved to be the effective alternative, hence, conservation and augmentation of potential biocontrol agents (predators and parasites) were being multiplied and used extensively. A number of villages in Mizoram (approximate area of 1000 ha.) is affected by rodent pests threatening bamboo cultivation. Due to incompetence of chemical pesticides, IPM strategies relying more on release of imported predators successfully helped in containing this polyphagous pest.[38]

Despite advancements in the field of plant breeding and molecular techniques a number of plant diseases (Table 2.2) continue to be threatening and capable of initiating severe harm in the event of epidemics as happened during the Irish potato and Bengal famine. Plant viruses are another group of biotic constraints for which to this date there are no economic management strategies. Some of the threatening viruses are as follows: Badnaviruses (Rice tungro), Geminiviruses (legumes, cotton, potato, tomato), Potyviruses (sugarcane, potato, cucurbits, papaya, grain legumes, and vegetables), Tospoviruses (vegetables and grain legumes), Ilar viruses (sunflower and legumes), and Nano viruses (Bunchy top of banana).

TABLE 2.2 Current Status of Key Plant Diseases of Economic Importance.

Disease	Current status
Banana leaf spot	1930 to date several epidemics
Botrytis gray mold	Frequent epidemics
Cereal smuts	Continuous loss
Karnal bunt	Spreading
Viral diseases of pulses	Continuous loss
Cotton leaf curl	Spreading
Mango malformation	Continuous loss
Coconut wilt	Spreading
Rhizoctonia root rot	Continuous loss
Downy mildews	Continuous loss
Red rot of sugarcane	Continuous loss
Gray mildews	Spreading
Tobacco streak virus	Spreading
Guava wilt	Continuous loss
Citrus die back & Gumosis	Spreading
Parawilt of cotton	Continuous loss
Vascular wilt of legumes	Continuous loss

2.3 HAZARDS ASSOCIATED WITH PESTICIDES

As a control tactic, over-reliance on pesticides and their injudicious use damages the natural balance of crop ecosystem. It disturbs the equilibrium of beneficial insect populations[39] and thus results in induced pest outbreaks due to the wrong use of pesticides.[40] The major hazards associated with the manufacture, distribution, and application of chemical pesticides include:

2.3.1 HAZARDOUS TO HUMAN HEALTH

The majority of the chemical pesticides are hazardous to human health, particularly pesticide users and the pesticide-treated foodstuffs consumers. The hazards posed to the users are high in the developing countries, as user protection measures are not implemented effectively leading to acute and constant chronic poisoning.

2.3.2 HAZARDOUS TO THE ENVIRONMENT

The application of chemical pesticides places a burden on the environment. This leads to contamination of the soil, groundwater, rivers and lakes, and the atmosphere. It is not only that pesticide contamination is confined to the immediate zone of application but also to distant regions.

2.3.3 RESURGENCE OF PESTS

The biotic natural enemies of pests can be wiped out with more use of pesticides, as their populations do not recover as quickly as those of insect pests. The use of selective action herbicides can in course of time lead to certain weed species becoming widespread. Changes in the biological diversity of agricultural systems and subsequently in ecological elements affect the pests.

2.3.4 RISK OF DEVELOPMENT OF PEST RESISTANCE TO CHEMICAL PESTICIDES

Numerous cases of development of resistance in insect pests, mites, weeds, fungi, and bacteria to certain chemical pesticides are known. At present, there is hardly either a single significant pest family or a group of chemicals already applied over a prolonged period, where resistant strains of the pests have not developed. The stronger the selection pressure of the pesticide and the shorter the generation period of the pest, greater are the chances of the development of resistant strains quickly. Cross-resistance has also been recognized, that is, the pests are unaffected to one chemical develops resistant to others with a similar mode of action. Some major pests have already been reported to have established resistance to an array of chemicals with different modes of action and chemistry.

2.4 THE RESPONSE TO THE PESTICIDE TREADMILL AND ASSOCIATED HAZARDS

IPM strategy, in an appropriate blend, is the natural response to the pesticide treadmill and associated hazards. Indian Government adopted IPM as the main plank of plant protection strategy in 1985 and signed the Agenda 21 of the UNCED.

Agricultural pests controlled by chemical means have increased, but its overutilization has adversarial effects on farm resources, consumer, farm workers' health, environment, and on international trade. IPM combining biological control, host plant resistance, appropriate farming practices, with judicious pesticides use, is the best choice for the future, as it promises yields, reduces costs, is eco-friendly, and contributes to the sustainable agriculture. The National Agricultural Policy[42] of the Government of India has given special emphasis on different components of IPM (Figure 2.1).

2.5 THREE PILLARS OF IPM STRATEGY

2.5.1 PRECAUTION

Numerous aspects of farm and crop management reduce sources of primary inoculum of pest and prevent the initial build-up of pest infestations and its secondary spread to adjoining crop and surrounding cropped areas. It is desirable to grow crops and varieties in appropriate locations where they fit to climate, soil, and topography. Local conditions provide the crop with optimal growing conditions from the start, which makes them strong enough to bear the injury to a tolerable level. Growing of nonlocale crop is to be avoided.

a. Crop Hygiene and Agronomic Practices

1. Deep summer plowing.

2. Cleaning of refuse and weed hosts
3. Crop rotation, mixed and alley cropping, intercropping
4. Selection of crop seed variety (resistant/tolerant) and seed treatment
5. Balanced integrated nutrient management (INM)
6. Water management (especially for rice), etc.,

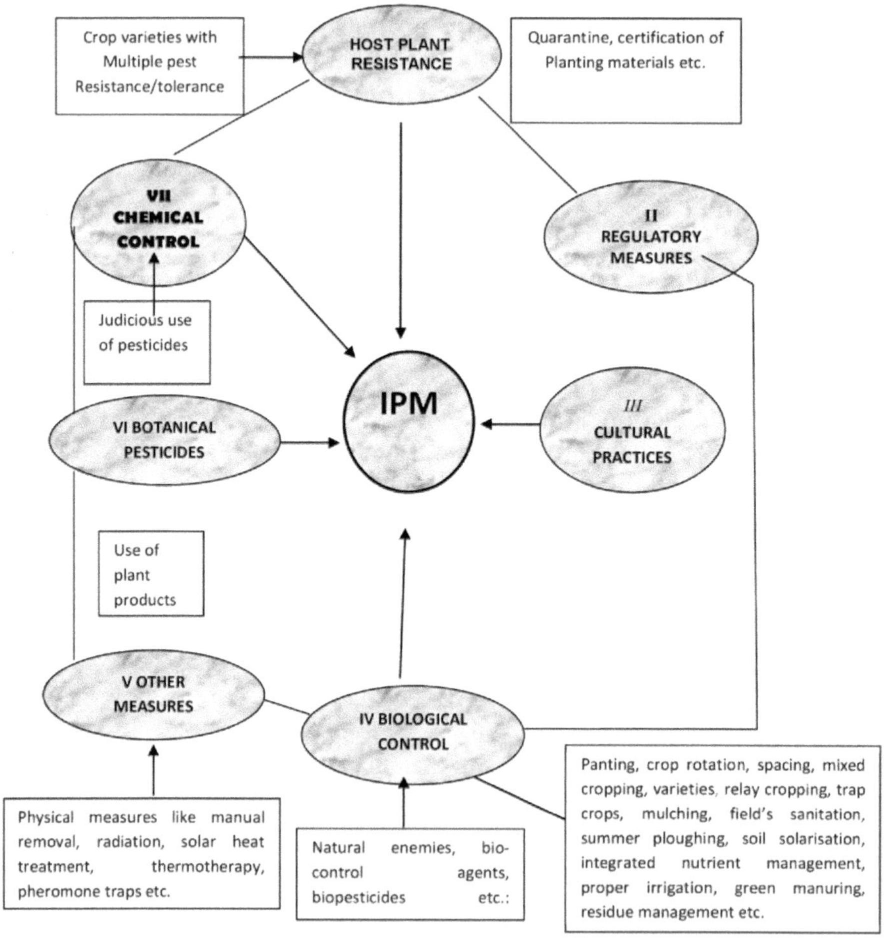

FIGURE 2.1 Different components of IPM.

Mechanical and physical crop protection methods are important in promoting good crop growth, and prevent or minimize weed, disease, and insect pest infestation. Traditional methods of pest management can be important components of IPM.[43] Plowing inverts, the soil with crop residue and weeds before the preparation of a seedbed for the next crop, which can too lead to increased erosion, so it should be combined with added conservation methods like the use of contour plowing and ridging. Changes in soil disturbance, location of plant residue, and weed ecology all impact the prevalence of diseases and insect pests need to be considered before designing IPM programs. Good crop hygiene is also vital in decreasing the build-up and carry-over of pest populations from one crop to another. Methods are labor-intensive (e.g., elimination and destruction of crop residue), but proved to an essentially effective technique of reducing the persistence of some pests to the subsequent season (e.g., in cotton, the leftover of pink bollworm from one crop to another can be reduced by eliminating and destroying cotton trash well in advance of sowing). Good fertilizer practice (INM) should be adopted to promote crop growth, but excessive use of nitrogen must be avoided—too much lush, vegetative growth encourages many diseases, insect pests, and weeds. Some examples of the agronomic practices which directly or indirectly influence the disease incidence and intensity have been listed below:

- Water supply to the crop can enhance or reduce pest incidence. Flooding of some crops, particularly lowland rice is important in stopping weed growth and reduction in root feeders. Though, flood irrigation can also undesirably disturb the survival of various soil-inhabiting natural enemies, for example, spiders. In vegetables, this can be decreased by growing crop on ridges or raised beds and should be taken into consideration in designing IPM systems.
- It is known that excess of nitrogen predisposes the host to rust and powdery mildew in wheat, blast in rice, tobacco mosaic, etc. Diseases like tomato wilt (*F. oxysporum* f. sp. *lycopersici*), damping-off (*Pythium* sp.), etc., are aggravated by low nitrogen. Excess of nitrogen attracts sucking pests, aphids in particular, which acts as vectors of many diseases.
- The application of potash and phosphate enhances the plant resistance to certain diseases. Application of K reduces the incidence of cotton wilt and stem rot in jute and rice. Application of superphosphate and potash helps to overcome root weevil attack in rice crop.
- Deficiency of some trace elements also impairs plant resistance, for example, deficiency of zinc predisposes maize plants to downy mildew.
- Application of micronutrients like molybdenum and boron checks black vein of Brassicae.

- Flooding controls *Fusarium* wilt of banana. Flooding of rice nurseries helps in eliminating the attack of armyworm and *Spodoptera mauritia*.

There are many other instances where soil fertility or nutrient status plays a role in pest incidence. Faulty mechanical sowing may cause seed injury and thereby predispose it to attack by soil fungi. By delaying emergence, deep sowing prolongs the susceptible phase between germination and emergence and favors many diseases and insect pests. So, sometimes shallow sowing favors the crop to escape disease infection. Sowing should always be as uniform as possible maintaining an optimum plant population. Planting geometry affects plant density which is related to competition for available nutrients, moisture, and sunlight and affects leaf wetness, etc. Plant density also affects natural enemies and results in greater pest control. *Nilaparvata lugens* incidence is always higher in closer plantings. Provision of alleys of 30 cm width after every 2–3 m has proved beneficial in reducing pest incidence in endemic areas. Sometimes, the magnitude of spacing influences the incidence of insect pests and diseases.

- Seedling blight of groundnut (*Aspergillus niger*), soybean root rot (*Macrophomina phaseolina*), etc., may be reduced by shallow sowing.
- Blast of rice, sheath blight of rice, damping-off, etc., can be reduced by wide spacing.
- *Verticillium* wilt of cotton bacterial leaf blight of rice, tomato spotted wilt virus in groundnut, etc., can be checked by close spacing. Closer spacing or higher plant density in chickpea had higher incidence of *H. armigera*.
- Adoption of increased seed rate in sorghum reduces shoot fly infestation.
- Nipping of leaves adjoining to grape bunches increases airflow preventing development of botrytis disease.

The nonavailability of preferred host to adults' insects and parasitic weeds certainly limits the survival chances and helps in the reduction of pest incidence. Some pests use the weeds or the wild hosts as the means of their survival during the absence of the main crop. With the availability of crop and conducive climate the pest mores to the crop and sometimes appear in the empirical fashion. Therefore, proper weed management may check the perpetuation of many diseases/insect pests.

Growing different crops in a rotation helps in reducing the build-up of some pests, especially those prevailing in the soil, such as root feeders, numerous fungal pathogens affecting the root system and nematodes.

Rotations apart from reducing the weed problems increase the population of predators and parasites. Planting inter as well as cover crops increases the biodiversity and helps in the conservation of biological agents in the field. To manage cotton leaf curl virus, the alternative host crops (cucurbits, solanaceous vegetables, and citrus orchards) should be avoided if possible.

Time of sowing/harvesting: in cases, where the susceptible stage in the crop's life or infesting stage in the pest's life is brief, altering timings of sowing and harvesting can be effective as a control measure. Early sowing of chickpea (Oct end) drastically reduces chances of occurrence of vascular wilt as well as *Ascochyta* blight disease.

- Cowpea wilt and wheat brown rust can be avoided by early sowing.
- Sugar beet damping-off *(Rhizoctonia* sp.) and wheat "take all" disease can be controlled by late sowing.

The mono-cropping aids in the control of pests by reducing population or breaking their life cycle, for example, parasitic weeds (Stigra and Orobanche). This is because the pest(s) of the preceding crop would not affect those of the following one. Rotation should necessarily involve different kinds of crops, whose pests are not common.

- Lady's finger (bhindi) following a cotton crop will suffer from increased pest infestation because of pests common to both crops.
- A similar situation exists in areas of monoculture, for example, Thanjavur delta where rice is the main crop season after season.
- Crop rotation generally affects insects having low host numbers and is relatively immovable in some stage of their development.
- Crop rotation is helpful in eradicating some soil-borne diseases; precisely those are soil transient but not soil-inhabiting (*Phoma* of Grapes). The crop rotation with unrelated plant species is detrimental to the survival of those pathogens, for example, vascular wilts.
- Crop rotation reduces the incidence of red rot of sugarcane, wilt of arhar (pigeon pea), wilt of pea and wheat mosaic, etc.
- Soil amendments: certain soil-borne pests and diseases can be checked through incorporation of some amendments. Diseases like scab of potato (*Streptomyces scabies*), *Verticillium* wilts, wilt of pea (*Fusarium oxysponum* f.sp. *pisi* race-I), etc., are favored by alkaline soil. Therefore lowering of soil pH through the application of elemental sulfur or gypsum reduces the incidence of these diseases.
- Clubroot of crucifers (*Plasmodiophora* sp.), bacterial wilt of potato (*Pseudomonas solanacearum*), cotton wilt (*Fusarium oxysporum* f. sp.

vasinfectum), etc., are favored by acidic soil. Therefore, liming helps in checking these diseases.

In cultivated fields normal cultural practices enable the addition of substantial quantities of organic, matter in the form of crop residues. In addition to that, many farming practices such as green manuring, use of oil cakes, straw composting, etc.

In "take all" disease of wheat, green manuring with mustard, rape, or pea creates such an antagonistic condition, which can destroy the pathogen.

- In potato scab, green manuring with soybean and pea gives good results.
- Potato black scurf (*R. solani*) can be checked by application of wheat straw compost.
- *Fusarium* wilt of banana can be checked by sugarcane residue compost. Nematodes can be controlled through the use of neem, cakes, and FYM.

b. Introducing Genetically Resistant Crops

Crop variety selection has continuously been a number one priority in crop protection, especially the growing of disease and insect pest-resistant varieties for the management of ubiquitous fungi. These may reduce the need for frequent treatments with crop protection products. The use of resistant varieties also encourages the survival of beneficial, both in terms of microbes (fungal and bacterial) and insects. Developments in genetic engineering will increase the potential for developing novel opportunities and choices for pest resilient and herbicide-tolerant crops as proved in the case of GM soybeans. A transgenic plant (*Bt* cotton) is a new means of delivering bioagents, where insecticidal proteins are introduced genetically into the plant. The insect pest is killed only when it feeds on the plant and nontarget organisms are unaffected.

c. Habitat Management

The food sources need to be provided by deliberate growth of certain wild vegetation (scented flowers) near main crop or on borders to support natural enemy populations. Careful management of the margins of farmland, as well as growing tree crops or hedges, are particularly important because they provide habitat, cover, and refuge for beneficial insects and other animals. In rice cropping systems field bunds with abundant grasses provide essential refuges for predatory spiders which help limit the population of rice pests, and for snakes which help control rats. The maize and cowpea are grown on borders of cotton field increases the population of coccinellids and these migrate to cotton in search of aphids and jassids and devour them.[43]

d. Trap Crops and Intercropping

Small plantings of a vulnerable or desired crop may be grown either near a main crop or as the separate row to serve as a "trap." It is intended for getting some return from an area, for when pests attack one crop; the other escapes and comes up well. But sometimes, one component of mixed cropping directly inhibits or checks some pest (s) of the other crop. Sometimes a pest can be attracted away from a valuable sensitive crop by another crop, which suffers less damage if attacked. In southern Maharashtra, linseed is sown in rows every 10–15 m across chickpea fields to attract bollworms during critical periods of crop development, scattered tobacco crop attracts *Helicoverpa* on their growing shoots which can be nipped away. Conventionally, some farmers sow different crops in alternate rows, or undersow a crop (e.g., maize) with a legume (e.g., beans) to help improve soil fertility and safeguard beneficial bioagents, for example, ladybird beetle. Astha experience shows that the growing of Setaria as 10 rows serve as live bird perch which attract the birds to predate on bollworms.

- Raising a crop of castor around chilies and tomatoes in citrus orchards for the control of fruit borer and sucking moths respectively.
- If *Crotolaria* is grown, it helps in checking the root-knot nematodes. Eggs are hatched into larvae, but adults do not emerge out.
- Mixed cropping of marigold in tomato helps in reducing the nematode infestation.
- Mixed cropping of sorghum and tobacco checks tobacco wilt. Intercropping of pigeon pea with sorghum reduces the incidence of pigeon pea wilt *(Fusarium udum)*.
- Mixed cropping *of* alfalfa and cocksfoot results in less incidence of Alfalfa mosaic.

2.5.2 MONITORING

Regular monitoring is essential to the timing of operational pest management. An efficient monitoring program can pay big dividends in depressing pest control costs. By detecting problems early, one can limit chemical use by treating at the primary inoculum site and/or using reduced rates. It is usually not desirable or possible to examine every plant or square inch of ground. Standardized sampling should be used in the representative areas to gain enough information to direct decisions for the whole field or location. An AESA, based on crop health at different stages of growth, pest and natural enemies population dynamics, soil condition, climatic factors, and farmer's past experiences are considered for decision making. Field scouting, use of sticky traps, pheromone traps, soil sample analysis for soil-borne pathogens and nematodes are usually employed. Diagnostic techniques, ETL, and pest forecasting models are available to give assistance in the proper timing of IPM interventions.

2.5.3 INTERVENTION

The aim is to minimize crop damage by pest populations to a tolerable level. Mechanical, biological, and chemical control measures are applied individually or in the grouping, taking into account costs, benefits, available labor force, timing, machines/tools, and control agents, along with ecological and environmental effects. Key IPM intervention measures available for farmers to reduce the effects of pests include the following.

a. Cultural and Physical Control

Before containing such methods in IPM approvals their effect on yields and requirements for labor (family labor) inputs need to be critically assessed.

Dry fallowing is generally followed where other methods of disinfestations of soil pathogens are not economical. The field is kept fallow for a period of one season to kill the pathogen by starvation. Wet fallowing involves frequent irrigation during the fallow period. It is usually practiced for some weeks. The main aim is to make pathogen germinate in the soil, which later dies due to lack of host plant. Wet fallowing reduces the pathogen population especially, of sclerotia, which retains viability in dry conditions for several years. It is also successful in reducing the population of *Pythium* and *Alternaria.* Flood fallowing is keeping the fallow land submerged. Under such conditions, the pathogens die due to a lack of oxygen. Deep plowing the hoeing help to bury resting propagules of pathogens present in soil or expose to harsh climate for inactivating as well as decay. Tillage operations tend to reduce soil pests by changing physical conditions of the soil, by eliminating the main or alternative host plants and by hastening the growth or increasing the vigor of crops. Plowing also helps in reduction of pest inoculums. Disc plowing does not open the furrows and, therefore, eggs/grubs are not exposed, to be picked up by predatory birds. However, the bullock-drawn desi plow opens furrows and exposes the different stages of white grubs, which are eaten away by the friendly birds (predatory birds). Further, the plowing with bullocks is generally started in the morning hours, when birds are awake and they are able to follow the track of the plow. Alternatively, the tractor plowing is similarly done in the night when the birds are at rest. The known practices are as follows:

- In brassica and millet downy mildew (*Peronospora* spp. & *S. graminicola*) and white rust (*Albugo candida*) oospores are harbored in the crop residues.
- Incidence of footrot of groundnut and chickpea (*Sclerotium rolfsii*) can be minimized with, deep plowing which destroys sclerotia in the stubble.
- Soil tillage with repeated plowing destroys the cysts of nematodes such as *Meloidogynae incognita.*

Sanitation and physical removal of inert plant materials, which are injurious to health of crop is the crucial factor governing disease buildup. Removal of crop residues tends to kill pests like beetles and caterpillars that hibernate or pupate in plant debris. After the harvest, the leftover should be destroyed by deep burying in the soil. It has been successful in reducing the infesting pests, borers in particular (of the preceding crop) in order to protect the new one.

Destruction of crop residues helps in reducing foot rot of groundnut and downy mildew of crucifers. Removal and destruction of infected plants check the secondary spread.

- Destruction of stubbles reduces the incidence of brown spot disease of rice caused by *Helminthosporium oryzae.*
- Deep burying of debris not only deprived the newly emerged nymphs of *Drosicha mangiferae* of their initial food and shelter but also exposed them to adverse conditions thus bringing their population to a manageable level.
- Similarly caging of gin trash kills hibernating larvae/emerging adults of pink bollworm (*Pectinophora gossypiella*) in cotton at the same time allow conservation of emerging parasitoids from affected larvae.
- Black scurf (*R. solani*) of potato can be controlled by soil solarization. In the summer months, the surface soil is exposed by light tillage/

interculture and then covered by thin transparent polypropylene sheets. The heat thus generated inside kills many soil-borne pests especially nematodes.

In the case of trees, severely affected plant parts should be pruned and destroyed by deep burying. (e.g., citrus canker and scale insects). Tool contamination especially the pruners needs to be avoided. During rainy season, physical injury to foliar parts needs to be avoided to reduce spread of bacterial diseases. Cutting knife is to be washed with formalin or disinfectant before cutting the fresh propagating materials. Composting of disease-infested materials for leaf manure should be avoided. Decomposing organic matter acts as breeding sits for Rhinoceros beetles (*O. rhinoceros*) in coconut.

Handpicking of egg masses, nymphs or larvae, and lethargic adults is sometimes practicable. The rice stem borer egg masses and groundnut red hairy caterpillar, gregarious larvae of *Amascta albistriga* and *S. litura* in their early stages, hairy caterpillars, snake gourd semilooper in their grown-up stages, the adults of *Epilachna* beetles on brinjal and cucurbits can be easily handpicked and killed. Insects like leafhoppers, ear head bugs, and grasshoppers can be netted and killed. Adults of Rhinoceros beetle are hooked out from the coconut crowns by means of an iron hook. Passing a rope across a rice crop is sometimes done so as to dislodge the case worms over the standing water.

Infection of *P. solanacearum* in potato and banana and that of *Colletotrichum falcatum* in sugarcane can be significantly prevented adopting above-mentioned sanitary measures.

- Seed solarization during the month of May and June is also effective in the smut of wheat (*Ustilago* sp.).
- In the case of pigeon pea shaking of plants results in dislodging of overgrown larvae which in turn are collected for use in preparation HaNPV.
- In case of orchards, 1 feet channels filled with water provide significant protection against ants and other beetles which not only cause the physical damage but also acts as vectors for viral disease.

The use of straw mulch or yellow polythene mulch repels certain insects, which are potential vectors of some viral diseases, and thus help in checking insect pests as well as insect vectors for transmission and spread of viral diseases. Underbalanced nutrition, a crop attains maximum resistance against pests and diseases. Due to excess application of nitrogenous fertilizers, plant foliage becomes sappy and it attracts insect pests. Adequate irrigation and proper drainage reduce the disease occurrence. Just in the case of vegetable crops, ridge planting is helpful to circumvent some diseases, especially causing stem or root rot. Nursery bed should also be raised. Tomato leaf curl can be controlled by repelling whiteflies with plastic mulching. Trash mulching or earthing up for one month or two after planting minimizes shoot borer attack in sugarcane.

b. Pheromones

With regards to insect pest management the population dynamics of a given species may not be precisely known but, long-term monitoring can help to elucidate patterns in population life cycles. These patterns can then be combined with the known biology of the insect to define the parameters of an effective control campaign. Over the last two or so decades, information collected from insect monitoring has led to the development of thumb rules of pest forecasting systems that allow farmers to begin spraying at precise time intervals for maximum impact. In the agricultural point of view, spraying may singly be effective for a short period, for example, when an insect stage is external on a crop plant, if this period is missed, and the larvae burrows into the plant then spraying is not cost-effective. This short period when spraying should take place is termed the "open window." The other method is mate disruption which involves the discharge of large quantities of synthetic female sex pheromone causing males to become confused and unable to locate females. The pheromones are also used in "lure and kill" strategies to attract the pest to contained insecticide deposits and reduce the requirement for whole crop spraying (as demonstrated against PBW in Punjab). Mating disruption to delay or diminish the population has been successfully demonstrated in the fruit fly.

c. Biopesticides

Biopesticides have the potential to manage phytophagous insect insects and plant diseases.

d. Biological Control

The selected organism can be a parasite, predator, or disease, which has a tendency to manage pests. Cost-effective technologies for a number of bioagents, such as the egg parasitoids like *Trichogramma japonicum* and

T. chilonis, larval parasitoids *Bracon hebetor, B. brevicornis, Goniozus* sp., *Brachymeria* sp, and predators, namely, *Chrysoperla carnea, Chrysoperla incarnata, Coccinella septumpunctata, Menochilus sexmaculatus* have been standardized and are now available commercially. Entomopathogenic fungi *Metarhizium anisopliae, Beauveria bassiana, Nemourea rileyi*, etc., is also widely used against several lepidopterous pests and specific success has been achieved in case of white grub. Viral pathogens, the GV and NPV are used to manage *Spilosoma obliqua, Amsacta albistriga S. litura, H. armigera.* Commercial preparations of *Bacillus thuringiensis* have become popular for the management of *P. xylostella* and *H. armigera.* The bioagents and microbial pesticides (Bt., fungal, and baculoviruses) have been widely used in the management of pests and avoiding the development of resistance. Entomopathogenic fungi and bacteria paralyze or kill their hosts by adversely affecting growth and development of host insects and their formulations are commercially available. In absence of popularity, it market is very negligible with total consumption of only 15–20 metric tonnes against the potential of 30 million tonnes. The production of biopesticides in the country is not adequate enough to cover 4–5% area under cultivation. Nevertheless, about 10–15% growth rate is envisaged for in the country. There is a sudden spurt and currently more than 760 units are involved in the production of numerous quality biocontrol agents.

However, the potential of the bioagents had not been exploited to the core. The natural enemies would keep 60% of the insect pest populations in nature under check, if there is no intervention by chemical pesticides. Diverse plant communities also provide shelter and alternate prey for many natural enemies.[44] Conservation and augmentation of bioagents hence is of significance in IPM. The present level of production of bioagents is not even sufficient to cover 4–5% area under any particular crop. The major constraints for the promotion of the bioagents include the difficulties in mass rearing technologies, temporal variations, the shorter life span, and delay in transit.

e. Botanical and Chemical Control

Botanicals sources of secondary metabolites or allelochemicals that provide for the fundamental or physiological or biochemical process of the plant. A number of plants such as Neem (*Azadiracta indica*), Pithraj or bullock's heart (*Amoora reticulate*), Mahua (*Madhuca longifolia*), Karanj (*Pongamia glabra*), Pinnae (*Calophyllum inophyllum*), Besharam (*Ipomoea sepiaria*), (*Polygonum hydropiper*), Gohkru (*Tribulus terristis*), Gandhigrass

(*Parthenium hysterophorus*), Haldi (*Curcuma longa*), Sweetflag (*Acorus calamus*), Kusum (*Schleichera trijuga*), Custard apple (*Annona squamosa*), Lantana (*Lantana kamara*), Periwinkle (*Vinca rosea*), Garlic (*Allium sativum*), *Artemisia absinthium*, citrus grass (*Cymbopogan citratus*), Dhatura (*Dhatura metal*) *Cassica auriculata*, and Kolingi (*Tephrosia purpurea*) have pesticidal properties. Amongst the botanical pesticides, six plants, namely, Neem, Pongamia, Cymbopogan, Annona, *Chrysenthemum*, Tobacco have been utilized and to some little extent poison vine or tuba root (*Derris elliptica*) as a source of Rotenone and Notchi (*Vitex negundo*) as fungicide have also been in use as botanical pesticides, though mostly in traditional pest control. Pure triterpenoids from neem oil has great potential as antifungal against *Dreschlera oryzae*, *F. oxysporum*, and *Alternaria tenuis*. However, only the products based on Azadirachtin, Cymbopogon, Pyrethrum, and Nicotine (currently banned for internal use) were registered with Central Insecticides Board. Around 735 botanical pesticide products from various companies stand registered with C.I.B. and these include 443 Azadirachtin based, 290 are pyrethrum based and one *Cymbopogan*-based products. In the case of cruciferous vegetables crops, the soap water of karanj soaps have proved significant over other chemical sprays. Other plants such as *Oscimum, melia*, *M. longifolia*, and *Calotropis* are being explored for their pesticidal properties.

The IPM works with the goal to reduce use of most chemical inputs, but they can still be helpful from time to time in case of emergency. Insecticides formed the key tools for the management of pests and diseases over the years. Their misuse and abuse incited several environmental hazards, the insecticides still remain to play a chief role in all pest management programs. However, with the growing concern about their ill effects, public demand for residue-free food, and policies to promote IPM with biorational methods, use of chemical-based pesticides would be reduced considerably in the years to come. Under the IPM strategy, the compounds that are compatible with bioagents to support production without degrading the ecological resource base are used.

2.6 DECISION SUPPORT AND AREA-WIDE MANAGEMENT SYSTEM

Farmers need assistance in deducing pest monitoring data. The mobile apps like "Expert systems" may be developed to farmers in different ways, including simple pegboards or charts, special booklets, radio, and television programs or more advanced aids such as prediction models and computer-based systems (available for cotton in India). Development and provision of the latest information is a key factor enabling farmers to implement IPM programs. IPM requires collaborative decisions within a national or more localized area to provide effective pest control. Some of these decisions need to be taken centrally by government with respect to restriction of pest and infested movement through implementation of quarantine regulations and legislation. There should be an adequate set-up for training of advisory services. The establishment of resistance management strategies against highly mobile pests and diseases on major crop systems is the need of the hour. Geographical information systems and remote sensing techniques are being used in IPM programs.

In India, ICAR-NCIPM, IPM program with 13 states institutes deliver, associate, and acclimatize new ecosystem-based technologies for a major pragmatic impact: 64% reduced insecticide sprays (from 9.8 to 3.5 per season), 98% reduced use of dust insecticides, increased use of real biopesticides like viruses to kill pod borers, botanicals like neem, and more selective insecticide molecules to conserve natural enemies, resulted 33–45% profit to smallholder farmers, 14–17% higher yields in pigeon peas and chickpeas, and 9–10.5% lower costs of production. What makes them remarkable is that they were implemented in the largest pulse production intensification program in the world, operating at the district level, and they are showing the way to make the system that successfully increased pulses production (from 14 to 19 million tonnes in 3 years making them more sustainable as well.[45]

2.7 CONCLUSION

To increase yields from the available land, which has not increased significantly during the last two decades, requires good crop protection against pre- and post-harvest losses. While on the one hand, productive land is contaminated with pesticide residues, on the other, substantial revenue is lost due to crop pests, which cannot be simply ignored as by reducing it can directly increase production. In many parts of the country, such as Uttar Pradesh, Punjab, and Haryana, human and soil health are in danger due to excessive use of pesticides. The ill-effects of chemical pesticides are not visible; however, they cannot be ignored due to lack of scientific manpower and infrastructure. Thus, pesticide residues in soil and resistance to pesticides, which have surpassed previous records, are the bane of the Indian farmer. The pesticide resistance developed due to multiple sprays of pesticides at times led to mass suicides in states like Andhra Pradesh and violent agitation in Punjab.

Contributions by Indian scientists to different constituents of IPM system for various field crops are well documented. The development of disease-resistant crop varieties, biological control of pests/diseases, manipulation of cultural practices to reduce pest/disease incidence, and the use of botanicals, such as neem are the practices largely adopted by farmers in general. Additionally, crop diversity, intercropping, favored the implementation of IPM at the farm level of small as well as medium farmers. The DAC&FW has been instrumental in providing financial assistance to various Department of Agriculture and Agricultural Universities for developing and producing biopesticides, biocontrol agents viz., NPV, GV, *Trichogramma, Chrysoperla,* and *Trichoderma.* Presently, 10 biopesticide production units at semicommercial level are able to cover an area of 1 million ha./annum in 10 crops against major insect pests in various regions of the country. Under this program, biopesticide production units and plant clinics at regional research stations have also been established to cater local needs. Microbial pesticides—stable formulations of (bacteria, viruses, nematodes) have been achieved. The technology has been transferred and being promoted by GOI, DST, and ICAR schemes.

IPM is built on detailed knowledge of pest biology, behavior, and ecology, not simply chemistry and toxicology. There after by the 1970s, adequate research was conducted to fill the technological gap and provide the knowledge to successfully implement IPM programs in economically important crops, such as rice, cotton, sugarcane, and few cash crop including vegetables. Salient IPM technologies for few economic crops have been shown in Table 2.3.

The IPM requires a major change in attitude about the way we currently remove pests and use pesticides, even at home and kitchen gardens. Managing pests requires more than reading a pesticide label and applying recommended dose. IPM seeks to decrease dependency on pesticides by using various practices for protecting the environment, for example, IPM for the kitchen garden includes planting well-adapted varieties that may naturally resist pests, keeping plants healthy and vigorous, encouraging natural pests' enemies like ladybugs, green lacewing, and spiders and, if required, use pesticides giving preference to botanical that is less lethal. IPM has made maximum utilization of conditions and methods that control pests naturally. NCIPM has field validated and perfected as how IPM can best be used in agriculture to get sustainable produce while maintaining the clean environment. While IPM can be a novel concept to many of us, it is an old practice to the farmers of India.

TABLE 2.3 Available IPM Technologies for Different Crops.

Crop	Key pest(s)	Management tactics
Rice	Brown plant hopper (BPH), gall midge, stem borer	Resistant varieties, pest monitoring, conservation and mass release of biotic agents, cultural methods
Sugarcane	Tissue borers, pyrilla, scale insects	Resistant varieties, augmentation and release of parasites (*Trichogramma, Epicarnia, Strumiopsis inference, Isotima javensis*, GV, etc.), need-based pesticides
Cotton	Boll worms, jassids, white flies	Mass release of biotic agents (NPV, *Chrysoperla, Trichogramma* spp., *Bracon* spp.), need-based pesticides
Groundnut	White grubs	Cultural and mechanical means, seed treatment with safe pesticides
Tobacco	*Helicoverpa*	Employing castor as a trap crop which acts as cover crop from harboring natural enemy. Spraying neem seed kernel extract on tobacco seedlings and crops. Spraying NPV
Cruciferous vegetables	Diamond back moth, other Lepidoptera	Biotic agents, trap cropping, need-based use of safe chemicals, neem B.T.
Mango	Leaf hoppers, mealy bugs	Tree banding, fungal pathogens
Citrus, Grapes, Guava	Mealy bug	Mess release of *Cryptolaemus montrouzieri* and *Laptomastrix dactylopii*
Coconut	Rhinocerous beetle	Use of baculovirus
Aquatic weeds	Water fern and water hyacinth	Introduction and mass release of exotic weevils, *Crytobagous salvinae* (for water fern), and *Neochitina* spp. (for water hyacinth)

The principles of IPM are crucial for sustainable agricultural production systems. Decision support software needs to be developed to allow IPM practitioners to estimate cost/benefit for a variety of management inputs and examine profitability of a system. Genetic engineering is being used to enhance potentiality of live modified organisms to increase productivity by reducing damage due to pest (insects, diseases, and weeds). Present examples include Bt corn, Bt cotton, Bt brinjal glyphosate-resistant soybeans, and transgenic wheat. Chemical pesticides are to be used wisely and judiciously in integrated pest control methods.[46,47] The IPM is the dynamic in nature and generates the local capacity for better sustainability.[48]

Different countries are now adopting different approaches to promote IPM as World Bank has made it mandatory to adopt 4.02 conditions for the projects financed by them.

To establish an IPM program, the following basic plans should be followed.

- IPM is knowledge-intensive making farm advisory services to play important role crop production in present time. Farm advisory service as proven under National Food Security Mission, under Government of India can be more effective in aiding to reduce dependency on pesticides.
- Understand the biology of the crop and identify the key pests as per the growing stage, know their biology, recognize the damage they inflict, and evaluate their economic status.
- Detect the basic environmental factors that affect pest and potential pest species in the ecosystem.
- Consider concepts, strategies, and materials, individually and combined, will help to perpetually suppress pest.
- Anticipate unexpected developments and move with attention to face the changes that can occur within it.
- Prevention is better than cure and considered as the formation of cropping systems which is less likely to experience substantial economic losses.
- Seek the weak associations in the life cycle of the key pest species, and direct control practices at these weak associations. Avoid broad influence on local resources prevailing in prevailing ecosystem and consider and develop methods, which preserve, complement, and increase the biotic and physical mortality factors that characterize the ecosystem.
- The collaboration among institutions and state officials to benefit from these human resources and experiences is the need of the hour. Communication and interaction with IPM specialists outside the region are missing due to language barriers and limited financial resources. Collaborative projects and networking activities to nurture interaction and exchange of knowledge and information can be of immense help.

Training farmers using the FFS approach is much simpler and the most comprehensive method of transferring IPM knowledge and skills to farmers. Regular interaction with farmers during the FFS meetings strengthens farmers' confidence in decisions making on crop management including pest management decision has proved a significant reduction in the use of chemical pesticides while increasing/conserving natural predators and parasitoids.[50,51]

KEYWORDS

- **integrated pest management**
- **biopesticides**
- **biological and chemical control**
- **insect pest and diseases**
- **regulatory measures**
- **pheromone**

REFERENCES

1. Carson, R.; Darling, L.; Darling, L. *Silent Spring.* Boston: Houghton Mifflin; Riverside Press: Cambridge, Mass, 1962.
2. Van den Bosch, R. *The Pesticide Conspiracy*; Doubleday: New York, 1978.
3. Pimentel, D.; Levitan, L. Pesticides: Amounts Applied and Amounts Reaching Pests. *BioScience* **1986,** *36* (2), 86–91.
4. Daxi R.; Kayserling, K. N. V.; Koch, C. K.; Link R.; Waibel, H. *Integrated Pest Management: Guidelines GTZ*; Federal Public of Germany, 1994; p 119.
5. Mehrotra, K. N. Pesticide Resistance in Insect Pests: Indian scenario. *Pesticide Res. J.* **1989,** *1* (2), 95–103.
6. Kishor, N. M. Pesticide Externalities in Coastal Andhra Pradesh. In *Natural Resource Economics: Theory and Application in India*; Kerr, J. M., Marothia, D. K., Singh, K., Ramaswamy, C., Bentley, W. R., Eds.; Oxford and IBH Publishing Company Pvt. Ltd.: New Delhi, 1997.
7. Alam, G. *A Study of Bio-Pesticides and Bio-Fertilizers in Haryana, India. Gatekeeper Series 93*; International Institute for Environment and Development: London, 2000; p 150.
8. Lewis W. J.; Lenteren, J. V.; Sharad C.; James, T. A Total System Approach to Sustainable Pest Management. *Proc. Natl. Acad. Sci.* 1997, *94* (23), 12243–12248.
9. Barlett, B. R. Natural Predators. Can Selective Insecticides Help to Preserve Biotic Control. *Agric. Chem.* 1956, *11*, 42–44.
10. Stern, V. M.; Smith, R. F.; van den Bosch, R.; Hagen, K. S. The Integration of Chemical and Biological Control of the Spotted Alfalfa Aphid (the Integrated Control Concept). *Hilgardia* **1959,** *29*, 81–101.
11. Ehi-Eromosele, C. O.; Nwinyi, O. C.; Ajani O. O. *Integrated Pest Management, Weed and Pest Control—Conventional and New Challenges, Sonia Soloneski and Marcelo Larramendy*: IntechOpen, 2013.
12. Hoskins, W. M.; Borden, A. D.; Michelbacher, A. E. Recommendations for a More Discriminating Use of Insecticides. *Proc. Sixth Pacific Sci. Congress* **1939,** *5*, 119–123.
13. Michelbacher, A. E.; Bacon, O. G. Walnut Insect and Spider Mite Control in Northern California. *J. Econ. Entomol.* **1952,** *45*, 1020–1027.

14. Smith, R. F.; Allen, W. W. Insect Control and the Balance of Nature. *Sci. Am.* **1954,** *190* (6), 38–42.
15. Geier, P. W.; Clark, L. R. An Ecological Approach to Pest Control. In *Proceedings of VIII Technical Meeting International Union for Conservation of Nature and National Resources*, 1961.
16. Geier, P. W. Management of Insect Pests. *Annu. Rev. Entomol.* **1966,** *11*, 471–490.
17. Smith, R. F. and Van Den Bosch, R. Integrated Control. In *Pest Control: Biological, Physical and Selected Chemical Methods*; 22. Kilgore and Doutt, R. L., Eds.; Academic Press: New York, 1967; pp 295–340.
18. Council on Environmental Quality. *Integrated Pest Management*; Council on Environmental Quality: Washington, DC, 1972; p 41.
19. Kogan, M. Perspective and Contemporary Developments. *Annu. Rev. Entomol.* **1998,** *43*, 243–270.
20. Bajwa, W. I.; Kogan, M. *Compendium of IPM Definitions (CID)—What Is IPM and How Is It Defined in the Worldwide Literature?* IPPC Publication No. 998, Integrated Plant Protection Center (IPPC), Oregon State University: Corvallis, OR, 2002.
21. FAO. *Report of the First Session of the FAO Panel of Experts on Integrated Pest Control.* Food and Agriculture Organization of the United Nations: Rome, 1967.
22. NAAS (National Academy of Science). Insect-Pest Management and Control. **1969,** *3*, 448–449. Principles of Plant and Animal Pest Control. *Natl. Acad. Sci. Pub*, 1695; p 508.
23. Singh, A.; Sharma, O. P.; Lavekar, R. C.; Bambawale, O. M.; Murthy, K. S.; Dhandapani, A. *IPM Technology for Rainfed Cotton*; National Centre for Integrated Pest Management: New Delhi, 2002.
24. Smith, R. F. History and Complexity of Integrated Pest Management. In *Pest Control Strategies*; Smith, E. H., Pimentel, D., Eds.; Academic Press: New York, 1978; pp 41–53.
25. Smith, R. I.; Apple, J. L.; Bottrell, D. G. The origins of Integrated Pest Management Concepts for Agricultural Crops. In *Integrated Pest Management*; Apple, J. L.; Smith, R. F., Eds.; Plenum Press: New York, 1976; pp 1–16.
26. Frisbie, R. E.; Adkission, P. L. IPM: Definitions and Current Status in U. S. Agriculture. In *Biological Control in Agricultural IPM Systems*; Hoy, M. A.; Herzog, D. C., Eds.; Academic Press: London, 1985, pp 41–51.
27. Kogan, M. Perspective and contemporary developments. *Annu. Rev. Entomol.* **1998,** *43*, 243–270.
28. Allen, W. A.; Rajotte, E. G. *Ann. Rev. Entomol.* **1990,** *35*, 379–397.
29. Pedigo, L. P. *Entomology and Pest Management*; Macmillan Publ. Comp.: New York, 1991, p 356.
30. Luckman, W. H.; Metcalf, R. L. The Pest Management Concept. **In** *Introduction to Insect Pest Management*; Metcalf, R. L., Luckman, W. H., Eds., 3rd ed.; John Wiley & Sons, Inc.: New York, 1994; pp 1–34.
31. Van Veldhuizen, L.; Hiemstra, W. Cutting back Cure or Prevent? *Ileia Newslett.* **1993,** *9* (2).
32. Dubey O. P.; Sharma, O. P. Policy and Perspective of Plant Protection Practices of Major Pulses. In *Proceedings of National symposium of Ind. Soc. of Pulses Research*, Ali, K. M., Chaturvedi, S. K., Gurha, S. N., Eds.; 2002; pp 87–91.

33. Jalali, B. L.; Sharma, O. P. Biocides and Non-Target Microorganisms: An Environmental Assessment. *Indian J. Microbiol.* **1993,** *33* (2), 83–92.
34. Sharma, O. P.; Lavekar, R. C.; Murthy, K. S.; Puri, S. N. *Habitat Diversity and Predatory Insects in Cotton IPM: A Case Study of Maharashtra Cotton Eco-System.* Radcliffe's IPM World Text Book, 2000.
35. Paroda, R. S. *Sustaining Our Food Security*; Konark Publishers Ltd.: Delhi, 2003; p 775.
36. Puri, S. N.; Murthy, K. S.; Sharma, O. P. Pest Problems in India-Current Status. *Indian J. Plant Protection* **1999,** *27* (1, 2), 20–31.
37. Dubey, O. P, Sharma, O. P. *Pest Management for Safer Environment*; DIPA, ICAR: New Delhi, 2005; p 64.
38. Muniappan, R.; Merle Shepard, B.; Watson, G. W.; Carner, G. R. First Report of the Papaya Mealybug, Paracoccus marginatus (Hemiptera: Pseudococcidae), in Indonesia and India. *J. Agric. Urban Entomol.* 2009, *25* (Jan 2008), 37–40. DOI: 10.3954/1523-5475-25.1.37
39. Lewis, W. J.; Joop van Lenteren, Sharad C.; James, T. A Total System Approach to Sustainable Pest Management. *Proc. Natl. Acad. Sci.* **1997,** *94* (23), 12243–12248.
40. Wood, B. J. Pest Control in Malaysia's Perennial Crops: A Half Century Perspective Tracking the Pathway to Integrated Pest Management. *Integr. Pest Manage. Rev.* **2002,** *7*, 173–190.
41. United Nations Conference on Environment and Development, Rio de Janeiro, Brazil, 3–14 June 1992.
42. National Agricultural Policy, Ministry of Agriculture, DAC, GOI. 2001. http://nlsenlaw.org/agriculture/domestic-legal-framework/national agricultural-policy/
43. Singh, A.; Sharma, O. P.; Jafri, A. A. Traditional Methods of Pest Management as Components of IPM. *IPM Mitra* **2001,** *2*, 34–44.
44. Wilkinson, T. K.; Landis, D. A. Habitat Diversification in Biological Control: The Role of Plant Resources. In *Plant-Provided Food for Carnivorous Insects*; Wackers, F. L., van Rijn, P. C. J., Bruin, J., Eds.; Cambridge University Press: Cambridge; New York, 2005.
45. Sharma, O. P.; Bambawale, O. M.; Yelshetty, S.; Patil, B. V.; Gopali, J. B.; Bhagat, S.; Khullar, M.; Neeraja, A.; Rachappa, V.; Patil, V. *Innovative Approach in Sustainable Production of Pigeonpea & Chickpea– Success Story of Gulbarga*; Published by ICAR-NCIPM: New Delhi, 2012, p 64.
46. Barzman, M.; Lamichhane, J. R.; Booij, K.; Boonekamp, P.; Desneux, N.; Huber, L.; Kudsk, P.; Langrell, S. R. H.; Ratnadass, A.; Ricci, P.; Sarah, J. L.; Messean, A. Research and Development Priorities in the Face of Climate Change and Rapidly Evolving Pests. *Sustain. Agric. Rev.* 2015, *17* (in press).
47. Lamichhane, J. R.; Barzman, M.; Booij, K.; Boonekamp, P.; Desneux, N.; Huber, L.; Kudsk, P.; Langrell, S. R. H.; Ratnadass, A.; Ricci, P.; Sarah, J. L.; Messean, A. Robust Cropping Systems to Tackle Pests Under Climate Change. A Review. *Agron. Sustain. Dev.* 2015, *35*, 443–459.
48. Reddy, P.; Lakshmi, Sharma, O. P.; Reddy, A.; Mallishwara, Joysthana, M. K.; Kumari, A. Implementation of IPM Practices in Farmers' Participatory Mode in Andhra Pradesh and Impact on Chickpea Pest Scenario, with Special Reference to Rust Disease. *Indian J. Plant Protection* 2015, *43* (1), 128–130.

49. Dachbrodt-Saaydeh, S. The Policy Perspective-How EU Member States Promote IPM Implementation? Paper Given at IPM innovation in Europe, Poznan, Poland, 2015.
50. Sharma, O. P.; Bantewad, S. D.; Patange, N. R.; Bhede, B. V.; Badgujar, A. G.; Ghante, P. H.; Kadam, M.; Bhagat, S.; and Kumari, A. Implementation of Integrated Pest Management in Pigeonpea and Chickpea Pests in Major Pulse-growing Areas of Maharashtra. *J. Integ. Pest Manage.* 2015, *15* (1), 12. DOI:10.1093/jipm/pmv011.
51. Bhede, B. V.; Bhosle, B. B.; Badgujar, A. G.; Sharma, O. P. Impact of Integrated Pest Management on Gram Pod Borer *Helicoverpa armigera* and Its Natural Enemies. *J. Entomol. Res.* 2014, *38* (4), 269–272.

CHAPTER 3

Innovative Integrated Pest Management Paradigm for Sustainable Crop Production with Special Reference to North East India

PRANAB DUTTA[1*], ASHOK BHATTACHARYYA[2], and ARTI KUMARI[3]

[1]*School of Crop Protection, College of Post Graduate Studies in Agricultural Sciences, Central Agricultural University, Umiam 793103, Meghalaya, India*

[2]*Former Vice Chancellor of Assam Agricultural University (AAU), Presently Director of Research (Agri.), AAU, Jorhat 785013, Assam, India*

[3]*Department of Plant Pathology, Assam Agricultural University, Jorhat 785013, Assam, India*

[*]*Corresponding author. E-mail: pranabdutta74@gmail.com*

ABSTRACT

The Northeastern (NE) region of India is bestowed with nature's blessing, rich biodiversity, and congenial environment for agriculture and horticulture. The NE states are well known for organic farming and IPM practices with limited use of chemical pesticides. However to enhance the agricultural production, it is very essential to understand the potential sources of growth, appropriate technologies, and investment accordingly. Thus to meet the demand of agricultural sector, Integrated Pest Management (IPM) is an economically viable and environmentally sustainable pest management alternative. IPM can be defined as a holistic strategy of pest management that amalgamates various control measures such as cultural, physical, mechanical, chemical,

Integrated Pest Management in Diverse Cropping Systems. Devendra Pal Singh, PhD (Ed.)

and biological measures in an effective manner to subdue pest population level below the level of economic threshold. It includes pest management techniques to diagnose, manage, and reduce risk on ecological health. Biological control measures have gained momentum in the last few decades over chemical control measures, which compromise the sustainable aspect of pest management. Tremendous research has been done at AAU, Jorhat, CAU, Umiam with the development of effective pest, nutrient management technologies having wider social acceptability, which proved to be a boon for farming community of the region. Nanotechnology is also emerging as a potential tool in effective management of pest, diseases, and weeds, and also helps in precise detection, diagnosis, monitoring of pest, and delivery of drugs. For successful implementation of IPM practices, advanced planning, knowledge of pest, resources, good agricultural practices, cohesive networking, interaction of all stakeholders, and cost-benefit management plays an important role. Thus, apart from management aspect, sustainability, business aspects are the two important pillars of new IPM paradigm which ensures profitability to the growers and affordability to consumers and sustainability to environment.

3.1 INTRODUCTION

The Northeastern (NE) region of India comprises of eight states namely Assam, Arunachal Pradesh, Meghalaya, Manipur, Mizoram, Sikkim, Nagaland, and Tripura and located between 22°05′, 29°30′N latitudes, 87°55′, 97°24′E longitudes. This region is bestowed with diverse geographical features comprising of 56% low altitude (valley or low land), 33% area is under mid-altitude flat upland and remaining under high altitude (upland terrace). Traditionally, the farmers in NE region practice organic farming since centuries. In several areas, jhum cultivation is being practiced for every 5, 10, and 20 years cycle with mixed cropping, leaving the cultivated field under bare fallow by default.

In the NEH region, Agriculture and Horticulture being the mainstay of the people has tremendous potential to flourish and enhance farm income by ensuring food and nutritional security, alleviating poverty, and thus accelerating economic growth of the region. Traditionally, agriculture in the NE states has been looked upon as a subsistence profession generating food for household consumption and ignoring its potential for commercialization, and growth. In view of its potential in spurring economic growth, and improving the livelihood of rural population, Agriculture and Horticulture started receiving increased attention of the policymakers in recent years. The NE Region Vision 2020 places considerable emphasis on enhancing agricultural and horticultural growth through technological change, area expansion, cropping intensity, and value-addition by attracting private investment in agro-processing. However, to unlock the growth potential of agriculture, it is essential to have comprehensive insight of the past and potential sources of its growth and target technologies and investment accordingly.

The major challenge of the 21st century is to feed the ever-increasing population throughout the world. To meet the increasing food demand food production system needs to be strengthened to its maximum efficiency. To maximize crop production per unit area, adoption of improved agricultural technologies in order to minimize crop losses should be encouraged. The use of chemical pesticides have contributed immensely toward making the country self-reliant in food grain production. However, the overuse of pesticides has caused an adverse impact on soil and ecological health. The negative outcomes of its overuse include deterioration of soil health, pesticide resistance, pest resurgence, elimination of natural enemies, negative effect on nontarget organisms, environmental toxicity, biomagnification of residues at higher trophic levels, and reduced crop yield. Thus, the central plank for protection of crops adopted by the Govt. of India was the Integrated Pest Management (IPM) approach since 1985.

Thus, to meet the ever-increasing demand of quality food products, IPM was adopted as economically viable, environmentally sustainable pest management strategy. Although, IPM components such as various cultural, mechanical, and physical control measures against numerous insect pests were developed by farmers a long ago, however, the IPM concept arouse due to ill effects caused by extensive use of pesticides. The term "Integrated Pest Control" was first coined by Barlett[1] as "integrated control" for the use of biological, cultural control measures to successfully manage pests of agricultural importance. IPM can be described as a comprehensive strategy that includes a combination of various control measures such as cultural, mechanical, physical, biological, and chemical to curtail population of insect pests below economic threshold. The United States Department of Agriculture—Agricultural Research Service[2] defined IPM as science-based, sustainable, and decision-making process, which aims at reducing the pest population level below economic threshold level by using numerous pest control techniques to identify, manage, and reduce risk posed by pests, pest management tools, approaches, and lessen the ecological and economical risks. The NE part of India is known for the production of safe and quality food products as they resort to traditional and IPM strategies for crop production. However, to increase the productivity of crops integration of management practices scientifically as per IPM programmes developed by researchers should be emphasized. The previous models of IPM were based on ecological and evolutionary aspects of management of insect pests.[3] With the development in agricultural technology, modern communication tools, globalization of trade, commerce, and increased awareness for quality food products among consumers emphasized the need to reconsider the earlier IPM paradigm and alter it as suitable for modern times. The new model is based on an earlier foundation based on ecological and economic aspects, is stretched to include management, business, sustainability aspects, also stresses the need to research, and extend the technologies to the ultimate beneficiaries. The new IPM paradigm is formulated including management and business and sustainability aspects with the objectives as cited below:[4]

- Reconfiguration of IPM concept as appropriate for recent times and encourage sustainable crop production system.
- To increase confidence, spread awareness on IPM-based sustainable system of crop production that fits different crops, agroclimatic conditions and ensures food security and social equality.
- To increase profitability of producers where informed consumers are given right to make food choices rather than vested interest groups.
- To minimize effect of conventional non-IPM-based practices on environment and tackles the challenges of managing certain insect pests.

3.2 MANAGEMENT ASPECT

The IPM model consists of four major components to address pest management challenge, which includes knowledge, resources available with the growers to manage pests, planning, organization of information to initiate management options at appropriate time, maintaining good communication channel to access, and disseminate knowledge effectively.

3.2.1 PEST MANAGEMENT

The concept of management of pests has changed over the years from complete eradication of pest to managing pest population level below those causing economic injury. The NE region of the country implements IPM strategy to effectively utilize natural resources with reduced dependence on chemical pesticides. The farmers of the region have evolved effective pest management practices due to their intimate experience with nature and natural resources. The various pest management tactics for sustainable crop production include host plant resistance, cultural, mechanical, physical, biological, and chemical control. With the advancement in science, nano-technology has demonstrated considerable potential to develop as one of the advanced tool for pest and disease management.

3.2.1.1 HOST PLANT RESISTANCE

The use of disease-resistant varieties for sustainable production of crops is the primary defense line in IPM. The pest resistant, tolerant varieties are developed through various conventional breeding, genetic engineering techniques. The disease-resistant gene makes the plant less attractive for the pest, pathogens to grow, develop, feed, and reproduce. It also increases the ability of plants to withstand the damage caused by pest during the growth period of plants.

3.2.1.2 CULTURAL CONTROL

The cultural control practices aim at making the crop environment hostile to pest and favorable to crops. Cultural control practices are preventative in action based on brief knowledge of prevailing crop pest, natural control mechanisms, and ecological relationships. The various cultural control practices include manipulation of planting, harvesting dates, spacing, crop rotation, fallowing, summer plowing, proper drainage, destruction of crop debris, planting cover crops, trap crops, intercropping, etc. Planting green manuring crop such as dhaincha (*Sesbania bispinosa*), incorporation into soil prior to planting main crops such as rice increases nutrient, organic matter content for the subsequent crop. Several green manuring crops also possess soil pest, disease suppressing abilities. Intercropping with crops such as sorghum, pigeon pea, maize, calotropis, various legume crops namely, Nepalese alder (*Alnus nepales*) also reduces spread and damage caused by insect pest.[5] Growing of border crops such as pumpkin, cucumber, okra, horticultural crops such as pineapple, banana attracts pest of vegetable crops, thus safeguards crop against the damage caused by pests. Also, maintenance of balanced nutrient supply and moisture content in soil is prerequisite for development of healthy plants. Crop rotation with nonhost crops helps to break the pest and disease cycle thereby reducing their build-up year after year. Ploughing exposes the soil-inhabiting pests and destruction of crop residues eliminates soil-borne inoculum and their breeding sites.

The cultural control methods of pest management include:

3.2.1.3 SELECTION OF SITE

Selection of suitable site is crucial in IPM to avoid pest infestation. An ideal planting site is suitable for the crop and natural enemies but unfavorable for insect pests. The factors that govern the selection process include history of planting site about occurrence of pest in the location, occurrence of potential overwintering sites and crops grown in the surrounding localities, soil conditions, presence of weeds, field borders, hedges, climate, topography, etc. The planting site should be well-drained, optimally fertilized, and manured. The emphasis must be laid on to avoid stressing the crop as stressed plants emit signals that attract pests and diseases.

3.2.1.4 CROP ISOLATION

The location of crops and the degree of isolation from other crops grown in the vicinity determines the probability of infestation of crops by insect pests. The crop must be isolated from older crops of same species, from closely related plant species that acts as source of inoculum and increases the spread diameter. This practice is most suitable for annual crops where predisposition factors are not congenial. Isolation of sequentially grown crops in time disrupts the continuity of host plants and thus prevents easy dispersal of pest species. For successful implementation of this technique, it requires cooperation from neighboring producers.

3.2.1.5 PLANTING DENSITY AND SPACING

To obtain maximum productivity with uncompromised crop quality, maintenance of optimum planting density, spacing is very crucial. The spacing affects the relative growth of the plant, also the pest population build up per unit of time, insect behavior, oviposition, etc. Some insects such as aphids on brassicas are attracted by low plant density, whereas others may be attracted by higher density of crops. Sometimes, closer spacing of plants may increase the effectiveness of natural enemies thereby resulting in greater control of insect pest.

3.2.1.6 TIME OF PLANTING

Planting time is crucial to allow the establishment of young plants to a tolerant stage before the pest attack, to reduce the susceptible period of pest attack and to avoid oviposition period of a particular pest. It also allows the crop to mature before a particular pest becomes abundant, provides sufficient time for the crop to compensate for the damage, and gap filling is very essential.

3.2.1.7 INTERCROPPING

Intercropping means "Growing multiple crops on the same land at the same time in definite rows." Intercropping with susceptible crops tend to attract the pest species thereby protecting the main crop. It also acts as a barrier in spread of the pest and also confers associational resistance by masking the odor of the main crop.

3.2.1.8 CROP ROTATION

Crop rotation is effective when crop belonging to a plant family is succeeded by another crop of different family. The common example of crop rotation includes rotation of cereals with legumes, root crops, and vegetable crops. However, crop rotation is effective against pest having limited range of host plants and depressiveness that cannot thrive for more than one season in absence of suitable host plant.

3.2.1.9 MANAGEMENT OF ALTERNATE, VOLUNTEER, AND TRAP CROPS

The weed plants, volunteer plants, collateral, and alternate hosts harbours many insect pests, diseases, and acts as focal point for future attack on crops.

Thus, its destruction is essential for reducing the pest attack in the main crop. Trap crops act as a more attractive food source and divert pest species away from the crop thus saving the crop. The trap crop later needs to be destroyed before the insect and pathogen reproduce.

3.2.1.10 TILLAGE

Tillage helps in controlling the soil-inhabiting crop pests by exposing the dormant structures of pests to heat, predators, birds and animals. Tillage also damages the soil-inhabiting phase of pests as well as crop residues, which might be the reservoir of several species of pest.

3.2.1.11 FERTLIZATION AND MANURING

Proper fertilization of plants enhances rapid growth, provides tolerance to, and ability to compensate for and damage caused by pest. The common fertilizer elements may have explicit and implicit effects on suppression of pests. Excess of nitrogenous fertilizer is known to increase pest incidence by making the plants more succulent. Phosphatic and Potassic fertilizers are well known to impede the level of pest and disease infestation making the plant tolerant. Thus, efforts should be made to ensure that plants are optimally fertilized avoiding nutrient stress and over-fertilization.

3.2.1.12 IRRIGATION AND DRAINAGE

Moisture is one of the key factors which affect the survival of some insect pest. Flooding of infested fields is sometimes used to control certain pest of crops like termite and nematodes, whereas some pest can be effectively controlled by drying out of soil.

3.2.1.13 MULCHING

Organic mulches are widely used, which permits the control of pest by encouraging the natural enemies present in the soil. Plastic mulches work by excluding the soil pests, whereas organic mulches provides suitable environment for the natural enemies.

3.2.1.14 TIME OF HARVESTING

Early harvesting helps in destruction of immature insects that are present in the foliage thus it disrupts its survival in the habitat.

3.2.2 PHYSICAL OR MECHANICAL CONTROL

Physical or mechanical control techniques include exclusion of pests by removal, destruction, and trapping. Pest exclusion is achieved by hand-picking, soil solarisation, netting, use of traps for insect pests, rodents, sterilization, mechanical tools to remove weeds, insect-pest, and bird deterrents like sonic devices and reflective materials. Tillage, slash and burn farming, hand weeding, and pruning of infested parts of trees are some of the agricultural practices that controls pest. Setting up of pheromone traps, light traps, baits such as rotten crab, frog to lure insects like Gandhi bug and placing gum prepared from fruits in possible rat runways for trapping helps controlling pest population in the main field. Baits containing toxic substances when placed in the field attract and kill pest. Pheromone lures confuse, attracts, and disrupts mating potential of insects thereby altering their behavioral response. To allow predatory birds to devour insect pests, perches for birds are also provided in the crop field.

The physical control methods are described as follows:

3.2.2.1 HANDPICKING

Handpicking, destruction of large sized, conspicuous insects is one of the ancient, effective methods of pest control. To control fruit flies, fruit borers, collection, and destruction of infested fruits are very commonly used. Insects such as tobacco caterpillar, sugarcane top borer, hairy caterpillar, etc., which lays eggs in masses can be easily eliminated by this method.

3.2.2.2 TRAPPING

Trapping is a popular mechanical method of pest control to lure insects to bait, light, color, pheromone and finally kill them. Several types of traps that are used to attract insects pests are yellow-pan traps, sticky traps, light traps, pitfall traps, and pheromone traps.

3.2.2.3 BARRIERS

Physical barrier debars the pest from attacking the crop. Nets and cloth screens protects the young, succulent plants from pests like hoppers, flea beetles, whitefly, etc. Trenches are built around trees to prevent attack of army worm, locust, etc. Metal, concrete barriers are used to control plants against termites, cutworms, etc. Sticky bands are applied around mango tree-trunks during the months of December-January to prevent the attack of mealy bugs.

3.2.2.4 PRUNING AND TOPPING

Removal, destruction of diseased, infested, dead branches, twigs, stem, and foliage can significantly reduce the dormant, overwintering stages of pest, thus their chance of spread in the successive year, e.g., mites, aphids, scale insects, etc.

3.2.2.5 SANITATION

Sanitation includes eradication of weeds, alternate hosts, destruction of crop residues, cleaning of field borders, bunds, removal of hibernating, and perpetuating sites of insect pests.

3.2.2.6 TEMPERATURE CONTROL

To control soil-borne pests in nursery bed, soil solarization is the suitable method to raise the temperature of soil to the level that is fatal to insect pest and pathogens. Low temperatures are utilized in storage and warehouses. The exposure of insects to subzero temperature for a period of 24 h is lethal to insect pests.

3.2.2.7 DRYING

Stored grains are dried to reduce the moisture content as low as 7–8% to control various pest of storage such as rice weevils, granary weevils, etc. Grains are dried under sun or by blowing dry air to manage storage pest.

3.2.2.8 RADIATION

Nonionizing radiations like gamma radiation kills all the stages of insect pests in storage conditions. This method is commonly employed to kill pests of fruits, vegetables, and grains during export and import of goods.

3.2.2.9 ULTRASONIC VIBRATIONS

To scare away lepidopterous insect pests from the crop fields, ultrasonic signals are used as they are sensitive to bats ultrasonic signals.

3.2.2.10 BIOLOGICAL CONTROL

Biological control refers to the use of living organisms for management of pests such as insects, weeds, plant diseases, and nematodes causing damage to crops. Biological control is one of the vital integral component of IPM that is safe to the environment. The biocontrol agents include predators, parasite, parasitoids and pathogens. The bioagent relies on predation, parasitism, competition, herbivory, or other mechanisms for their natural action but also involves active human management role. The effectiveness of biocontrol agents depends on species, biology, time of release, and environmental parameters. Biological control is delineated into three groups namely, importation, augmentation, and conservation.

3.2.2.11 IMPORTATION

Importation previously practiced as classical biological control involves the introduction of natural enemies of a particular pest species into new areas where their natural occurrence is not known. The natural enemies are generally collected from their native region and are released into new areas to control insect-pest that have been interposed fortuitously. In order to be effective against pest, the introduced natural enemies must possess good adaptability and colonizing ability under varied environmental conditions. The highest control by bioagents is seen if they have persistence in relation to time which enables them to survive during the interim unavailability of the host species. The most classical example of importation was in controlling cottony cushion scale (*Icerya purchasi*) in Australia using Vedalia beetle (*Rodolia cardinalis*).

Alligator weed introduced into the United States from South America was controlled by alligator weed flea beetle released in Florida.

3.2.2.12 AUGMENTATION

Augmentation means enhancement of population of existing natural enemies in an area by timely release of great numbers of natural enemies in that particular area, which have additive action on existing population. The augmentative release is based on the assumption that natural enemy population to provide biological control is less than optimal in the natural environment, but the number can be enhanced by periodic release. Augmentation is further classified into groups namely, inoculative and inundative release. In inoculative release, control agents are liberated in fewer numbers to allow them to establish and propagate with the objective to provide long-term control. On the other hand, inundative release refers to release of large number of bioagents with the objective of rapidly reducing the damaging pest population level in the crop field. The inoculative release of parasitoidal wasp, *Encarsia farmosa* to control whitefly is one of the common example. The most widely used example of inundatively released parasite is *Trichogramma* spp. to control lepidopteran pest like rice stem borer. Similarly, microbial pesticides such as *Bacillus thuringiensis*, *Trichoderma* spp., *Beauveria bassiana*, etc., entomopathogenic nematodes are applied in large quantities for a rapid knock down effect.

3.2.2.13 CONSERVATION

Conservation of natural enemies existing in a particular area to increase their survival and activity at the expense of harmful pest population is the third method of biological control. Conservation can be achieved by creating ecological strips consisting of noncrop plants to act as source of food, shelter, and protection of natural enemies from deliberate human interventions. For example, growing of nectar producing plants in the rice field at borders is practiced to support parasite, predators, and parasitoids of harmful plant hoppers.

3.2.2.14 BOTANICALS

The plant parts and plant-derived products have now emerged as effective plant protection agent to safeguard crops and its products. Botanicals are highly biodegradable, environment-friendly, safe to nontarget organisms, and compatible with other IPM components. The botanicals commonly used are neem (*Azadirachta indica*), pyrethrum, rotenone, *Cympopogan khasianus*, *Cannabis sativa*, etc. In northeast India, several botanicals are used such as Calotropis leaf powder to control mustard saw fly, stored grain pest, root-knot nematodes, stem decoction of neem (*Azadirachta indica*) to control rice stem borer, leaves of *Munromia wllichi*, *Croton caudatus*, *Adhatod avessica,* and *Artemisia vulgaris* to control several pests of rice, leaves of Heigri (*Dillenia indica* L.), *Artemisia vulgaris* L., *Melia composite,* etc. to control several stored grain pests.

Bio-pesticides market showed steady pattern of growth over the years. Its market share (as per 2014) has been reported as 10% of total pesticides market globally (http://www.adas.uk/News/counterfeit-and-illegal-pesticides-in-food-supply-chains-what-should-businesses-be-doing-to-minimise-the-risk and accessed on 08 Nov 2015). A wide array of beneficial microbes such as fungi, bacteria, nematodes, and viruses have been developed into bio formulations for safeguarding field crops, horticultural crops, protected cultivation, garden crops, etc. The important bio-pesticides produced and utilized in the country are *Bacillus thuringiensis* (Bt), *Trichoderma* species, Nuclear Polyhedrosis Virus, and botanical such as Neem (*Azadirachta indica*). The regulatory mechanisms are more flexible for biopesticides unlike chemical pesticides. The target specificity and environmental compatibility renders it more popular among end users. They are benign to beneficial insects and residue development is minimal. These can be effectively utilized in organic cultivation and have no chances of resistance development in insect pests. A number of products have been developed by small companies as well as multinational companies namely, Syngenta, Bayer, and Novozymes.

In the NE region of India, the study on biological control was initiated during 1970–1980s at Assam Agricultural University. The study was done to biologically control *Rhizoctonia solani* using *Trichoderma viride.* Thereafter, many effective bioformulations have been developed by the institute till date (Figure 3.1).

Integrated Nutrient Management practices and proper nutrient scheduling against major, commercial crops in Assam have been formulated and incorporated into package of practices. Three indigenous strains of Azotobacter (Azo52), Azospirillum (Az71), and PSB have been identified by the researchers working on management of natural resources at AAU and passed on the technology to several companies for mass production to cater the needs of biofertilizers under organic cultivation. With the aim to promote organic cultivation in the region, AAU has developed several bioformulations namely, Org-Trichojal, Org-Metajal, Org-Vertijal, Org-Beauverijal, and talc-based bioformulations namely, Biofor-PF, Biozin-PTB popularly called as Jaiva Kiran, which are found effective against diseases of vegetables such as wilt of solanaceous crops, cabbage, ginger, etc.[6,7] and sucking pests of crops such as thrips, whiteflies, aphids, etc. (Table 3.1). For commercialization and popularization of above-mentioned liquid bioformulations, the technology was supplied to NE-based company M/s. Green Biotech Eco-solution Pvt. A bioformulation prepared from mutant isolate of *Trichoderma viride* was developed to control white rot pathogen *Sclerotinia sclerotiorum.* Also, an effective and cheap mass production technique using solid substrate was developed for a local *Beauveria bassiana* strain isolated from rice hispa (*Dicladispa armigera*). Besides, AAU has done tremendous research, recommended, and developed mass production techniques for bio-control agents like *Trichogramma chilonis* and *T. japonicum* for management of dreaded pests of rice.

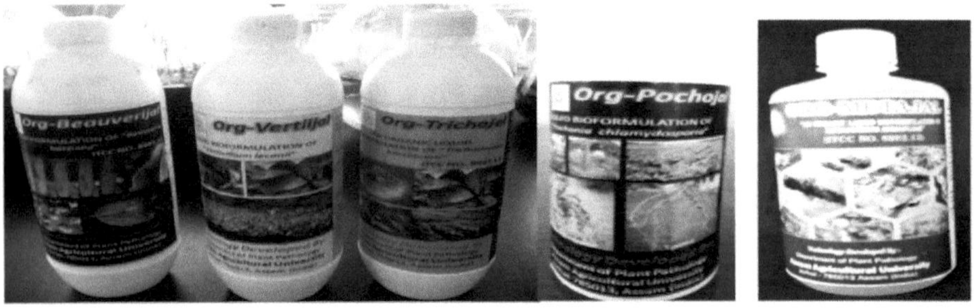

FIGURE 3.1 Biopesticides developed at Assam Agricultural University, Jorhat, Assam, India.

The classical biological control work for the management of native invasive as well as alien invasive weeds was initiated at Assam Agricultural University, Jorhat in collaboration with CABI, DFID, UK, etc. The Trinidad isolate (W1761) of *Puccinia spegazzinii* was tested for host specificity to *Mikania micrantha* using centrifugal phylogenetic testing at quarantine centers of NBPGR, New Delhi, which revealed that no infection of the fungus was present on 74 economically essential species of plants. Thus, the fungus *P. spegazzinii* was released for the first time as a classical biocontrol agent in India as well as the continent of Asia to effectively manage the invasive weed of tea ecosystem, *M. micrantha.* With this work, India became the eighth country in the entire globe to control weeds using pathogenic fungus. Another study identified *Puccinia cacao* from NE region as promising biocontrol agent against *Hygrophila polysperma.*

TABLE 3.1 Some of the Example of Biopesticides with the Targeted Pests and Diseases.

Name of formulation	Biocontrol agent	Target pests/diseases
Liquid		
Org-Metajal	*Metarhizium anisopliae*	Termite, aphid, and scale insect; Also act as endophyte and give induced resistance
Org-Trichojal	*Trichoderma harzianum*	Soil-borne plant pathogen like *Rhizoctonia solani*, *Sclerotium rolfsii*, *Sclerotinia sclerotiorum*, *Fusarium*, etc. Enhances plant growth and yield
Org-Vertijal	*Verticillium leccanii*	Red spider mite, Scale insect
Org-Beauverijal	*Beauveria bassiana*	Rice hispa, Tea Mosquito bug, Pathogens namely, Fusarium etc.
Org-Cilliumjal	*Purpurocillium lilacinus*	Root knot nematode
Org-Pochojal	*Pochonia chlamydosporia*	Root Knot nematode
Org-Metahim	*Metarhizium anisopliae*	Termite, aphid, scale insect
Solid		
Biofor-PF-2	*Pseudomonas fluorescens, Trichoderma harzianum*	Fungal, bacterial soil borne plant pathogen
Biozin-PTB	*Pseudomonas fluorescens, Bacillus brevis, Trichoderma viride*	The species of plant pathogens like *Pythium, Fusarium, Colletotrichum, Rhizoctonia, Macrophomina, Sclerotinia, Xanthomonas, Ralstonia*
Bio-Llium	*Verticillium lecannii*	whitefly, mites, aphids, jassids, thrips, ants, Nematodes
Bio-Meta	*Metarhizium anisopliae*	Red ants, termites, mosquito larvae, planthoppers, cattle ticks
Bio-Sona	*Beauveria bassiana*	Rice hispa, *Bemisia tabaci,* Helicoverpa, mites, coffee borer
Bioveer	*Trichoderma viride*	Bacterial, fungal wilt, anthracnose, root, stem rots
Biogreen-5	*Beauveria bassiana, Metarhizium anisopliae, Pseudomonas fluorescens, Trichoderma viride, Bacillus thuringiensis*	Wilt disease of solanaceous vegetables, wilt of turmeric& ginger, Rhizome rot, Tea mosquito bug, tea looper, red spider mites, stem borer, rice hispa
Biotime	*Metarhizium anisopliae, Pseudomonas fluorescens, Trichoderma harzianum*	Bacterial, fungal wilt, anthracnose, root, stem rots, diseases of tea, Red ants, Termites, coconut leaf beetle, stem borer, leaf folder
Biosona	*Beauveria bassiana*	Rice hispa, aphids, Tea mosquito bug, stem borer, thrips, whitefly, looper, mites
Biozium	*Trichoderma harzianum*	Bacterial, fungal wilt, anthracnose, root, stem rots

Similarly, Successful control of water hyacinth, *Eichornia crassipes* has been achieved by the exotic weevil *Neochetina eichhorniae, N. bruchi* @50,000 per ha, and the weevil dispersal was seen in eight different districts of Assam. The probable route of dispersal may be aerial migration, Brahmaputra river and its tributaries. For management of various lepidopteran pests of field and horticultural crops like sugarcane borer, *Chilo* spp., *tomato fruit borer, Helicoverpa armigera*; cut worms, *Agrotis* spp.; cotton boll worm, *Pectinophora gossypiella, Ear* spp., maize bore, *Chilo partellus* rice stem borer, *Scirpophaga incertulus,* egg parasitoid Trichogramma has been identified as potential bioagents which suppress various caterpillar pests damaging many agricultural crops. Trichocards bearing parasitoid eggs are liberated in the fields 30–45 days after sowing or transplanting @50,000–100,000 eggs/ha/release. Overall trichocards are released three to six times based on population build up of pest and subsequent development of eggs of target insect pest.

The suitable environmental conditions of NE India favors BCA i.e., RH (>80%), temperature (12°C–35°C), low pH, Rainfall (300–3000 mm) favorable for BCA, wide diversity in terms of species, strains of biopesticidal resources i.e., plants, fungi, bacteria, virus, parasitoids, predators, and good potential market for biopesticides for large scale production of rice and tea as organic.

The trainings are imparted to the end-users as well faculties and Scientists of SAU and ICAR institutes and line departments for the technical aspects of bioformulations. The organic package of practices for the state of Assam was a long-felt need for organic growers of the state. Considering the need and an Organic Package of Practices of selected crops of Assam and 2019 has been prepared on organic cultivation of selected crops by the specialists/researchers of Assam Agricultural University, state department of Agriculture and Horticulture & Food Processing through the coordination of the Directorate of Research (Agri.). The recommendations in the Organic Package of Practices were made in a workshop jointly organized by the Department of Agriculture and Department of Horticulture & Food Processing, Govt of Assam, Assam Agricultural University, Jorhat, on May 29, 2019 at Guwahati.

At Central Agricultural University (Imphal), Umiam Campus indigenous biocontrol agents based liquid bioformulation (namely, Um-Tricho, Um-Bir, Um-Met, Um-Comb, Um-Pseudo, Um-Bac, UmRaj, UmLec) has been developed under the leadership of Dr. P. Dutta of School of Crop Protection (Figure 3.2). During farmers field trial, as a component of IPM, these products showed successful results in different districts of Meghalaya. The said formulations, the packages developed for IPM can be utilized by the farmers of other NE states of India for cultivation of crops like cereals, vegetables and spices.

FIGURE 3.2 Biopesticides developed at CPGSAS, Central Agricultural University (Imphal), Umiam, Meghalaya, India.

Apart from biopesticides and biofertilizers, bioherbicides are also the integral component of IPM. Microbes are extensively used to develop bioherbicides among which the notable ones are developed by Marrone Bio Innovations. The company has produced two new products namely, MBI-005(OpportuneTM) consisting of *Streptomyces* sp i.e. a nonpathogenic strain *Burkholderia* A396. The product was registered under EPA (2012) and its mode of action is both systemic as well as contact. Second product MBI011 is in pipeline, which is developed using the aqueous extracts of long pepper, *Piper longum* wherein sarmentine acts as active ingredient. The Scotts Company commercialized *Phoma macrostoma*-based bioherbicides, which was registered in California, USA. For management of turf weeds, Phoma TM could be evolutionary. *Sclerotinia minor* (SarritorTM)-based bioherbicide is available commercially in Canada.[8]

The bio-pesticides emerged as preferred plant protection agent over chemical pesticides due to many reasons such as sale of forged pesticides in the country, which prevents realization of its ability to effectively manage pest population. It was reported that 10% of global pesticides traded are spurious amounting US$ 4.73 million annually. The illegal trade of pesticides is a lavishing area of crime due to high net profit ratio. The risk associated with simulated pesticides are: (1) health risk on animals and human due to potential residue build-up on food products, (2) ecological and environmental hazards such as soil and water pollution, (3) crop security if crop failure occurs due to simulated pesticides, and (4) social risk associated with defamation caused by use of illegal pesticides. The types of illegal pesticides found in the market are: (1) Feigned pesticides produced and packaged to resemble original products, but the active ingredients may not be at par as indicated on labels – it may contain toxic materials or might have subdued active ingredient; (2) Feigned pesticides asserting to be exact counterpart of "product X"; and (3) Illicit parallel imports – wherein products produced and approved in a nation are utilized in another country.

3.2.3 CHEMICAL CONTROL

The use of chemical pesticides in IPM is the ultimate option when other components of IPM fail to subdue the population of pest below economic threshold. Chemical pesticides are used only when needed along with other approaches to reduce the pesticide usage, its harmful impact on environment. In IPM, pesticides are selected, applied in a way that diminishes its possible harmful effect on nontarget organisms, human, and environment. Chemical control includes synthetic as well as microbe-based (Spinosad and Avermectin) and botanical (Azadirachtin and Pyrethrins) pesticides. The use of natural pesticides should be encouraged as they are relatively safe. Chemical pesticides are generally categorized based on its mechanism of action. Rotation of chemicals having different mode of action is essential to reduce the threat of resistance development by the pest. Thus, judicious use of chemical pesticides based on strict pest surveillance as well as economic threshold level should be done in combination with other pest management tactics. Before the application of chemicals, we must have thorough understanding of chemicals to be sprayed and its application protocol.

3.2.4 NANOTECHNOLOGY

To resolve current challenges in agriculture and pest management, nanotechnology have evolved as an emerging science to supplement the conventional approaches and can potentially be integrated with IPM programs. Attempts are made to produce potential pesticides at nanoscale, that is, nanopesticides that can increase the effectiveness of pesticides.

Encapsulation of pesticides by polymeric nanomaterials also ensures targeted delivery of these nanopesticides thereby ensuring reduced pesticide load in the environment.[9,10] Similarly, encapsulated fertilizers and other agrochemicals allow slow release of nutrients for its proficient use. For controlling pests and plant diseases, nanoparticles such as Silver (Ag), Gold (Au), and Zinc oxide (ZnO) nanoparticles have been extensively studied and are found effective against number of pests and diseases. AgNPs are effective against insect pests like papaya mealybug, mustard aphid, brinjal fruit, shoot borer, cabbage worm, rugose spiraling whitefly, gram pod borer, rice moth, etc. Nanoparticles are also effective in suppressing phytopathogens namely, *Sclerotium rolfsii*, *Rhizoctonia solani*, *Fusarium oxysporum*, *Colletotrichum musea*, *Xanthomonas oryzae* pv. *Oryzae,* etc.[11–13] A nano-bioformulation synthesized from chitosan was found effective as seed treating agent against soil-borne plant pathogens with enhanced plant growth parameters[14] but without any harmful effect on soil properties up to some concentration.[15]

Nanosensors can play crucial role in successful management of various crop pathogens such as bacteria, virus, fungi, etc by its early detection and diagnosis. The viral diseases are not easy to diagnose and control under field conditions, however, smartsensors show potential to detect various strains of virus and stage of application of therapeutic measures to control diseases.[16] The nanosensors could also be used to monitor soil conditions, nutrient status, and crop growth conditions to increase agricultural productivity.

Nanomaterials can also be developed as carrier for delivery of pest-resistant genetic material into target crops for development of pest-resistant varieties. Nanobiotechnology allows integration of specific resistant traits to deter pest from feeding, oviposition, reproduction, thereby making the host plant hostile for its survival. For controlling weeds, parasitic plants, encapsulated herbicides can be an effective alternative. Encapsulation increases the effectiveness of herbicides, allows its gradual release into the soil thereby reducing the amount of herbicides required for effective control of parasitic plants.

As per the demand of crop, pest infestation level, one among the above mentioned pest management options can be utilized for effective management of pest and diseases throughout the production period.

3.3 KNOWLEDGE AND RESOURCES

The knowledge, awareness of various control tactics, biology of pest, damage potential and availability of resources allows the grower to make decision appropriate for the cropping situation.

3.3.1 PEST

For identification of effective control strategy, it is imperative for the farmers to recognise pests and their most damaging stages. Understanding of population dynamics of pest, seasonal fluctuations, nature of damage, economic significance, preference of host, alternate, collateral host, and other related information is essential for formulation of effective management strategies.

3.3.2 AVAILABLE MANAGEMENT OPTIONS

The selection of most effective control option ideal for the situation is very essential as management option of different pest varies. Some insect pests, diseases can be effectively controlled using biological control, whereas some other requires application of chemical pesticides. Pests that become abundant during a particular season can be managed by altering sowing and harvesting dates. Commercially available microbial pesticides such as bacteria, fungi, viruses, and entomopathogenic nematodes can be used against various soil pathogens, phytopathogens and against pests having chewing mouthparts and fungi effective against pests having suctorial mouthparts.

3.3.3 AVAILABLE RESOURCES

Availability of resources is the key factor for management of pests. A particular pest may be controlled by certain option but absence of resources at the right place and time hampers the management process. Again, the feasibility of control measures depend on acreage of field crop, value of crop grown and available financial means. For example, release of biological control agents may be very effective in controlling several pests, diseases but its availability in certain areas is the main hindrance that needs immediate attention. Again, application of chemical pesticides increases the cost of cultivation, is unsuitable for application in low-value crops. Use of netting can be very effective in controlling pests, but the expense of netting limits its availability to small and marginal farming community. Moreover, regulatory guidelines that prevents the use of certain pesticides or promotes the use of others may have profound impact on implementation of IPM programmes. A number of broad-spectrum pesticides have been banned in several countries including India. The recent concern for environmental health has led to such restrictions. With increasing regulations on use of chemical pesticides and there is a greater need for better understanding of available control options for its use in most effective way.[17]

The factors including knowledge and resource component covers the implementation stage of IPM, while presence of ample resources to implement the various components of IPM is very critical, which is often overlooked. Limited availability of IPM specialists, monetary resources and time-consuming agricultural research are some of the challenges in development and dissemination of IPM strategies. Moreover and IPM technologies developed in some cases fails to reach the stage of implementation due to various obstacles such as greater cost of commercialization or lack of social adaptability.

3.4 PLANNING AND ORGANIZATION

This component includes management aspect of IPM model for collection of data, planning, and actual actions to be taken against pest infestations.

3.4.1 PEST MONITORING

Regular monitoring and surveillance of crop fields for pest occurrence and its population density is the preliminary stage for protection of crops. Prompt detection of pest and diseases enables effective management of pest by precise application of treatment and removal of infested plant parts. When the level of pest population continues to aggravate, regular monitoring becomes essential to assess the damage and to determine the threshold level to initiate pest control measures. Monitoring is also essential to avoid application of pesticides in the crop field at low pest populations that do not require treatment. In such situation, improved sanitation and appropriate cultural, physical control measures provides the desired control action. However, decision-making based on mere physical scouting is not possible due to lack of essential resources. Drone-assisted observation of areas that are exposed to various biotic, abiotic stresses, detection, and identification of various pests can improve the monitoring efficiency and promote precision application of pesticides.[18]

In the digital era, surveillance system needs to be digitized for more precise and effective results. GAP or IPM programmes coupled with e-pest surveillance systems are essential for training, commercial manufacturing system and to keep track of the dispersal and migration of invasive pest.

As invasion of dreaded pest can cause huge loss to the economy, thus it necessitates linking of management action to IPM. Preborder checks are desirable and more effective instead of delaying till pest gains entry into the country. Thus, altered bio-safety policies could ensure protection of biological resources from exotic threats.

The present world challenge is to grow enough food from limited land resources to feed ever-expanding population.[19] Global warming and climate change leading to changing dynamics of pests further aggravates the challenges for the growers and breeders. IPM strategies provide a beam of hope for achieving sustainability in crop production wherein more emphasis must be laid on precise diagnosis of pest, skill, and capacity development of growers, extension personnel and scientists and prepare them for future climate challenges. It can be achieved through Private-Public-Partnership. Human resource development with thorough knowledge on pros and cons of IPM needs to be more emphasized by NARS system of education. The environmental parameters such as increased temperature and level of CO_2 may pose risk of enhanced vulnerability of biological control agents, diminished chemical management efficacy, migration of invasive pest, and breakdown of durable resistance in plants. Thus, it is very essential to formulate a management strategy that is compatible with environmental and ecological balance.

3.4.2 REGULATION OF INFORMATION

To efficiently manage information of an agricultural farm, develop institutional knowhow, and recordkeeping proves to be an effective tool. A good recordkeeping must record information about pests, seasonal fluctuations, extent of pest damage, interaction of pest, plants with environmental parameters, soil conditions, plant nutrient status, irrigation practices, and other related information that builds the knowledge base of growers, prepares them to take adequate preventive and curative actions.

3.4.3 CORRECTIVE ACTIONS

The most important aspect of IPM is taking corrective action at right time. Even with the existence of knowledge about pests, utilization of resources, the damage caused by pest can be controlled only when preventive and curative measures are initiated at the right time. Timely actions are required to be taken by all the farmers of a particular area in a coordinated manner to ensure successful control of pest and diseases. If overlooked in an individual's farm, insect pests may spread in the entire locality within a short span of time with larger social, regulatory, and economic implications.

3.5 COMMUNICATION

Communication is the key component of IPM to transfer collective knowledge to the farming community for their benefit. The technologies developed by extensive research can be disseminated to the stakeholders by various modern and traditional communication tools. Also, the information about types of pest (Endemic and invasive pests), new emerging threats, and control measures can be channeled to the end users through good communication network. Farmers and other stakeholders should stay well updated about existing, potential pests, and their remedial measures. In the era of digital world, it has now become easier to gain information through attending trainings, workshops, and webinars. The newsletters, scientific or trade journal articles, agricultural magazines and package of practices, etc also act as a good source of information on latest updates of IPM. The innumerable online resources from various universities across states, other government, nongovernment organizations, and smartphone applications also provides frequent updates.[20] Well trained and informed growers are competent to address pest issues effectively. Regular updates on IPM are also essential for agricultural scientists and extension workers since they are involved in the development and extension of new strategies.

3.5.1 COMMUNICATION AMONG GROWERS

Good communication channel aids in educating farming communities about pests and control measures. The crop production and crop protection strategies are considered to be an individual affair but pests are beyond the limitations of geographical boundaries and can spread to larger areas when not effectively managed. Sharing of information and resources within the farming community can enhance the pest control efficiency and benefit the entire farming community.

In addition to the above-mentioned components of an IPM model and elements that determine sustainability and social acceptability of products, economic profitability at a wider scale and their impact on global food security must also be included. Thus, business and sustainability aspects are the two most important layers surrounding the four components of IPM.

3.6 BUSINESS ASPECT

Business aspect includes the interest of growers, sellers and consumers. The sole interest of producers is to meet the demands of market with the objective of making minimal profit while maintaining ecological and human safety. Sellers assess the market situation and make strategies to meet the demand of consumers, making enough profit to stay in the system. Consumers are interested in nutritious and healthy produce at reasonable prices. Thus, growers, sellers, and consumers maintain harmonious balance of food demand and supply in an ideal system. In such as balanced system, global food security is ensured, products are safe, affordable to both growers and sellers make sufficient profit with reduced risk to the environment. However, such an ideal system is difficult to achieve in real world situation. This balance is often disrupted due to various reasons such as consumers misconception about production systems, their tendency to pay premium prices for food products that are perceivably safe, their demand for uniformly shaped, colored food products, growers battling with challenges of producing high-quality food under continuous threat posed by pests and diseases while sellers exploit the producers to earn maximum profit from the products. The desired balance can be achieved if growers implement IPM practices to produce healthy, safe products, consumers are well informed, aware of IPM strategies and then sellers would be bound to market IPM-based products to meet consumers demand.

In northeast India, the use of chemical pesticides is very limited, farmers resort to traditional and eco-friendly crop protection measures for growing crops. Thus, food products from northeast India have great potential to reach the consumers across the world but there is a dire need to enhance productivity of crops by promoting scientific IPM strategies.

3.7 SUSTAINABILITY ASPECT

IPM practices ensure economic viability, environmental sustainability, and social acceptability of the food produce. Economic viability is ensured at both producer and consumer level as food products are offered at affordable prices and sellers are expected to earn a reasonable margin of profit from it. Again, IPM practices maintain environmental safety by balanced application of available pest, disease management tactics, and social acceptability is enhanced since IPM-based products are safe and affordable. Over the past decade with the rising concern on human, environmental health, demand for safe, and healthy food products has risen exponentially. Organic food products are generally considered as safe and sustainable but organic farming is always associated with costs and risks. The organic food products are costlier and can be purchased by upper-class consumers only, whereas low-income groups cannot afford it. Organic products can be grown and marketed only by certified growers, whereas IPM-based products can be grown by all the interested growers. The crop production practices followed by organic farmers based on USDA National Organic Program follows the same IPM strategies as used by non-organic producers. The comprehensibility and profitability are ensured by IPM-included production system. IPM technique relies on experience and keen observation and knowledge, integration of multiple control tactics and thus qualifies IPM (excluding chemical pesticides) as appropriate for organic farming also. IPM approach focuses on cultural, physical, mechanical, biological, microbial control, and advocates pesticide usage only when extremely desired to suppress pest population level below economic threshold level.

Thus, IPM practices have evolved as safe, profitable and practically feasible alternative.[21] The IPM-based products lead to social equality and ensures global food security by optimizing cost of inputs, reducing waste generation, growing confidence among consumers, and enhancing adoption of sustainable practices.

3.8 ADVANTAGES OF IPM

3.8.1 MAINTENANCE OF ECOLOGICAL BALANCE

IPM allows eradication of pest and diseases by application of blend of pest management practices, which are safer with reduced pesticide load in the environment. The IPM addresses environmental safety issues, minimizes effect on nontarget organisms, pesticide residue hazards. The IPM also minimizes harmful impact on soil, air, and water thereby maintaining ecological balance.

3.8.2 DELAYED PESTICIDE-RESISTANCE DEVELOPMENT

In conventional pest management practices, chemicals are applied repeatedly on crops, which enhance the speed of development of pesticide resistance. The insect pests that survive the pesticides and transfer the genes of resistance to the next generation lead to development of super-pests. The use of IPM has the potential to resist development of pesticide-resistant species of pest and pathogens.

3.8.3 REDUCED COST OF MANAGEMENT

Conventional pest management strategies allowed recurrent application of chemical pesticides to control pests. The IPM reduces the frequency, dosage of pesticide application, thus reduces the cost of cultivation. The initial investment in IPM practices appears to be cost-intensive but proves to be cost-effective in long run.

3.8.4 SOCIAL ACCEPTABILITY

The social acceptability of IPM-based products is high as compared with chemical-intensive methods. The IPM-based product reduces risk of exceeding the standards of residue limit set by government. This builds up the confidence of consumers on food products developed using IPM-based cultivation practices.

3.8.5 FLEXIBILITY

The IPM practices increase knowledge of farmers on pest, management tactics, allows the grower to modify the management practices as per required to meet the specific demand.

The IPM is everybody's business, which can be done with cooperation of farmers and government and cohesive networking among all the stakeholders is essential. Krishi Vigyan Kendras, under ICAR, SAU's and nongovernmental organizations plays a central role in providing advisories related to crop production and protection as well as enhancing awareness level among farmers in IPM. It requires prior planning, good agricultural practices ensuring environment safety and higher benefit-cost ratio. Regular monitoring of crops linked with precise diagnosis, timely decisions, prompt action, and expert advices can prevent pest outbreak in future. Thus, pest dynamics monitoring via e-pest surveillance can forecast information on pest emergence, abundance, and distribution pattern based on changes in environmental parameters along with dissemination of smart-phone based advisories through SMS or different apps. However, in India short to medium-term weather forecasting using numerical weather prediction models have been least harnessed. Also, future climate projections of sleeper pests have not been done in India for various regional services of crop protection. The use of state-of-the-art technology through innovative, tactical research allows Integrated Decision Support System for overall protection of crops via network of KVKs. Real time surveillance of pest through Information Communication Technology has played tremendous role in acquisition and dissemination of data from rural places to the main station via internet. The compiled data are uploaded in the website comprehensively from which SAU's can directly access it and via State Agricultural Department (West Bengal, Maharashtra, and Odisha) issue advisories through mobile phones and extension workers, which amplifies preparedness for future pest outbreak.

3.9 CONSTRAINTS IN ADOPTION OF IPM

The corroboration trials established IPM as the most efficacious crop protection approach. However, it has not reached desired level of adoption by the farming community due to number of limitations. Some important constraints are as follows:

1. Higher cost of hybrid seeds of vegetable crops than the traditional varieties.
2. Farmers' lack cognizance about IPM.
3. Intensive input of labor is required.
4. Unavailability of various inputs such as traps, biocontrol agents, resistant varieties are main drawbacks in IPM implementation of vegetable crops. Also, mass multiplication of BCA requires trained professionals.
5. Difficult to convince end-users about IPM practices and their productiveness.

6. IPM technology is relatively complex due to integration of various components.
7. The IPM practices are time consuming and implementation is tedious.
8. Credit acquisition is one of the major drawback.

3.10 CONCLUSION

IPM is an age-old and environmentally sound concept of pest management. IPM amalgamates different components of management namely, cultural, physical, biological, and chemical control measures in an effective manner to sustain crop production and minimize the effect on environment. Preventive measures such as the use of resistant varieties, soil solarization, field sanitation, manipulation of microenvironment of crop, and many more ensures reduced inoculum of pests and pathogens for attack on susceptible stage of crop. Curative measures such as application of biological control agents, botanicals and chemical pesticides helps in reducing the population density of pest below economic threshold level. The IPM promotes enhanced application of blend of management practices with reduced usage of chemical pesticides. With the advancement in technology and nanotechnology has emerged as one of the advanced science of 21st century. Nano-based materials exhibit tremendous scope for utilization in plant disease diagnosis and detection and precise application of pesticides. The various components of IPM include pest management, knowledge on pests and resources, planning and organization for effective usage of resources and proper communication among stakeholders. Proper understanding of pests helps in identification of suitable pest management strategy and is crucial for timely application of control options. Regular monitoring is also a very crucial component of IPM, which helps in determination of economic threshold level. Proper communication among growers and IPM specialist helps in creating awareness and enhances social adaptability of IPM practices. In the modern times, IPM is redefined where latest agricultural technologies and communication tools plays crucial role in economic viability and social adaptability. Apart from the management aspect, sustainability and business aspects are the two important pillars of new IPM paradigm. The new outlook of IPM paradigm ensures profitability to growers, affordability to consumers and sustainability to environment and food security to the expanding population in the entire globe.

KEYWORDS

- **integrated pest management**
- **cultural control**
- **biological control**
- **nanotechnology**
- **north east India**
- **social acceptability**

REFERENCES

1. Barlett, B. R. Natural predators. Can Selective Insecticides Help to Preserve Biotic Control? *Agric. Chem.* **1956,** *11*, 42–44.
2. USDA-ARS (United States Department of Agricultural Research Service). A National Road Map for Integrated Pest Management. 2018. https://www.ars.usda.gov/ARSUserFiles/OPMP/IPM%20Road%20Map%20FINAL.pdf
3. Peterson, R. K. D.; Higley, L. G.; Pedigo, L. P. Whatever Happened to IPM? *Am Entomol.* **2018,** *64*, 146–150.
4. Dara, S. K. The New Integrated Pest Management Paradigm for the Modern Age. *J. Integr. Pest Manage.* **2019,** *10* (1), 1–9.
5. Chhetry, G. K. N. Indigenous Pest, Disease Management Practices in Traditional Farming Systems in North East India: A Review. *J. Plant Breed. Crop Sci.* **2009,** *1*, 28–38.
6. Bhattacharyya, A.; Dutta, P. Achievement of Assam Agricultural University on Biological Control. In *Biocontrol, Utilization of Insects for North East India*; Haldhar, S. M., Dutta P., Saha, R. K., Nagesh, M., Selvaraj, K., Singh, M. P., Eds.; DEE, CAU: Imphal, 2020; pp 17–23. ISBN: 978-81-938078-3-5.
7. Dutta, A.; Boarh, A.; Mahanta, B.; Dutta, P. Integration of Biocontrol Agents, Chemical for the Management of *Meloidogyne incognita and Sclerotinia sclerotiorum* Complex on French Bean. *J. Entomol. Zool. Stud.* **2020,** *8* (2), 1546–1551. E- ISSN: 2320-7078; P-ISSN: 2349-6800.
8. Sardana, H. R. Development, Validation of IPM Technology for Selected Vegetable Crops for NEH Region. In *Orientation Course on Important Agricultural, Horticultural Crops of NEH Region*; Sehgal, M., Balodi, R., Tripathi, A. K., Deka, B. C., Sardana, H. R., Eds.; ICAR-NRCIPM, 2020; pp 10–24.
9. Sahayaraj, K. Nanotechnology, Plant Biopesticides: An Overview. In *Advances in Plant Biopesticides*; Singh, D., Ed.; Springer India: New Delhi, India, 2014a; pp 279–293. ISBN: 978-81-322-2006-0.
10. Sahayaraj, K. Novel Biosilver Nanoparticles, Their Biological Utility, Overview. *Int. J. Pharm,* **2014b,** *4*, 26–39.

11. Das, A.; Dutta, P. Antifungal Activity Biogenically Synthesized Silver, Gold Nanoparticles Against Sheath Blight of Rice. *J. Nanosci. Nanotechnol.* **2021,** *21* (6), 3547–3555(9). doi:10.1166/jnn.2021.18996.
12. Goswami, R.; Bhattacharyya, A.; Dutta, P. Nanotechnological Approach for Management of Anthracnose and Crown Rot Diseases of Banana. *J. Mycol. Plant Pathol.* **2020,** *50* (4): 370–381.
13. Kaman, P. K.; Dutta, P. Synthesis, Characterization, Antifungal Activity of Biosynthesized Silver Nanoparticle. *Indian Phytopathol.* **2019,** *72*, 79–88.
14. Baruah, S.; Dutta, P. Fungus Mediated Synthesis, Characterization of Chitosan Nanoparticles, Its Combine Effect with *Trichoderma asperallum* Against *Fusarium oxysporum, Sclerotium rolfsii, Rhizoctonia solani. Indian Phytopathol.* **2020.** DOI 10.1007/s42360-020-00289-w.
15. Ahmed, A. A.; Dutta, P. Effect of Green Synthesized Silver Nanoparticles on Soil Properties. *J. Med. Plant Stud.* **2020,** *8* (1), 34–37.
16. Mousavi, S. R.; Rezaei, M. Nanotechnology in Agriculture, Food Production. *J. Appl. Environ. Biol. Sci.* **2011,** *1*, 414–419.
17. Hillocks, R. J. Farming with Fewer Pesticides: EU Pesticide Review, Resulting Challenges for UK Agriculture. *Crop Protection* **2012,** *31* (1), 85–93.
18. Yue, Y.; Cheng, X.; Zhang, D.; Wu, Y.; Zhao, Y.; Chen, Y.; Fan, G.; Zhang, Y. Deep Recursive Super Resolution Network with Laplacian Pyramid for Better Agricultural Pest Surveillance, Detection. *Comput. Electr. Agric.* **2018,** *150*, 26–32.
19. Sharma, S. B.; Wightman, J. Vision Infinity for Food Security: Some Whys, Why Nots, Hows! SpringerBriefs in Agriculture, 2015, pp XVII, 98. DOI: 10.1007/978-3-319-23249-2.
20. Dara, S. K. Android Version of the IPM Info App with New Features Just Released. University of California Agriculture, Natural Resources. eJournal Strawberries, Vegetables. 2016. https://ucanredu/blogs/blogcore/postdetail.cfm?postnum=14030.
21. Dara, S. K. Safe, profitable, and Practical Label for Sustainable Production, Food Security. *Progr. Crop Consultant* **2018,** *3*, 20–23.

CHAPTER 4

Concept and Practices of IPM in Relation to Different Crops and Cropping Systems

DEVENDRA PAL SINGH*

ICAR-Indian Institute of Wheat and Barley Research, Karnal 132001, India

**E-mail: dpkarnal@gmail.com*

ABSTRACT

The cropping systems are used for managing associated pests in crops over years and emphasis is given on the integration of different tools and methods of integrated pest management (IPM) in a cropping system mode. Different concepts and approaches in IPM include pest assessments and diagnosis, decision making in IPM and case examples of pest management in cropping systems involving cereals, food legumes, oilseeds, and horticultural crops are highlighted. The IPM technology development and the pest management based on ecology are discussed in different agroecosystems. The cropping systems approach includes soil and crop health as well as specific IPM and production goals. The development of IPM in different cropping systems is a dynamic in nature and requires regular revisions based on crops, cultivars, agriculture practices, associated pests, and the environment.

The use of pesticides resulted increase in yields and simplified cropping systems. However, it is associated with contamination of ecosystems with undesirable effects on human health. The emergence of pest resistance as well as reduced availability of active substances of such pesticides is posing problems in their uses. Therefore the cropping systems with less dependent pesticides are good for better sustainability of food production.

Integrated Pest Management in Diverse Cropping Systems. Devendra Pal Singh, PhD (Ed.)

4.1 INTRODUCTION

Pests are defined as "Organisms which damage or interfere with crops in fields, orchards and landscapes, pests of crops include insect pests, nematodes, and plant pathogens, which cause diseases, or other unwanted harms to crops." The integrated pest management (IPM) includes all the available pest control techniques and measures that are discouraging the build-up of pest populations along with minimum risks to health of human and ecosystem. The IPM includes all the best combinations of the cultural, biological, and chemical methods to manage diseases, insect pests, and weeds. The IPM measures should be locally available in cost-effective manner. The IPM may vary from one site to another site in a socially acceptable way. However, the implementation of IPM at the level of farmers depends how best they adopt practices. The farmers view the IPM practices based on their practical use and value. The IPM is considered as an ecosystem-based strategy of pest control. It gives long-term prevention of pests or their damage. In IPM, the use of techniques like biological control, manipulation of habitat, modification in cultural practices, and deployment of resistant varieties is done in a combined manner. The use of pesticides is only advised under emergency based on proper monitoring of pest according to established guidelines to remove only the target organism. The other advantage of IPM is that it minimizes risks posed by pesticides to human health, beneficial insects, and nontarget organisms as well as the environment. The IPM is of great use in rice agroecosystems. The healthy agroecosystems resulted higher rice yields in case of "BRRI dhan 29" and yielded 7.4 tons per ha in the IPM treated plots.[1]

4.2 CONCEPT OF IPM

Through IPM, the damage of crops caused by pests is minimized on long-term basis by managing the ecosystem. The crop health is maintained by growing disease and insect pests resistant plants. However a cultivar may not have resistance to all the diseases and insect pests and therefore other methods and techniques are used to manage rest of these under IPM. The emphasis in IPM is not simply eliminating the pests but look into those factors of environmental that are responsible to develop pest and its ability to thrive. The phenomenon is understood properly and the unfavorable conditions are created for the pest. It is important to monitor and identify pest correctly to decide whether IPM is needed or not. It is also the key in knowing whether a pest may become a problem and which best management strategy would be best to manage it.

The IPM is the integral part of systems like "Integrated Crop Management (ICM)" or Farming Systems, "Good Agricultural Practices," and "Sustainable Agriculture." Under ICM practice, the whole farm is managed strategically using IPM as a component, which results in to minimum pollution, safeguards natural assets of farm, improves energy efficiency and crop profitability".[2] The 'Good Agricultural Practices' is "efficient production of good-quality food, feed, and fiber with minimum negative impact on natural resources, environment, and ensure proper health and safety of farmers".[2]

4.3 PRACTICES OF IPM

Under IPM programs, pest management approaches are combined for greater effectiveness. The practices for managing pests may be categorized as follows:

4.3.1 BIOLOGICAL CONTROL

Under it, the use of natural enemies like, predators, parasites, pathogens, and competitors is done to manage pest damage. The insect pests and disease-causing agents have many natural enemies.

4.3.2 CULTURAL CONTROLS

The cultural controls are practices to aim to reduce pest establishment, reproduction, and dispersal as well as its survival. Too much application of irrigation can increase root diseases and weeds. The common practices used are as follows:

- Removal of plant debris, cleaning of bunds, and summer ploughing of soil may be beneficial to kills various pests.
- Soil testing for nutrients deficiencies.
- Use of treated and certified seeds.
- Pest resistant/tolerant varieties.
- Alteration of time of sowing and harvesting that may help in escape of peak season of pest attack.
- Use of nonhost crops in crop rotation.

- Optimum use of fertilizer with plant spacing and use of FYM as well as bio-fertilizers.
- Proper water management. Proper weed management.
- Setting sticky traps to manage white flies and aphids.
- Trap crops on the borders of the fields, intercropping, and harvesting of crop close to ground level.
- Spraying and dipping of seedlings using copper fungicide and biopesticide solutions to manage soil-borne diseases.
- Pruning of crowded, dead, broken, and diseased branches and burning of these to reduce inocula of pathogens. The wounds may be covered with Bordeaux paste or paint.

4.3.3 MECHANICAL AND PHYSICAL CONTROLS

Under mechanical and physical controls the pest is killed directly or blocked by making environment unsuitable for it. These may include using pest traps, steam sterilization of the soil, or use of barriers to keep birds or insects away. The major thrust is on:

- Remove and destroy the egg masses, larvae, pupae, and adults of insect pests as well as diseased plant parts.
- Use of bird perches.
- Trapping of insects using light traps.
- Dislodging leaf feeding larvae of caseworm and leaf folders using rope.
- Erection of bird scares in the field.
- Pheromones use to disrupt mating and creation of kill zone as well as for mass trapping.

4.3.4 CHEMICAL CONTROL

The use of chemicals should only be done when needed most and these are applied along with other approaches for long-term control of pests. Such selected pesticides are applied in a way to minimize their possible harm to people, nontarget organisms, and the environment. The IPM components are pest identification, monitoring, counting pest numbers and its damage, prevention of pest, application of biological, cultural, physical, mechanical, and chemical management techniques along with estimation of effect of pest management. The society and environment are benefited most by application of IPM. The use of safer pesticides like neem based and biopesticides may be promoted. The strips or isolated patches may be sprayed in case of infection or infestation is isolated and not the whole field.

4.3.5 BENEFITS OF IPM

The advantages of IPM are:

- Enhanced crop profitability due to no or little losses to yields.
- Assured yields and produce quality.
- Low severity of infestation of pests.
- The pesticide resistance or resurgence of pests is decreased.
- Safer produce.
- Better market potential of produce, sustained market share, and access.
- Low risk of restrictions or deregistration.
- The novel products, techniques, and services are facilitated.
- The product lifecycle is extended.

The proper knowledge of crop protection areas like prevention, monitoring, and intervention is prime requirement to synthesize IPM module of a crop. The build-up of pests is prevented. The monitoring of pests and bioagents is done through scouting for insect pests and diseases to decide the ways of interventions (physical, biological, and chemical methods) to get maximum economic value of crops with the use of minimal effects on the environment. The crop is grown in locations best suited climatically with proper soil and its topography. The cultivation of similar crops side by side of mixed cropping may substantially increase build-up of pest population and therefore, should be avoided. Maize is cultivated with a legume (Cowpea) helps in improvement in soil fertility and reduces weeds. These cropping systems also help to reduce pest damage. The crop rotation with nonhost crops reduce root-feeding insects and fungi.

Following principles are used to make IPM successful:

1. *PREVENTION AND SUPPRESSION:* It may be achieved by using crop rotation, adequate cultivation techniques, deployment of resistant and tolerant cultivars, certified seed and planting material, balanced fertilization, liming and proper irrigation and drainage practices, hygiene measures, conservation and use of beneficial organisms.
2. *MONITORING:* Observations of pests in the fields, scientifically sound warning, forecasting and early diagnosis, and advices from professionally qualified advisors.

3. *DECISION-MAKING:* Based on monitoring the decision on application of plant protection measures in time and space is made. For this, a robust and technically sound threshold values are essential components. The threshold levels mentioned for the region, specific areas, and crops in a particular climatic condition may be taken into account before treatments, wherever possible.
4. *NONCHEMICAL METHODS:* These are biological, physical, and other nonchemical methods but these should provide satisfactory pest control.
5. *SELECTION OF PESTICIDES AND DOSES:* The pesticides should be specific for the target along with least side effects on health of human population, nontarget organisms, and ecosystem. The measures like reduced doses and reduced application frequency may be preferred to avoid risk for development of resistance in populations of insect pests and pathogens.
6. *PESTICIDE RESISTANCE MANAGEMENT:* The antiresistance strategies may be used in case of pesticides known to cause resistance in insect pests so that the effectiveness of the products is maintained for longer period. For this purpose, multiple pesticides with different modes of action are used in a specific manner in crops like cotton and vegetables.
7. *EVALUATION:* The success of plant protection measures is assessed using volume of pesticides applied.

4.4 IPM IN DIFFERENT CROPS AND CROPPING SYSTEMS IN INDIA

The Govt. of India has adopted IPM in the overall Crop Production Programme from 1985 with following details:

- Minimum losses caused by pests and diseases in crops.
- Emphasis in ecologically sustainable IPM approaches over chemical pesticides at farmers fields.
- Promotion of bio-control for managing pests in crops.
- Conservation of the diverse agro-ecosystem to facilitate desired population of various natural enemies of pests.
- Making farmers aware on safe and judicious use of pesticides with label claims and their recommended doses.
- Survey and surveillance for insect pests and diseases aiming to forewarn the farmers for their potential epidemics.
- Human resource development in IPM.

The plant pathogens are known to cause serious diseases in different crops in India. These are blast and blight of paddy; leaf blight, rust, wilt, stem, and cob rots in maize; rust and Karnal bunt in wheat; wilt, blights, and root rots in food legumes; stem rot, pod rots, and foliar diseases in case of groundnut; gray mold, downy, and powdery mildews, *Alternaria* and bacterial blights in oilseeds; downy and powdery mildews in case of mango, oranges, and grapes; wilt and leaf spots in cotton; red rot and smut in sugarcane; damping-off in vegetable and tobacco; wilt and powdery mildew in vegetables. The species of *Alternaria, Aspergillus,* and *Fusarium* are important in storage, which is known to spoils the quality and viability of grains, fruits, and seeds.

The main advantages of using cropping system in IPM are better management of resistance in a pest, suppression of pests, and increase in biodiversity. A good cropping system, however, needs good understanding of nature of pest, nonhost crops to be rotated, intercropping, and understanding interaction crop-pest-NE and secondary plant.[3]

4.4.1 IPM PACKAGES

About 87 crops that include **Cereals crops** like, Maize, Rice, Wheat, Sorghum, and Pearl Millet; **Food legumes** namely, Black gram, Green gram, Gram, Horse gram, Red gram, Lentil, Moth bean, Rajmah, Pea, **Oilseeds** like, Rapeseed & Mustard, Sesame, Groundnut, Soybean, Safflower, Sunflower; Oil Palm, **Vegetables** crops- Potato, Onion, Tapioca, Tomato, Vegetables (Cruciferous, Leguminous, Cucurbitaceous crops) Okra, Brinjal, Chillies, Jack fruit, Spinach, Broccoli, Lablab bean; **Cash crops-** Cotton, Sugarcane, Tobacco, **Fruit crops-**Citrus, Apple, Pineapple, Raspberry, Sapota, Pomegranate, Grapes, Banana, Litchi, Papaya, Apricot, Peach, Pear, Fig, Phalsa, Saffron, Custard apple, Mango, Guava, Kiwi, Passion fruit, Cherry, Coconut, Walnut, Ber, Amla, Loquat, Strawberry, Watermelon, **Spices and beverages-** Cardamom, Coriander, Cumin, Black Pepper, Fennel, Ginger, Turmeric, Clove, Cashew, Curry Leaf, Arecanut, Fenugreek, Tea, Olive, Garlic, Betelvine, Coffee, Curry Leaf, Drumstick, and Mint.[4]

4.4.2 IMPACT OF IPM IN INDIA

About 93% of the farmers in India use chemical control and 51% of these get their plant protection advice from dealers. A total of 22% from extension officials and 73% take plant protection measures based on the first appearance of the pest. The plant protection on various crops amounts to the range of 7–40% of the total crop production cost. In spite of advocating IPM for the past two decades, only 3.2% of the farmers had adopted it in different crops. The IPM resulted 20–100% reduction in the use of pesticide in different crops in India.

The crop yields in IPM fields increased in the range from 6.72 to 40.14% in rice and 22.7–26.63% in cotton as compared to non-IPM fields. The use of pesticides is reduced in the range of 50–100% in case of paddy whereas it was 29.96–50.5% in cotton with the adoption of IPM. The use of biopesticides and neem-based pesticides went from 123 MT during to 7682 MT (Tech. Grade) in recent years. The total consumption of chemical pesticide also reduced from 61357 MT (1994–1995) to 49438 MT during 2018–2019.[4]

4.4.3 IPM IN DIFFERENT CROPS AND CROPPING SYSTEMS

Cropping systems may vary from one another in terms of intensity, inputs, and acreage systems. Accordingly pest management efforts and inputs also vary along this continuum. Under low value and extensive cropping systems, the inexpensive management strategies/tactics like cultural practices and host-plant resistance are generally followed. It is more applicable in cereals, whereas pest management in high-value crops like vegetables, fruits, and cash crops in general is far more intensive due to high value of these and comparatively low pest resistance. The use of pesticides is typically highly efficacious but often expensive with temporary effect on crops. There are differences in terms of use of inputs in IPM in developed and developing countries in case of similar commodity like potato.

Bragg et al.[5] outlined IPM practices in cereal-based cropping system in tropics. However, it needs experts to take decisions on IPM strategies based on monitoring of insect pests, diseases and pathotypes, deployment of host resistance, and prevailing weather conditions and advisories are given to farmers using modern communication technologies. The inputs are made available by state governments in India. The survey and surveillance are done to categorize different regions and relative importance of insect pests and diseases is assessed. The monitoring of key pests like Helicoverpa, Spodoptera, Pectinophora, Rhinoceros, Aproaerema, and Scirpophaga populations in different cropping systems is done using sex pheromones.[6] The use of transgenic plants having insecticidal genes impacted positively in reducing the need of insecticide inputs, increased safety to human beings and nontarget, natural enemy, low pesticide residues, and a safer environment.[7]

4.4.3.1 WHEAT

Fusarium graminearum and related species in wheat may cause seedling blights, foot rots, and scab. The wheat-maize cropping system favors the pathogen to overwinter in corn stalks and therefore becomes more severe. In wheat, it causes scab on ear heads and reduces wheat yields, as well as infected grain may have mycotoxins, which may impact the health of human and cattle. Sharma and Singh[8] gave wheat health management or IPM in India. The IPM practices involve survey and surveillance of pests and diseases, demarcation of important agroecological zones for different diseases, strategic deployment of resistant varieties, choosing high disease-free seed, use of seed treatment, and foliar sprays, which in turn minimized environmental and nutrient stresses; maintain beneficial insects and microorganisms. The adoption of conservation agriculture practices in the rice-wheat cropping system and its impact on soil microbes has been also studied.[9] The cereal cyst nematode (*Heterodera avenae*) is suppressed in soils having higher organic matter contents and reduced tillage[10] and also enhanced populations of beneficial organisms, including earth worms.[11] The foliage pathogens like rusts, powdery mildew, spot blotch, etc. are, however, managed well using host resistance in wheat.

4.4.3.2 BARLEY

The cereal cyst nematode of barley (*Heterodera avenae)* is well managed using crop rotation with crucifers like sarson, toria, raya, taramira, and other crops like gram, berseem, carrot, and coriander.[12]

4.4.3.3 SOYBEAN

Collar rot or Sclerotium blight caused by *Sclerotium rolfsii* is common pathogen in soybean-wheat cropping system.

4.4.3.4 CORN

During years with droughts aflatoxin production in corn is favoured due to infection of *Aspergillus* spp. whereas rice fields with corn intercrops had 26–48% less plant hoppers population than in rice monoculture.

4.4.3.5 COTTON

The intercropping of cotton with crops like cowpea, soybean, groundnut, and green gram has been recommended in central and southern part of India for management of insect pests. The use of trap crops namely okra, canabinus, castor, marigold (Tagetes), pigeon pea, coriander, jowar, and maize is recommended.

4.4.3.6 GROUNDNUT

The trap crop like sunflower is used in groundnut to suppress the population of the two most important defoliators, *S. litura,* and *H. armigera* and pests prefer sunflower for oviposition and larval feeding more than groundnut crop. A newly hatched larvae disperse immediately from the egg sites on groundnut crop whereas they stay longer than a week on sunflower plant.

4.4.3.7 VEGETABLES

The bio-pesticides such as neem-based products and NPV against *Spodoptera* spp. and *Helicoverpa armigera* are largely used in IPM strategy in vegetable crops. In cotton and vegetables, *Bacillus thuringiensis* (Bt) has importance in suppressing pest populations.[12] The use of trap crops is also done in managing insect pests in vegetables.[13]

The detailed information of diseases and insect pests and existing management practices in different crops in India is in Table 4.1.

Way Forward

- The plant protection measures should be developed in the manner so that pest build-up is avoided in the entire cropping system along with reduced residues of pesticides in produce and water bodies.
- Capacity building at farm level for giving better knowledge of soil, water, nutrient and pest management in an integrated manner.
- For monitoring of the crops at different critical stages using pheromones and weather based advisory system.
- Pests and diseases surveys to get an update on their incidence, distribution, and losses in different geographic regions.

TABLE 4.1 Economically Important Insect Pests and Diseases of Major Crops and Their Management in India.

Crop	Existing control methods	Common name	Scientific name	Economic thresh hold levels (ETLs)
Cereals				
Rice	Integrated Pest Management (IPM)	Stem borer	*Scirpophaga incertulus*	One egg mass per sqm-or white ears up to 5%
		Brown plant hopper	*Nilaparvata lugens*	10 hoppers per clump
	Resistant varieties	Gall Midge	*Orseolia oryzae*	5–10% silver shoots
		Leaf folder	*Cnaphalocrocis medinalis*	10–15% webbed foliage
	Integrated Disease Management (IDM)	Blast	*Pyricularia oryzae*	
		Sheath blight	*Rhizoctonia solani*	
		Bacterial leaf blight	*Xanthomonas oryzae*	
Wheat	Resistant varieties	Aphid	*Schizaphis graminum*	5–10% of plants with infestation
	Resistant varieties	Brown rust	*Puccinia recondita* f. sp. *tritici*	
		Black rust	*Puccinia graminis* f. sp. *tritici*	
	IDM	Karnal bunt	*Neovossia indica*	
	Chemical	Loose smut	*Ustilago segetum*	
Maize	Chemical	Stem borer	*Chilo partellus*	5–10% infestation
		Shoot fly	*Atherigona* spp.	Dead hearts up to the range of 5–10%
		Earworm	*Helicoverpa armigera*	25–30% damage to cobs
	Resistant varieties & Chemical	Maydis leaf blight	*Cochliobolus heterostrophus*	
		Common rust	*Puccinia sorghi*	
	Chemical	Downy mildew	*Peronosclerospora* sp.	
	Resistant varieties	Stalk rot	*Fusarium moniliforme*	
	Resistant varieties	Charcoal rot	*Macrophomina phaseolina*	

TABLE 4.1 *(Continued)*

Crop	Existing control methods	Common name	Scientific name	Economic thresh hold levels (ETLs)
Legumes				
Pigeonpea	IPM	Pod borer	*Helicoverpa armigera*	Three larvae/ 2 plants or 2–3 eggs 5 twigs
	Chemical	Pod fly	*Melanagromyza obtusa*	In all endemic locations
		Leaf webber	*Maruca vitrata*	Five webs per plant
		Pod sucking bugs	*Clavigralla gibbosa*	One egg mass per plant
	Resistant varieties	Wilt	*Fusarium udum*	
	IDM	Phytophthora blight	*Phytophthora drechsleri* f. sp. *cajani*	
	Resistant varieties	Sterility mosaic	*Aceria cajani* transmitted sterility mosaic virus	
Chickpea	IPM	Pod borer	*Helicoverpa armigera*	Two larvae/plant or three eggs per plant
	Chemical	Cutworm	*Agrotis ipsilon*	5% plant mortality
	Resistant varieties	Wilt	*Fusarium oxysporum* f.sp. *ciceri*	
		Dry root rot	*Rhizoctinia bataticola*	
		Collar rot	*Sclerotium rolfsii*	
	IDM	Ascochyta blight	*Ascochyta rabiei*	
Soybean		Botrytis gray mold	*Botrytis cinerea*	5% plant infestation
		Stunt	Bean leaf roll virus	5% incidence
		Hairy caterpillar	*Spilosoma obliqua*	Five larvae meter row
	Chemical and Resistant varieties	Pod blight of soybean	*Colletotrichum dematium* f. sp. *truncata*	

TABLE 4.1 *(Continued)*

Crop	Existing control methods	Common name	Scientific name	Economic thresh hold levels (ETLs)
	Resistant varieties	Bacterial pustule	*Xanthomonas campestris*	
	Cultural and Resistant varieties	Bacterial Blight	*Pseudomonas* spp.	
		Charcoal rot	*Macrophomina phaseolina*	
	Resistant varieties	Collar rot	*Sclerotium rolfsii*	
Oilseeds				
Groundnut	IPM	Leaf miner	*Aproaerema midicella*	Five mines/plant at around 30 days of crop age
		Tobacco caterpillar	*Spodoptera litura*	20–25% defoliation at 40 days
		Aphids	*Aphis craccivora*	Aphids (5–10 in numbers)/terminal at seedling stage
	Chemical	Thrips	*Scirtothrips dorsalis*	Five thrips per terminal at seedling stage
		Crown rot	*Aspergillus niger*	
	Resistant varieties & cultural	Stem and pod rots	*Sclerotium rolfsii*	
		Aflatoxin	*Aspergillus flavus*	
	IDM	Early leaf spot	*Cercospora arachidicola*	
		Late leaf spot	*Phaeoisariopsis personata*	
	Resistant varieties & IDM	Rust	*Puccinia arachidis*	
Sunflower	Chemical	Gram pod borer	*Helicoverpa armigera*	One larva per head
		Gray mold	*Botrytis cinerea*	
		Alternaria blight	*Alternaria helianthi*	

TABLE 4.1 *(Continued)*

Crop	Existing control methods	Common name	Scientific name	Economic thresh hold levels (ETLs)
	Resistant varieties	Wilt	*Verticillum dahliae*	
		Scorch	*Maacrophomina phaseoli*	
Sesame	Chemical	Leaf webber	*Antigastra catalaunalis*	Two to five webbs per plant
	Chemical	Phytophthora blight	*Phytophthora parasitica*	
	Resistant varieties	Charcoal rot	*Macrophomina phaseolina*	
		Wilt	*Fusarium oxysporum* f. sp. *sesami*	
		Cercospora leaf spot	*Cercospora sesami*	
		Alternaria leaf spot	*Alternaria sesami*	
		Bacterial blight	*Xanthomonas campestris*	
Rapeseed	Chemical	Aphids	*Lipaphis erysimi*	5–10 aphids per plant
	Resistant varieties	Alternaria blight	*Alternaria brassicae*	
		Downy mildew	*Peronospora parasitica*	
		Powdery mildew	*Erysiphe cruciferarum*	
Vegetables				
Brinjal	IPM	Borer of fruit and stem	*Leucinodes orbanalis*	Shoot and fruit infestation in the range of 1–5%
	Chemical	Damping-off	*Phytophthora* or *Pythium* sp.	
	Resistant varieties	Wilt	*Fusarium ozonium*	
		Phomopsis blight	*Phomopsis vexans*	

TABLE 4.1 *(Continued)*

Crop	Existing control methods	Common name	Scientific name	Economic thresh hold levels (ETLs)
Cabbage & Cauli-flower	IPM	Dimond back moth	*Plutella xylostella*	The pest incidence in the range of 1–5%
		Tobacco caterpillar	*Spodoptera litura*	1–5% incidence
	Chemical	Downy mildew	*Perenospora parasitica*	
		Alternaria blight	*Alternaria solani*	
		Black rot	*Xanthomonas campestris*	
		Stalk rot	*Sclerotinia sclerotiorum*	
Tomato	IPM	Fruit worm	*Helicoverpa armigera*	1–5% fruit damage
		Late blight	*Phytophthora infestans*	
		Leaf blight	*Septoria lycopersici*	
	Resistant varieties and cultural	Tomato spotted wilt Viral disease	Tomato spotted wilt virus	
	Resistant varieties	Wilt	*Psuedomonas solanacearum*	
Fruits				
Apple	IPM	San Jose Scale	*Quadraspidiotus perniciosus*	5% trees are infested
		Mites	*Panonychus ulmi*	5–10% foliage infestation
	Resistant varieties and Chemical	Apple Scab	*Venturia inaequalis*	
Grapes	Chemical	Flea beetle	*Scelodonta stricollis*	20% foliar damage
		Thrips	*Retithrips syriacus*	Five thrips/young leaf
		Mealy bugs	*Maconellicoccus hirstutus*	1% bunch infestation
		Fruit flies	*Carpomyia vesuviana*	Oranges 1–2% incidence

TABLE 4.1 *(Continued)*

Crop	Existing control methods	Common name	Scientific name	Economic thresh hold levels (ETLs)
		Defoliators	*Papilio demoleus*	20–30% foliar damage
		Mango Hopper	*Amritodes atkinsoni*	Two to five hoppers per inflorescence
		Leaf webber	*Orthaga exvinacea*	10% incidence
		Stem borer	*Batocera rufomaculatus*	Appearance of the pest
		Anthracnose	*Gloeosporium ampelophagum*	
		Downy mildew	*Plasmopara viticola*	
		Powdery mildew	*Uncinula necator*	
Cash Crops				
Cotton	IPM	American bollworm	*Helicoverpa armigera*	5–10 % boll infestation
		Pink bollworm	*Pectinophora gossipiella*	5–10% boll infestation
		Whitefly	*Bemisia tabaci*	8–10 adults/leaf
		Spoted bollworm	*Earias insulana*	5–10% boll infestation
	Resistant varieties	Verticillium wilt	*Verticillium dahliae*	
		Root rot	*Rhizoctonia* spp.	
	IDM	Alternaria leaf spot	*Alternaria macrospora*	
	Chemical	Anthracnose	*Colletotrichum gossypii*	
Sugarcane	IPM	Stem borer	*Chilo sacchariphagus*	Up to 10% damage at tillering phase
		Scale insect	*Melanapsis glomerata*	20–30% canes with scale incidence
	Resistant varieties	Red rot	*Colletotrichum falcatum*	
		Smut	*Ustilago scitaminea*	

TABLE 4.1 *(Continued)*

Crop	Existing control methods	Common name	Scientific name	Economic thresh hold levels (ETLs)
		Wilt	*Fusarium sacchari*	
Tobacco	IPM	Tobacco caterpillar	*Spodoptera litura*	5–10% leaves with damage
		Whiteflies	*Bemisia tabaci*	5–10 flies/leaf
	Chemical	Damping-off	*Pythium aphanidermatum*	
		Frog-eye spot	*Cercospora nicotianae*	

- Deployment of resistant varieties of crops to biotic stresses through farmers networks.
- Altered agronomic practices for supporting natural enemies.
- Up scaling of use of botanicals and bioagents available locally as an alternative to chemicals.
- Participation of community in IPM.
- Participatory research with farmers.
- To update farmers on input and output market intelligence.
- Promotion of farm clinics.

4.5 CONCLUSION

Through IPM, negative impacts of crop protection activities on the environment are minimized. It is a technique to provide safer food to growing population. The IPM is a dynamic in nature and requires constant updating in approach, its components as well as tactics with time and space. It impacted different cropping systems quite positively in an ecosystem and will go in a long way to provide sustainable crop production under changing environmental conditions and business scenario. The transfer of IPM technology to the farmers is a challenging task and training of farmers in crop production and protection involving methods like identifying pests, their natural enemies, economic injury levels, agro-ecosystem, and its analysis are crucial to farmers in their decision making.

KEYWORDS

- **sustainable**
- **cropping systems**
- **non-chemical pest management**
- **organic farming**
- **exotic pests**
- **agroecosystem management**
- **biological control**

REFERENCES

1. Alam, M. Z.; Haque, Md. M.; Islam, Md. S.; Hossain, E.; Hasan, S. B.; Hasan, S. B.; Hossain, Md. S. Comparative Study of Integrated Pest Management and Farmers Practices on Sustainable Environment in the Rice Ecosystem. *Int. J. Zool.* **2016,** Article ID 7286040, 1–12 http://dx.doi.org/10.1155/2016/7286040.
2. Carroll, B. N. Development in Transition: Crop protection and More. In: *Emerging Technologies for Integrated Pest Management–Concepts, Research, and Implementation*, Kennedy, G. G.; Sutton, T. B.; Eds.; APS Press, American Phytopathological Society, St. Paul, MN, 2000; pp 339–354.
3. Poddar, N.; Yele, Y.; Kumari, A. Cropping System Approach in IPM. Adaptive Crop Protection Management Strategies Editors: D. Prasad, Gopal Lal and Imtiyaz Ahmad. ISBN: 978-93-88317-08-5 Write and Print Publications, New Delhi, 2019; pp. 358–371.
4. DAC&FW. 2021. https://agricoop.nic.in
5. Bragg, D. E.; Rondon, Silvia, I. J.; Gavloski, U. S.; Abrol, D. P. *Integrated Pest Management in the Tropics*; Abrol, D. P., Ed.; New India Publishing Agency: New Delhi, 2016; pp. 249–273.
6. Ranga Rao, G. V.; Kumari, B. R.; Sahrawat, K. L.; Wani, S. P. Integrated Pest Management (IPM) for Reducing Pesticide Residues in Crops and Natural Resources. In *New Horizons in Insect Science: Towards Sustainable Pest Management*; Chakravarthy, A. K., Ed.; Springer, India, 2015; pp 397–412.
7. Sharma, A. K.; Singh, D. P. *Integrated Pest Management in Wheat Based Cropping Systems*; DWR: Karnal, 2008; 258p.
8. Rani, A.; Singh, D. P.; Sharma, R. K.; Chhokar, R. S. Resource Conservation Agricultural Practices, Rhizosphere and Diseases of Wheat Under Wheat-Rice Cropping System. *Int. J. Curr. Microbiol. Appl. Sci.* **2017,** *6* (11), 1290–1298.
9. El Titi, A.; Ipach, U. Soil Fauna in Sustainable Agriculture: Results of an Integrated Farming System at Lautenbach, F.R.G. Agriculture. *Ecosyst. Environ.* **1989,** *27*, 561–572.
10. Zwart, K. B.; Bloem, J.; Burgers, S. L. G. E.; Bouman, L. A.; Brussaard, L.; Lebbink, G.; Didden, J. C. Y.; Marinissen, W. A. M.; Vreeken-Buys, J. J.; de Ruiter, P. C. Population Dynamics in the Below-Ground Food Webs in Two Different Agricultural Systems. *Agric. Ecosyst. Environ.* **1994,** *51*, 187–198.
11. Khan, M. R. Nematode Diseases of Crops in India. In *Recent Advances in the Diagnosis and Management of Plant Diseases*; Awasthi, L. P., Ed.; Springer India, 2015; pp 183–224.
12. Rupela, O. P.; Rahman, S. J.; Ranga Rao, G. V.; Rama, J.; Sasi Jyothsna, J. S.; Humayun, P. Protecting Vegetables from Insect-Pests Using Low-Cost and Biological Options-a Two-Year on-Farm Experience. Presented at the International Conference on Indigenous Vegetables and Legumes, 2006. Jointly Organized by AVRDC-The World Vegetables Center, Regional Center for South Asia and ICRISAT, at ICRISAT Campus, 12–15 Dec 2006.
13. Sarkar, S.; Wang, E.; Wu, S.; Lei, Z. Application of Trap Cropping as Companion Plants for the Management of Agricultural Pests: A Review. *Insects* **2018,** *9*, 128.

CHAPTER 5

Integrated Pest Management in Rice-Based Cropping Systems in India

GURURAJ KATTI*, CH. PADMAVATHI, and R. MAHENDER KUMAR

ICAR-Indian Institute of Rice Research, Rajendranagar, Hyderabad, India

**Corresponding author. E-mail: gururajkatti59@gmail.com*

ABSTRACT

Rice is the major staple food crop cultivated widely across diverse eco systems throughout India. In recent years, increasing scarcity of resource inputs coupled with other socio-economic constraints have posed serious challenges in achieving sustainable food production in India. Improved productivity and profitability through the identification of appropriate rice-based cropping systems seems to be the possible solution to mitigate these challenges. However, insect pests are the key biotic stresses posing hurdles in realizing the potential benefits of rice-based cropping systems. Five pests of national significance affecting rice crop yields along with impacts on climate change and its concomitant effects on the agronomies of rice-based cropping systems need to be tackled to ensure ecological, economic, and social benefits for rice farmers. A holistic approach of integrated pest management, with emphasis on biointensive inputs, offers a long-term, eco-friendly solution to the pest problems of this most important and prevalent cropping system of the country.

5.1 INTRODUCTION

India is one of the leading rice-growing nation in the world with largest agricultural area of 43.8 m ha and the second highest production of 118 million

Integrated Pest Management in Diverse Cropping Systems. Devendra Pal Singh, PhD (Ed.)

tonnes. Rice and its products constitute the dietary composition of more than 65% of the people of India. The crop is cultivated in diverse environments, *viz*., waterlogged, deep water, hilly area, high humidity, high temperatures, and areas regularly subjected to uncertain situations like salinity, alkalinity, and floods. The cropping intensity involves cultivation of three crops in a year in fertile riverine deltas with assured irrigation facilities. It, however, varies from one environment to the other depending upon weather, soil, and food consumption pattern (Figures 5.1–5.4).

FIGURE 5.1 Hill rice ecosystem.

FIGURE 5.2 Deep water rice ecosystem.

FIGURE 5.3 Upland rice ecosystem.

FIGURE 5.4 Irrigated rice ecosystem.

In view of the dominance and impact of rice-based cropping systems in Indian agriculture, increased productivity of these systems is crucial for ensuring poverty alleviation and livelihood security through generating employment and higher income.[1] More than 250 cropping systems have been identified in India based on crop spread, of which 30 cropping systems are important and commonly adopted across the country. The predominant cropping systems have been delineated into 15 different agro-climatic zones based on soil, water, rainfall, temperature, and rice-based cropping systems, which are predominant in eleven zones. The rice-based cropping systems, including both low land and upland crops, range from rice–rice–rice to rice followed by different cereals, pulses, oilseeds, vegetables, and fiber crops (Table 5.1). Rice-based cropping systems may include lowland and upland crops. Among these, rice-wheat cropping system is the major system in the country occupying 10.5 m ha^2 followed by rice–rice (6.0 m ha) and rice fallow (4.4 m ha). There is a wide diversity in insect pest profiles in these cropping systems with variation in resource utilization patterns, which may undergo more dynamic alterations due to the impact of the global climate change scenario.

TABLE 5.1 Distribution of Rice-Based Cropping Systems in the Country.

Cropping system/States	Area ('000 ha)
Rice–wheat—West Bengal, Uttar Pradesh, Gujarat, Punjab, Madhya Pradesh, Jammu, Kashmir, Himachal Pradesh, Haryana, Bihar, Maharashtra, Assam	**9847.3**
Rice–rice—Andhra Pradesh, Kerala, Tamil Naidu, Assam, Gujarat, Odisha, Karnataka	**5894.0**
Rice–fallow—Bihar, Karnataka, Madhya Pradesh, Jammu, Kashmir, Maharashtra	**4419.9**
Rice–vegetable—Odisha, West Bengal, Gujarat, Tamil Naidu, Maharashtra	**1243.5**
Rice–ground nut—Tamil Nadu, Maharashtra, Andhra Pradesh, Odisha, Karnataka	**1022.9**
Rice–lathyrus—Madhya Pradesh, Bihar	**950.0**
Rice–urdbean—Andhra Pradesh, Odisha	**598.9**
Rice–mungbean—Karnataka, Andhra Pradesh, Odisha	**597.5**
Rice–chickpea—Uttar Pradesh, Odisha, Maharashtra, Madhya Pradesh	**512.9**
Rice–mustard—Uttar Pradesh, Madhya Pradesh, Bihar, Odisha, West Bengal	**480.7**
Rice–potato—West Bengal	**462.4**
Rice–sugarcane—Gujarat, Tamil Naidu, Bihar	**432.6**
Rice–horse gram—Odisha	**245.9**
Rice–jute—West Bengal	**120.1**

5.2 RICE-BASED CROPPING SYSTEMS IN INDIA AND INSECT PEST SCENARIO

5.2.1 RICE–WHEAT SYSTEM

The rice–wheat cropping sequence has a major role to play due to its substantial contribution to the cereal production. However, efficiency and productivity of this cropping system is under threat due to indiscriminate adoption of rice–wheat sequence, thereby leading to adverse shift in pest situations as well as decline in factor-productivity. The continuous cultivation of rice–wheat cropping system has created many problems related to low water-use efficiency and land degradation problems. Indiscriminate exploitation of groundwater, inefficient land use, imbalances in fertilizer use, and indiscriminate application of pesticides have led to the build-up of insect pests and diseases.

Among the insect pests in rice, yellow stem borer (YSB) is dominant in this cropping system; but, in the last few years, higher levels of pink stem borer (PSB) populations have been observed at the reproductive phase, while white stem borer (WSB) is also prevalent, but only in vegetative phase.[1] In the rice–wheat system, planthoppers are another group of insect pests which have gained serious pest status in the last decade mainly due to the imbalanced use of fertilizers and widespread abuse of pesticides, even leading to serious environmental and human health concerns. In the hill regions of Himachal Pradesh, five species of stem borers have been reported, though YSB, PSB, and WSB are more prevalent (Table 5.2).

5.2.2 RICE–RICE SYSTEM

Rice–rice is also an extensively cultivated cropping system in India covering an area of about 6 m ha area. Among the cropping systems, this is the second-most widely prevalent system extending to Eastern and Southern region including the states of Odisha, West Bengal, Telangana, Andhra Pradesh, Karnataka, Tamil Nadu, and Kerala. Deficient resource use and poor soil health are major issues for sustaining productivity of rice–rice cropping system.

The insect pest problems are relatively more, both in terms of occurrence and intensity in rice–rice cropping system compared to the other rice-based cropping systems. This is because of a continuous supply of host resources for the multiplication and survival of insect pests affecting rice. YSB being a monophagous pest of rice crop is a longer surviving and continuous pest in this cropping system. In addition, leaf folder (LF), brown planthopper (BPH), white-backed planthopper (WBPH), and gall midge (GM) are widely prevalent. Among the pests of regional significance, rice hispa is a recurring pest in these cropping system regions.

In some of the areas of West Bengal, the insect pest incidence in these cropping systems is mainly restricted to stem borers, particularly YSB, which has emerged as a serious biotic constraint in boro rice. In the other rice-based cropping systems of the North Eastern region and particularly Assam, yellow stem borer is common in both wet season and boro rice. However,

TABLE 5.2 Major Insect Pests in Rice-Based Cropping Systems.

Cropping system	Major insect pests	States	Agro-climatic zone
Rice–wheat, Rice–potato–potato	Yellow stem borer, pink stem borer, planthoppers, leaf folder, hispa	Jammu & Kashmir, Himachal Pradesh, Uttarakhand	Western Himalayan Region
Rice–fallows, rice–rice, rice–potato/oilseeds	Stem borers, leaf folder, hispa	Assam, North Eastern States, West Bengal	Eastern Himalayan Region
Rice–rice, rice–wheat, rice–potato–jute/vegetables	Stem borers, planthoppers	West Bengal	Lower Gangetic Plains
Rice–wheat, rice–maize, rice–potato–sunflower	Stem borers, planthoppers, leaf folder	Eastern Uttar Pradesh, Bihar	Middle Gangetic Plains
Rice–wheat, sugarcane ratoon–wheat	Yellow stem borer, pink stem borer, planthoppers, leaf folder	Central, Western Uttar Pradesh	Upper Gangetic Plains
Rice–wheat	Yellow stem borer, pink stem borer, planthoppers, leaf folder	Punjab, Haryana	Trans-gangetic Plains
Rice—urdbean, rice–niger/linseed, rice–vegetables, rice–fallow	Stem borers, leaf folder, planthoppers, gall midge, ear head bug	Chhattisgarh, Jharkhand, Madhya Pradesh, Maharashtra	Eastern Plateau, Hills Region
Rice–rice, rice–mung bean/urdbean, rice–peanut–mung bean	Stem borers, leaf folder, planthoppers, gall midge	Andhra Pradesh, Odisha, Tamil Nadu, Puducherry	East Coast Plains and Hills Region
Rice–rice	Stem borers, leaf folder, planthoppers, gall midge	Andhra Pradesh, Tamil Nadu	Southern plateau & Hills Region
Rice–rice	Stem borers, leaf folder, planthoppers, gall midge	Goa, Maharashtra, Karnataka, Kerala, Tamil Nadu	West Coast Plains & Hills Region
Rice–fallows	Stem borers	Andaman & Nicobar Region	Island Region

gall midge, leaf folder, and planthoppers have also been reported but only in the wet season. Although their incidence has been sporadic, occasionally pest outbreaks occur, particularly those of planthoppers and leaf folder, as observed in recent times.

5.2.3 RICE—PULSES/OILSEEDS/FALLOW/VEGETABLES SYSTEM

Rice–pulses/oilseeds cropping sequences are mainly prevalent in the Eastern plateau & Hills, Chhattisgarh, and Andaman Nicobar Islands and are considered as valuable for food, nutritional security, and ideal to maintain balanced soil health.

Under the rice–pulses cropping system, two sequences, viz., rice–urdbean and rice–mung bean are key to the rain-fed lowland rice ecosystem of Odisha state as urdbean and mung bean are grown as winter crops following rice. In the rice–oilseeds cropping system of eastern India, groundnut and rapeseed/mustard are also cultivated as winter crops in harvested rice fields. These have replaced the commonly prevalent practice of rice–fallow system, which no longer is economically sustainable for the farmers in this region. With the introduction of different-duration crop varieties, rice–fallow areas are gradually getting reduced. However, Odisha accounts for around 10% of the total rice fallow area of India.

In the remaining areas, rice crop followed by vegetables/potato is also quite popular among farmers because it is beneficial in terms of better utilization of land compared to leaving it fallow. Also, the emerging rice–maize cropping systems in South Asia are increasing in states of South India. Although the yield potential of this system is very high, the nutrient demand of this system has led to declining soil fertility in those regions.

In all these cropping system regimes, the major pests reported are yellow stem borer, leaf folder, gall midge, and ear head bug. However, in the past few years, planthoppers have also emerged as one of the key biotic stresses affecting rice crop in these regions. In the case of rice–maize system, pink stem borer has gained importance because the pest can attack both hosts, similar to the rice–wheat cropping system.

5.3 INSECT PEST SCENARIO IN RICE

Insect pests are the most challenging constraints for farmers to manage in the country because of the constantly prevailing warm and humid tropical weather conducive to pest incidence and multiplication. Particularly in rice-based cropping systems, the uniqueness and diversity of the cultivated agroecosystems makes the pest scenario more difficult to handle in an economically and environmentally gainful way.

Over 100 species of insects attack and feed rice crop from nursery to maturity, also in storage. Of these, about a dozen are consistently reported as major pests of economic importance causing yield losses up to 25%[3] and the damage caused by them varies in time and space. Multilocation studies on insect monitoring under All India Coordinated Rice Improvement Project (AICRIP) revealed that in the last five decades, there has been an increase in the number of important insect pests from 3 to 19 with planthoppers leading the list of key pests, due to their spread in intensity across the country.[4] However, there are five major insect pests of rice found in different rice-based cropping systems in India as described as follows:

5.3.1 STEM BORERS

Five species of stem borers have been recorded to seriously damage rice crop. Among these, yellow stem borer (YSB), *Scirpophaga incertulas* Wlk., is spread across and severe in all the prevalent rice ecosystems. The other major borers are pink stem borer (PSB), *Sesamia inferens*, confined to north-western regions and white stem borer (WSB), *Scirpophaga innotata*, common in southern and northern hill regions. The other two species, dark-headed stem borer, *Chilo polychrysus* Meyr., and striped stem borer, *Chilo suppressalis* Meyr., are prevalent in eastern and north-eastern states of West Bengal and Assam, respectively.

The stem borer has the ability to cause damage in both vegetative and reproductive phases of the rice crop. However, its incidence during vegetative stages does not have serious impact on yield, though heavy infestation during nursery and early tillering stage can result in economic losses for the farmers. At the tillering stage, the neonate stem borer larvae enter and feed inside the tiller resulting in central leaf whorl drying, often turning into a "dead heart." Such affected tillers can be easily pulled out by hand and do not bear panicles. During the reproductive stage, the affected panicles turn whitish with ill-filled grains or sometimes there is complete loss of grains. The "white ear" damage is very difficult to avoid as the farmer comes to know of it only after its occurrence and also has a direct impact on yield (Figures 5.5–5.7).

FIGURE 5.5 Adult male and female.

FIGURE 5.6 Dead heart.

FIGURE 5.7 White ears.

5.3.2 GALL MIDGE

Gall midge (*Orseolia oryzae* Wood-Mason) has been observed to be an important pest confined mainly to irrigate or rainfed rice including shallow, upland, and deep-water rice. Similar to stem borer, gall midge is also one of the key hidden pests of rice as most of the pest life cycle is completed within the rice plant. In fact, rice farmers identify this pest in the field only through its damage symptoms after its occurrence. The maggots emerging from eggs laid on leaf surfaces or leaf sheaths cause damage by feeding on the growing tip resulting in an elongation of the leaf sheath transforming it into a gall. The galls are commonly referred to as silver shoots as they are similar to onion silvery leaves glistening in the field. In some cases, there would be no gall development but necrosis of the growing tip is noticed. There is also heavy tillering and formation of stunted seedlings associated with gall formation. Gall midge infestation can cause significant loss in yields, if its incidence is heavy up to 50 days after transplanting (Figures 5.8 and 5.9).

FIGURE 5.8 Adult midge.

FIGURE 5.9 Silver shoots.

5.3.3 PLANTHOPPERS

There are two types of planthoppers commonly observed in India with brown planthopper (BPH) being more dominant than WBPH in occurrence and distribution. BPH (*Nilaparvata lugens* Stal) is widely distributed across all the rice ecosystems but is more prevalent in rainfed and irrigated wetland ecologies. The pest incidence starts during the active tillering stage and continues up to harvest. The other species, WBPH (*Sogatella furcifera* Horvath), hitherto not serious, has been increasingly recorded in areas where rice varieties resistant to BPH are grown.

Both species of planthoppers occur as mixed populations in fields though BPH dominates the other species in most of the rice ecosystems. Dimorphic, winged as well as half-winged males and females, along with wingless nymphs confined to the base of the rice plants, suck the sap from the tillers. The affected plants initially turn greenish yellow, spread further, and lead to yellowing and drying of the plants. Due to confinement and slow mobility of the pest, the patches of infestation spread slowly in concentric circles among the hills within the field. However, due to the rapid multiplication potential of the pest, the infestation in severe cases results in the affected field getting a burnt appearance. Hence, this condition is referred as "hopper burn." The pest populations migrate over long distances causing a widespread infestation in short time. BPH is also a vector of grassy stunt virus (Figures 5.10–5.12).

FIGURE 5.10 Field showing Hopper burn.

FIGURE 5.11 BPH.

FIGURE 5.12 WBPH.

5.3.4 LEAF FOLDER

Among the foliage pests of rice, leaf folder (*Cnaphalocrocis medinalis* Guenee) is an important one having the ability to cause severe defoliation. However, since rice plants have the innate ability to compensate for foliar damage by producing more tillers, leaf folder infestation can result in yield loss only when flag leaf is severely affected during early reproductive stage of rice crop. Earlier considered as minor pest, leaf folder has now made a resurgent niche across the rice-based cropping systems because of expansion of irrigated rice area, input responsive vigorously growing high yielding varieties, and indiscriminate use of nitrogenous fertilizers and insecticides by the farmers.

The leaf folder larvae feed on the leaves remaining within the longitudinal folds made by stitching together of leaf edges causing linear pale white strips on damaged leaves. Severely infested fields present a whitish appearance due to the drying up of leaf margins and tips providing triggering point for initiation of insecticide sprays by the farmers (Figures 5.13 and 5.14).

FIGURE 5.13 Leaf folder adult moth.

5.3.5 GUNDHI BUG

Gundhi bug (GB) is considered a serious pest of rice as it attacks crucial milky stage during reproductive phase of the crop. There are three species reported to be prevalent of which, *Leptocorisa oratorius* Thunberg and *L. acuta* Thunberg are important. The pest survives on alternate hosts such as grasses and associated dicotyledonous weeds before migrating to paddy crop. Among the grasses, *Echinocloa colona* serves as potential alternate host before the pest shifts to unequally ripened paddy fields planted with early to late flowering varieties.

FIGURE 5.14 Leaf folder damage.

The developing milky grains are the feeding targets of the pest, which results in affected panicles showing characteristic blackish spots at the point of feeding. The infestation leads to panicle discoloration and malformation of grains in the panicles.

In addition to the aforementioned discussion, there are a few pests of regional significance which are sporadic but can cause considerable losses depending upon time and place of occurrence.

5.3.6 RICE HISPA

Rice hispa (*Dicladispa armigera* Olivier), though a minor pest, can assume the status of an economically damaging insect when its infestation is heavy during nursery or early tillering stage just after transplanting. In India, this pest earlier was more common in north-eastern parts, however, now is relatively well-distributed across eastern, central, and parts of southern India. In case of serious infestation, intense foliar feeding by adults and grubs results in leaf drying and fields with scorched appearance. Rice hispa can multiply and migrate fast to adjoining fields quickly spreading damage over short distances.

5.3.7 CASE WORM

Rice case worm (*Parapoynx stagnalis* Zeller) is a regularly occurring pest mainly in rainfed rice areas as well as under irrigated conditions particularly where there is continuous standing water in fields for a long period. Low land and flooded deep water areas provide favorable conditions for the buildup of this pest during the early stages of the crop therby adversely affecting crop growth. The larvae cut the leaf blades and form tubular cases, which help them to survive and feed on the crop. Prolonged pestiferous conditions result in patches of voracious defoliation, stunted crop growth, and loss of plant population.

5.3.8 CUT WORM/EAR-CUTTING CATERPILLAR

It is a sporadic pest (*Mythimna separata* Walker) reported to cause heavy losses during sudden outbreaks in the states of Bihar, Uttar Pradesh, Odisha, and Madhya Pradesh. Commonly occurring weeds, such as *Echinocloa colonum* and *Eleusine coracana* in rice fields act, as alternate hosts enabling the pest survival during early stages of crop growth.

Cut worm damage in the early crop stages involves excessive foliage feeding sometimes leading to complete removal of leaves and severe plant loss. However, higher populations during reproductive stages can play more havoc resulting in cutting of ear heads with total yield loss.

5.4 INTEGRATED PEST MANAGEMENT IN RICE-BASED CROPPING SYSTEMS

Integrated pest management (IPM) with its holistic approach remains a sustainable and realistic option for rice farmers to manage the pests across different rice-based cropping systems. IPM provides the ideal ecologically sound framework for managing pests under which crop-management specialists can intelligently integrate available pest-management components such as the use of pest-resistant varieties, biological suppression or natural regulation, modification of agronomic practices, and habitat manipulation or deploying only as last resort eco-friendly insecticides to tackle the pests in times of unexpected outbreaks or emergency situations. This long-term strategy also has the potential to cause minimum risk to human health and environment as well as ensure economic and social gains for the farmers.

IPM adoption by farmers across ecosystems, crops, and countries relies on the operating circumstances, options, and capabilities; nevertheless, it remains the most viable approach toward realization of profitable, yet sustainable, rice production. In the last decade, there has been a paradigm shift in IPM from being considered as a separate entity to the present context of its inclusion as a component of integrated crop management (ICM) under the wider realm of farming system management.[5] Also, IPM in a broader sense involves managing not only "pests" but also pesticides, parasites, predators, pathogens and all the key stakeholders linked to farming.

Innovation is the key to any progress and refinement. Similarly, IPM strategies have also evolved over the years through the synergy of developments in scientific research and discoveries related to biotechnology and other fields. The newer technological innovations leading to concomitant changes on market and the policy environment fronts have offered novel opportunities to wean away from the need for chemical pesticides toward the evolution of a holistic refinement of IPM for sustainable rice production. The latest motive trend has been to make IPM more biointensive and provide safe and gainful solutions for the farmers. The different components of IPM available for adoption in different rice-based cropping systems are enumerated as follows:

5.4.1 PEST DIAGNOSTICS AND SURVEILLANCE

Pest diagnostics along with surveillance forms the most vital cog of rice IPM strategy. Conventionally, farmers have been habituated to identify or diagnose pest problem or incidence mainly through damage symptoms and that too most of the time after the pest has made its appearance. Also, several of the farmers fail to differentiate among the damage symptoms arising out of an insect pest, disease attack, or nutrient deficiency. Abiotic stresses such as droughts and floods further complicate the farmers' decision making to undertake suitable management measures. Hence, simple, cheap, and rapid pest diagnostic tools should be made available to farmers.

Recent scientific research developments have helped in the development of innovative technologies such as remote microscopy diagnosis (RMD) and barcoding techniques for easy, accurate identification of pest or disease infestation and demarcation from nutrient deficiency disorders. Reliable and immediate identification of all the rice pests is now possible through the coordinated efforts of the National Agricultural System organizations, agricultural universities, as well as state departments of agriculture aided by farmers' active cooperation and response. The amalgamation of cutting-edge research in biotechnology and nanotechnology in sensors leading to equipment of increased sensitivity allows rapid tackling of climate change induced stresses. Traditionally, light traps have been the major means for the farmers to monitor the onset and development of the pest population dynamics in the field. However, advanced studies on semi-chemicals have provided novel means to accurately monitor pest populations. Sex pheromones are species-specific and offer great promise as monitoring tools to take timely actions for preventing yellow stem borer in rice. Pheromone traps with 5 mg lure placed in a sleeve trap with a yellow canopy lid @ 3 traps/acre can be used to monitor adult male moths of yellow stem borer throughout the crop growth. The lure needs to be replaced every three weeks to maintain its effectiveness and a threshold level of 30–35 male moths per trap per week is considered for pest-management decisions. Further, efforts are under way to provide similar tools for reliable information on other species of stem borers as well as leaf folder, caseworm, hispa, and earhead bug.[6]

Although prevention of attack is the best approach to manage a pest, sometimes regular monitoring of pest populations after their appearance becomes unavoidable and even essential to avert recurrences of pest outbreak particularly in view of unpredictable disruptions due to climate change scenario. Understanding of the shifts in pest profiles over time and space needs accurate and reliable tools of pest surveillance. In India, *e-surveillance*, a novel web-based monitoring system of using tools of information and communication technology has enabled real-time monitoring of the swarming caterpillar populations in rice in 13 districts of Odisha. This has been extended to all the rice pests across the rice ecosystems of the country. Integration of a large amount of data related to changes in pest populations, soil profile, plant phenology, and other phyto factors along with weather parameters over diverse ecosystems for a long period is now possible through the application of Information Technology (IT) tools. Pest-forewarning systems are now in place aided by geographical information system (GIS) and weather data-driven pest distribution maps to provide action advisories to farmers as part of decision support system.[7]

5.4.2 HOST PLANT RESISTANCE (HPR)

Resistant varieties are the most efficient, economical, and practical tools for encountering the pest problems and are ideally compatible with other components of IPM. In India, rice IPM has been strengthened with the development and inclusion of cultivars resistant to gall midge, brown planthopper, whitebacked planthopper, and green leafhopper, along with other components of insect pest management.[4] However, the pest populations are quite capable of evolving adaptive biotypes to overcome the effect of resistant variety as observed in the case of gall midge evidenced by reports on identification of six biotypes across the cropping systems. Hence, concerted efforts are continuously in place to refine and develop newer varieties to not only withstand insect pest attacks but also be resistant to multiple stresses. Also, due to changing pest scenario and situations of altered pest profiles in different cropping systems, there is always the need for multiple pest-resistant varieties to tackle two or three pests occurring simultaneously along with diseases in rice. Multiple pest resistant varieties, such as, Suraksha, Vikramarya, Shaktiman, Rasmi, and Daya, are already being cultivated by farmers in states like Andhra Pradesh, Odisha, West Bengal, Kerala, and Madhya Pradesh (Table 5.3).

TABLE 5.3 Varieties of Rice Resistant to Insect Pests.

Insect pest	Released varieties
1. Gall midge	Sneha, Pothana, Kakatiya Erramallelu, Kavya, Rajendradhan 202, Karna, Ruchi, Samridhi, Usha, Asha, MDU 3, Bhuban, Samalei, Orugallu, Abhaya, Shakti, Suraksha, Daya, Pratap, Udaya, IR 36, Shaktiman, Tara, Kshira, Sarasa, Neela, Lalat, Phalguna, Mahaveer, Vibhava, Divya, Dhanya Lakshmi, Surekha, Vikram, Kunti, Triguna, Sita, Samleswari, Karma Mahsuri, Dhanarasi, Mahamaya, Jyothi
2. Brown planthopper	Chaitanya, Krishnaveni, Vajram, Pratibha, Makom, Pavizham, Manasarovar, Co-42, Chandana, Nagarjuna, Sonasali, Rasmi, Jyothi, Bhadra, Neela Annanga, Daya, Aruna, Kanaka, Remya, Bharatidasan, Karthika, Vijeta, Cotton Dora Sannalu, KRH2, PA6201, Hybrid 6129, ADT37
3. White-backed planthopper	Haryana Basmati, HKR120, Laatha, Narendra 2002, Jitendra, Satyaranjan

5.4.3 CULTURAL MANAGEMENT

Appropriate manipulation of cultural practices offers the viable means for the resource deficient small and marginal farmers to indirectly suppress pest populations through resource use efficient techniques, particularly in rainfed rice. Simple practices like early and synchronous planting can help in escaping damage from major pests like stem borer, planthopper, and gall midge. Simultaneous harvesting over large areas helps in the destruction of diapausing larvae and pupae of stem borers. Leaving alley ways by providing 30 cm width after every 2–3 m during transplanting in the main field make the crop habitat unsuitable by allowing adequate sunlight and aeration not preferred by planthoppers. Similarly, water management in the field by draining or flooding can help in suitably tackling pests such as caseworm and planthoppers. Field sanitation measures can take care of leaf folders, hispa, and earhead bug through the removal of alternate hosts like weeds and preventing off-season survival and spread across the fields.[8]

Integrated soil health and plant nutrient management strengthens plant system through induced resistance to withstand the insect populations preventing them from either reaching to "pest" level status or enabling the plants to yield well despite stress impacts. Vigorous plant health also results in a substantial reduction of pesticide use ensuring cost optimization. Application of nitrogen in splits along with use of slow release fertilizers such as Neem-coated urea helps in checking the pest incidence aiding improvement in plant health toward achieving higher yields. Use of leaf color charts is recommended for optimized use of nitrogenous fertilizers (Figure 5.15). Application of recommended doses of other key primary and secondary nutrients through site-specific nutrient management is another practical means available to rice farmers to get direct benefits of pest suppression at no additional cost. Studies carried out under AICRIP have enabled the development of nutrient management packages for the major rice-based cropping systems of the country.[4] Application of organic manures like FYM or vermicompost facilitates build-up of beneficial natural enemies of pests by enhancing higher populations of detrivorous, plankton feeders, and also higher populations of predators—both below and above the water.

Precision farming based on intensive grid-sampled information obtained by GIS and global positioning system (GPS) is another new dimension initiative for efficient soil and plant health management.

5.4.4 NATURAL ENEMY CONSERVATION

Rice ecosystem is endowed with rich and diverse wealth of natural enemies such as predators, parasitoids, insect pathogens, and their in situ interactions, which can be conserved for exploitation under IPM.[10] Biointensive pest-management strategies (BIPM) can strengthen the natural regulatory mechanisms for in situ management of pest populations with least disturbance to the balance of nature. The BIPM focuses more on measures to restructure the agricultural ecosystem toward conservation of natural enemies and to the disadvantage of a pest.

FIGURE 5.15 Leaf color chart.[9]

Habitat manipulation through naturally innovative strategies such as use of trap crop and ecological engineering for the management of stem borer and planthoppers, respectively, is one of the recent eco-friendly attempts designed to protect rice crop. Growing aromatic rice, preferably Pusa Basmati 1, as a trap crop can effectively manage yellow stem borer as this pest prefers scented rice as host over the non-scented varieties. This practice has lot of potential in rice–rice cropping system of the peninsular region. The recommendation is to plant one row of trap crop for every 2.5–3 m of main crop and preferably in the east–west direction, and for effective results (Figure 5.16).[4,10]

Ecological engineering for natural enemy impact/conservation biological control has also been successful for the management of planthoppers. Growing flowering plants such as marigold and pulses like cowpea soybean etc. on bunds surrounding paddy fields can effectively help in reducing planthopper pests in paddy fields (Figure 5.17). These crops serve as reservoirs of natural enemies by acting as pollen and nectar sources to attract natural enemies of planthoppers.[11]

FIGURE 5.16 Rice field with trap crop.

Inundative releases of natural enemies can also be made to supplement the efforts for their natural conservation. Release of egg parasitoids, *Trichogramma japonicum* adults, against yellow stem borer and *T. chilonis* against leaf folder, 5–6 times @ 1,00,000 ha^{-1} starting from 15 days after transplanting, are effective and useful. These releases help in the supplementation of already existing natural populations of these parasitoids and thus increase the parasitism percentage in rice fields.[10]

FIGURE 5.17a, b Ecological engineering with flowering plants (marigold and pulses).

5.4.5 BEHAVIORISTIC MANIPULATIONS USING SEX PHEROMONES

Pheromones are the chemicals regulating the intra specific communication among members of the same species. Pheromones have no adverse effects on the biota or the environment, are unaffected by abiotic factors, and hence can be harmoniously integrated with other components to manage rice pests.

Sex pheromones have been found promising for the management of yellow stem borer (YSB) in monitoring as well as direct control through male annihilation by mass trapping. The rationale of pheromone-mediated mass trapping technique is to place enough traps to concentrate pest insects into a restricted space (catch enough males) and leave the females of the species without mates. Mass-trapping technique offers great promise against monophagous pests like YSB on rice with crop often cultivated extensively and contiguously. Multilocation studies conducted with 3:1 blend of (Z) 11-hexadecenal, (Z) 9-hexadecenal @ 5 mg/dispenser have led to the recommendation of trap density of 20 traps/ha for optimum management of YSB. Under AICRIP, location-specific IPM modules with pheromone mass trapping as one of the important components have been validated across rice ecosystems.[4]

5.4.6 CHEMICAL MANAGEMENT

Intelligent use of chemicals can be an easy, effective, and sometimes even economical choice for the farmers to manage insect pest populations in rice. Under AICRIP, regular intensive screening and evaluation of newer insecticide molecules for their efficacy against rice pests has helped in the identification of suitable chemical options in different cropping system regimes.[4] Environment friendly chemicals with novel modes of action and effectiveness at very low doses are now ready for use by the farmers. However, use of pesticides has always been recommended only in situations of pest outbreaks or resurgences or only as a last resort when alternate options do not yield results.

Effective chemical use can be ensured by right choice of active ingredient, suitable formulation, precise application time, and technique based on pest biology and crop phenology. Nano-pesticides or Nano-encapsulated pesticides provide opportunities for accurate and targeted delivery of pesticides. Agricultural research on the uses of unmanned aircraft (drones) for various agricultural activities including surveying fields, crop health, watering, and application of pesticides, fertilizers, or other beneficial inputs provides the futuristic possibility of using chemicals in more effective and environment friendly manner.

5.5 CONCLUSION

The potential use of biopesticides including botanical pesticides needs a fresh relook in the light of new developments in advanced chemistry and formulation technology.[12] Similarly, indigenous technical knowhow's (ITKs) can also be exploited in specific farmer-friendly situations. Use of smashed locally available crabs placed on pointed bamboo sticks is one of the most widely used traditional pest management practices in North-Eastern region for effective management of earhead bug.[10] However, ITKs have promise and potential for future application if they can be fully investigated using newer analytical standardization tools to make them consistent in performance as effective, cheap, and widely available alternative products ready for use by farmers.

As a futuristic strategy, the versatile tool of biotechnology has progressively provided novel and more-powerful alternatives to the cumbersome and time-taking conventional resistance breeding. Development of transgenic through the incorporation of novel genes from the soil bacterium, *Bacillus thuringiensis (Bt)*, producing *cry* insecticidal toxins into commercial rice varieties for resistance to yellow stem borer, is ready for extensive field testing in India. Introgression of multiple toxins in transgenic plant varieties through gene stacking and DNA marker technology to tag and map several major resistance genes is now possible in case of hopper pests and gall midge. Gene silencing through RNAi and recently established CRISPR/Cas9-based gene editing system also offers exciting biotechnological means of exploiting genomic intelligence for the development of novel pest-management tactics.

KEYWORDS

- **rice-based cropping systems**
- **insect pest scenario**
- **rice–wheat system**
- **rice–rice system**
- **rice–pulses system**
- **insect pest scenario in rice**
- **stem borers**

REFERENCES

1. Kumar, R. M.; Sreedevi, B.; Tuti, M. D.; Aarthi Singh, Sowmya Saha, Arun, M. N.; Gangaiah, B.; Surekha, K.; Brajendra, P.; Vidhan Singh, T.; Prasad, M. S.; Prakasham, V.; Katti, G.; Padmavathi, Ch.; Somasekhar, N.; Nirmala, B.; Subbarrao, L. V.; Muthuraman, P.; Voleti, S. R. Sustainable Rice, Rice Based Cropping Systems (All India Coordinated Rice Improvement Programme (AICRIP)-Experiences). Technical Bulletin No. 105/2019. ICAR-Indian Institute of Rice Research, Rajendranagar, Hyderabad, 2019, p 198.
2. Panda, B. B.; Satapathy, B. S.; Nayak, A. K.; Tripathi, R.; Md. Shahid; Mohanty, S.; Bhaduri, D.; Khanam, R.; Nayak, P. K. Agro Ecological Intensification of Rice Based Cropping System. In *Rice Research for Enhancing Productivity, Profitability, Climate Resilience*; Pathak, H., Nayak, A. K., Jena, M., Singh, O. N., Samal, P., Sharma, S. G., Eds.; 2018, ISBN: 81-88409-04-09.
3. Dhaliwal, G. S.; Jindal, V.; Mohindru, B. Crop Losses Due to Insect Pests: Global, Indian Scenario. *Indian J. Entomol.* **2015,** *77* (2), 165.
4. Ravindra Babu, V.; Padmavathi, Ch.; Neeraja, C. N.; Krishnaveni, D.; Subba Rao, L. V.; Raghuveer Rao, P.; Prasad, G. S. V.; Katti, G.; Ram, T.; Subrahmanyam, D.; Kumar, R. M.; Hari Prasad, A. S.; Surekha, K.; Prasad, M. S.; Chaitanya, U. 50 Years of AICRIP… Way Forward. Technical Bulletin No. 92/2016. ICAR–Indian Institute of Rice Research, Rajendranagar, Hyderabad, Telangana State, India, 2016, 292pp. ISBN: 978-81-928249-8.
5. Padmakumari, A. P.; Katti, G.; Rajender, K. R. Aromatic Rice as Trap Crop for Yellow Stem Borer Management in Rice. Paper Presented in the National Conference on Technological Challenges in Social, Environmental, Agricultural Reforms (TECH SEAR-2017) on 9th–10th September 2017. By Green Reap Welfare Society, ICAR–IIRR at Sri SVS Sastry Auditorium, IIRR, Hyderabad 30, 2017.
6. Sampathkumar, M.; Shanker, C.; Padmavathi, Ch.; Katti, G. Binary Sex Attractant Blends for Caseworm, *Nymphula depunctalis*. *Indian J Plant Protection.* **2015,** *43* (3), 364–365.
7. Sailaja, B.; Padmavathi, Ch.; Krishnaveni, D.; Katti, G. Subrahmanyam, D.; Prasad, M. S.; Gayatri, S.; Voleti, S. R. Decision Support Systems for Pest Monitoring, Pest Management. In *Improving Data Management, Decision Support Systems in Agriculture*; Armstrong, L., Ed.; BurleighDodds Science Publishing: Cambridge, UK, 2020. http://dx.doi.org/10.19103/AS.2020.0069.18
8. Padmavathi, Ch; Katti, G.; Vallabuni, S.; Yenumala, G. P. Physiological Age Status of Female Adults, Off-Season Survival of Rice Leaffolder *Cnaphalocrocis medinalis* in India. *Rice Sci.* **2015,** *22* (5), 237–244.
9. Padmavathi, Ch.; Sreedevi, B.; Krishnaveni, D.; Surekha, K.; Padmavathi, G.; Prasad, M. S.; Mahender, K. R.; Katti, G.; Ravinder Babu, V. Vari Pocketbook (Telugu). Indian Institute of Rice Research, Rajendranagar, Hyderabad, India. 2016, pp. 64.
10. Padmavathi Ch, Chitra Shanker, Gururaj Katti. Advances in rice IPM -Indian Scenario. *J Res, PJTSAU.* **2017,** *45* (1&2), 1–13.
11. Shanker, C.; Katti, G. Ecological Engineering for Biointensive Management of Rice Pests. *Indian Farming*. **2017,** *67* (5), 31–34.
12. Katti, G. Current Status, Potential of Biopesticides in Pest Management in Rice. In *Biopesticides in Sustainable Agriculture -Progress, Potential*; Koul, O., Dhaliwal, G. S., Khokhar, S., Singh, R., Eds.; Scientific Publishers (India): Jodhpur, India, 2014; p. 474.

CHAPTER 6

IPM in Rice-Wheat Cropping Systems

RAMJI SINGH[1*], H. S. VISWANATH[1], AJAY TOMER[2], DURGA PRASAD[3], and U. P. SHAHI[4]

[1]*Department of Plant Pathology, Sardar Vallabhbhai Patel University of Agriculture and Technology (SVPUAT)-Meerut, India*

[2]*Institute of Plant and Microbial Biology (IPMB), Academia Sinica, Taipei, Taiwan*

[3]*Department of Plant Pathology, Banda University of Agriculture and Technology (BUAT), Banda, Uttar Pradesh, India*

[4]*Department of Soil Science and Agricultural Chemistry, Sardar Vallabhbhai Patel University of Agriculture and Technology (SVPUAT)-Meerut, Uttar Pradesh, India*

[*]*Corresponding author. E-mail: singh.ramji@gmail.com*

ABSTRACT

Rice–wheat cropping system is one of the most important production systems, which contributes a major share in the production of both the crops in Indo-Gangetic plains. This covers the area of 13.5 m ha of cultivated land covering across four countries, namely, Bangladesh, India, Nepal, and Pakistan. Among the several productivity constraints, namely, poor water management, nutrient deficiencies, the occurrence of several pests, diseases, and weeds are of utmost importance. Since R–W system contribute major share in the food bowl of the country hence constrains related to productivity need to be resolved with a minimum adverse effect on the quality of products, food chain, and biodiversity. Rice wheat cultivation in a continuous sequence; year after year reflects a regular increase in the

Integrated Pest Management in Diverse Cropping Systems. Devendra Pal Singh, PhD (Ed.)

population dynamics of several pests, pathogens, and ultimately yields loss of both the crops. In addition, this system also favors a heavy population of selective weeds. Adopting the practices based on principles of integrated pest management (IPM) and INM with a target of minimizing pests, diseases, and weeds thereby the quantity of agrochemicals can be minimized. Several insect pests, diseases, and weeds, which need to be monitored frequently, harm rice and wheat both. Based on monitoring and economic threshold level of specific crop enemy, strategic decision has to be taken for the application of specific management practices. The management practices should be decided in a long-term perspective and also in the context of sustainability of the environment and biodiversity as well. Several cultural practices and antagonists and natural enemies have been developed/identified, tested, and validated for the management of insect pests, diseases, and weeds, which can be effectively used for managing the insect pests, diverse pathogens, and weeds in both the crops. Several adopted cultural practices have been found to be quite effective in minimizing weed population and are commonly being used by farmers. Pest and disease specific resistant cultivars have also been developed, notified, and released in both the crops for different agroclimatic zones against important insect pests and diseases, which should be adopted with an aim to manage these enemies using host resistance. However, need-based chemicals may also be applied if there is no option and crop losses are supposed to be very big. Several IPM modules for rice, wheat individually, also as a system approach have been mentioned and discussed in this chapter.

6.1 INTRODUCTION

Rice and wheat are the most important staple food crops which form an integral part of the human diet for more than 655 million people of Southeast Asia which is about 8.5% of the world's population. Rice–wheat cropping system (RWCS) is one of the most important production systems, spreading across the world to a larger extent of area, thus acting as a food source to the vast population. Indo-Gangetic plains (IGPs) contribute a major share in the rice and wheat production not only in India rather cover an area of 13.5 m ha of cultivated land across four countries, namely, Bangladesh, India, Nepal, and Pakistan. About one-third of the rice and wheat growing areas in South Asia produce staple grains for more than 1 billion people (20% of world's population). This level of yield accounts for 40% of rice and wheat production in India. This cropping system also serves a major source of employment, livelihood, and income for larger chunk of rural and urban people. It fulfills the food and nutritional requirement of 80 and 60%, respectively, of the Indian population. Planting of rice crop during Kharif (wet) season followed by wheat crop in rabi (dry) season is the main characteristic of RWCS. The weather requirements range from warm wet summers with heavy to normal rain and cool dry winters with intermittent light rains. Favorable conditions of RWCS with high productivity prevail in the western parts of IGP due to irrigated areas with higher mechanization and input-intensiveness apart from conjunctive use of ground and surface water. Weather in rain-fed areas is less favorable with low productivity in the eastern zones due to less mechanization and low-input use with more constraints of poor drainage and rainwater conservation. The Green Revolution technologies adopted during the 1960s and 1970s significantly helped to increase the yield of both the crops and system as well. Thereafter, declining growth rates of production had become a serious concern and being noticed that increased production in the future may not keep pace with the human population growth and is expected to continue in the future too. However, many researchers were perplexed about the state of sustainability of RWCS in near future with the gradual reduction in yields and degradation of the natural resources and higher yield losses as this system is more prone to biotic and abiotic stresses like insect pests, diseases, multinutrient deficiencies, weeds, nematodes, drought flood, etc.[1,2] Several biotic and abiotic stresses incur a significant reduction in the rice–wheat system productivity. Among the several stresses and productivity constraints, poor crop stand, poor water management, nutrient deficiencies, occurrence of several pest and diseases are some of the most important ones.

Among the diseases, blast, sheath blight, bacterial leaf blight, whereas among the insect pests, brown plant hoppers and stem borers have a major damaging effect in rice crop. In the wheat, foliar blight, spot blotch, three different types of rusts and Karnal bunt are the most damaging ones.[3] Very recently, a new pathogen causing wheat blast has been reported in neighboring country, that is, Bangladesh which is expected to arrive any time in India also and great losses are anticipated. In RWCS, the biotic factors usually are responsible to bring in a yield loss ranging from 20 to 30% in both the crops. Continued RWCS has led to the development of such a potential pest dynamics that the insect pests that attacked only either of the crops earlier are now attacking both the crops, for example, *Sesamia inferens*, rice pink stem borer, shoot fly (*Atherigona soccata*), etc.[4] In case of the pathogenic organisms, transmission of soil and seed-borne pathogens can occur from crop-to-crop. A repeated rice–wheat system may lead to the build-up of soil-borne pathogens like Genus *Fusarium, Rhizoctonia*, and *Sclerotinia.* Generally, pest infestation is higher in hot and humid climates during rice cultivation as compared to that of in the cold season during wheat cultivation.[5] Apparent concerns that require quick attention in rice–wheat system are the increasing risks of biotic stresses due to increased losses of yields due to various pests, which include insect pest and plant pathogens. This has happened especially due to the emergence of new races and biotypes of insect pests, fungal and bacterial pathogens, and plant-parasitic nematodes. Besides these biotic risks, multinutrient deficiencies, droughts, flooding, and early temperature increase during the winter season are major abiotic risks. Building up of soil pathogens, weed flora, reducing inherent fertility of the soil, and depletion of the ground water table are also some of the production constraints of the cropping system. The dependence on chemicals for the management of these major biotic stresses (diseases and insect-pests) has become imperative in most of the crops. Reduction in the market value of produced grains in the global market due to the presence of pesticide residues, resistance development among pests against multiple pesticides, serious health implications, and negative deteriorating effects on the ecosystem due to accumulation of toxic residues of pesticides in the environment, contamination of entire food chain, groundwater toxicity have led a shifting from chemical pesticides, and inorganic fertilizer applications toward integrated pest management (IPM) and INM practices for sustainable RWCS. Therefore, applying the practices based on IPM strategies to minimize the dependence on these chemicals to manage these pests and plant pathogens has become mandatory. IPM is defined as a management strategy that utilizes and integrates all available appropriate techniques/practices optimally in order to maintain the pest population at levels below those causing economically unacceptable damage or from reaching the economic injury level by maintaining the balance between the environment and population dynamics of the pest species by keeping the damage caused by chemical pesticides to the environment as low as possible.[6]

6.2 IPM IN RICE

India is one of the leading countries in rice production and ranks second (116.42 million MT) next only to China with the largest area under cultivation of rice crop in the world (44.6 m ha). Rice crop is grown under wide range of agroecological habitats in diverse weather and soil conditions in India. In RWCS, rice is grown during the khaki season which is constantly prone to stress throughout the crop season due to the attack of various harmful biotic agents like disease-causing pathogens, insect pests, nematodes, weeds, and rodents. Implementation of IPM strategies helps in checking the increase of pest population, suppression of dormant pathogens, weeds, and soil insects along with maintaining the fertility of soil, thereby decreasing nutrient deficiencies and disorders. IPM strategies in rice production primarily focus upon the optimal use of pesticides, complementary weed control methods, alternative practices for cultural and biological control measures in order to keep the pest population to the minimum and thereby reducing the yield losses. Since IPM is a dynamic, knowledge-intensive practice and therefore, it needs continuous up-gradation of the technology from time to time with the changing pest dynamics. In order to reach the goal of increasing productivity in the near future, it requires the implementation of modern innovative and advanced agricultural strategies by the farmers. However, apart from the practice of innovative/intensive agriculture, there is the aggravation of biotic constraints like diseases, insect pests, and weeds. Reports of more than 100 species of insect pests have been recorded as rice pests and of which about a dozen are found to be of economic significance in India. It is estimated that 10% yield losses occurring annually in rice production are mainly due to infection with various diseases. Even if the loss caused by diseases of rice in a year is reduced to 5% on an average to the total rice production in the country, it accounts a lot in the production increase and food security, which can be achieved by proper implementation of IPM practices in rice.

Major insect pests, diseases, nematodes, weeds, rodents, etc., affecting rice cultivation in the IGPs are given in Tables 6.1–6.5

TABLE 6.1 Major Insect Pests in Rice.

Name of the pest	Scientific name
Brown plant hopper	*Nilaparvata lugens*
Yellow stem borer	*Scirpophaga incertulas*
White-backed plant hopper	*Sogatella furcifera*
Gall-midge	*Orseolia oryzae*
Rice Leaf folder	*Cnaphalocrocis medinalis*
Rice Gundhi bug	*Leptocorisa acuta*

TABLE 6.2 Major Diseases, Nematodes, and Weeds of Rice.

Disease	Scientific name
Sheath blight	*Rhizoctonia solani*
Brown spot	*Helminthosporium oryzae (syn: Bipolaris oryzae)*
Rice blast	*Pyricularia oryzae*
False smut	*Ustilaginoidea virens*
Bacterial leaf blight	*Xanthomonas campestris* pv. *oryzae*
Nematode	
Root-knot nematode	*Meloidogyne graminicola*
White tip nematode	*Aphelehchoides besseyi*
Weed	
Barnyard grass	*Echinochloa crusgalli*
Nut grass	*Cyperus rotundus*
Weedy rice	*Oryzae* spp. (wild rice)

TABLE 6.3 Major Diseases and Nematodes of Regional Significance.

Disease	Scientific name	Areas in which pests are prominent
Sheath rot	*Sarocladium oryzae*	Widely distributed in many regions
Bakanae	*Gibberella fujikuroi*	Widely distributed in many regions
Stem rot	*Sclerotium oryzae*	Widely distributed in many regions
Rice Tungro Virus		Confined to few regions
Nematode		
Root lesion nematode	*Pratylenchus indicus*	Upland rice cultivated areas in SRI system of Jharkhand and Orissa
Cyst nematode	*Heterodera oryzae*	Kerala only
Ufra	*Dilylenchus angustus*	Prominent in Tripura, West Bengal, and Assam,
Rice root nematode	*Hirschmanniella oryzae*	Mostly found in Orissa, West Bengal, Bihar, Andhra Pradesh

Major practices which constitute the practices of IPM for rice in RWCS involve the following.

6.2.1 PEST MONITORING AND SURVEY

The field survey or field/pest scouting is aimed at determining the economic threshold level (ETL) of different pests and pathogens in the field. The management practices of any disease or insect-pests using different applications of bioagent or chemical should be considered only when the population dynamics of insect pest and incidence/severity of disease reach to its ETL.

TABLE 6.4 Major Weeds/Rodents of Regional Significance.

Weed	Scientific name
Benghal day flower	*Commelina bengalensis*
Spiny amaranth	*Amaranthus spinosus*
False daisy	*Eclipta alba*
Saramolla grass	*Ischaemum rugosum*
Indian goose grass	*Eleusine indica*
Hairy crabgrass	*Digitarias anguinalis*
Wood sorrel	*Oxalis latifolia*
Lesser fimbry	*Fimbristylis littoralis*
Swamp rice grass	*Leersia hexandra*
Water hyacinth	*Monochria vaginalis*
Crown grass	*Paspalum* spp.
Para grass	*Brachiaria* spp.
Asian sprangle top	*Leptochloa chinenesis*
Panic grass	*Panicum* spp.
Four leaf clover	*Marsilea quadrifoliata*
Rodent	
Smaller bandicoot	*Bandicota bengalensis*
Indian gerbil	*Tatera indica*
Field mice	*Mus* spp.
Soft furred field rat	*Millardia meltada*

6.2.2 MONITORING PEST POPULATIONS USING PHEROMONE TRAPS/LIGHT TRAPS, ETC.

Populations of many insect pests can be monitored through such traps, which can give an idea about the extent of damage, these insect species can cause based on the assumptions of their population dynamics being prevailed in the crop field.

6.2.3 AGROECOSYSTEM ANALYSIS (AESA)

This is an approach that aims at understanding the interactions between different components in an agroecosystem, thereby analyzing the different factors that predispose the crop to be affected by various pests and pathogens.

6.2.4 DETERMINING THE ETL FOR DIFFERENT PEST AT DIFFERENT GROWTH STAGES OF THE CROP PLANT

The understanding of the ETLs of the pest population becomes essential in order to take protective measures so that cost of cultivation does not exceed the profit incurred by the farmer. ETLs of different pests of paddy crop have been presented in Table 6.5.

For the success of an IPM program, pest monitoring and prevention (Cultural, genetic), timely interventions at appropriate time are the essential components.

6.2.5 CULTURAL PRACTICES

6.2.5.1 COMMON CULTURAL PRACTICES

1. During land preparation prior to transplanting in Kharif, 45 days old crop of Sesbania or sun-hemp should be incorporated in the soil wherever possible.
2. Selection of suitable resistant or moderately resistant variety against important diseases for cultivation.
3. Use of disease, insect-free certified seed.
4. Seed treatment with *Trichoderma* spp. or *Pseudomonas* spp. @ 5–10 g/kg seed for management of soil and seed-borne diseases or seed dressing with carbendazim 50% WP @ 2 g/kg seed for seed- or soil-borne diseases.
5. Seed treatment with imidacloprid 200 SL (20%) @ 0.25 L/100 kg or Chlorpyriphos 20% EC @ 10000 mL/ha (seed mixed with 10% solution of gum Arabica prior to sowing).
6. Seed treatment with Carbosulfan 2 g/kg seed for root nematodes.
7. Timely planting/sowing should be ensured.

TABLE 6.5 ETLS for Different Pests of Paddy Crop in Its Different Crop Growth Stages.

Stage of crop	Pest/disease	ETL
Nursery	BLB, Kresek phase	Two to three young plants/m^2
	Yellow stem borer	One egg mass/m^2
	Root-knot nematode	One nematode/gall
Early vegetative to late tillering phase	Rice Hispa	Two adult insects or dead leaves/hill
	BPH/WBPH	10–15 hoppers/hill
	Rice leaf folder	Two infested (folded) leaves with larva/hill
	Stem borer	10% dead hearts observed or two egg masses/m^2 or one moth/m^2
	Sheath blight	Wavy irregular Lesions of 5–6 mm in length, two to three infected plants/m^2
	Sheath rot	Lesion of 2–3 mm length on sheath, three to five infected leaves/m^2
	Bacterial leaf blight	Two to three infected leaves/m^2
	Leaf blast of rice	Three to five spindle-shaped lesions/leaf
	Tungro virus disease	One infected/diseased plant/m^2
	Brown leaf spot of rice	Two to three infected plants/m^2 or two to three spots/leaf
	BPH/WBPH	15–20 hoppers/hill
Panicle emergence to booting phase	Leaf folders	Two leaf folders or folded leaves/hill
	Stem borer	25 moths/trap/week or two egg masses/m^2 or one adult moth/m^2
	Neck blast	Two to five neck infected plants/m^2
	Sheath rot	Five infected plants/m^2

6.2.5.2 PRESOWING IRRIGATION

1. Many weeds will be induced to germinate by applying presowing irrigation and thereby the destruction of these weeds, through ploughing in the area where nursery or seedlings are to be transplanted.
2. Raising healthy nursery: nursery bed should be exposed for soil solarization using thin and transparent polythene sheets for at least 45 days prior to nursery seeding. Solarization is highly effective in the destruction of several soils borne pathogen, nematodes, pest's

eggs/pupae, and weed seed. Solarization will benefit the crop vigor by improving the level of soil fertility.

3. Ensuring the absence of weed plant from rice seedlings by manually removing them at the time of transplanting.
4. Removal of weeds from field, destruction of leftover nursery, cleaning of bunds to ensure that they should not be a shelter for survival of pathogen, pests to check their subsequent transfer to main field/crop.
5. Depending on the duration of the variety normal spacing is maintained having 30–36 hills/m^2.
6. Judicial use of fertilizers, micronutrients by soil application or through foliar sprays.
7. In endemic areas of plant hopper, bacterial leaf blight and stem rot affected fields, alternate wetting, and drying of field are done to avoid water logging conditions. Weed growth can be kept under check by maintaining a thin film of water on the surface of the soil.
8. Avoid application of nitrogenous fertilizers in the BLB endemic areas or when symptoms of BLB are noticed to be started.
9. In plant hopper and sheath blight endemic areas, formation of alley spacing of 30 cm at every 2.5–3 m distance should be made.
10. Seed is sowed in lines by maintaining proper spacing in direct-sown rice which makes intercultural operations easy for the weed management. Mechanical methods of weeding should be practiced after 2–3 weeks of transplanting and one more weeding, if necessary, after 4–6 weeks of sowing.
11. Harvesting of crop is done close to the ground level in order to destroy insect pests hidden in the stubbles and thereby exposing the insects to birds. In the standing crop, bird perches should be erected so that the predatory birds may get sheltered over these perches and can catch their prey insects from the standing crop field.
12. After harvesting, the fields should be flooded with water, and with the help of rotavators the field should be ploughed to kill dormant hibernating larvae of insect pests, especially of yellow stem borer present in the crop stubbles. Summer ploughing of fields should be practiced to expose and kill the hidden larvae and pupae of rice swarming caterpillar present in the soil. This will facilitate the predatory birds to catch the exposed eggs and pupae or adult.
13. Summer hot weather or sunny heat may also adversely affect the hidden eggs and larvae and dormant structures of pathogens and nematodes.

6.3 MANAGEMENT OF RICE INSECTS THROUGH CULTURAL PRACTICES

Important cultural practices to minimize the infestation of major insect pests of rice are given below.

6.3.1 RICE HISPA (DICLADISPA ARMIGERA)

1. Detopping/clipping and destruction of the top one-third portion of the leaves during the early vegetative stage of the crop having a high infestation of eggs and grubs.
2. Collection of adults by sweep nets and their destruction.
3. In case of heavy infestation of the pest, top dressing with nitrogenous fertilizer should be skipped.
4. Removal of grassy weeds to destroy the alternate and collateral hosts of the pests.
5. Early sowing/planting of rice at the beginning of the rainy season.
6. Hand-picking of damaged leaves to remove adults from the field.

6.3.2 RICE LEAF FOLDER (CNAPHALOCROSIS MEDINALIS)

1. Early planting of crops to avoid greater damage from the pest.
2. Wider spacing and low use of nitrogenous fertilizers.
3. Removing the cause of shade within the field.
4. Proper sanitation of the field before transplanting
5. Use of insect tolerant varieties.

6.3.3 YELLOW STEM BORER (SCIRPOPHAGA INCERTULAS)

1. Clipping off/detopping the leaf tips before transplanting. Use of resistant varieties.
2. Use of short duration and early maturing varieties.
3. Destruction of larvae residing in stubbles through tillage operation after harvest followed by flooding.
4. Delay in planting and seeding time to avoid peak emergence period of the moth.

6.3.4 GREEN LEAF HOPPERS (NEPHOTETTIX VIRESCENS)

1. *N. virescens* serves as the vector for rice tungro virus disease
2. Transplanting older seedlings (>3 weeks old) reduces the susceptibility to the rice tungro disease.
3. Early planting especially in the dry period reduces the risk of insect-vector borne disease.
4. Application of nitrogenous fertilizers should be done in optimum levels.
5. Proper weed management in the field and removal of bunds that may be preferred by hoppers for grassy hosts.

6.3.5 BROWN PLANT HOPPERS (NILAPARVATA LUGENS)

1. Avoid the use of high doses of nitrogenous fertilizers and dense planting to check the closer spacing of the plants. Use of fertilizer in split doses.
2. Drain out the water from the field for 3–4 days in times of heavy infestation.
3. Planting early maturing varieties and maintaining synchronous planting is effective for the pest management.

6.3.6 WHITE BACKED PLANT HOPPER (WBPH) (SOGATELLA FURCIFERA)

1. Drain out the water from the rice fields. Intercropping of dry land rice with cotton or pigeonpea.
2. Use of moderately resistant varieties like IR48, IR52, IR60, IR62, etc.

6.3.7 RICE CASE WORM (NYMPHULA DEPUNCTALIS)

1. Adoption of wider hill spacing (30 × 20 cm) and early planting during the cropping season usually reduces the damage of case worm by escaping the peak insect pest activity.
2. Selection of older and hardened seedlings for transplanting by reducing the duration of susceptible crop growth stage.

3. Judicial use of nitrogenous fertilizers in split applications reduces the rice case worm's abundance.
4. Field draining should be maintained to destroy the larvae.

6.3.8 WHORL MAGGOT (HYDRELLIA PHILIPPINA)

1. Field draining should be maintained at every 3–4 days interval during the first month of transplantation. This will prevent the attraction of adult flies to the standing water and thereby reducing egg laying by adults.
2. Maintaining higher plant density may reduce the damage.
3. Growing water ferns like Azolla and *Salvinia molesta* by forming a thin film on the top surface prevent rice whorl maggot infestation.
4. Direct-sown rice is not congenial for pest development as adults are not attracted to direct-seeded rice.
5. Dense planting with less spacing helps to prevent oviposition by the adult and further damage.

6.3.9 GALL MIDGE (ORSEOLIA ORYZAE)

1. Ploughing under the ratoon of previous crops can reduce infestation.
2. Removal of alternate and collateral hosts around the field can decrease gall midge infestation.
3. Field water draining for 5–7 days greatly affects gall midge populations.
4. Use of early maturing varieties may escape high infestations.
5. Optimum application of nitrogen and potassium fertilizers in split dosages reduces the insect populations.
6. Avoiding staggered planting (complete planting without delay in an area within 3 weeks) reduces gall midge infestation.

6.3.10 GREEN STINK BUG (NEZARA VIRIDULA)

1. In order to protect late-maturing varieties (main crop) which give yield, early maturing varieties can be selected as trap crops, and finally trap crops are sprayed by insecticides.
2. Changes in the planting date can reduce the *N. viridula* infestation.

3. Use of trap crop which can serve the purpose of green-manuring like *Sesbania rostrata*, intercropping with soybean is also effective in reducing stink bug population

6.4 MANAGEMENT OF RICE DISEASES

6.4.1 BACTERIAL LEAF BLIGHT (XANTHOMONAS ORYZAE PV. ORYZAE)

1. Practice clean cultivation and destroy the weeds inside and around the rice field.
2. Avoid water inundation or flooding with water to a depth in the nursery.
3. Large quantities of nitrogenous fertilizers should be avoided and balanced fertilization should be practiced.
4. Use resistant varieties
5. Disease-free seed should be used for planting.

6.4.2 RICE BLAST (PYRICULARIA ORYZAE) (SEHGALETAL.2001).

1. Disease-free seed should be used for planting
2. Use of resistant or tolerant cultivars.
3. Sowing should be done at the proper and appropriate time.
4. Nitrogenous and phosphate fertilizers should be used judiciously.
5. Once initial symptoms are noticed, affected, and surrounding plants should be sprayed with Carbendazim @ 0.1% to check the further spread and to confine and destroy the plant pathogenic organisms at the initiation point only.

6.4.3 SHEATH BLIGHT OF RICE (RHIZOCTONIA SOLANI)

1. Draining of the field helps in controlling the disease. Destruction of collateral host.
2. Burning of crop residues to destroy the sclerotia of the pathogen.
3. Balanced application of NPK fertilizers.
4. Green manuring with *Sesbania aculeate* to check the survival of sclerotia.

5. Use of soil amendments like oil cake and saw dust and rice husks to ensure high antagonist population and activity.
6. Grow resistant varieties.
7. Once initial symptoms are noticed, the affected and surrounding plants should be sprayed with Carbendazim @ 0.1% to check the further spread, to confine and destroy the plant pathogenic organisms at the initiation point only.

6.4.4 *BROWN SPOT OF RICE (BIPOLARIS ORYZAE)*

1. Sowing should be done with healthy disease-free seeds.
2. Destruction of collateral hosts of the crop.
3. Removal of stubbles along with proper fertilization.
4. Use of resistant varieties for planting.

6.5 MANAGEMENT OF RICE NEMATODES

1. Use nematode free seed for sowing
2. Deep summer ploughing helps in reducing the nematode incidence.
3. Burning of stubbles and weeds after harvesting the crop.
4. Soil solarization of the nursery field.
5. Flooding of the fields.

IPM in any crop consists of using predators, parasitoids, and parasites of different insect pests, resistant varieties, and spray schedules of chemical pesticides synergistically in such a way that all these practices should be compatible with each other.

A detailed account of major parasitoids and predators, resistant varieties of rice against different insect pests, and schedules of chemical application as a component of IPM modules for different crop stages have been given in Tables 6.6–6.8.

6.6 IPM IN WHEAT

Wheat is the world's most important grain crop. Wheat ranks first position in the world among the cereals with respect to both area and production. It is a crop that primarily grown in temperate regions. It constitutes the staple food in at least 43 countries. The Levant region of the near east is

TABLE 6.6 Major Parasitoids and Predators of Insect Pests in Rice.

Egg parasitoids		
Type of insect egg parasitized	**Parasitoid**	**Number of eggs parasitized**
Leaf and plant hoppers	*Anagrus* spp.	15–30 eggs/day
Leaf and plant hoppers	*Gonatocerus* spp.	Eight eggs/day
Leaf folder and case worm and YSB	*Trichogramma chilonis*	–
Leaf folder and case worm and YSB	*Trichogramma chilonis*	–
Yellow and white stem borer	*Tetrastichus schoenobii*	Three borer eggs/wasp for adult development
Yellow stem borer (YSB)	*Trichogramma japonicum*	–
Yellow and white stem borer	*Telenomus rowani*	A female parasitizes 20–40 eggs
	Tetrastichus and *Telenomus*	Both parasitize the same eggs of the pest species but not the same egg
Leaf folder	*Copidosomopsis nacoleiae*	A few eggs can yield about 200–300 wasps
Rice leaf and plant hoppers	*Oligosita* spp.	Two to eight eggs/day
Larval parasitoids		
Type of insect larva parasitized	**Parasitoid**	**Adult features**
Leaf folder	*Elasmus* spp.	small elongated adults with pointed abdomen
Yellow stem borer and rice hispa	*Bracon* spp.	Females-wingless with pincher like front claws
Yellow and white stem borer	*Amauromorpha accepta*	Adult is of red or black in color having a white band at the tip of abdomen
Leaf hoppers, plant hoppers	*Haplogonatopus* spp., *Pseudogonatopus* spp.	Parasites and predators, brown or black in color
Stem borer and semilooper	*Cotesia flavipes*	Base of hind legs has yellow-brown to red markings
Leaf folder	*Stenobracon nicevillei*	Black head having orange-brown body, Forewings have three pairs of black spots, ovipositor with twice the length of its body is black in color

TABLE 6.6 *(Continued)*

	Larval and pupal parasitoids	
Type of insect larva parasitized	**Parasitoid**	**Adult features**
Larvae of whorl maggot	*Opius* spp.	Adult is small brownish orange in color with black ovipositor, long antennae and yellow legs
Stem borer	*Xanthopimpla flavolineata*	Medium-sized yellow-orange color adult with black ovipositor and transparent wings
Leaf folder, rice skipper, green horned caterpillar	*Brachymeria excarinata*, *Brachymeria iasus*	Adults of *Brachymeria excarinata* are black with yellow markings on both ends of hind tibia having no cheek whereas *Brachymeria iasus* have, a triangular cheek, yellow markings on tip of femur and on ventral half of tibia
	Predators	
Type of prey	**Predator common name**	**Scientific name**
Leaf folder larvae	Carabic beetle	*Ophionea nigrofasciata*
Leaf-and plant hoppers, stem borer, whorl maggot flies and leaf folder moths	Wolf spider	*Pardosa psuedoannulata,*
Moths/adults of rice pests, namely, grasshoppers	Orb spider	*Argiope catenulata*
Small hoppers, small larvae, exposed eggs	Coccinellid beetles	*Micraspis hirashimai, Harmonia octamaculata*
Leaf, plant hoppers, eggs, and small moths	Rove beetle	*Paederus fuscipes*
Stem borer, leaf-and plant hoppers, leaf folder moths	Long-jawed spider	*Tetragnatha maxillosa*
Adults of whorl maggots, plant hoppers, leaf hoppers, and moths of rice pests	Lynx spider	*Oxyopes javanus*
Plant hoppers	Ground beetle	

TABLE 6.7 Resistant Varieties of Rice Against Different Insect-Pests and Diseases.

Pest	Resistant/tolerant varieties
Stem borer	CR Dhan 802, CR Dhan 309, Vikas, Satyabhama (CR 2340-11), Abhishek
Brown plant hopper	CR Dhan 802, VL Dhan 86, Virender, IR69726-29-1-2-2-2, IR70454- 144-1-1-3-2, CR 2711-114, CR 2711-76, CR 2711-139, CR 2711-149 and CR2712-2
Leaf folder	CR Dhan 802, CR Dhan 309, Satyabhama (CR 2340-11), CR Dhan 500 (CR 2285-6-6-31), CR Dhan 300, Abhishek
White Backed plant hopper	Virender, JKRH 401(hybrid), IC568061, AC 111, AC 1066, AC 1073, AC 124, AC 1418
Green leaf hopper	Vikramarya, Nidhi, IR 64, CR 1009, PY 3, Co 46
Gall midge	DRR Dhan 54, DRR Dhan 55, Gangavati Ageti, CR Sugandh Dhan 907 (CR 2616-3-3-31), Purnabhog (CRM 2203-4), Shusk Samrat
Rice blast	DRR Dhan 54, DRR Dhan 55, Sahyadri megha, Sasyasree, Vani, CR Sugandh Dhan 907 (CR 2616-3-3-31), Purnabhog (CRM 2203-4)
Sheath blight	CRDhan-601, Manasarovar, Geetanjali, CR 1014, IET 17886, IET 20443, Gangavati Ageti, CR Dhan 300
Bacterial blight	DRR Dhan 53, BR-4-39-51-2, BR-51-49-6, IR3796-14-2, ARC-5925, ARC-5943
Brown spot	Gangavati Ageti, VL Dhan 65, CNM529, CNM540, Shusk Samrat
Bakanae	CR Dhan 311 (Mukul), Khira, Sadabahar
Sheath rot	MDR40049, CN1035-61, OR-090-3-158, CR Dhan 300, Shusk Samrat
False smut	Harsha, Vaishak, Ranjit, Luna Suvarna
Rice-root nematode	RP1155-128-1, CR52, N136, CR44-32, Mtu.28, CR320, Tin Pakia
Root-knot nematode	Swarnadhan and Dhanrasi, IRIOF 571, IRIOF 602, CR3047 and CR-Dhan 303
White tip nematode	Gurmatia, PNB, Indira, Basumati, IGPI-37, RTN68, Yamuna, SSM591
Stem nematode	Rayada16-06, Rayada16-08

TABLE 6.8 Scheduled Chemical Sprays as a Component of IPM Module for Different Crop Growth Stages.

Insect pest	Scheduled chemical sprays (nursery stage)
Gall midge	Carbosulfan 25% EC @ 0.8–1 L/ha/Lambda cyhalothrin 5% EC 12.5 L a.i./ha or Carbosulfan 6% G @ 16.7 kg/ha or Ethofenoprox 10% EC 50–75 mL a.i./ha/ or Fipronil 0.3% GR 50–75 g a.i./ha
Stemborer	Carbosulfan 6% G 1 kg a.i/ha or, Cartap hydrochloride 50% SP @ 1 kg/ha, or Cartap hydrochloride 4% granules @ 18.7 kg/ha
Thrips	Fipronil 80% WG 40–50 g/ha/ Lambda cyhalothrin 5% EC 12.5 L ai./ha
Green leaf hopper	Fipronil 5% SC 50–75 g ai./ha
	Scheduled chemical sprays (vegetative stage)
Leaf folder	Spraying Cartap hydrochloride 4% granules @ 18.75–25 kg/ha/Bifenthrin 10% EC 50 g ai./ha/Cartap hydrochloride 50% SP @ 1 kg/ha/ Chlorantraniliprole 18.5%SC 30 g ai./ha/ Carbosulfan 6% G 1 kg ai./ha
Stemborer	Cartaphydrochloride50% SP @1 kg/ha/Bifenthrin 10% EC 50 g ai./ha/ Cartap hydrochloride 4% granules @18.7 kg/ha/Chlorantraniliprole 18.5% SC 30 g ai./ha
Gall midge	Foliar spray of Lambda cyhalothrin 5%EC 12.5 L ai./ha/Fipronil 0.3%GR @16.6–25 kg/ha at 20 days after transplanting
Brown plant hopper, white-backed plant hopper	Spraying of Imidacloprid 70% WG @ 30–35 g/ha or Ethofenoprox 10% EC@ 0.5–0.75 L/ha or Buprofezin 25%SC @ 0.8 L/ha or Imidacloprid 30.5% SC@ 60–75 mL/ha
Hispa	Foliar Spray with Triazophos 40% EC 2.5–5 kg/ha, Lambda cyhalothrin 5% EC 12.5 L ai./ha
Caseworm	Foliar spray with phenthoate 50% EC 500 g ai./ha.
	Scheduled chemical sprays (panicle initiation to booting stage)
Leaf folder	Foliar spray of Cartap hydrochloride 50% SP @ 1 kg/ha or Bifenthrin 10% EC 50 g ai./ha/ Chlorantraniliprole 18.5% SC 30 g ai./ha. or Cartap hydrochloride 4% granules @ 18.7–25 kg/ha can be done
Stem borer	Cartap hydrochloride 50% SP @ 1 kg/ha/ Chlorantraniliprole 18.5% SC 30 g ai./ha or Bifenthrin 10% EC 50 g ai./ha or Cartap hydrochloride 4% granules @ 18.7–25 kg/ha/
Brown plant hopper/white backed plant hopper	Spraying of Ethofenoprox10% EC @ 0.5–0.75 L/ha or Imidacloprid 70% WG@ 30–35 mL/ha or Buprofezin 25% SC @ 0.8 L/ha or Imidacloprid 30.5% m/m SC @ 60–75 mL/ha

TABLE 6.8 *(Continued)*

Insect pest	**Scheduled chemical sprays (nursery stage)**
	Flowering
BPH/WBPH	Spraying of Imidacloprid 70% WG @30–35 mL/ha or Fipronil 0.3% GR @ 16.6–25 kg or Ethofenoprox 10% EC @ 500–750 mL/ha or Triazophos 40% EC 2.5–5 kg/ha or Buprofezin 25% SC @ 800 mL/ha
Diseases	**Scheduled chemical sprays (nursery stage)**
Bacterial leaf blight	Spraying of 100–150 ppm of Streptomycin sulfate 9% + Tetracycline hydrochloride1% SP
Rice blast	Tricyclazole 70% WG at the rate of 0.3 kg/ha or 75% WP at the rate of 0.3–0.4 kg/ha or isoprothiolane 40% EC @ 0.75 L/ha
	Scheduled chemical sprays (vegetative stage)
Sheath blight	Apply Hexaconazole 5% EC@ 1 L/ha or Propiconazole 25% EC @ 0.75 L/ha or Propiconazole10.7%+Tricyclazole 34.2% SE @ 0.5 L/ha or Validamycin 3% @ 2 kg/ha
Bacterial leaf blight	Spraying of Streptocycline at the rate of 100–150 ppm solution at the early vegetative stage, if necessary Second spraying should be done before grain set. Nitrogen application must be avoided to the maximum extent, incase if it is required minor split doses should be given. Other chemicals as mentioned earlier
Rice blast	Isoprothiolane 40% EC @ 0.75 mL/ha or Tricyclazole 75% WP @ 0.3–0.4 kg/ha or Tricyclazole 70% WG @ 0.3 kg/ha
	Scheduled chemical sprays (panicle initiation to booting stage)
Sheath blight	Spraying of chemical fungicides like Propiconazole 10.7% + Tricyclazole 34.2% SE @ 0.5 L/ha or Hexaconazole 5% EC @ 1 L/ha or Propiconazole 25% EC @ 0.75 L/ha or Validamycin 3% L @ 2 L/ha which are highly efficient against sheath blight
Bacterial leaf blight	Reduce nitrogen application and apply if needed only small doses of N in more split doses and chemicals are sprayed as mentioned earlier
	Scheduled chemical sprays (flowering stage)
Sheath blight	Foliar spray of fungicides like Hexaconazole 5%EC @ 1 lit/ha/Propiconazole 10.7%+Tricyclazole 34.2% SE @0.5 L/ha or Validamycin 3%L @ 0.2 kg/ha or Propiconazole 25% EC @ 0.75 L/ha can be done

TABLE 6.8 *(Continued)*

Insect pest	Scheduled chemical sprays (nursery stage)
Rice Blast	Spray Tricyclazole 75% WP @ 0.3–0.4 kg/ha/Isoprothiolane 40% EC@ 750 mL/ha or Ediphenphos 50% EC @0.5–0.6 mL/ha. Apply nitrogen only if needed or in small doses
Weed management	➢ In direct-seeded grassy weeds rice after 18–20 days sowing, application of Cyhalofop-butyl 10% EC @ 0.75–0.80 L/ha. If needed one more hand weeding can be given after 4–6 weeks
	➢ At 3–10 days after transplanting, application of Anilophos 24%+ 2,4-D-ethylester32% EC @ 1–1.5 L/ha or Metsu furonmethyl 10% + Chlorimuronmethyl 10% WP @ 20 g/ha can be done
	➢ Within 4–6 days after transplanting, Pretilachlor 50% EC @ 1.0–1.5 L/ha, Oxadiargyl 80% WP @ 0.125 kg/ha, Ethoxysulfuron 15% WDG @ 83.3–100 g/ha, Chlorimuronethyl 25% WP @ 24 g a.i./ha, Anilophos 2% G @ 20–25 kg/ha, Cinmethalin 10% EC @ 0.75–1.0 L/ha, Metsufuran-methyl 20% WG @ 20 g/L can be applied as Pre-emergence application
Rodent control	➢ Community-based management approach is required
	➢ After one and half month of transplantation, Bromodiolone (0.005% ai.) is applied in baits (ETL): 15 burrows/ha
	➢ Setting up of indigenous traps one month after transplantation
	➢ The second application may be given to residual live burrows with Bromodiolone (0.005%)
	➢ The above control practices may be repeated except zinc phosphide (as bait shyness is developed in rodents). If the rodent population exceeds ETL (Economic threshold level) Zinc phosphide (2.5%) in baits can be applied to check them efficiently

the source of origin for wheat but now being cultivated across the world. The most important wheat-growing countries are the USSR, USA, China, India, Canada, Argentina, Australia, and a number of European countries. The total area of cultivation under wheat crop throughout the world is about 225.62 m ha with the production of 685.6 million tonnes. It is the second most important food crop being next to rice and contributes to the total food grain production of the country to the extent of about 25%. In the total cultivated area of wheat in the world, 12.40% of world's wheat area lies in India which accounts to 11.77% of total world wheat production. If we talk of India in terms of production, the state of UP again occupies first place followed by other states like Punjab, Haryana, and Madhya Pradesh. The subsequent states followed next to them are Rajasthan, Bihar, Maharashtra, Gujarat, West Bengal, Uttarakhand, Himachal Pradesh, Jammu & Kashmir, and Karnataka. The contribution of these states in the production is about 99.5% at the national level. Wheat straw is used for feeding the cattle. The introduction of Mexican dwarf wheat varieties in the mid-1960s revolutionized the wheat productivity as well as its total production in India. Wheat contributed to the lion's share in the Green Revolution.

In the case of wheat, area of cultivation is greater than any other commercial crop. Global trade in wheat crop is greater than rest of the crops. Due to its high protein content among the major cereal crops, it is considered as leading protein source for human diet globally than maize (corn) or rice. It is next to rice as main human food crop in terms of total production ahead of maize remaining in the produce after its extensive use in animal feeds.

Wheat is a winter season rabi crop that needs cool climatic conditions, hot summers where the plains of the Indo-Gangetic region which are highly fertile are the most suitable areas for its cultivation. Although well-drained clay loams, loamy soils are considered best for wheat cultivation, it is also grown well on sandy loam soils. Wheat sowing takes place in the months of October to December and harvesting is done during the months of February to May. Based on agroclimatic conditions, India is broadly divided into five agroclimatic zones for wheat cultivation.

1. The north-western plains zone: it is a highly important zone among all others, which comprises of plains of the states of western Uttar Pradesh, Haryana, Punjab, Rajasthan, and Jammu. The sowing of wheat s takes place during late October–November and harvesting usually start by mid-April.
2. The central zone: it comprises of Bundelkhand area of Uttar Pradesh, Madhya Pradesh, Gujarat, and southeastern part of Rajasthan. This

zone of wheat cultivation largely depends (nearly 75%) on rainfed irrigation. Durum wheat of the best quality is produced in this zone.

3. The north eastern plains zone: it comprises of the eastern part of Uttar Pradesh, and states of Bihar, Sikkim, West Bengal, Assam, Manipur, Tripura, Mizoram, Arunachal Pradesh, Meghalaya, Nagaland, and Orissa. Sowing of wheat takes place in late November, early December, or even from late December to early January also after sugar cane ratoon is harvested. Since rice crop is usually harvested a little late in these areas, harvesting takes place to the end of March–April, sometimes goes even up to the first fortnight of May also.
4. The northern hill zone: it comprises the hilly areas of states of Uttarakhand, Jammu &Kashmir, Sikkim, Himachal Pradesh, West Bengal, and Assam. Sowing takes place in October, remains dormant in the cold months of November to March, and starts growing as the temperature rises in April and gets harvested in May–June.
5. The peninsular zone: southern states of India, that is, Maharashtra, Andhra Pradesh, Karnataka, and Tamil Nadu fall under this zone. Sowing is usually completed by early November, harvesting begins in the second half of February and production is very early in this zone.

A detailed account of insect pests of wheat and natural enemies of different insect pests have been given in Tables 6.9–6.11

TABLE 6.9 Pests of National Significance in Wheat.

Insectandmites	**Scientific name**
Pink stem borer	*Sesamia inferens*
Army worm/cutworm	*Mythimna separata*
Wheat aphid	*Sitobian avenae*
American pod borer	*Helicoverpa armigera*
Brown mite	*Petrobia lateens*
Termite	*Odontotermis obesus, Microtermes obesi*
Shoot fly	*Atherigona naqvii*
Diseases	
Yellow/stripe rust	*Puccinia striiformis* var. *striiformis*
Alternaria leaf blight	*Alternaria triticina*
Brown rust	*Puccinia recondita* f. sp.*tritici*
Spot blotch	*Bipolaris sorokiniana*
Loose smut	*Ustilago tritici*
Karnal bunt	*Tilletia indica*

TABLE 6.9 *(Continued)*

Insectandmites	Scientific name
Black rust	*Puccinia graminis tritici*
Foot rot	*Pythium graminicolum*
Flag smut	*Urocystis agropyri*
Powdery mildew	*Blumeria graminis*
Hill bunt	*Tilletia tritici*
Fusarium head blight	*Fusarium graminearum*
Nematodes	
Cereal cyst Nematode	*Heterodera avenae*
Root-knot Nematode	*Meloidogyne* spp.
Yellow ear rot or Tundu	*Rathaybacter tritici* + *Anguina tritici*
Seed gall Nematode	*Anguina tritici*
Weeds	
Broad-leaved/dicot weeds	
Jangali Palak	*Rumex dentatus*
Field bind weed	*Convolvulus arvensis*
Yellow pea	*Lathyrus aphaca* L.
Corn spurry:	*Spergula arvensis*
Fine leaf fumitory	*Fumaria parviflora*
Swine cress	*Coronopus didymus*
Thistle weed	*Cirsium arvense*
Scarlet pimpernel	*Anagallis arvensis*
Onion weed	*Asphodelus tenuifolius*
Sweet clover:	*Melilotus indica*
Lambs quarter	*Chenopodium* spp.
Grasses	
Blue grass	*Poa annua*
Canary grass	*Phalaris minor*
Rye grass	*Lolium spp*
Wild oat	*Avena ludoviciana*
Sedge	
Purple nut sedge	*Cyperus rotundus*
Rodents	
Field mouse	*Musbo oduga*
Soft fur red field rat	*Millardia* spp.
Indian Gerbil	*Tatera indica*
Lesser bandicoot	*Bandicota bengalensis*
Black Rat	*Rattus*

TABLE 6.10 Pest of Regional Significance in Wheat.

Insectpests	Botanical name	Important regions
Wheat thrips	*Anaphothrips favicinctus, Haplothrips tritici*	Different regions of Uttar Pradesh
Ghujhia Weevil	*Tanymecus indicus*	
Diseases		
Seedling blight	*Rhizoctonia solani, Fusarium* spp.	Different regions of Himachal Pradesh, Tamil Nadu, and Punjab
Spot blotch	*Bipolaris sorokiniana*	West Bengal, Eastern areas of Uttar Pradesh and different regions of Bihar

TABLE 6.11 Natural Enemies of Wheat Insect and Mite Pests.

Parasitoids	Examples
Nymphal and adult parasitoids	*Thripobius semiluteus*, *Aphelinus* spp., *Aphidius* spp.
Pupal parasitoid	*Ichneumon* spp.
Egg parasitoid	*Trichogramma* spp.
Larval parasitoids	*Campoletis* spp., *Bracon* spp., *Carcelia* spp.
Predators	Preying mantis, ladybird beetles, spiders, big-eyed bug, earwigs, ground beetles, fire ants, dicyphus, predatory mites, *Orius* spp., hover fly, mirid bugs, predatory thrips, robber flies, reduviid bugs, pentatomid bugs, black drongo, lacewings, common mynahs

6.6.1 IPM STRATEGIES DURING PRESOWING STAGE OF WHEAT

6.6.1.1 COMMON CULTURAL STRATEGIES

1. Well-decomposed FYM application helps to prevent termite infestation.
2. Timely sowing of the crop and avoiding delay in sowing.
3. Deep ploughing of fields during summer to control nematodes, by killing eggs, pupae, and dormant resting structures of soil-borne pathogens by exposing them to high temperatures.
4. Soil solarization is an effective cultural practice to destroy inoculum of soil-borne pathogens, larvae, pupae, and eggs of insect pests, weed seed, rhizomes, bulbs, etc.
5. Timely sowing without any delay should be followed.
6. Rouging diseased plants and proper field sanitation should be followed.
7. Destruction of collateral and alternate host plants which may harbor disease inoculum.

root of infected plant remains short and bushy. The gelatinous egg masses are produced, but most of these remain inside the aerenchymatous cavities in root cortex (Figure 7.11a, b). The eggs hatch and the emerged J2 infect new sites within the root, or come out and invade new roots when soil moisture levels are low and J2 gets enough grips to move to new feeding sites. The biology and ecology of these nematodes have been described in detail.[2] Although several species of *Meloidogyne* have been found infecting rice roots and the more specialized ones are *M. graminicola, M. hainanensis, Magnaporthe oryzae, Magnaporthe oryzicola, M. salasi,* and *M. triticoryzae.* These species show a greater preference to monocotyledonous hosts, more commonly the graminaceous plants in family Poacae. These nematodes may belong to the subgeneric Hypsoperine group in morphology, meiotic parthenogenesis, and differences in the ITS of mitochondrial DNA from mitotic parthenogenic *M. incognita, M. javanica,* and *Meloidogyne arenaria.*[3]

FIGURE 7.3 A close up of rice root-galls.

7.4.2 SURVIVAL STRATEGY

In order to understand the population behavior, it is important to understand the survival strategy of these nematodes, especially *M. graminicola* and *M. triticoryzae* that have been described.[2] These nematodes can survive long

8. Use of trap crops like marigold or pea for the management of leaf miner and border tall crops like mustard to control aphid population.[11,12]
9. Growing pest attractant, repellant, or trap crops to avoid severe damage by pests to the main crop around the field bunds and destroying them.[13,14]

6.6.1.2 *NUTRIENT MANAGEMENT*

1. Application of FYM @ 4.0 t/acre at the time of field preparation 2–3 week before sowing or vermicompost @2.0 t/acre 1 week before sowing.
2. *Trichoderma harzianum* or *Pseudomonas flourescens* colonized FYM or vermicompost should be applied to ensure better crop health and also for induced systemic resistance against broad spectrum of pests and diseases.
3. Cultivation of green manure crops to improve soil health to ensure better root growth and crop health and vigor.

6.6.1.3 *WEED MANAGEMENT*

1. Adoption of stale seed bed technique at the time of field preparation to minimize the damage caused by the weed to the crop.
2. Adopting zero tillage also minimizes the weed seed germination.

6.6.2 *IPM STRATEGIES AT THE TIME OF SOWING*

6.6.2.1 *NUTRIENT MANAGEMENT*

1. Seed treatment with Azotobacter at the rate of 0.2 kg/acre.
2. Fertilizers should be applied based on testing report of the soil, general standard recommendation per acre for wheat is 60 kg N, 25 kg P_2O_5, 20 kg K_2O.
3. At the time of sowing one-third portion of N, full dose of P&K must be applied.

6.6.2.2 *WEED MANAGEMENT*

1. Hand weeding is done whenever required.

6.6.3 IPM STRATEGIES AT VEGETATIVE STAGE

6.6.3.1 NUTRIENT MANAGEMENT

1. Apply one-third portion of N as top dressing with first irrigation and remaining one-third portion of N with second irrigation.

6.6.3.2 WEED MANAGEMENT

1. Postemergence application of 2,4-D-dimethyl amine salt 58% SL @ 0.3–0.4 L by dissolving in 200–240 L of water per acre or metsulfuron-methyl 20% WP @ 20 g in 500–600 L of water+ surfactant (Iso-Octyl Phenoxyl Poloxethanol 12.5%) @ 200 mL/acre to control wide range of weeds in broader leaf category.
2. To control grasses like wild oat, *Phalaris* spp., and also many broad-leaved weeds, application of weedicide Isoproturon 50% WP @ 0.8 kg dissolved in 300 L of water per acre can be applied as both pre and post-emergence spray.

6.6.4 IPM THROUGH COMMON CULTURAL STRATEGIES

1. Well decomposed FYM application helps to prevent termite infestation.
2. Timely sowing of the crop and avoiding delay in sowing.
3. Compulsory irrigation at critical growth stages of crop.
4. Drought stress or water logging conditions should be avoided.
5. Instead of unnecessary application of chemicals needful chemical sprays should be done in order not to disturb the parasitic activity of natural enemies, when larval parasitoids are noticed in the field.

6.6.5 IPM THROUGH COMMON MECHANICAL STRATEGIES

1. Insect's eggs and larvae mass must be collected and destroyed.
2. Diseased and insect-infested plant parts must be collected and destroyed.
3. Setting up of yellow and blue sticky traps for aphids and thrips, respectively, at the rate of 4–5 traps/acre. Set one light trap per acre. This practice should be operated between 6 and 10 pm

4. Pheromone traps should be installed at the rate of 4–5/acre for monitoring adult moth's activity (lures are changed for every 2–3 weeks)
5. In order to attract predatory birds like King Crow and common mynah, etc., set bird perches at the rate of 20/acre.
6. Make bonfire during evening hours between 7 and 8 pm in order to attract pests toward fire light in a way to get them killed.

6.6.6 IPM THROUGH COMMON BIOLOGICAL STRATEGIES

1. Conservation of natural enemies and implementation of augmentative release of natural enemies.

6.6.7 IPM STRATEGIES AT EAR HEAD STAGE

6.6.7.1 NUTRIENT MANAGEMENT

1. Foliar spraying of micronutrients should be given whenever deficiency symptoms were noticed in the field.

6.6.7.2 WEED MANAGEMENT

1. Uproot the weeds present in the field to check the weed seed spread.

6.6.8 CULTURAL CONTROL STRATEGIES

1. Low-lying water stagnant areas of the field are more prone to Karnal bunt infection. Field drainage and uniform land leveling can reduce disease incidence.
2. Increased row spacing, decreased seed rate, avoiding delay in the sowing must be followed.
3. Checking irrigation at the end of flowering or during awn emergence period may check disease development to some extent. Balanced use of nitrogen and potash fertilizers to avoid lodging of the crop.
4. Soil solarization or mulching with polythene sheets prevents the germination of teliospore present in the stubbles or soil.

5. After harvesting, destroy the crop stubbles by burning.
6. Crop rotation with nonhost crop.

6.6.9 CHEMICAL CONTROL STRATEGIES

1. Spraying of Propiconazole 25% EC @ 200 mL mixed in 300 L of water/acre or Bitertanol 25% WP @ 0.86 kg dissolved in 300 L of water to spray for 1 acre or Thiram 75% WS dissolved 10–12 g in 400 mL of water to be applied for one acre. Different strategies for IPM in wheat have been summarized in Table 6.12.

6.7 IPM MODULE FOR WHEAT IN THE RICE–WHEAT CROPPING SYSTEM

6.7.1 STRATEGIES FOR PRESOWING STAGE

1. Identify the termite hills (termitaria) in the wheat field and elimination of queen should be ensured by spraying with chlorpyriphos.
2. Use only well decomposed FYM manure in the field.
3. Two days before wheat sowing, seed must be treated with chlorpyriphos.
4. Follow normal timely sowing and avoid delayed sowing of wheat on a community basis.

6.7.2 STRATEGIES FOR POSTSOWING STAGE

1. Foliar spraying with chlorpyriphos@ 0.8 L active ingredient per 1 ha followed by irrigation.
2. For the late sown crop, seed rate should be increased by 5–7%.
3. Periodical removal of dead hearts formed due to shoot fly.
4. For the higher ETL level of shoot fly pest (more than10% incidence), two times spraying must be carried out with insecticide, cypermethrin @ 50 g a.i./ha at 15 days interval.
5. Sprays with formohian and oxydementon-methyl or phosphamidon @160, 160, 250 mL a.i./ha effectively control sucking pests and these sprays can be repeated at fortnight intervals.

TABLE 6.12 Different IPM Strategies Against Insect Pests, Diseases, and Nematodes in Wheat.

Physical and mechanical control	Cultural and biological control	Chemical control
Aphids		
	Monitoring for pest & natural enemies regularly; Growing trap or barrier crops like mustard around the field. Planting tall border crops like sorghum/millet/maize to reduce pest density. Following common cultural practices. Conserving the natural enemies	Thiamethoxam 25% WG @ 20 g in 200 L of water per acre
Shoot fly		
	Regular field monitoring for observation of pest and defender population and barrier crops like mustard crop around the field should be grown. Plant tall border crops like maize and sorghum or millet to reduce pest population. Fish meal traps can be used for monitoring the pest. Use of resistant genotypes like HP-19123, KRL-213	Foliar spray with cypermethrin 10% EC at the rate of 0.2 L dissolved in 200–300 L of water/acre when dead hearts are noticed. Soil application of Phorate10% G at the rate of 7.5 kg per acre
Termites		
1. Dismantle termitaria (termite hills) around the field and try to find and kill the termite queen	At 10 days interval three summers ploughing reduces juvenile population. Well decomposed FYM application helps to prevent termite infestation. Avoid late sowing of crops. Attracting the termite population by placing one kg of raw cow dung at 10 different places in the field which act as an attractant for termites. Destruct the crop residues which serve as a source of infestation by termites. Use of kerosene oil or any crude emulsion oil to destroy the termite colony in the termatorium. Incorporation of neem cake to soil @ 80 kg/acre. Application of fungal bioagents *Trichoderma viride/harzianum*, bacterial bioagent *Pseudomonas fluorescens* as seed treatment and soil application	Thiamethoxam 30% FS @ 1.32 kg per 40 kg seeds. Soil drenching with Chlorpyrifos 20% EC @ 0.8–1.2 L/acre

TABLE 6.12 *(Continued)*

Physical and mechanical control	Cultural and biological control	Chemical control
Armyworm/cutworm		
	Field observations must be taken regularly for pest and natural enemy populations. Barrier crops like mustard should be grown around the field. Setting up of Pheromone traps at the rate of 4–5/acre and bird perches at the rate of 10/acre can be installed for monitoring fruit borer activity and attracting predatory birds. Lure replacement with fresh lures after every 20–25 days interval. Economic threshold level is 8–10 moths/day/trap for fruit borer	Carbaryl 50% WP @ 800 g in 200 L of water/acre. Trichlorfon 5% GR at the rate 0.3 kg per acre. Trichlorfon 50% EC @ 0.3 L per acre dissolved in 300 L of water per acre
Pink borer	Conserve natural enemies. Use of trichocard with 50,000–100,000 Trichogramma per ha. Establish bird perches @ 10/acre in order to attract predatory birds. Others control measures same as in the case of aphid. Ploughing and flooding the field for killing the larvae	Foliar spray of Dichlorvos 76% EC @ 112.8 150.4 mL dissolved in 200–400 L of water/acre. Field showing more than 5% dead hearts should be sprayed with 0.8 L Triazophos 40 EC per ha
Helicoverpa (American pod borer)		
	Raising intercrops in 1:2 ratio such as coriander, onion, maize, cowpea, urd bean. Grow border guard crop like sorghum or maize all around wheat in four rows. Erect pheromone traps @ 10/ha to monitor the pest. Light traps during. November–December; Erect bird perches @ 50/ha to attract predatory birds like Drongo. When eggs and early instar larvae are noticed spray NSKE 5% or neem-based Insecticides. Use of microbial insecticides, namely, NPV 200 LE/ha. *B.t. formulation* 400 g or 400 mL/ac thrice at weekly interval in evenings in winter. Conserving the natural enemies like coccinellid beetles, spiders, dragonfly, lacewing, *Cotesia* spp., syrphid flies, etc. Release *T. pretiosum* 3–4 times @ 0.4 lakh/acre from flower initiation stage at weekly intervals inundatively. Release of larval parasitoids like *Chrysoperla zastrowii*, *Campoletis chloridae* in the field	Foliar spray of indoxacarb 1 mL/L or spinosad 0.3 mL/L per acre

TABLE 6.12 *(Continued)*

Physical and mechanical control	Cultural and biological control	Chemical control
Brown mite		
	Selection of tolerant varieties such as C 306 for brown mite. Follow common cultural practices. Conserve these natural enemies by growing the plants as mentioned earlier: Foliar spray with 2% of Neem oil or NSKE (5%)	Foliar spray of Quinalphos 25% EC at the rate of 0.6 L mixed in 200–400 L of water per acre
Wheat bug, ghujhia weevil, and wheat thrips	Irrigation is a must at critical crop growth stages. Avoid drought stress or water logging conditions. Limit chemical sprays to enhance natural enemy populations (larval parasitoids)	
Stem, leaf and yellow rust of wheat	Seed biopriming with bioagents like *Trichoderma viride*, *Trichoderma harzianum*, and *Pseudomonas flourescens* to induce resistance and plant growth promotion in wheat. Adopting mixed cropping and crop rotation. Avoid overuse of N fertilizers	Foliar spray with Propiconazole 25% EC or Tebuconazole 25% EC @ 0.2% of 200 L of water/acre should be applied when initial symptoms are noticed
Powdery mildew of wheat	Disease is confined to specific locations mostly and even some portions of particular locations. New varieties mostly do not have adequate resistance	Foliar spray with Triadimefon 25%WP @ 200 g dissolved in 300 L of water/acre
Seedling blight, foliar blight, leaf blight, and spot blotch	Use certified seeds of resistant varieties only. Crop rotation and timely sowing, Growing tall crop plants as a barrier to check air borne pathogens/diseases	Foliar spray of Propiconazole @ 0.1% or Tebuconazole @ 0.1% of about 200 L/acre. Foliar spray with Thiophanate Methyl 70% WP @ 0.28 kg in 200 L of water/acre
Loose smut of wheat		
Cold water treatment of seeds by soaking during hot summer months in the morning hours and exposing them to the hot sun from 8 am to	Selection of resistant or tolerant varieties	Seed treatment with Tebuconazole 2% DS @ 0.2/10 kg seed. Seed treatment with Carboxin 37.5%+ Thiram 37.5% DS @ 3.0 g/kg.

TABLE 6.12 *(Continued)*

Physical and mechanical control	Cultural and biological control	Chemical control
12 noon and kept for drying in the afternoon. This will ensure killing the dormant mycelium of causal organism residing inside the seed and provides a proper management of disease reducing and limiting the fungicides use. However, care should be taken to prevent loss of seeds viability		Seed treatment with Carboxin 75% WP @ 2–2.5 g/kg seeds
For seed gall nematode		
Use of only pure seed without any contamination by sieving or providing 2% brine solution treatment to remove galls in order to prevent ear cockle disease. Dry cleaning: Galls can be separated by a coarse sieve from the healthy seeds. Winnowing or fanning to separate the galls from the seeds. Brine floatation: seeds soaked in 2% salt solution in place of plain water may remove almost 100% galls	Use certified seed of resistant varieties. Seed treatment @ 5 g/kg seed and foliar spray with *Pseudomonas fluorescens* @ 5 g/L. Apply neem cake by incorporation in the soil at the rate of 80 kg/acre	Seed treatment with Carbofuran 3% CG @ 26.6 kg/acre

6. Use of zinc phosphide for rodent management.
7. Spraying of insecticide and fenitrothion @ 500 g ai./ha can effectively reduce cutworm, semilooper, and pyrilla population.

6.8 PRACTICING DRY-WET-DRY TRANSITION IN RICE-WHEAT CROPPING SYSTEM MAY RESULT BIOLOGICAL CHANGES IN SOIL

Puddling the soil is the traditional method of rice culture to form a saturated root zone above a compacted subsoil layer for the quick establishment of transplanted seedlings that reduces sewpage of standing water from the rice field during the cropping period. Generally, crops like wheat cannot tolerate the anaerobic conditions to which rice is well adapted in this cropping system. So, the significant dominating feature of R–W systems is the conversion of soil from aerobic to anaerobic conditions when wheat is followed by rice and then again reverted back when it is followed by wheat again annually. These conversions have significant impact on biological status that is responsible for the change of growing conditions for different crops in this system. Only a few organisms are likely to be adapted to both aerobic and anaerobic conditions and the continuous transition between these conditions limits the survival of biotic stress-causing organisms in R–W systems. Insect pests, weeds, and disease problems are most likely to be caused by mobile organisms which are readily reintroduced by the means of various dispersing agents like wind, water, or other biotic organisms as well. These prevailing conditions also affect and may influence the survival of other beneficial microorganisms, especially among them the most important and farmer's friendly are those which play significant role in atmospheric nitrogen fixation and its transformations in the soil. Unlike other nutrients, major proportion of nitrogen exists in the organic fraction from where it continuously got converted from mineralized to immobilized state and again converted to mineralized form (nitrate or ammonium forms) by the action of different microorganisms. In strong reducing conditions, nitrate is converted to ammonium, whereas under anaerobic conditions, it is lost in the form of gaseous through denitrification. Mineralization changes rapidly due to drying and wetting of the soil. Unless plant roots utilize its available nitrate form, it is at a risk of loss by leaching from the rhizosphere zones. Thus, under RWCSs, frequent transitions of wet, dry conditions, and there is potential for significant loss of nitrogen by leaching and denitrification. Further, since nitrite is an intermediate in both the reduction of nitrate, the oxidation of ammonia, aerobic denitrification via nitrate may be more substantial, widespread than previously realized, especially on soils that are alternatively subjected to wetting and drying.[15] There has been much research in rice and wheat crops on the N dynamics separately, but so far, almost none of the studies have been undertaken altogether kept in a combined manner on the N dynamics in R–W systems. From that vast literature, however, it is possible, despite the complexity of the chemical interchanges, the various microorganisms involved, and identifying as simple distinction that assists in the management of N dynamics in these systems. The anaerobic conditions required for the growth of rice crop have economies of ammonium whereas aerobic conditions of wheat and other crops have nitrate economies. This is true even though rice does utilize nitrate form in the rhizosphere, especially at high N level, during the reproductive stage[16,17] wheat can utilize ammonium, although it is in frequently available in aerobic soils.

6.9 EFFECT OF CROP RESIDUES ON MICROORGANISMS IN RICE–WHEAT CROPPING SYSTEMS

Carbon and nitrogen sources supplied from the crop residues act as substrate for the change in the activity of soil microorganisms in the formation of soil organic matter (SOM). But those microorganisms also equally compete for available nutrients in the soil with plants which also include those nutrients/organic matters which get released due to the decomposition of crop residues. Pronounced effects are observed when residues are incorporated into the soil.[18,19] Nutrient content of residues and the rapid release of some nutrients and their immobilization and fixing of nutrients by the soil microflora and subsequent release in available form or fixation in resistant SOM provide a strong base for residue management in crop nutrition. Residue management is important in R–W systems as combine harvesters are now most popularly used for crop harvesting. One undesired option for the disposal of crop residue waste is burning the stubbles or crop residues. But this lead to great losses of nutrients (up to 80%) of N,[20] 25% of P, 21% of K,[18] 4–60% of S,[21] and also causes significant air pollution and elimination of beneficial soil microorganisms and insects. Despite these negative effects, burning has one advantage where it helps to destroy insect pests and diseases present in the leftover crop residues in the soil. A second option is incorporation of these crop residues into the soil after harvest but then decomposition process takes much time and immobilization of soil nitrates occurs[21] thereby reducing the N uptake and yield of succeeding wheat crops by about 40%.[22,23] The integrated utilization of rice or wheat straw along with the added inorganic fertilizer can enhance the yield of rice and wheat in R–W systems along with efficient use of inorganic fertilizers in fewer quantities.[24] In Mahapatra's study, straw containing 29 kg N/ha with the added urea at the rate of 58 kg N/ha substituted for 87 kg N/ha of urea in the production of rice, and it was also found that soil mineral N was maintained to the same level as urea when applied alone. Retention of some or all of the crop residues on the surface may be the best option in many situations. Crop residues decompose very slowly on the surface thereby increasing the organic carbon and total nitrogen content in the top soil profile of 5–15 cm, apart from the firm binding of top soil and preventing the losses caused due to erosion.[25] Surface retention of residues had increased nitrogen uptake by 29%, soil nitrate NO_3 concentration by 46%, and yield by37%, as compared to burning.[22,26,27] Retention too had some disadvantages, however, providing shelter for both harmful as well as useful microorganisms may reduce the detrimental effects of harmful microorganisms. The patterns of nutrient release from the residues of individual crops can vary greatly depending upon weather and management and the condition of the soil and its microflora.

Yoneyama and Yoshida[28] observed that although only 10% of N from rice straw was recovered by the first crop followed by a further 10% by the second, it significantly improved the uptake of nutrients which coincided with peak crop demands. Retention of rice straw in the field is the source of carbon substrate for increased activity of soil microorganisms, heterotrophic nitrogen fixation, increased soil carbon and nitrogen, and reduced N fertilizer requirements for rice.[18,19] Rice straw, however, decomposes at a faster rate and release N to soil if it is treated with urea and incorporated in to soil during field preparation.[29] An important feature of irrigated rice-cropping systems is the initial faster decomposition but slow rate of decomposition in the long-run because of the prevailing flooded conditions as compared to aerobic conditions where decomposition takes place at a faster rate.[30] This probably explains why soil organic carbon levels are relatively stable in two- or three-crop rice systems compared to R–W systems where decomposition is faster.[31]

6.10 MANAGEMENT OF INSECT-PESTS, DISEASES, NEMATODES, AND WEEDS IN RICE–WHEAT CROPPING SYSTEMS

The productivity of component crops in the R–W systems may be reduced significantly by damage caused by insects, diseases, rodents, nematodes, termites, and competition from weeds. The important issues for R–W cropping systems are how the pressure shift occurs among the pest species from rice to wheat crop and how this affects the susceptibility reaction of these pests on these crops and the population changes either by increasing or reducing their density and allowing some control measures to be taken. Savary et al.[32] used cluster and correspondence analyses to characterize relationships to cropping practices and crop status and injuries due to specific insects and weeds and diseases. They also used principal component analysis and step-wise regression analyses to generate yield loss estimates for rice. They reported that in northern India, damage caused by weeds, dead hearts caused by yellow stem borer, yield loss caused by brown spot (*B. oryzae*), and sheath blight (*R. solani*) were most devastating to rice. They also found that brown spot and sheath blight mostly prevailed in heavily cropped areas to a greater extent. They noted that many problems were reduced where fallow was included and implicating pest problems with increased agricultural intensification.

6.11 MANAGEMENT OF INSECTS

Continuous rice–wheat cropping has resulted in greater incidence of some insects. For example, the rice pink stem borer, caused by *S. inferens* and shoot fly, by *Antherigona oryzae,* became great threats in India where they normally used to attack only rice crop but now started attacking both crops in R–W cropping systems.[4] Similarly increasing threats of green leaf hopper (*Nyphotettix nigropictus*) and dark-headed stemborer (*Chilo suppressalis*) due to R–W systems in Nepal is another example of this.[33] Management of residue also affects the incidence of stem borers in rice. For example, stem borer larvae get survive in the residues if crop residues of rice are incorporated by ploughing following sequential cultivation of wheat until the next rice crop. On the contrary, if residues are retained on the surface of the soil where wheat crop is grown with irrigation and under the influence of fertilizers, these crop residues get decomposed and stem borer larvae do not survive.[34] With periodic flooding and drying, R–W systems suffer little damage due to termite also.[35]

6.12 MANAGEMENT OF DISEASES

Due to the existence of continuous anaerobic conditions prevailing in rice crop, some of the surviving soil and seed-borne pathogens can be transmitted from one season crop to other season crops in R–W systems. Repeated R–W cropping may allow build-up of soil-borne pathogens, such as *Fusarium,*

Rhizoctonia, and *Sclerotium*. In both crops, *R. solani* and *S. rolfsii* usually found to cause sheath blight and stem rot, respectively. In the case of sheath blight, if infection starts before flag leaf emergence, yield loss can reach upto 20% of total yield.[4,36] Spot blotch (*Bipolaris sorokiniana*) is also considered as one of the most important diseases of eastern IGPs of RWCSs. Sheath blight pathogen prevail and survive best under conditions of high temperature, humidity, while the spot blotch caused by *B. sorokiniana* is well adapted to lower temperature and maximum incidence will be noticed in the low lying (Tarai) areas of the IGP.[37] *Bipolaris* spp. can well establish and proliferate on crop residues and are likely to become problematic in zero tillage systems.[34] Likewise, northwest and western IGP is mainly dominated by yellow rust (*Puccinia striiformis* f. sp. *tritici*) by occurring in severe form apart from *B. sorokiniana* (spot blotch) and *Pyrenophora tritici-repentis* (tanspot) in wheat. In the eastern Gangetic Plains, yellow rust has recently become a major problem, whereas spot blotch is more problematic under reduced tillage where the pathogen survives on alternative hosts and crop residues.[38] There are several factors and practices of crop management that also influence incidence of the disease. Thus, delaying the harvest of the rice crop may make wheat crop more prone and vulnerable to the diseases like loose smut (*Ustilago tritici*), leaf rust (*Puccinia recondita* f. sp. *tritici*), and yellow rust by causing significant damage. At the same time, early sowing of wheat crop reduces the impact of those diseases which makes utilization of short-season rice, more significant, and its timely harvesting is subsequently followed by rapid establishment of the wheat crop. Broadening the range of crops and diversification of different crops can serve as disease breaking and become an important part of R–W cropping management and probably a large number of cropping combinations came into existence. That is why rotation of R–W with sugarcane (plus ratoon) every few years is commonly followed by farmers in north India. Similarly in Northern Bangladesh, jute (*Corchorus* spp.) is frequently grown before the main-season monsoon rice. Singh[39] also reported a yield benefit to the rice of 0.4 t/ha in R–W sequences that include rapeseed and sugarcane and sugarcane ratoon when compared to R–W cropping pattern alone. Reduced pressure of pests and diseases is also one of the important reasons for coming up with this response in crop diversification.

6.13 MANAGEMENT OF WEEDS

Although weeds such as *Echinochloa colonum* and *E. crus-galli* are commonly present and built up in rice crop field. However, these weeds are not of great significance in transplanted rice where seedlings are transplanted into puddled, weed-free soil, even though presowing FYM incorporation is a source of weed seed except for a few weed species which can withstand even flooded conditions. Even these weeds that are emerged can be kept under check by the farmers by doing at least one-hand weeding. However, in wheat crop, canary grass (*Phalaris minor*) has become a serious problem in R–W systems because its dormant seeds are able to survive even flooded conditions in the soil of rice cropping duration.[40] In Nepal, Pire (*Polygonum hydropiper*), another weed of wheat crops is also becoming a big threat to the R–W systems.[41]

Overall, it is estimated that a yield loss of around 15–25% in rice and 15–30% in wheat crop can occur in R–W cropping system when these crops are grown in sequence.[36] Although weed control is highly challenging task in unpuddled soils, there are some merits also where planting can be done by zero tillage without much disturbance to the soil on raised beds where only a few weeds get germinated or very little establishment of weed takes place. Apart from that, planting on the raised beds allows mechanical weeding even during the early stage of crop growth and reducing high costs on selective herbicides and allowing the use of nonselective herbicides which are available at cheaper costs. Apart from all these practices, crop rotations are one of prominent cultural practices for the management of weeds by creating a pressure of competitive forces against weeds and also by creating additional opportunities or advantages in selecting a specific array of useful and effective herbicides. For example, replacing wheat crop for 1–2 years by sunflower (*Helianthus annus*), sugarcane or barseem (*Trifolium alexandrium*), in raised bed planting systems may drastically reduce the population of *P. minor*.[40] However, the greater increase in the risk of weed is expected with direct seeding rice on bed systems.

6.14 MANAGEMENT OF NEMATODES

The majority of nematodes were found to be polyphagous in nature affecting wide range of crops other than rice and wheat except some species that are specifically associated with either of them or both in R–W cropping system.[42] For example, *Hirschmanniela oryzae* (rice root nematode, RRN) and *Tylenchorhynchus* spp. are two groups of plant-parasitic nematodes that are currently predominant in R–W cropping areas of Uttar Pradesh, India.[43] *H. oryzae* is found to be predominant in flooded-rice fields while the *Tylenchorhynchus spp* is found to be dominant in wheat fields and even some observations and initial studies also suggest that *H. oryzae* could also be a possible dominant nematode apart from the serious problem of root-knot nematode, *Meloidogyne graminicola* in the R–W systems of Nepal.[44] Likewise, apart from the species of *M. graminicola*, other species of *Meloidogyne* such as *M. incognita* or *M. javanica* could be associated with several crops by posing a severe threat to crops in R–W-legume systems of the entire IGP because *Meloidogyne* spp. can parasitize all the component crops.[42] On the contrary to the conclusions and findings of Dwivedi and Upadhyay,[43] Pokhrel et al.[44] found that the introduction of wheat into continuous rice cropping in Nepal reduced the incidence of *H. oryzae* to which wheat is found to be resistant. Likewise, by rotating the wheat and barley crops with legume crops like mung bean and black gram reduces the incidence of *Pratylenchu sindicus*[45] or replacing early rice by wheat and broad bean or linseed reduces the incidence of RRN in the ensuing main season of rice.[44] Also, the replacement of maize crop by rice crop in the Indian Punjab had decreased the populations of the cereal cyst nematode *(Heterodera avenae)* substantially which is a major predominant nematode of wheat in maize–wheat cropping system. This replacement, however, also resulted in the substantial increase of *H. oryzae*.[46] Soil solarization has shown many positive results in some experiments of R–W system in countries like Bangladesh and Nepal where heating of soil in-situ by covering with thinnest plastic sheet was found to drastically reduce nematode population along with many soil-borne pathogens, weed seed germination, and insect pest's eggs and pupae by increasing the availability of nutrients, antagonist activity which could act as an useful diagnostic tool in the assessment of soil health.[47]

6.15 CONCLUSION

RWCS is a very important and major cropping system in South Asia. Both these crops constitute a staple food for more than 1 billion people (20% of world's population) and thus a major source of employment, livelihood, and income to many rural and urban people. In this system, the biotic factors alone are responsible to bring in a yield loss of 20–30% in both the crops. Dependence on chemicals for the management of several biotic stresses (diseases, insect pests, and weeds) has become imperative in these two crops. Reduction in the market value of produced grains, pesticide resistance in insect-pests, health hazard complications, negative effects on ecosystem due to pesticide residues, and contamination of food chain and groundwater has necessitated a shifting from using chemical pesticides and inorganic fertilizers toward IPM and INM for sustainable production of both the crops. It will be more appropriate if a term integrated crop management may be adopted in practice for IPM and INM together. In RWCS, certain pests and diseases are of global importance while some are of regional importance. Regular survey and surveillance along with pest scouting for monitoring the population dynamics, onset of diseases and their progress, agroecosystem analysis (AESA) to determine the predisposing factors for a crop to be affected by various pests and pathogens, and also to determine the ETL for different insect pests at different growth stages of the crops are of utmost importance. On the basis of survey and monitoring, different strategies based on cultural practices and mechanical methods have to be applied to check the insect pests and diseases to minimize the crop losses. Presowing irrigation can be a very effective cultural operation for weed management, where nursery or seedlings are to be transplanted. Similarly, soil solarization of nursery plots/beds may help in raising healthy nursery of rice crop with no soil-born pathogen or nematodes and also to ensure the freedom from weed also. Balance application of fertilizers with the need-based application of micronutrients can also be effective in bringing down the level of pests, diseases along with enhancing the level of resistance in plants against pests and diseases. Pests and disease specific different effective IPM modules have been developed and validated which need to be strategically applied. Similarly, specific natural enemies of important insect pests, namely, insect's parasites, parasitoids, and predators are commercially available which need to be strategically released/applied in the area where the insect pests are supposed to be a problem. Similarly, some fungal and bacterial antagonists like *Trichoderma* sp. and *P. flourescens*, other plant growth-promoting rhizobacteria are well known and commercially available hence they can be effectively applied either as seed biopriming or seedling root dipping for their direct effect against several seed- and soil-born plant pathogens. Applications of these fungal and bacterial bioagents also act as potential resistance inducers against several airborne pathogens in rice and wheat both. Plant-based natural products such as neem and karanja cakes, their seed extract, leaves extracts, etc., have also some pest deterrent properties and accordingly same should be used for pest management in both the crops. The use of resistant cultivars is another feasible option without any additional expense with no toxic effects to either crop or environment. In both the crops, disease and pest specific resistant cultivars have been developed, notified, and released by competent authorities for different agroclimatic zones which need to be adopted in cultivation. However, a regular vigilance will be necessary to monitor the breakdown of resistance in these cultivars. The use of multiline cultivars and varietal mixtures approach may ensure the longevity of any resistant cultivars in a particular geographical location. However, crop stages specific schedules for chemical sprays against important pests and diseases can also be adopted as a component of the IPM module to ensure effective management of diseases and insect pests in both the crops.

KEYWORDS

- **rice-wheat cropping system**
- **IPM**
- **diseases**
- **insects**
- **weeds**
- **nematodes and soil microbial changes**

REFERENCES

1. Gill, M. S.; Pal, S. S.; Ahlawat, I. P. S. Approaches for Sustainability of Rice (*Oryza sativa*)- Wheat (*Triticum aestivum*) Cropping System in Indo-Gangetic Plains of India–A Review. *Indian J Agron.* **2008,** *53* (2), 81–96.
2. Ladha, J. K.; Fischer, K. S.; Hossain, M.; Hobbs, P. R.; Hardy, B. Progress Towards Improving the Productivity, Sustainability of Rice-Wheat Systems: A Contribution by the Consortium Members. IRRI Discussion Paper No. 40, 2000.
3. Timsina, J.; Connor, D. J. Productivity, Management of Rice-Wheat Cropping Systems: Issues, Challenges. *Field Crops Res.* **2001,** *69*, 93–132.
4. Nagarajan, S. Plant Protection Problems in Rice-Wheat Rotation System- a Perspective. *Oryza* **1989,** *26*, 329–333.
5. Dubin, H. J.; VanGinkel, M. The Status of Wheat Disease Research in Warmer Areas. In *Wheat for the Nontraditional Warm Areas*, Saunders, D. A., Ed.; CIMMYT: Mexico, DF, 1991; pp 125–145.
6. FAO. *FAO Production Yearbook*, Vol. 53. FAO: Rome, 1999.
7. Prakash, J. S. Bentur, S.; Prasad, M.; Tanwar, R. K.; Sharma, O. P.; Bhagat, S.; Sehgal, M.; Singh, S. P.; Singh, M.; Chattopadhyay, C.; Sushil, S. N.; Sinha, A. K.; Asre, R.; Kapoor, K. S.; Satyagopal, K.; Jeyakumar, P. Integrated Pest Management for Rice. Director National Centre for Integrated Pest Management, LBS Building, IARI Campus: New Delhi, 2014; p 43.
8. Sehgal, M.; Jeswani, M. D.; Kalra, N. Management of Insect, Disease and Nematode Pests of Rice, Wheat in the Indo-Gangetic Plains. *Jf Crop Prod* **2001,** *4* (1), 167–226.

9. Sendhil, R.; Singh, R.; Singh, S.; Kumar, A.; Sharma, I. An Exploration into Changing Food Consumption Pattern in India. *Crop Improv.* (Special Issue), **2012,** 1315–1316.
10. Gurr Men, S. D.; Altieri, M. A. *Ecological Engineering for Pest Management Advances in Habitat Manipulation for Arthropods*; CSIRO Publishing: Collingwood, AUS, 2004.
11. Sharrow, S. H. Defoliation Effects on Biomass Yield Components of Winter Wheat. *Can. J. Plant Sci.* **1990,** *70*, 1191–1194.
12. Walmsley, M. R.; Capinera, J. L.; Detling, J. K.; Dyer.; M. I. Growth of Blue Grama, Western Wheatgrass Following Grasshopper Defoliation, Mechanical Clipping. *J. Kansas Entomol. Soc.* **1987,** *60*, 51–57.
13. Capinera, J. L.; Roltsch, W. J. Response of Wheat Seedling to Actual, Simulated Migratory Grasshopper Defoliation. *J. Econ. Entomo.* **1980,** *73*, 258–261.
14. Zuckerman, E.; Eshal, A.; Eyal, Z. Physiological Aspects Related to Tolerance of Spring Wheat Cultivars to Septoria Tritici Blotch. *Phytopathology* **1997,** *87*, 60–65.
15. Ponnamperuma, F. N. The Chemistry of Submerged Soils. *Adv. Agron.* **1972,** *24*, 29–96.
16. Luo, A. C.; Xu, J. M.; Yang, X. Effect of Nitrogen (NH4NO3) Supply on Absorption of Ammonium, Nitrate by Conventional Hybrid Rice During Reproductive Growth. *Plant Soil* **1993,** *155/156*, 395–398.
17. Yang, X.; Zhang, J. Q.; Ni, W. Z.; Dobermann, A. Characteristics of Nitrogen Nutrition in Hybrid Rice. *Int. Rice Res. Notes* **1999,** *24*, 5–8.
18. Ponnamperuma, F. N. Straw as a Source of Nutrients for Wetland Rice. In *Organic Matter, Rice*; Banta, S., Mendoza, C. V., Eds.; IRRI: Los Banos, Philippines, 1984; pp. 117–136.
19. Ladha, J. K.; Tirol, A. C.; Daroy, M. L. G.; Caldo, G.; Ventura, W.; Watanabe, I. Plant Associative N2 Fixation (C_2H_2 Reduction) by Five Rice Varieties and Relationship with Plant Growth Characters as Affected by Straw Incorporation. *Soil Sci. Plant Nutr.* **1986,** *32*, 91–106.
20. Raison, R. J. Modification of the Soil Environment by Vegetation Fires, with Particular Reference to Nitrogen Transformation: A Review. *Plant Soil.* **1979,** *51*, 73–108.
21. Lefroy, R. D. B.; Chaitep, W.; Blair, G. J. Release of Sulfur from Rice Residue Under Flooded, Nonflooded Soil Conditions. *Aust. J. Agric. Res.* **1994,** *45*, 657–667.
22. Bacon, P. E. Effect of Nitrogen Fertilization, Rice Stubble Management Techniques on Soil Moisture Content, Soil Nitrogen Status and Nitrogen Uptake by Wheat. *Field Crops Res.* **1987,** *17*, 75–90.
23. Sidhu, B. S.; Beri, V. Effect of Crop Residue Management on the Yields of Different Crops, Soil Properties. *Biol. Wastes.* **1989,** *27*, 15–27.
24. Mahapatra, B. S.; Sharma, G. L.; Singh, N. Integrated Management of Straw, Urea Nitrogen in Lowland Rice Under a Rice-Wheat Rotation. *J. Agric. Sci. Camb.* **1991,** *116*, 217–222.
25. Rasmussen, P. E.; Collins, H. P. Long-Term Impacts of Tillage Fertilizer, Crop Residues on Soil Organic Matter in Temperate Semi-Arid Regions. *Adv. Agron.* **1991,** *45*, 93–134.
26. Bacon, P. E.; Cooper, J. L.; Effect of Rice Stubble, Fertilizer Management on Wheat Growth After Rice. *Field Crops Res.* **1985a,** *10*, 229–240.
27. Bacon, P. E.; Cooper, J. L. Effect of Rice Stubble, Fertilizer Management Techniques on Yield of Wheat Sown After Rice. *Field Crops Res.* **1985b,** *10*, 241–250.
28. Yoneyama, T.; Yoshida, T. Decomposition of Rice Residue in Tropical Soils. I. Nitrogen Uptake by Rice Plants from Straw Incorporated, Fertilizer (Ammonium Sulfate), Soil. *Soil Sci. Plant Nutr.* **1977,** *23*, 33–40.
29. Sharma, M. P, Bali, S. V. Effect of Rice Residue Management in Wheat Yield, Soil Properties in Rice-Wheat Cropping System. *Ind. J. Agric. Sci.* **1998,** *68*, 695–696.

30. Neue, H. U.; Scharpenseel, H. W. Decomposition Pattern of 14C-Labeled Rice Straw in Aerobic, Submerged Rice Soils of the Philippines. *Sci. Total Environ.* **1987,** *62*, 431–434.
31. Bronson, K. F.; Cassman, K. G.; Wassmann, R.; Olk, D. C.; Noordwijk, M.; Garrity, D. P. Soil Carbon Dynamics in Different Cropping Systems in Principal Ecoregions of Asia. In Lal, R., Follet, R. F., Stewart, B. A., Eds.; *Management of Carbon Sequestration in Soil*; CRC Press: Boca Raton, FL, 1998; pp 35–57.
32. Savary, S.; Srivastava, R. K.; Singh, H. M.; Elazegui, F. A. A Characterization of Rice Pests, Quantification of Yield Losses. *Crop Protection* **1997,** *16*, 387–398.
33. Gyawali, B. K.; Ghimire, M. N.; Shrestha, K. P. Entomology Studies in the Naldung Rice-Wheat Area, 1992/1993–1996/1997. In *Proceedings of the Rice-Wheat Research End-of-Project Workshop*; Hobbs, P. R., Rajbhandary, N. P., Eds.; NARC/CIMMYT/RWC: Kathmandu, Nepal, 1997; pp. 39–44.
34. Hobbs, P.; Giri, G. S.; Grace, P. Reduced, Zero Tillage Options for the Establishment of Wheat After Rice in South Asia. Rice-Wheat Consortium Paper Series 2. Rice-Wheat Consortium for the Indo-Gangetic Plains; CIMMYT: New Delhi, 1997; 18 p.
35. Chander, S.; Garg, R. N.; Singh, G.; Chander, S.; Singh, G. Termite Infestation in Relation to Soil Physical Properties in Wheat Crop Under Rice-Wheat Cropping System. *Ann. Agric. Res.* **1997,** *18*, 348–350.
36. Yadav, R. L.; Gangwar, K. S.; Prasad, K. Dynamics of Rice- Wheat Cropping Systems in India. Technical Bulletin. Project Directorate for Cropping Systems Research, Modipuram, India, 1998; 90 pp.
37. Rajaram, S. Breeding, Testing Strategies to Develop Wheat for Rice-Wheat Rotation Areas. In *Wheat Production Constraints in Tropical Environments*; Klatt, A. R., Ed.; CIMMYT: Mexico DF, 1998; pp 187–196.
38. Duveiller, E.; Dubin, H. J.; Reeves, J.; McNab, A., Eds. *Helminthosporium Blights of Wheat: Spot Blotch, Tan Spot. Proceedings of the International Workshop*; CIMMYT: Mexico City, Mexico, 1998; 337 pp.
39. Singh, Y. *Rice-Wheat Cropping Systems: NARP Basic Research Sub-Project. Progress Report 1994–1995*; Pantnagar University: Pantnagar, India, 1995.
40. Malik, R. K.; Gill, G.; Hobbs, P. R. Herbicide Resistance a Major Issue for Sustaining Wheat Productivity in Rice-Wheat Cropping Systems in the Indo-Gangetic Plains. Rice-Wheat Consortium Paper Series 3. Rice-Wheat Consortium for the Indo-Gangetic Plains: New Delhi, 1998, 36 pp.
41. Ranjit, J. D. Weeds, Weed Management in the Rice-Wheat System. In *Proceedings of the Rice-Wheat Research End-of-Project Workshop*; Hobbs, P. R., Rajbhandary, N. P., Eds.; NARC/CIMMYT/RWC: Kathmandu, Nepal, 1997; pp 13–22.
42. Sharma, S. B.; Johansen, C.; Midha, S. K., Eds. Nematode Pests in Rice-Wheat-Legume Cropping Systems. In *Proceedings of the Regional Training Course. Rice-Wheat Consortium Paper Series 4. Rice-Wheat Consortium for the Indo-Gangetic Plains*; New Delhi, 1998; 105 p.
43. Dwivedi, K.; Upadhyay, K. D. Nematode Pests in Rice- Wheat-Legume Cropping Systems in Central Uttar Pradesh. In *Nematode Pests in Rice-Wheat-Legume Cropping Systems. Proceedings of the Regional Training Course. Rice-Wheat Consortium Paper Series 4*; Sharma, S. B., Johansen, C., Midha, S. K., Eds.; Rice-Wheat Consortium for the Indo-Gangetic Plains: New Delhi, 1998; pp 60–62.

44. Pokhrel, R. R.; Khan, S.; Hobbs, P. R. Plant Parasitic Nematodes Associated with the Rice-Wheat Ecosystems of Nepal. In *Proceedings of the Rice-Wheat Research End-of-Project Workshop*; Hobbs, P. R., Rajbhandary, N. P., Eds.; NARC/CIMMYT/RWC: Kathmandu, Nepal, 1997; pp 81–86.
45. Rao, Y. S.; Prasad, J. S.; Yadav, C. P.; Padalia, C. R. Influence of Rotation Crops in Rice Soils on the Dynamics of Plant Parasitic Nematode Populations. *Biol. Agric. Hort.* **1984,** *2*, 69–78.
46. Sakhuja, P. K.; Singh, I.; Kaur, D. J.; Randhawa, N. Nematode Pests in Rice-Wheat-Legume Cropping Systems in Punjab State of India. In *Nematode Pests in Rice-Wheat-Legume Cropping Systems. Proceedings of the Regional Training Course. Rice-Wheat Consortium Paper Series 4*; Sharma, S. B., Johansen, C., Midha, S. K., Eds.; Rice-Wheat Consortium for the Indo-Gangetic Plains: New Delhi, 1998; pp. 67–68.
47. Katan, J.; DeVay, J. E., Eds. *Soil Solarization*; CRC Press: Boca Raton, FL, 1991; 267 pp.

CHAPTER 7

Root-Knot Nematode Disease: An Expanding Problem in Rice-Based Cropping Systems

H. S. GAUR*

School of Agricultural Sciences, Sharda University, Greater Noida 201306, India

**E-mail: hs.gaur@sharda.ac.in, hsgaur2@gmail.com*

ABSTRACT

Rice, *Oryza sativa*, is the second most important staple foodgrain for humans. From 162 million ha, most of it in Asia and under this crop about 496.6 million mT is produced. Average productivity is low at <3 metric tonnes per hectare than China and USA and some other countries, which could be due to differences in rice genotypes, soil, water, climate, and many pests and diseases to which rice is susceptible. Attention has been on the management of insects and diseases caused by various fungi, bacteria, and viruses, but plant-parasitic nematodes have remained overlooked. Several species of nematodes infect the aboveground and belowground parts of plants and vary in their adaptations to edaphic conditions. The root-knot nematode, *Meloidogyne graminicola* is a widespread serious pest wherever rice is grown while *Meloidogyne incognita* and *Meloidogyne triticoryzae* also cause yield and quality losses in rice in some parts of India. Although adapted to all soil types and rice ecologies they multiply more and cause greater crop losses in lighter soils and low water irrigated and upland conditions. Decreasing water availability and practices of direct seeding and raised-bed planting, etc., and more than one crop of rice in a year favors the root-knot nematodes. Damage

Integrated Pest Management in Diverse Cropping Systems. Devendra Pal Singh, PhD (Ed.)

is more in direct-seeded crops and nurseries. In the popular rice–wheat cropping systems, the wheat crops suffer less due to low winter temperatures but support these nematodes until the next rice crop. Alternative cropping systems with nonhost crops are needed to give a break to the continuous availability of host crops. Nematode resistant varieties are rare and effective nematicides are either not available or are expensive and hazardous to humans and environment. These nematodes multiply on weeds and alternate host crops and also can survive in dry-infested roots for several months. African *Oryza glaberrima* and some other germplasm have resistance genes that may be may be introduced into commercially preferred varieties through traditional and marker-assisted breeding and biotechnological tools including the CRISPER-Cas in times to come. Presently, management of these nematodes is mainly based on tillage, water and organic matter, crop rotations, solar heat, and biocontrol agents.

7.1 INTRODUCTION

Rice (*Oryza sativa*) is one of the most important food crops for humans and it covers approx. 162 million hectare (M ha) in the world producing about 760 million metric tonnes (M mT) of paddy or 496.6 M mT of milled rice in 2019–2020. This was 18.92% of total foodgrain production of 2624.6 M mT that included 730.9 M mT wheat and 1397.1 M mT coarse grains and expected to go upto 501 M mT in 2020–2021 and if the Covid-19 pandemic does not upset. Much of the coarse grains are used to feed livestock. Rice is the staple food for over half of the human population and more so in Asia where more than 90% of the world's rice is produced and consumed. About 53.5% of total rice came from 45.5% area under rice in China and India put together in 2018–2019. Nine other countries, namely, Bangladesh, Indonesia, Vietnam, Thailand, Myanmar, Philippines, Japan, Pakistan, and Cambodia together produced 32.2% of the milled rice from 37.3% of the total area under rice in the same year. Some countries in Africa, Australia, Russia, Europe, and Latin and Caribbean America as well as USA and Canada also cultivate rice to a small scale (Figure 7.1). The productivity of rice, generally reported in terms of paddy grain varies from less than 3 mT/ha to as high as 6.7 mT/ha in China and 8.38 mT/ha in the USA. There may be various reasons for this vast variation in productivity such as the kind and variety of rice genotypes, soil, water, and climate. Two other reasons are (1) production technology and (2) management of diseases and pests. The seasonal warm-humid weather, besides its biological characteristics render rice susceptible to many kinds of pests and diseases including nematodes.

FIGURE 7.1 Rice root and tillering affected by increasing population densities (left to right) of root-knot nematodes.

Rice germplasms exhibit much variability in genetics, grain length, thickness, color, stickiness, aroma, growing practices, and hence many cultivars. Of the various kinds of germplasm, the Asian rice, *O. sativa* subspecies *indica* and *japonica* are the prominent types with over 40,000 varieties. African rice, *Oryza glaberrima* is also cultivated in some areas. Wild rice, *Zizania* spp. has four species, namely, northern wild rice (*Zizania palustris*), wild rice (*Zizania aquatica*), Texas wild rice (*Zizania texana*), and manchurian wild rice (*Z. latifolia*). Wild type rice germplasm is not of much commercial significance but can be source of genes for various abiotic and biotic stresses. The African rice, *O. glaberrima* germplasm too is known to offer resistance to several pests including nematodes.

The *indica* rice grains are long and elongate on cooking and remain unsticky while the *japonica* rice grains are short-grained and sticky on cooking (glutinous). Some *indica* rice varieties and like basmati are scented due to some volatile organic compounds and are richer in 2-acetyl 1-pyrroline, proline and methylglycxal contents and compared to nonscented rice types.[1] The scented cultivars such as basmati considered premium rice are relatively lower yielding but fetch higher prices and hence an important item of national and international trade in India and Pakistan who have rights as geographical indicators for some parts. The specific sets of agroecological conditions of this region promote the typical basmati aroma.

7.2 CROP ECOLOGY

Rice is, in general, a water-loving crop although it is quite versatile. Many types of rice are adapted to various ecologies and varied cultivation practices are popular. In brief, there are five major rice ecologies, namely, (1) irrigated/inundated (53%)—flooded throughout the season, (2) rainfed lowland (31%)—bunded, rainfall dependent variable depth, and duration of standing water, (3) deep-water (3%)—river delta, flooding during a part of the season with water up to 3 m deep, (4) tidal wetlands—near sea coasts and inland estuaries, and (5) upland (13%)—without surface water accumulation and rainfed without water control. Most of the rice is grown in warm-humid agroclimates, hence it is a summer-rainy season crop. The crop duration varies from 2 to 6 months. Plants may be less than a meter to 2–3 m tall and produce many tillers aboveground and large number of fine roots in soil. The roots have large aerenchymatous cavities serving as reservoir of oxygen even if rhizosphere has anaerobic conditions.

Rice is grown in a range of soil types. Clay and clay loam soils are able to hold water for a longer period. In such soils aeration is poor and anaerobic conditions prevail for most part of the season in lowlands. In the upland sandy loams and generally, a practice of puddling is followed to reduce percolation and hold standing more water for longer time and for the first half or more of the season, and aerobic conditions prevail for a major part of the season. River deltas and coastal zones usually have sandy soil with standing water for a part or whole season. The number of rice crops in a year varies with local climate; in warmer regions, two or three crops of rice are grown, one after the other, whereas in areas with well defined winter season, only one crop is grown during the summer–rainy season.

7.3 PLANTING METHODS

Rice is sown using paddy seed. Different methods are practiced for raising a crop:

i) Raising young seeding in a nursery over about 5% of the area, while the main field is tilled, irrigated, and puddled using bullock or tractor-mounted puddlers. In total, 3–4 week seedlings are removed and individually transplanted in the puddled field manually or using rice transplanting machines and standing water is retained for a part or most of the season until a couple of weeks before harvest.
ii) Seedlings are raised and uprooted when they are about 15 days old. The main field is tilled and puddled as in (i) above and young seedlings are thrown onto the muddy soil where they establish.
iii) Seed is soaked and allowed to sprout. The young sprouts are thrown onto the puddled soil.
iv) Dry seed is sown directly in moist soil without standing water. Rainfall is the major source of water and that may provide intermittent dry and wet conditions of varying durations or flooded for long periods depending upon rainfall.
v) In deep-water conditions, rice is sown when standing water is absent or shallow. As the water levels rise and the rice also grows tall with part of the plant above water.

The species and population densities of different plant-parasitic nematodes prevailing in the rhizosphere of rice as well as methods of their management also depend on the kind of growing conditions of rice.

7.4 NEMATODE PARASITES OF RICE

The above-ground parasites of rice are the stem nematode, *Ditylenchus angustus* mainly in lowland and deep water crops and the white-tip nematode, *Aphelenchoides besseyi* found in upland, irrigated, lowland, and deep water crops. Both these nematodes can survive in anhydobiotic conditions in crop residues while *A. besseyi* can survive under the seed-coat of paddy grains. The root-parasites of rice include the root-knot nematodes, *Meloidogyne graminicola, Meloidogyne hainanensis, Meloidogyne lini, Meloidogyne incognita, Meloidogyne javanica, Magnaporthe oryzae, Meloidogyne salasi*, and *Meloidogyne triticoryzae;* cyst nematodes, *Heterodera oryzae, Heterodera oryzicola,* and *Heterodera sacchari;* rice–root nematode, *Hirschmanniella* species., *Hirschmanniella belli, Hirschmanniella gracilis, Hirschmanniella imamuri, Hirschmanniella mexicana, Hirschmanniella mucronata, Hirschmanniella oryzae,* and *Hirschmanniella spinicaudata;* root-lesion nematode, *Pratylenchus indicus, Pratylenchus pseudopratensis,* and *Pratylenchus zeae.* Some other nematodes infecting rice roots include *Criconemoides onoensis, Hoplolaimus indicus, Paralongidorus australis,* and *Xiphinema ifacolum.*[2]

There is an ecological gradient in plant-parasitic nematodes infecting rice in relation to the four major ecological conditions (Table 7.1). Whereas *D. angustus* is limited to high water levels in deep water and lowlands, the *Hirschmanniella* spp. also prevail in irrigated rice. *A. besseyi* and *C. opoensis* and *M. graminicola* have been reported from the wide range of high and low water levels, indicating their greater adaptation, but are more damaging in lower water regimes in irrigated and upland rice. The root-knot nematodes and *M. graminicola* are the most widespread and highly damaging to rice, at times resulting even in total crop failures. Usually, population densities of root-knot nematodes are lower in soils that harbor high population densities of *Hirschmanniella* spp. due to divergent soil ecology preference. The other species of root-knot and cyst nematodes as well as *Hoplolaimus* and *Paralongidorus* prevail in irrigated and upland rice, whereas the loot lesion nematode, *Pratylenchus* spp. are confined to low water regimes in upland rice. Therefore, the alternative crops, cultural practices, and other methods of nematode control or management would depend upon considerations related to the specific rice ecology being addressed. In view of the intercontinental occurrence and predominance, the further discussion is focused more on *M. graminicola*, which was described as late as 1965, this species has attained much economic significance, since it causes serious losses, sometimes total crop failures of rice, the world's most important staple food crop.

TABLE 7.1 Occurrence of Plant Parasitic Nematodes of Rice in Different Rice Ecologies.

Deep water	Lowland	Irrigated	Upland
Ditylenchus angustus			
Hirschmanniella H. belli, H. gracilis, H. imamuri, H. mexicana, H. mucronata, H. oryzae & H. spinicaudata			
Aphelenchoides besseyi, Meloidogyne graminicola			
	Criconemoides onoensis		
		M. oryzae	
		Heterodera elachiata, H. oryzae, H. oryzicola, H. sacchari Meloidogyne hainanensis, M. lini, M. incognita, M. javanica, and M. salasi & M. triticoryzae Hoplolaimus indicus, Paralongidorus australis	
			Pratylenchus brachyurus P. indicus, P. pseudopratensis, P. zeae Xiphinema ifacolum

7.4.1 SYMPTOMS AND BIOLOGY

ABOVEGROUND: The rice crops affected by root-knot nematodes exhibit patches of yellowing, stunted plants with poor tillering, partly of fully sterile panicles (Figure 7.2). The patches are more prominent in rice than in wheat. The patches widen in the area over the years until small fields are fully affected. The wheat crops, especially when early sown, shown stunting and chlorosis (Figure 7.5) in early stages but symptoms become less prominent in later stages, due to lesser nematode activity at low temperature.

FIGURE 7.2 Rice root galled due to root-knot nematodes.

BELOWGROUND (Figures 7.3–7.9): The roots of young plants in nursery, or in field in direct-seeded crops, are infected by the second-stage juveniles (J2), induce syncytium, and tiny galls due to the hyperplasia and hypertrophy typical of root-knot nematodes. The nematode develops, producing third and fourth juvenile stages and adult females and males within the root. The root tips exhibit swelling behind the root tip due to which the root-tip gets hooked or clubbed with many females and other stages in these (Figure 7.10). The galls are also formed on other parts of the primary and secondary roots, which are produced in larger numbers and these are also infected. Thus the

FIGURE 7.4 Egg matrix inside root tissue.

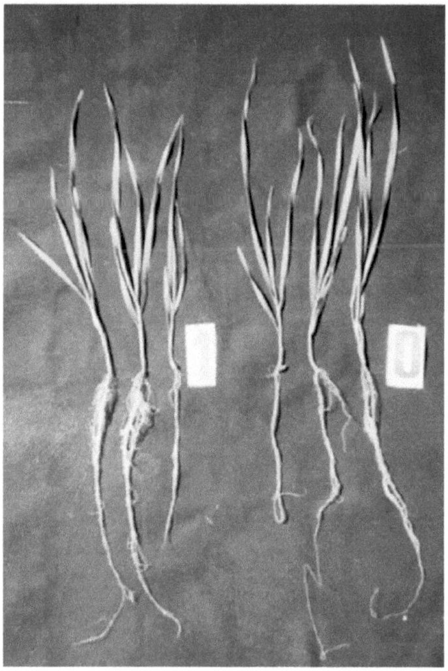

FIGURE 7.5 Wheat seedlings with roots infected by root-knot nematode, *M. triticoryzae*.

FIGURE 7.6 Close up of galled roots of wheat seedlings.

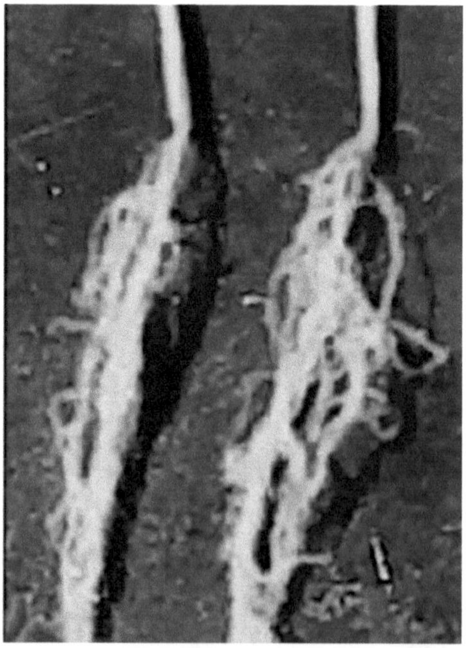

FIGURE 7.7 Close up of a root gall on young wheat seedling with egg matrix emerging on root surface.

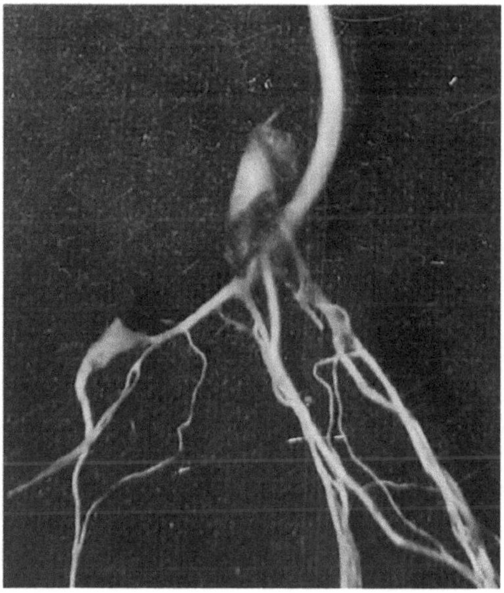

FIGURE 7.8 Growth of wheat plants (Left—in nematode-free soil) as affected by *M. triticoryzae* @ 4 J2 per cm^3 soil.

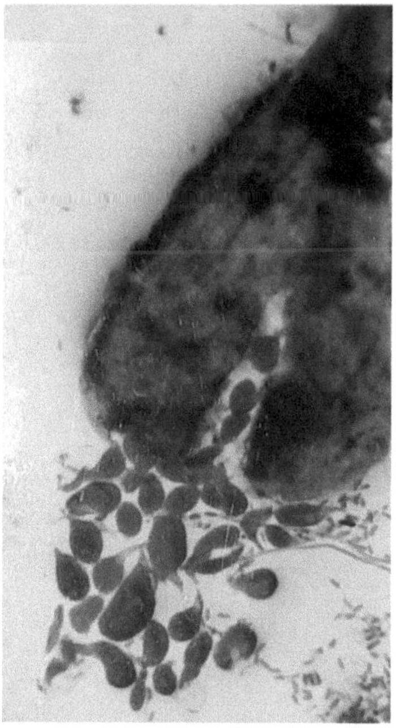

FIGURE 7.9 Large number of females, eggs, juveniles, and males in a single rice root tip gall.

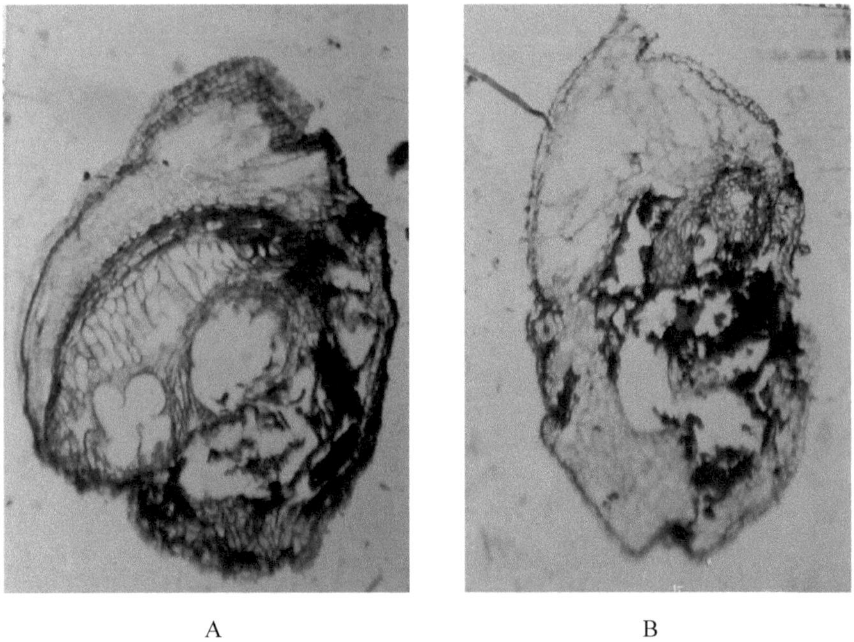

FIGURE 7.10 A. Transverse section of rice root galls showing damage to tissues. B. Transverse section of wheat root galls showing damage to tissues.

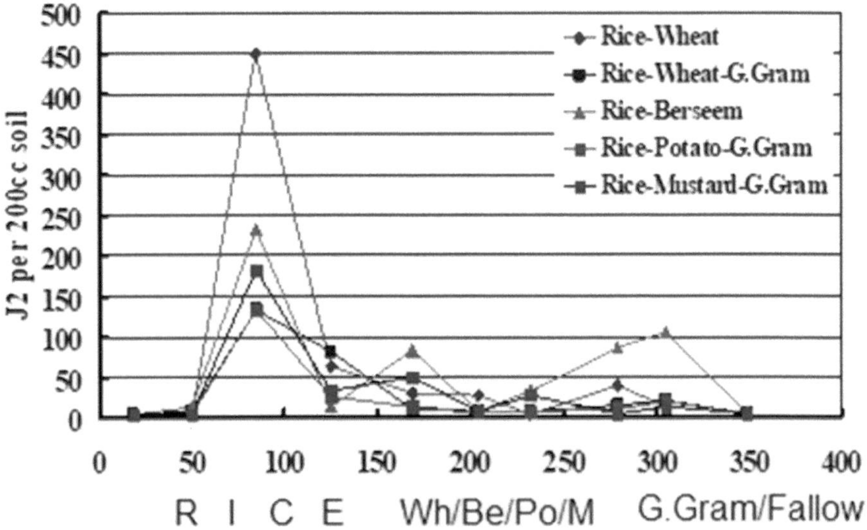

FIGURE 7.11 Population dynamics of *Meloidogyne triticoryzae* under rice-based cropping systems.

periods in dry soil in unhatched eggs embedded in pieces of dried galled roots or soil clods wherein they are dried slowly and undergo a kind of quiescence, that responds to host root diffusates. Gaur et al.[4] reported three types of eggs of *M. triticoryzae,* (1) hatching freely in water, (2) hatching when exposed to host root diffusate, and (3) not hatching for long periods. The eggs produced by the later generations on the senesced host produced larger proportion of the third kind of eggs. Therefore, bioassay using hosts rice and wheat is more effective than normal extraction techniques in detecting these nematodes in soil during off-season summer and winter months. The use of sieving and decanting or elutriators fails to detect infestation at times when free J2 are not present in the soil.

7.4.3 EFFECTS OF SEASON AND POPULATION BEHAVIOR

Even in the good monocot host crops like wheat, that is equally susceptible to rice at temperatures between 25°C and 35°C, the population growth is lesser owing to the lower winter temperatures. Yet, these nematodes are able to complete at least one generation on early sown wheat and one generation on late sown wheat. This is sufficient to maintain population densities to provide enough initial population densities for the following rice crops. This has been found to be one of the major reasons for the fast buildup of population densities and declining productivity of the rice crop in the popular rice–wheat cropping system in parts of Nepal and the Indo-Gangetic region of India. The farmers are not ready to stop cultivating rice for one or two seasons, but if they can replace wheat with mustard or other economically important dicot crops, these nematodes can be managed to a good extent. However, good control of legume and grassy weeds is necessary. The *Phalaris minor*, a problem weed in wheat in winter season and the weeds like *Echinocloa* spp. and wild rice, common in normal rice crops are very good hosts of *M. graminicola* and can serve as indicator plants. The duration of crop rotation is an important consideration. The population density of J2 in soil declines rapidly after harvest of the host crop, but a proportion of unhatched eggs can survive more than a year in slowly dried root galls or soil clods, explaining the adaptation of the rice root-knot nematodes to survive for long periods and due to which 1-year crop rotations are not sufficient.

At similar temperature, many crops are infected, but at relatively lower temperatures in the winter season the infection rates and population growth are much lower even on monocot crops. Singh et al.[5] demonstrated that early sown wheat with more cumulative degree hours above a basal of 10°C during the first 50 days supported more reproduction and residual population for the subsequent rice crop than the late sown wheat. The alteration in the date of sowing of the winter crops with suitable late sown cultivars could, therefore, be helpful in reducing the population growth of *M. graminicola*.

7.4.4 MANAGEMENT

As for most other nematodes, a number of cultural, physical, chemical, biological, and genetic methods are deployed with varying degrees of success. Use of more than one of these methods in an integrated mode is more successful and economical. Since nematodes are not the only pests concerning a farmer, the nematode management methods must also be integrated with the practices against other pests and diseases. Also some methods of reducing nematode infection and damage are also useful against other pests and *vice-versa*. Hence, nematode management practices should be designed as one of the components of the integrated pest management package against nematodes. The problem of root-knot nematode in rice has attained international importance as one of the serious threats to rice and some other staple food crops. Various approaches for its management have been extensively reviewed in the recent decade.[2,6–10]

7.4.4.1 PREVENTION OF SPREAD

The spread of root-knot nematodes within a rice field is rapid because (1) the plantation/sowing practices involve mixing and moving of soil with the plough and the flooding providing a continuum disseminating the infection more rapidly than in other crops, and (2) the introduction of nematode infection with the infected roots of rice seedlings that are grown in infested nursery-bed in the same of other fields. The infested seedlings are not only careers of the nematode but also have lesser vigor than the healthy ones for the very fact that younger stages of the plant have lower tolerance levels. Producing rice seedlings on uninfested or properly treated nursery-beds is very important to have healthier seedlings and prevent dispersal of the nematode. Since the area of nursery-beds is small, the use of biological control agents and nematicides and/or soil solarization can be economical. Proper mapping/ demarcation of the infested areas of the field as evident from the patches of poorly growing crop and some area with mild or no aboveground symptoms encircling the patch in the previous rice season, and isolating these areas with good bunding and avoiding movement of soil and water from these areas to other areas can reduce the spread of infestation. The control measures applied to thus demarcated areas can serve as a precision farming practice to reduce costs considerably, as well as reduce spread.

7.4.4.2 IMPACT OF CROPPING SYSTEMS

In areas where multiple rice crops are grown in a year on the same land, the nematode problems are more serious. Rice farmers are so used to this crop, due to technical familiarity and staple food requirements, that they are reluctant to give up rice farming for a few years or seasons. It is necessary to find good remunerative alternative nonhost crops with local acceptance. Wheat is an equally good host of *M. graminicola* and *M. trticoryzae* under similar temperature conditions. Since the wheat crop season includes lower temperature regimes of 2–3 months, especially during vegetative growth phase, the reproduction and development of these nematodes are slowed down. At least one generation can still be completed depending upon the sowing date of wheat and temperature.[11] Therefore, late sown wheat with suitable varieties may be a possible way to lower the inoculum for the following rice crop. The host range of *M. graminicola* is broad, but not as large as that of *M. incognita* or *M. javanica*. Although graminaceous plants are preferred by *M. graminicola*, the plants belong to several other families, including the popularly grown monocot and dicot crops and commonly occurring monocots and some dicot weeds in other families.[45] Crop rotations with mustard, chickpea, carrots, radish, maize, spinach, blackgram, brinjal, groundnut, tomato, okra, potato, sweet potato, cowpea, sesame, castor, sunflower, marigold, soybean, turnip, garlic, ginger, cauliflower, jute, etc., have been reported helpful in reducing population growth.[11–13] Effect of five cropping systems on the average population density of *M. triticoryzae* is shown in Figure 7.11. Replacing wheat with mustard or potato and inclusion of mungbean and incorporation of organic matter into soil could reduce population densities. There can be differences among cultivars with respect to host status, as well as differences in the infectivity of different geographical populations; therefore, testing is necessary before recommending a crop in a particular area. Win et al.[14] obtained the highest rice yield in the two-season crop rotation sequences green gram–rice–black gram–rice, sesame–rice–cowpea–rice, and sunflower–rice–sesame–rice. These yields were double of the one-season crop rotation sequences and three times of the yield in continuous rice cropping. Crops like sorghum, pearl millet, oats, and common grasses support populations.

Onions in rotation with rice suffer serious infection and damage due to *M. graminicola.* Continuous rice–wheat cropping system gradually leads to increased nematode infestations and this may be one of the important factors, besides nutrient loss, damage to soil structure, decreased soil organic matter, etc., contributing to the gradual decrease in factor productivity over the years. Judiciously planned and executed cropping systems and weed management can be the most practically feasible method for these nematodes.

7.4.4.3 ORGANIC MANURES AND BIOFERTILIZERS

Application of organic manures, or incorporation of crop residues into soil, was found to significantly reduce infection in the subsequent rice crop. One such study was done by growing a short duration (60 days) grain legume crop of mungbean, *Vigna radiata*, harvesting matured pods, and incorporation of green stover into soil reduced root-knot nematode infection in rice.[15] The use of cyanobaterial biofertilizers, the blue-green algae, like *Nostoc muscorum, Aulosira fertilissima*, etc., was also found to reduce infection levels. Gaur[16] found that the culture filtrates of these blue-green algae were nematostatic and inhibited hatching, whereas the extracts of ultrasonicated cells were nematicidal in nature. Similar findings have been made latter for these and some other species of cyanobacteria.[17] Since the distribution of root-knot nematodes in the field is aggregated generally following a negative binomial distribution, it implies that, generally, larger area of the field is uninfested or having low population densities of nematodes, than the area having high population densities above the economic damage threshold. Hence, any nematicides or other control agents applied on the entire field would not have any beneficial effect on the former larger area and thus a waste. On the other hand, a nematode-suppressive biofertilizer will have the dual beneficial effect of nematode suppression and fertility improvement in the entire field. Therefore, if we can select cyanobacteria with strong nematicidal/nematostatic properties in addition to the fertility improvement, this will be doubly beneficial and more profitable.

7.4.4.4 BOTANICALS

Application of oilseed cakes of neem, *Azadirachta indica*, castor (*Ricinus commis*), simarouba (*Simarouba glauca*), etc., @ 2.5–5.0 g/kg soil have been tried and found to reduce *M. graminicola*; however, large quantities required render their use economically unviable. Seed treatment with 8% Econeem resulted in improved root and shoot growth.[18] Soaking the seed or dipping the roots of seedlings in the 2.5–5% aqueous extracts of neem seeds have been reported to reduce infection. Application of organic amendments like biomass of sesame, buckwheat, neem leaves, chinaberry leaves, or chicken manure to soil can reduce the infection and damage.[19] Patil et al.[16] found that rice plants supplied nitrogen in nitrate form were infected more by *M. graminicola* than the plants equal amount of ammoniac form. The aerobic rice, with soil having nitrate abundance, suffers more damage than the crop in submerged soil where ammoniac form is more abundant. Therefore, organic manures wherever available should be applied and mineral fertilizers should be used only as supplements if required.

7.4.4.5 SOIL SOLARIZATION

Rice is a crop of warmer regions that normally have some periods in the year with good sunshine and high temperature. This factor can be very useful in adopting soil solarization of the area intended for preparation of nursery-beds. The soil is tilled, lightly irrigated, if not already moist, and covered with colorless, thin (25–40 μm thick) polyethylene sheets, margins buried in soil, and left for exposure to hot sun for 3–6 weeks. The soil temperature in the top about 15 cm soil layer is raised by about 10°C above the ambient for some hours daily. This reduces nematode populations due to death of soil-borne infective J2 by starvation and or heat; the eggs in root-debris also hatch and the J2 die. There are additional advantages of the reduction of weeds, fungi, and bacteria in the solarized or pasteurized soil.[20] The period of solarization depends on the prevailing day temperature and available solar heat in the area. Thereafter, the polyethylene sheet is removed and rice seeds can be sown to raise seedlings. Usually, this treatment is sufficient to raise nematode-free seedlings. Efficacy can be increased by the application of some systemic organophosphate or carbamate nematicide at the time of sowing or about a week before uprooting the seedlings for transplantation. The presence of systemic nematicide in the tissues of seedlings may protect the seedlings from early infection in the field.

7.4.4.6 WATER REGIMES

The rice root-knot nematodes like M. *graminicola* can thrive in a whole range of water regimes; however, the damage to the crop is lesser in rice crops with soil submerged for the first half or lingers in the season.[21] Since most of the rice of the world is cultivated in the Asia-Pacific region in high water regimes in lowland or irrigated conditions, the pertinent issues of global climate change and water deficits raise increasing demands for reducing (1) water use and (2) methane gas emissions. Therefore, attempts are being made to promote practices like aerobic rice, raised-beds, direct seeded, and upland rice. However, these conditions have been found to accentuate the infestation and damage due to rice root-knot nematodes.[5] Nie et al.[22] also reported decline in yield after continuous cultivation of aerobic rice. It is necessary to grow only highly resistant varieties of rice in aerobic conditions[23] or under the system of rice intensification.

7.4.4.7 TILLAGE PRACTICES

Cultural practices can play a big role in managing the root-knot nematodes in rice. There have been a number of trials on reduced or zero tillage and raised-bed planting as a part of the conservation agriculture drive to save soil structure, water, and energy. However, the root-knot nematode infection and population growth as well as crop damage due to nematodes are more compared to the tradition practice of summer ploughing and puddling before transplanting.[24] The damage is more in direct-seeded crops in moist nonsubmerged soil, since the young seedling roots are immediately infected. Therefore, nematode management either by resistant varieties, induced resistance, organic amendments, and soil solarization with or without nematicides is essential if reduced tillage practices advocated as part of conservation agriculture are to be adopted.

7.4.4.8 BIOLOGICAL CONTROL

A number of mutualistic and endophytic bacteria and fungi have been found to inhibit nematode infection and damage to the crop. *Psedomonas fluorecens*, generally regarded as plant growth promoting bacteria (PGPR) and nonpathogenic strains of *Catenaria anguillulae*, *Fusarium oxysporum*, *Trichoderma harzianum*, *Gliocladium virens*, etc., have been reported by many to suppress root galling and nematode population densities in many crops including rice by several workers. Anitha and Rajendran[25] reported 60% reduction in *M. graminicola* population density by seed treatment with *P. fluorecens* and its soil application. However, most reports are of experimental trials and demonstrations, the use of biological control agents by farmers on large scale is rare.

7.4.4.9 NEMATICIDES

A number of nematicides have been tried to control rice root-knot nematodes with variable success. Methyl bromide was used to assess yield potential, almost eradicated the nematode and rice yield by three times.[12] Carbofuran and many other nematicide-insecticides have been found to improve yields. Nursery-bed treatment with carbofuran at 0.3 g a.i./m^2 and its field application at 1 kg a.i./ha 40 days after transplanting has been found effective. Soaking of rice seeds in 0.1% solution of carbosulfan or 0.2% phosphomidan, for up to 12 h reduced egg mass production and root and increasing grain yield in many trials in India.[26,27] Volatile (fumigants) and nonvolatile nematicides applied to soil and/or for seedling root dips or seed soaking can reduce nematode infections and population growth. Seeds can be treated with carbofuran or carbosulfan. Systemic nematicides such as oxamyl or fensulfothion, phorate, carbofuran, and DBCP can be useful as soil treatments or for seedling root dips. Telone (1, 3-dichloropropene), a fumigant, can be injected into the soil before the crop is planted. The effects of chemical treatment last only one season. *Meloidogune graminicola* completes a life cycle in about 22–25 days and fecundity is high, hence the high population densities are resumed in a few months. The nematicide applied in nursery-bed is unlikely to result in pesticide residues in grain due to the long period between transplanting and harvest. Most nematicides have been banned due to local and global regulations, including the UN protocol.[28] Before recommending the use of any nematicide, one must enquire if it is officially permitted to be used in a particular state or country. Many importing countries put up trade barriers with very low MRLs, resulting in the rejection of exported rice. Therefore, alternative methods are necessary to reduce the losses due to serious pests like *M. graminicola*. Application of nematicides limited to the treatment of seedbeds, seed, or seeding roots is not likely to result in pesticide residues in the grain.

7.4.4.10 INTEGRATION

Anyone of the above-said practices is individually only partially effective in reducing the nematode infection and averting the crop damage. Many studies have been carried out to achieve better management by combining two or more practices in integrated approach. Combinations of crop rotation with nonhost crops and application of nematicides, or application of biological control agents and PGPRs such as *Pseudomonas fluoresecns* or *T. harzianum* have been reported more effective than either of these alone. The treatment of *P. fluorescens* or *T. viride* @20 g/m² + Carbofuran 3G has been found to improve grain yield and reduced nematode population density.[8,9] Combination of deoiled seed cakes and PGPRs/biocontrol agents has also been found effective in some experimental trials. The organic materials provide good substrates for the microorganisms and enable better colonization of the rhizosphere. In a number of trials carried out by the author, the integration of soil solarization of nursery-beds with these practices in main fields was quite effective and economically viable in north Indian conditions. Soil solarization of area for nursery-beds and application of nematicide was very effective in producing nematode-free rice seedlings.[29]

Gaur[15] gave an integrated package to manage root-knot nematodes in rice–wheat cropping system:

i) The area of the main field of rice is much larger than the nursery-beds; therefore, soil solarization of the entire area is not cost-effective.
ii) The field may be ploughed a 2–3 times in summer and exposed to solar heat and desiccation OR grow a short duration leguminous crop, for exmaple, mungbean, and plough back the depodded green haulm into soil as mentioned above, before preparing the field for planting rice.
iii) During the same period, the area intended to be used to raise rice seedlings should be lightly irrigated, ploughed, and covered with 40 μm thick colorless polyethylene sheet, with edges burrowed, exposed to summer heat for 3–6 weeks.
iv) After removing polyethylene sheet, apply carbofuran @ 1 g a.i./ha or deoiled neem seed cake @ 50 g/m² in infestation is high and thereafter sow the seed.
v) After 2–3 weeks prepare main field soil with water (puddling) and transplant nematode-free rice seedlings.
vi) In addition to these, replacing wheat crop in winter season with mustard of chickpea or some other nonhost crop and control of grassy weeds in alternate years, can minimize root-knot nematode damage in rice.

These practices fit in very well with the agronomy and agroecology of the region under rice–wheat cropping systems. These are adjustable with and add to the efficacy of other methods of management of weeds and other pests and pathogens of rice.

In the deep-water and lowland rice areas, the above practices may not be feasible. Crop rotations with *Crotalaria spectabilis*, *Crotalaria juncea*, etc., have been found to reduce root-knot nematodes in subsequent rice. Farmers need to be guided against taking multiple successive crops of rice on the same field, a common practice in eastern India, Bangladesh, Myanmar, and other Asian countries.

7.4.4.11 HOST PLANT RESISTANCE

There are two approaches to imparting resistance in the crop plant against nematodes and other pests or pathogens: (1) physiological induced resistance by application of certain regulatory or inhibitory chemicals, and (2) Genetic resistance by introducing certain genes from related plants by traditional plant breeding, or from unrelated sources as developing transgenic or genetically modified plants.

a. Induced Resistance

Increasing the levels of jasmonic acid, ethylene and salicylic acid can increase tolerance to nematodes. On the other hand, rise in abscisic acid (ABA) increases tolerance to abiotic stresses like drought, cold, and salinity, but decreases the tolerance to root-knot nematodes.[30] Thus, lower ABA levels may result in increased resistance to pathogens. The basal plant defense system, for example, potentiated H_2O_2 generation, lignin formation, and callose deposition were noted in the galls of BABA-treated plants. However, defense activators may reduce plant growth, because of the trade-off between growth and defense.[31] Priming using thiamine root dip treatment[32] or silicon[33] decreased infection.

b. Resistant Varieties

The ideal method to manage root-knot nematodes would be to grow nematode-resistant varieties. A lot of effort has been made in this direction and some resistant varieties have been developed, especially at the IRRI, Philippines, ICAR-National Rice Research Institute, India, and some Institutes in China. The use of resistant genotypes such as ARC-12620, INRC-2002, CR-94-CCRP-51, Khao Pahk Maw, and LD2 which have shown resistance against *M. graminicola* can be taken up in the affected areas. Some related rice species such as some accessions of

Oryza longistaminata Chev. et Roehr. and *O. glaberrima* Steud are also resistant.[12,34] Unfortunately, several genes causing an interspecific *O. sativa* and *O. glaberrima* sterility barrier were also identified and, among them, the S1 locus had the strongest effect.[35] The hybrid nonviability and sterility may occur in crosses with *O. glaberrima.* However, successive backcrosses, followed by selfing and produced fertile plants derived from O. *sativa* and *O. glaberrima* crosses.[36] Certain semiochemicals including small lipophilic molecules in root exudates could be responsible for the differential attraction and parasitism.[6] The resistant varieties of *Oryza glabberima* have lesser root and stele diameter and cortical thickness compared to the susceptible *O. sativa* cultivars.[37] The incompatible reaction of *O. glabberima* was attributed to the plant hormone pathway involving jasmonates; the effecters from the infecting nematodes may suppress plant defense.[38]

Research is also underway on the use of molecular techniques, for example, (1) development and use of molecular markers for nematode resistance, (2) marker-assisted plant breeding, (3) RNAi techniques to induce resistance, and (4) development of transgenic nematode-resistant rice germplasm or use of CRISPR-Cas9-based targeted genome editing.[39] The inhibition of nematode infestation has been successful using the ectopic expression of cystatin protease inhibitors (reviewed by Lilley et al.[40]). In transgenic approach, overexpression of a rice thionin gene OsThi7 of the PR13 gene family did not reduce galling but reduced number of females and reproduction in susceptible rice cultivar, Nipponbare.[31] This also reduced root colonization of *Pythium*, another pathogen associated with "soil sickness" syndrome.[41] Thionins can be toxic to mammalian cells if its expression is not restricted to root.[42] The success so far has been limited but there are possibilities of success, however, public resistance to genetically modified crops is a big question in many producing and importing countries.

7.5 CONCLUSION

The awareness of the economic significance of managing root-knot nematodes in rice has increased and these nematodes are now recognized as one of the nationally important problems in rice cultivation in most rice-growing countries. This problem will increase further with the decrease in available water and organic matter content of soil. The need to grow more food from limited land will increase cropping intensity and further accentuate this nematode problem. A number of cultural practices through management of cropping system, crop rotation, tillage of soil, water and organic matter management, soil solarization, and on a limited scale use of rhizosphere bacteria and fungi as biocontrol agents or plant growth promotion improving tolerance of the plant, as well as a few resistant varieties have become available. The previous decade has seen much research effort to develop resistant varieties with a shift toward the application of biotechnology tools for marker-assisted plant breeding, RNAi, CRISPER-cas, and transgenic approaches. It is expected that more potent biological control agents and dependable resistant varieties will become available. Hopefully, public resistance to genetically modified crops will subside. However, the traditional cultural practices of managing soil–water-cropping system, soil solarization, and growing any available resistant varieties will ever remain the widely practiced practical method at the farmer's level.

KEYWORDS

- **rice-wheat cropping system**
- ***Meloidogyne graminicola***
- ***Meloidogyne triticoryzae***
- **nematode management**
- ***Oryza sativa***
- **rice**
- **wheat**

REFERENCES

1. Maraval, I.; Sen, K.; Agrebi, A. Meenut, C.; Morere, A.; Boulanger, R.; Gay, F.; Mestres, C.; Gunata, Z. Quantification of 2-acetyl-1-pyrroline in Rice by Stable Isotope Assay Through Headspace Solid-Phase Microextraction Coupled to Gas Chromatography-Tandem Mass Spectrometry. *Analytica Chimica Acta.* **2010,** *675,* 148–155.
2. Peng, D-L.; Gaur, H. S.; Bridge, J. Nematode Parasites of Rice. In *Plant Parasitic Nematicides in Subtropical and Tropical Agriculture*; Sikora, R. A., Coyne, D., Hallman, J., Timper, P., Eds., 3rd ed.; CABI International & USDA, 2018; pp 120–162, 876 pp.
3. Gaur, H. S.; Mende, N. von; Perry, R. N. Differentiation of Two Groups of Species of the Genus *Meloidogyne* by Polymerase Chain Reaction and Restriction Fragment Length Polymorphism of Ribosomal DNA. *Afro-Asian J. Nematol.* **1996,** *6,* 50–54.

4. Gaur, H. S.; Perry, R. N.; Beane, J. The Influence of Root Diffusate, Host Age and Water Regimes on Hatching of the Root-Knot Nematode, *Meloidogyne triticoryzae*. *Nematology* **2000,** *2*, 191–199.
5. Singh, I.; Gaur, H. S.; Briar, S. K.; Sharma, S. K.; Sakhuja, P. K. Role of Wheat in Sustaining *Meloidoigyne graminicola* in Rice-Wheat Cropping System. *Int. J. Nematol.* **2003,** *13*, 79–86.
6. Dutta, T. K.; Ajoy, K.; Ganguly, Gaur H. S. Global Status of Rice Root-Knot Nematode, *Meloidogyne graminicola. Afr. J. Microbiol. Res.* **2012,** *6*, 6016–6021.
7. Mantelin, S.; Bellafiore, S.; Kyndt, T. *Meloidogyne graminicola*: A Major Threat to Rice Agriculture. *Mol. Plant Pathol.* **2017,** *18*, 3–15.
8. Narasimhamurthy, H. B.; Ravindra, H.; Sehgal, M. Management of Rice Root-Knot Nematode, *Meloidogyne graminicola. Int. J. Pure Appl. Biosci.* **2017,** *5*, 268–276.
9. Pankaj; Sharma, H. K.; Prasad, J. S. The Rice Root-Knot Nematode, *Meloidogyne graminicola*: An Emerging Problem in Rice-Wheat Cropping System. *Indian J. Nematol.* **2010,** *40*, 1–11.
10. Pankaj; Sharma, H. K.; Singh, K.; Lal, J. Management of Rice Root-Knot Nematode, *Meloidogyne graminicola* in Rice (Oryza *sativa*). *Indian J. Agric. Sci.* **2015,** *15*, 701–704.
11. Gaur, H. S.; Sharma, S. N. Influence of Date of Sowing Wheat on the Population Growth of Plant Parasitic Nematodes, *Meloidogyne triticoryzae* and *Tylenchorhynchus mashhoodi* and Saprozoic Nematodes and Relationship with Temperature. *Indian J. Nematol.* **2000,** *30*, 117–123.
12. Soriano, I. R. S.; Reversat, G. Management of *Meloidogyne graminicola* and Yield of Upland Rice in South-Luzon, Philippines. *Nematology* **2003,** *5*, 879–884.
13. Win, P. P.; Kyi, P. P.; Maung, Z. T. Z.; Myint, Y. Y.; Cabasan, M. T. N.; Waele, D. D. Host Status of Rotation Crops in Asian Rice-Based Cropping Systems to the Rice Root-Knot Nematode *Meloidogyne graminicola. Trop. Plant Pathol.* **2016,** *41*, 312–319.
14. Win, P. P.; Pyine, P. K.; Maung, Z. T. Z.; Myint, Y. Y.; Cabasan, M. T. N. Crop Rotation Sequencing to Minimize Yield Losses of Summer Irrigated Lowland Rice in Myanmar Caused by the Rice Root-Knot Nematode, *Meloidogyne graminicola. Int. J. Pest Manage.* **2020,** *66*, 319–331.
15. Gaur, H. S. *Root-Knot Disease of Rice and Wheat: Problem and Management*; Technical Bulletin, Indian Agricultural Research Institute: New Delhi, 2003; 20 p.
16. Gaur, H. S. Studies on the Hatch Inhibition and Nematicidal Action of Two Cyanobacteria and a Green Alga. *Ann. Plant Protection Sci.* **1995,** 3, 158–163.
17. Holajjer P.; Kamra, Anju; Gaur, H. S.; Manjunath, M. Potential of Cyanobacteria for Biorational Management of Plant Parasitic Nematodes: A Review. *Crop Protection* **2013,** *53*, 147–151.
18. Das, P.; Deka, B. C. Efficacy of Neem Based Pesticides Against *Meloidogyne graminicola* as Seed Treatment. *Indian J. Nematol.* **2002,** *32*, 204–205.
19. Dangal, N. K.; Sharma-Poudyal, D.; Shrestha, S. M.; Adhikari, C.; Duxbury, L. M.; Lauren, J. G. Evaluation of Organic Amendments Against Rice Root-Knot Nematode at Seedling Stage of Rice. *Nepal J. Sci. Tech.* **2008,** *9*, 21–27.
20. Gaur, H. S.; Perry, R. N. The Use of Soil Solarization for Control of Plant Parasitic Nematodes. *Nematological Abstracts* **1991,** *60*, 153–167.
21. Padgham, J. L.; Duxbury, J. M.; Mazid, A. M.; Abawi, G. S.; Hossain, M. Yield Loss Caused by *Meloidogyne graminicola* on Lowland Rainfed Rice in Bangladesh. *J. Nematol.* **2004,** *36*, 42–48.

22. Nie, L. X.; Xiang, J.; Peng, S. B.; Bouman, B. A. M.; Huang, J. L.; Cui, R.; Visperas, R. M. Alleviating Soil Sickness Caused by Aerobic Monocropping: Responses of Aerobic Rice to Fallow, Flooding and Crop Rotation. *J. Food, Agric. Environ.* **2009,** *7*, 723–727.
23. Waele. D. De.; Das, K.; Zhao, D.; Tiwari, R. K. S.; Shrivastava, D. K.; Vera-Cruz, C.; Kumar, A. Host Response of Rice Genotypes to the Rice Root-Knot Nematode (*Meloidogyne graminicola*) Under Aerobic Soil Conditions. *Arch. Phytopathol Plant Protection* **2013,** *46*, 670–681.
24. Chandel, S. T.; Gaur, H. S.; Alam, M. M. Effect of Tillage and Water Management on the Population Behaviour of *Meloidogyne triticoryzae.* Root-Knot Nematode in Rice Crop. *Arch. Phytopathol. Plant Protection* **2002,** *35*, 195–200.
25. Anitha, B.; Rajendran G. Integrated Management of Root-Knot Nematode Meloidogyne graminicola Infecting Rice in Tamil Nadu. *J. Plant Protection Environ.* **2005,** *2*, 108–114.
26. AICRPNematodes. 2020. https://www.iari.res.in/index.php?option=com_content&view =article&id =184& Itemid=932
27. Deka, B. C.; Das, P. Efficacy of Certain Chemicals as Seedling Root-Dip Treatment Against Rice Root-Knot Nematode, *Meloidogyne graminicola. J. Agric. Sci. Soc. North East India* **2002,** *15*, 112–114.
28. MBTOC. Report of the Methyl Bromide Technical Options Committee. Montreal Protocol on Substances That Deplete the Ozone Layer. MBTOC Assessment Report, UNEP, 2010; 387p.
29. Chandel, S. T.; Gaur, H. S.; Alam, M. M. Effect of Soil Solarization and Nematicidal Treatment of Nursery Beds and Summer Ploughing of Main Field on Root-Knot Nematodes and Performance of Rice Crop. *Arch. Phytopathol. Plant Protection* **2002,** *35*, 291–297.
30. Asselbergh, B.; De-Vleeschauwer, D.; Hofte, M. Global Switches and Fine-Tuning-ABA Modulates Plant Pathogen Defense. *Mol. Plant-Microbe Interact.* **2008,** *21*, 709–719.
31. Ji, H.; Kyndt, T.; He, W.; Vanholme, B.; Gheysen, G. b-Aminobutyric Acid-Induced Resistance Against Root-Knot Nematodes in Rice is Based on Increased Basal Defense. *Mol. Plant Microbe Interact.* **2015,** *28*, 519–533.
32. Huang, W. K.; Ji, H. L.; Gheysen, G.; Kyndt, T. Thiamine-Induced Priming Against Root-Knot Nematode Infection in Rice Involves Lignification and Hydrogen Peroxide Generation. *Mol. Plant Pathol.* **2015,** *17*, 614–624.
33. Li-Ping, Z.; De-Liang, P.; Xu-Li, W.; Ling-An, K.; Huan, P.; Shi-Ming, L.; Ying, L.; Wen-Kun, H. Priming Effect of Root-Applied Silicon on the Enhancement of Induced Resistance to the Root-Knot Nematode *Meloidogyne graminicola* in Rice. *BMC Plant Biol.* **2018,** *18*, 50.
34. Soriano, I. R. S.; Prot, J. C.; Matias, D. M. Expression of Tolerance for *Meloidogyne graminicola* in Rice Cultivars as Affected by Soil Type and Flooding. *J. Nematol.* **2000,** *32*, 309–317.
35. Garavito, A.; Guyot, R.; Lozano, J.; Gavory, F.; Samain, S.; Panaud, O.; Tohme, J.; Ghesquie`re, A.; Lorieux, M. A Genetic Model for the Female Sterility Barrier Between Asian and African Cultivated Rice Species. *Genetics* **2010,** *185*, 1425–1440.
36. Ghesquie`re, A.; Sequier, J.; Second, G.; Lorieux, M. First Steps Towards a Rational Use of African Rice, *Oryza glaberrima*, in Rice Breeding Through a 'Contig Line' Concept. *Euphytica.* **1997,** *96*, 31–39.
37. Cabasan, M. T. N.; Kumar, A.; Bellafiore, S.; De Waele, D. Histopathology of the Rice Root-Knot Nematode, *Meloidogyne graminicola*, on *Oryza sativa* and *O. glaberrima. Nematology* **2014,** *16*, 73–81.

38. Kyndt, T.; Nahar, K.; Haegeman, A.; De Vleesschauwer, D.; Hofte, M.; Gheysen, G. Comparing Systemic Defence Related Gene Expression Changes upon Migratory and Sedentary Nematode Attack in Rice. *Plant Biol.* **2012,** *14*, 73–82.
39. Haegeman, A.; Bauters, L.; Kyndt, T.; Rahman, M. M.; Gheysen, G. Identification of Candidate Effector Genes in the Transcriptome of the Rice Root Knot Nematode *Meloidogyne graminicola. Mol. Plant Pathol.* **2013,** *14*, 379–390.
40. Lilley, C. J.; Kyndt, T.; Gheysen, G. Nematode Resistant GM Crops in Industrialized and Developing Countries. In *Genomics and Molecular Genetics of Plant– Nematode Interactions*; Jones, J. T., Gheysen, G., Fenoll, C., Eds.; Springer: Heidelberg, 2011; pp 517–541.
41. Van Buyten, E.; Hofte, M. *Pythium* Species from Rice Roots Differ in Virulence, Host Colonization and Nutritional Profile. *BMC Plant Biol.* **2013,** *13*, 203.
42. Stec, B. Plant Thionins—The Structural Perspective. *Cell Mol. Life Sci.* **2006,** *63*, 1370–1385.

CHAPTER 8

Defense Mechanisms of Maize: Implications for Integrated Pest Management

P. LAKSHMI SOUJANYA[1], J. C. SEKHAR[1*], S. B. SUBY[2], K. R. YATHISH[1], K. SRINIVASA BABU[3], and G. SHYAM PRASAD[3]

[1]*Winter Nursery Centre, ICAR-Indian Institute of Maize Research, Rajendranagar, Hyderabad-500030, India*

[2]*ICAR-Indian Institute of Maize Research, Pusa Campus, New Delhi-110012, India*

[3]*ICAR-Indian Institute of Millets Research, Rajendranagar, Hyderabad-500030, India*

[*]*Corresponding author. E-mail: jcswnc@rediffmail.com*

ABSTRACT

Maize (*Zea mays* L.) is the most demanded cereal crop globally highest genetic yield potential, grown in diverse environmental conditions. As the damage caused by insect pests is economically significant in maize with effective and sustainable pest management strategies are the need of the hour. Plant resistance is a complex phenomenon involving several traits that determine the categorization of resistance/susceptibility to insect pests. Maize contains a wide array of constitutive and inducible defense mechanisms including structural and biochemical traits to protect from insect pests attack. Maize produces Herbivore-induced plant volatiles ($HIPV_S$) and specific volatile organic compounds to activate systemic plant defenses, in neighboring plants. In maize, several phytohormones and their derivatives

Integrated Pest Management in Diverse Cropping Systems. Devendra Pal Singh, PhD (Ed.)

play an important role in modulating plant defenses. In this book chapter the role of morphological characters, biochemical constituents associated with maize resistance to insect pests, and induced defenses by various external stimuli have been discussed. Besides, the utilization of HIPVs in attracting natural enemies, herbivore-associated molecular patterns, phytohormones in plant signaling, and the role of endosymbionts are deeply elaborated. Furthermore, implications of maize IPM, and novel strategies for enhancing maize defenses are highlighted.

8.1 INTRODUCTION

Maize is a potential cereal to ensure food security and also for doubling farmers' income. It is grown over an area of 192 m ha in 165 countries across the world with a production of 1186.86 mt. India represents nearly 4% of the global maize area, 2% of global production. It requires a moderate climate for its growth and gives higher yield in a short period. Maize is widely used as poultry feed (47%), fodder (13%), food (13%), starch industry (14%), processed food (7%), export, and other purposes (6%). Maize is a source for many value-added products including specialty corn such as QPM, baby corn, popcorn, and sweet corn. The production target of maize is 50–60 mt by 2025. At present, maize is grown in an area of 9.2 m ha with a production of 31.51 Million Tonnes (2021). Approaching this target is very much challenging as insect pests is one of the major reasons for low production in maize. In tropical and subtropical regions, the climate is highly favorable for the rapid multiplication of a wide range of insect pests. Maize is attacked by about 139 insect pests during its growth.[1] Spotted stem borer (SSB) *Chilo partellus* (Swinhoe) and pink stem borer (PSB) *Sesamia inferens* Walker and shoot fly *Atherigona* spp. are the predominant insect pests of maize till 2018. However, during May 2018, the invasion of fall armyworm (FAW) (*Spodoptera frugiperda*) was observed for the first time on maize in Karnataka. FAW has invaded major maize growing states of the country within 1 year.[2]

The SSB attacks maize mostly during *the kharif*/rainy season causing 26–80% yield losses in different agro-climatic regions of India,[3] whereas PSB is the major pest during *rabi*/winter season. The favorable conditions for rapid multiplication are warmer temperatures, and high relative humidity, and abundance of host plants. The extensive feeding by the larvae results in "dead heart" formation.[4] The FAW has become the predominant pest of maize causing substantial damage resulting in 73% yield losses.[5] Early instars scrape the upper epidermis of leaves resulting in papery spots. The damage by late instars (fourth instar onwards) results in ragged holes with the presence of large amounts of fecal pellets in the plant whorl. During the reproductive stage, the larvae infest tassels, cobs, and feed on kernels making its control very difficult.

Even though maize insect pests can be controlled by chemical insecticides extensive use is not recommended as it leads to ecological imbalance human health hazards and toxicity to nontarget organisms. In this sense, Host Plant Resistance (HPR) is one of the best alternatives for sustainable pest management. From the historical perspective, HPR to manage insect pests is highly successful in several crops. Moreover, plant resistance is an integral component of IPM in maize. Plants try to escape from insect pests damage and at the same time suitable adaptations are developed by insect pests too. This intricate interaction results in the development of morphological and biochemical plant defensive traits. Understanding the plant defensive traits is necessary for the development of safe and sustainable integrated pest management (IPM) strategies. Keeping in view of the discussed concerns, the current state of knowledge of various direct and indirect defense mechanisms and induced defenses, phytohormones, involved in maize resistance has been discussed for sustainable management of insect pests of maize.

8.2 MECHANISMS OF PLANT DEFENSE

The plant defense responses are categorized into direct and indirect plant defenses as per the trophic level. Irrespective of external stimuli the constitutive plant defense is present in plants whereas the induced defense is triggered by elicitor application, oviposition, and insect attack.[6,7] Direct defenses affect the herbivore's performance or behavior while indirect defenses contribute to traits affecting the herbivore's performance or by the release of attractants through the beneficial insects.[8]

8.2.1 DIRECT DEFENSES

Morphological and biochemical defenses in plants are considered as direct defenses and play a major role in imparting resistance to insect pests of maize. To tolerate insect pests attack, plants activate various structural, biochemical, and molecular mechanisms. Morphological traits are initially utilized by the plants to withstand insect damage. However, insect growth and development are directly affected by biochemical-based defense.[9] Secondary metabolites play an important role in protecting plants from subsequent insect attack.

8.2.1.1 MORPHOLOGICAL TRAITS

Various researchers identified structural and anatomical plant traits imparting resistance to maize.[10–13] Morphological/structural traits include trichomes, spines, thorns, epicuticular wax, tissue toughness, which play predominant role in hindering insect pests attack. The structures of trichomes are unicellular or multicellular, simple or branched, scale-like or glandular, and cover both surfaces of leaves, stems, shoots, and roots. The movement of insects on the plant surface is interfered by trichomes and insects can mechanically damaged. The shape of the trichome varies from the hooked, straight, spiral, branched, or unbranched, can be glandular or nonglandular. Glandular trichomes secrete secondary metabolites, which form a combination of structural, chemical defenses.

In the past, several researchers reported that trichomes inhibit the oviposition of SSB.[14–16] The presence of trichomes on the upper leaf surfaces of the resistant genotypes resulted in oviposition reduction by SSB.[17,18] In maize, an important morphological trait resistant to FAW is leaf toughness.[19] It was reported that the thickness of the cuticle and the epidermal cell wall is the most consistent difference between the resistant and susceptible maize inbreds.[20] The cell wall complex of the resistant inbreds derived from exotic germplasm originating in the Caribbean islands and primarily Antigua was 1.7 times thicker than that of the susceptible inbreds. Morphological traits attributed to FAW resistance include higher amounts of leaf cuticular lipids,[21–23] thicker cell wall complex of epidermal layer[20] inhibits herbivore attack. Apart from trichome densities, leaf toughness[16,24] also hinders the oviposition and the development of SSB.

8.2.1.2 BIOCHEMICAL TRAITS

Several biochemicals in maize impart resistance/susceptibility to insect pests.[25,26] Different categories of plant chemicals imparting resistance to maize insect pests have been elaborated.

8.2.1.3 PLANT NUTRIENTS

Resistant maize genotypes contain lower reducing and total sugars and were positively related to infestation by stem borer.[27,28] It was reported that protein content was positively correlated with stem borer susceptibility.[28] Further, it was observed significantly low protein content (13.21%) in resistant genotype (X2(Y) 5089) of maize when compared to highly susceptible CM 202 (16.42%). It was observed significant low carotenoid, crude protein, nitrogen, leaf chlorophyll, and moisture contents in resistant maize genotypes.[25] Different nutrient concentrations in stem tissues of maize genotypes, CM 137, and HY 464 impart resistance/susceptibility to SSB.[29] Significantly higher stem sugar content was observed in susceptible maize genotypes.[24] It was observed that nitrogen, protein, total sugars, reducing sugars, and phosphorus were negatively correlated with the total life span of SSB.[30] Higher defensive proteins were reported in Maize Antiguan inbred Mp708 resistant to FAW.[31] Sugars present in maize genotypes play role in imparting resistance to SSB.[32] A higher concentration of amino acids, namely, aspartic acid and tyrosine, abundance of crude and acid detergent fiber imparts resistance to FAW.[33]

8.2.1.4 SECONDARY METABOLITES

Plants produce a large and wide variety of organic compounds known as secondary metabolites, which can either be constitutive or triggered due to herbivore attack. Different secondary metabolites include Benzoxazinoides, phenolics, flavonoids, terpenoids, alkaloids, etc. impart resistance to insect pests. Secondary metabolites also improve the plant fitness apart from defending them from different biotic stresses. Phenolic compounds, hydroxamic acids are the most common group of secondary metabolites and play a major role in imparting resistance to insect pests.[34–37] Among phenolic compounds, maysin imparts resistance to FAW, Corn earworm *Helicoverpa zea*.[35,38] through cell wall fortification, lignifications, phenolic constituents, ferulic acid, and *p*-coumaric acid provide resistance to *Sesamia nonagrioides*.[39] Cell wall-bound *p*-CA, FA also contribute resistance to *Sesamia inferens* infesting maize.[40] However, apart from phenolics, a wider array of biochemical traits may play role in imparting resistance to PSB. DIMBOA, the major hydroxamic acid (2, 4-dihydroxy-7-methoxy-(2H)-1, 4-benzoxazin-3-(4H)-one),[34] its degradation products, MBOA, HMBOA, have deleterious impact on *Ostrinia nubilalis*,[41] corn borer *Sesamia nonagrioides*,[42] while minimum effect was observed on FAW and southwestern corn borer.[43] In *Spodoptera exigua*, *Spodoptera littoralis*, *S. frugiperda*, and *Diatraea grandiosella*, resistance have been associated with induced benzoxazinoid accumulation, methylation of DIMBOA-Glc to produce HDMBOA-Glc in response to larval feeding.[44–47] It was reported that diterpenoid phytoalexins exhibited significant antifeedant activity against European corn borer.[48]

For corn earworm C-glycosyl flavone maysin and chlorogenic acid present in maize impart resistance. Total phenols and tannin content were higher in maize lines resistant to SSB.[49] The cell wall components including cellulose, hemicellulose, and lignin also exhibit resistance to stem borers.[50,51] Higher levels of JA,[52] (E)-β-caryophyllene JA responsive indirect defense volatiles were observed in genotypes resistant to FAW.[53] Protease inhibitors (PIs) are important plant defense proteins against insect herbivory.[54] By binding to the digestive enzymes in the insect gut, the PIs decrease protein digestion of the insects. *Maize insect resistance1* (*mir1*) gene was identified in Mp708 and which encodes a cysteine (Cys) proteinase and Mir1-Cys protease and which disrupts the peritrophic matrix of *FAW*.[55] *In FAW*, the synthesis of new proteases and degradation of PIs were observed.[56] The Mir1-Cys protease is also toxic to corn leaf aphid *Rhopalosiphum maidis* when fed with an artificial diet containing Mir1-Cys protease.[57] Significant dose-dependent inhibition of SSB growth and development to recombinant CanPI-7 and PIs from *Capsicum annuum* (Pin-II type PIs) was observed.[58] Mir1-Cys protease accumulation can be reduced by blocking ETH synthesis or perception in Mp708 resulting in susceptibility of maize to *FAW* and *corn leaf aphid*.[57,59] The Poly phenol oxidases are important in plant defense, regulate feeding, growth, and development of insects. Due to insect infestation or wounding the PPO genes are differentially induced by signaling molecules. Expression of tobacco anionic peroxidase in maize resulted in 100% mortality of *Ostrinia nubilalis*. Besides, sublethal effects were observed in FAW.[60] Enhanced defense mechanisms including genes encoding jasmonic acid (JA) biosynthetic enzymes, proteinase inhibitors, and defense-related transcription factors were observed due to Mediterranean corn borer attack in maize.[61] Similarly, an increase of 1, 3-benzoxazine-4-ones, phospholipids, *N*-hydroxycinnamoyltyramines and azelaic acid and tryptophan was observed on maize seedlings due to leaf worm infestation.[62] It was observed that elevated levels of the hormones JA, its derivative JA-Ile (JA-isoleucine conjugate) when oral secretions of armyworm- fatty acid amino conjugates were applied after mechanical wounding.[63]

8.2.2 INDIRECT DEFENSES

8.2.2.1 HERBIVORE-INDUCED PLANT VOLATILES (HIPVS)

HIPVs defend plants with the help of beneficial insects.[64,65] Volatiles are also released by secondary metabolites that attract the insect's natural enemies.[66,67] It was reported that maize induces volatiles after applying synthetic volicitin, caeliferin, and inceptin, which inturn attract natural enemies of insect pests.[68] In some maize landraces, oviposition by SSB induces the release of volatiles, which attract egg parasitoid *Trichogramma* spp. and larval parasitoids (*Cotesia sesamiae*).[69] After *SSB* oviposition and (*E*)-β-caryophyllene and other terpenes were released by maize landraces C-2101, B-3016, and H-2034.[69] Eightfold (*E*)-β-caryophyllene was released by south American maize landrace Braz1006 resulting in attraction towards egg parasitoid (*C. sesamiae*) of SSB.[70] It was reported that sesquiterpene EβC is emitted in response to above,[71] below-ground injury[72] in maize, which serves as an attractant for natural enemies.[72,73] Further, sesquiterpenes protect from insects attack.[73] It was reported that western corn rootworm damage attracts its natural enemy nematode due to the production of volatile (E)-β-caryophyllene in maize roots resulting in minimum damage.[72] Oviposition by SSB induces the release of volatiles that attract egg parasitoid *Trichogramma* spp. and larval parasitoids *Cotesia sesamiae* in landraces of maize.[69]

Several defense mechanisms in maize can also be induced by egg deposition. Due to the presence of elicitors in ovipositional fluids plant defenses are activated. HIPV production was observed in the oviposited plants in the "Nyamula" landrace. Further, compared with control plants, maximum amounts of EAG-active biogenic volatiles were emitted from these plants.[74] In triggered defense responses, genetic variability was investigated with respect to stemborer egg-laying in maize.[75]

8.2.2.2 HERBIVORE ASSOCIATED MOLECULAR PATTERNS (HAMPS) AND EFFECTORS

Many insects, microbial, and plant-derived molecules in insect frass could be identified by plants to elicit defensive responses.[76] It was demonstrated that wound-responsive defense genes in maize were induced by proteins from FAW frass.[77] Besides, accumulation of Herbivore Induced Gene Transcripts, Defense Protein RIP2 was suppressed in maize leaves by FAW frass proteins when fed with larvae of FAW. In the oral secretions of FAW, phospholipase

C was identified which could modulate defense responses in various host plants.[78] Elevated levels of plant growth hormones such as auxins were identified in oral secretions of European corn borer.[79] The ECB utilizes the OS-derived auxin to increase the nutritional quality, impact subsequent herbivory by promoting plant susceptibility.

The saliva in lepidopteran insect pests was predominantly secreted by salivary glands.[80] Salivary proteomic analysis of two strains of FAW (C &R rice strain) was performed on different host plants. It was revealed that several categories of salivary proteins of FAW modulate plant defenses in maize. Indeed, glucose oxidase (GOX) and apyrases were identified in the corn earworm and FAW saliva regulates plant defenses.[81] Regurgitant of beet armyworm, leafworm, and armyworm induced accumulation of JA, defense-related genes.[82]

8.3 ROLE OF PHYTOHORMONES IN INDUCED RESISTANCE

Plant hormones are involved in regulating plant growth and defense mechanisms. Salicylic acid (SA), JA mediate Phenylpropanoid, octadecanoid pathways, respectively, activate induced resistance in plants.[83–85] Defense-related genes are triggered due to mechanical wounding, insect regurgitation, exogenous application ofphytohormones, or by insect feeding. The production of benzoxazinoids is triggered by JA, which protect from insect herbivores. Through these pathways and plant defensive secondary metabolites are produced, which impact insect development. Maize plant resistance to corn leaf aphid can be enhanced by increasing *mir1* transcription levels through ethephon (an ETH donor) treatment.[57] Indole a volatile aromatic compound elevates herbivory-induced JA and JA-Ile contents in maize. Further, it was also reported that have shown that airborne indole can also prime neighboring maize plants through indole biosynthesis mutants to release herbivory-induced mono-, homoterpenes.[86] Methyl jasmonate (MeJA) triggers defense signaling when infested by Asian corn borer (*Ostrinia furnacalis*) attack in maize.[87] Higher levels of oxylipin 12-oxo-phytodienoic acid (OPDA), an important signal transduction molecule in maize (Mp708) contributed to corn leaf aphid (CLA) resistance. CLA feeding significantly induced the expression of the transcripts encoding Maize insect resistance1-Cysteine Protease, which is an important defensive protein.[88] Flooding stress and FAW larval release in maize resulted in increased production of the SA and C-glycosyl flavones (maysins) and the herbivore-induced volatile phenolics resulting in additively and synergistically extra protection to maize.[89]

8.4 ROLE OF INSECT ENDOSYMBIONTS IN INDUCED RESISTANCE

Earlier, several workers studied on insect endosymbionts in relation to plant defenses.[78,90] Western corn rootworm accumulated more defense genes when treated with antibiotics to eliminate endosymbionts than untreated WCR.[91] Similarly, several species of symbiotic bacteria have been identified in the oral secretions of FAW and these herbivore-associated bacteria can mediate insect–plant interactions differently in multiple hosts.[78] Induced defense response in maize was also observed by the increased content of salivary GOX through gut-associated bacteria in corn earworm.[92]

8.5 IPM INTERVENTIONS IN MAIZE

The concept of IPM emphasizes to utilize multiple tactics to keep pest population levels below action thresholds. Excessive dependence on chemical insecticides leads to the evolution of resistance, resurgence, and toxicity to natural enemies and environmental pollution. Therefore, it is not advisable to depend too much on one tactic, which leads to the intensification of selection pressure. Pest-specific IPM strategies in maize[93] are discussed in the below subheadings.

8.5.1 IPM PRACTICES FOR SSB

- Destruction of the stubbles.
- Deep summer plowing (not applicable to conservation agriculture).
- Planting of two to three rows of Napier grass as trap crop in field boundary.
- Maize intercropped with cowpea in 2:1 ratio.
- *Trichogramma chilonis* 8 cards/ha (150,000 parasitized eggs/ha) in two releases at 7 and 15 days after germination.
- Pulling out of dead hearts.
- Spraying of Chlorantraniliprole 18.5 SC @150 mL/ha when infestation crosses 10% of plants, that is, 10% of plants show visible leaf feeding damage.

8.5.2 IPM PRACTICES FOR PSB

- Removal of dead/dried maize stems.
- Thorough plowing in summer.

- Pulling out of dead hearts.
- Spraying of chlorantraniliprole 18.5 SC @150 mL/ha when infestation crosses 10%, that is, 10% of plants show visible leaf feeding damage.

8.5.3 IPM PRACTICES FOR FAW

8.5.3.1 PREPLANTING PRACTICES

- Intensive plowing in summer.
- Soil application of neem cake @ 200 kg/acre.
- Planting of flowering plants around the bunds to attract beneficial insects.

8.5.3.2 SOWING TO V6 STAGE

- Proper time of sowing (Not too late or early).
- Ridge and furrow planting method to be followed.
- Application of optimum dosage of fertilizers.
- Planting of three to four rows of napier grass/hybrid napier as trap crops.
- Maize intercropped with legumes.
- Bird perches erection @10/acre.
- Installation of pheromone traps @ 4/acre for monitoring.
- Removal of weeds and other alternate hosts.
- Destruction of egg masses and FAW caterpillars.
- Application of sand or soil mixed with lime in 9: 1 ratio into whorl of maize plants.
- Foliar application with 5% neem seed kernel extract (NSKE) or azadiractin @ 1500 ppm (1 L/acre) @ 5 mL/L when 5% plant infestation was observed.
- After observing one moth/trap/day, release of egg parasitoids, viz., *Telenomus remus* @ 4000/acre or *Trichogramma pretiosum* @ 16,000/acre.
- Whorl application of *Bacillus thuringiensis* v. *Kurstaki* formulations (400 g/acre) @ 2 g/L or *Metarhizium anisopliae* or *Beauveria bassiana* with spore count of 1×10^8cfu/g (1 kg/acre) @ 5 g/L or SfNPV (600 mL/

acre) @ 3 mL/L or entomopathogenic nematode (EPN) (4 kg/acre) @ 20 g/L of water at 5–10% infestation.

- Whorl application of above mentioned insecticides for FAW, namely, Chlorantraniliprole 18.5 SC (80 mL/acre) @ 0.4 mL/L; Spinetoram 11.7 % SC (100 mL/acre) @ 0.5 mL/L; Emamectin benzoate 5% SG (80 g/acre) @ 0.4 g/L when more than 10% infestation is observed.

8.5.3.3 V7 STAGE TO FLOWERING STAGE

- Installation of pheromone traps @ 4/acre should be continued.
- Foliar application of 5% NSKE or azadiractin, 1500 ppm (1 L/acre) @ 5 mL/L after observation of one moth/trap/day or 5% of fresh FAW infestation.
- Whorl application of *Bacillus thuringiensis* v. *Kurstaki* formulations (400 g/acre) @ 2 g/L or *Metarhizium anisopliae* or *Beauveria bassiana* with spore count of 1 × 10^8 cfu/g (1 kg/acre) @ 5 g/L or SfNPV (600 mL/acre) @ 3 mL/L or entomopathogenic nematode (EPN) (4 kg/acre) @ 20 g/L of water when more than 10% infestation is observed.
- Whorl application of above mentioned insecticides for FAW, namely, Chlorantraniliprole 18.5 SC (80 mL/acre) @ 0.4 mL/L; Spinetoram 11.7 % SC (100 mL/acre) @ 0.5 mL/L; Emamectin benzoate 5% SG (80 g/acre) @ 0.4 g/L when more than 20% infestation is observed.

8.5.3.4 FLOWERING TO HARVEST

- Hand picking, destruction of larvae that bore into ears.
- Application of *Bacillus thuringiensis* v. *kurstaki* formulations (400 g/acre) @ 2 g/L or *Metarhizium anisopliae* or *Beauveria bassiana* with spore count of 1×10^8 cfu/g (1 kg/acre) @ 5 g/L or SfNPV (600 mL/acre) @ 3 mL/L or entomopathogenic nematode (EPN) (4 kg/acre) @ 20 g/L of water is recommended at 10% ear damage.

8.5.3.5 PRECAUTIONS

Interval of minimum 1 week should be there between parasitoid release, application of neem, or chemical insecticides.

Under the present scenario, the most resilient solution is the utilization of sustainable alternatives that are compatible with IPM strategies. The natural association of insect pests to host plants is disrupted due to inbuilt protection present in resistant plants. Therefore, exploration of defense mechanisms in maize is beneficial to develop new interventions in insect pests management. Utilizing structural and biochemical plant traits in breeding programs is a potential approach to combat yield losses incurred by insect pests. Further, induction of plant defenses is a new concept in IPM to stimulate defense reactions against insect pests. Moreover, the plant defense activators are compatible with other IPM strategies and can also promote attractiveness to natural enemies.

8.6 FUTURE PROMISES AND CHALLENGES

Utilization of modern breeding techniques to transfer target traits in cultivated maize lines harbor the potential for sustainable agriculture. Utilization of wild relatives for identifying new insect-resistant genes such as *R* genes is also essential for sustainable pest management. This is because defense-related genes exhibit higher expression levels in teosinte (resistance to *FAW)* compared to maize. Secondary metabolites responsible for insect resistance can be enhanced by recapturing from wild relatives of maize through conventional breeding techniques. Maize genotypes with high trichome density can be developed for defense against stem borers. Genotypes with higher capacities of tritrophic interactions with their natural enemies can be utilized in breeding programs. Quantitative trait loci mapping for cell wall-bound components can be done to enhance resistance against stem borers. However, the accumulation of biochemical compounds in plants is very complex and is highly dependent on genotype, plant tissue, and plant stage. Maize defense can also be enhanced by engineering the production of secondary metabolites through the cross-species transfer of metabolic pathways. A proper, deeper understanding of the molecular mechanisms, and ecological consequences of induced resistance in maize are important for designing new pest control strategies. Novel molecules involved in plant signaling can be explored against maize insect pests through novel tools. The exploitation of biotechnological tools including gene stacking using different combinations of Proteinase Inhibitors, RNAi-based approaches, and CRISPR/Cas9-mediated genome editing may enlighten newer dimensions in IPM.

8.7 CONCLUSION

In maize, stem borers and FAW are one of the major obstacles for increased maize production. Nowadays tremendous emphasis is on sustainable agriculture, which reduces the dependency on chemical insecticides. Under these circumstances and novel plant resistance-based approaches can be deployed as a part of IPM to revolutionize the management of maize insect pests. Using plant natural defenses, that is, morphological and biochemical defensive traits are the most promising way for crop improvement. Apart from utilizing natural defenses induction of plant defenses is another strategy that is currently being explored in which plants prepare to respond quickly and efficiently to activate their defense responses through external stimuli. Also, exploiting plant volatiles to attract natural enemies is advantageous for sustainable maize protection.

KEYWORDS

- **maize defenses**
- **morphological traits**
- **secondary metabolites**
- **insect pests**
- **induced defense responses**
- **phytohormones**
- **IPM**

REFERENCES

1. Dhillon, M. K.; Kalia, V. K.; Gujar, G. T. Insect-Pests and Their Management: Current Status, Future Need of Research in Quality Maize. In *Maize: Nutrition dynamics, Novel Uses*; Choudhary, D. P., Kumar, S., Langyan, S., Eds.; Springer: New Delhi, 2014; pp 95–103.
2. Rakshit, S.; Ballal, C. R.; Prasad, Y. G.; Sekhar, J. C.; Lakshmi Soujanya, P.; Suby, S. B.; Jat, S. L.; Siva Kumar, G.; Prasad, J. V. Fight Against Fall Armyworm *Spodoptera frugiperda* (Editor: J. E. Smith); ICAR-Indian Institute of Maize Research: Ludhiana, Punjab, 2019; p 52.

3. Panwar, V. P. S.; Management of Maize Stalk Borer, *Chilo partellus* (Swinhoe) in Maize. Stresses on Maize in the Tropics (Editor: P. H. Zaidi, N. N. Singh); Directorate of Maize Research: New Delhi, 2005; pp 376–395.
4. Kumar, H.; Asino, G. O. Resistance of Maize to *Chilo partellus* (Lepidoptera: Pyralidae): Effect of Plant Phenology. *J. Econ. Entomol.* **1993,** *86*, 969–973.
5. Murua, G.; Molina-Ochoa, J.; Coviella, C. Population Dynamics of the Fall Armyworm, *Spodoptera frugiperda* (Lepidoptera: Noctuidae), Its Parasitoids in Northwestern Argentina. *Florida Entomol.* **2006,** *89*, 175–182.
6. Sharma, H. C.; Sujana, G.; Rao, D. M. Morphological, Chemical Components of Resistance to Pod Borer, *Helicoverpa armigera* in Wild Relatives of Pigeonpea. *Arthropod-Plant Interact.* **2009,** *3*, 151–161.
7. War, A. R.; Paulraj, M. G.; Ahmad, T.; Buhroo, A. A.; Hussain, B.; Ignacimuthu, S.; Sharma, H. C. Mechanisms of Plant Defense Against Insect Herbivores. *Plant Signal. Behav.* **2012,** *7*, 1306–1320.
8. Schulze, E. D.; Beck, E.; Müller-Hohenstein, K. *Pflanzenökologie*; Springer: Heidelberg, 2002.
9. Kariyat, R. R.; Balogh, C. M.; Moraski, R. P.; De Moraes, C. M.; Mescher, M. C.; Stephenson, A. G. Constitutive, Herbivore-Induced Structural Defenses Are Compromised by Inbreeding in *Solanum carolinense* (Solanaceae). *Am. J. Bot.* **2013,** *100*, 1014–1021.
10. Kumar, H. Resistance in Maize to *Chilo partellus* (Swinhoe) (Lepidoptera: Pyralidae): An Overview. *Crop Protection* **1997,** *16*, 243–250.
11. Rao, C. N.; Panwar, V. P. S. Anatomical Plant Characters Affecting Resistance to *Chilo partellus* (Swinhoe) in Maize. *Ann. Plant Protection Sci.* **2001,** *9*, 304–306.
12. Sharma, H. C.; Dhillon, M. K.; Pampapathy, G.; Reddy, B. V. S. Inheritance of Resistance to Spotted Stem Borer, *Chilo partellus*, in Sorghum, *Sorghum bicolor*. *Euphytica* **2007,** *156*, 117–128.
13. Dhillon, M. K.; Gujar, G. T. Maize Genotypes Identified with Resistance to Spotted Stem Borer, *Chilo partellus and* Favorable Agronomic Traits. *Ann. Plant Protection Sci.* **2013,** *21*, 224–228.
14. Kumar, H.; Saxena, K. N. Ovipositional Responses of *Chilo partellus* (Swinhoe) to Certain Susceptible, Resistant Maize Genotypes. *Int. J. Trop. Insect Sci.* **1985,** *6*, 331–335.
15. Van den Berg J. Oviposition Preference, Larval Survival of *Chilo partellus* (Lepidoptera: Pyralidae) on Napier Grass (*Pennisetum purpureum*) Trap Crops. *Int. J. Pest Manage.* **2006,** *52*, 39–44.
16. Munyiri, S. W.; Mugo, S. N.; Otim, M.; Tefera, T.; Beyene, Y.; Mwololo, J. K.; Okori, P. Responses of Tropical Maize Landraces to Damage by *Chilo partellus* Stem Borer. *Afr J. Biotechnol.* **2013,** *12*, 1229–1235.
17. Dubey, S. L.; Sarup, P. Ovipositional Responses of Moths of *Chile partellus* (Swinhoe) on Different Maize Germplasms. *J. Entomol. Res.* **1982,** *6*, 1–9.
18. Ampofo, J. K. *Chilo partellus* (Swinhoe) Oviposition on Susceptible, Resistant Genotypes. *Insect Sci. App.* **1980,** *6*, 323–330.
19. Bergvinson, D. J.; Arnason, J. T.; Hamilton, R.; Mihm, J. A.; Jewell, D. C. Determining Leaf Toughness, Its Role in Maize Resistance to the European Com Borer (Lepidoptera: Pyralidae). *J. Econ. Entomol.* **1994,** *87*, 1743–1748.
20. Davis, F. M.; Baker, G. T.; Williams, W. P. Anatomical Characteristics of Maize Resistant to Leaf Feeding by Southwestern Corn Borer (Lepidoptera: Pyralidae), Fall Armyworm. *J. Agric. Entomol.* **1995,** *12*, 55–65.

21. Yang, G.; Wiseman, B. R.; Isenhour, D. J.; Espelie, K. E.; Chemical, Ultrastructural Analysis of Corn Cuticular Lipids, Their Effect on Feeding by Fall Armyworm Larvae. *J. Chem. Ecol.* **1993,** *19*, 2055–2074.
22. Bernal, J. S.; Melancon, J. E.; Zhu-Salzman, K. Clear Advantages for Fall Armyworm Larvae from Feeding on Maize Relative to Its Ancestor Balsas Teosinte May Not Be Reflected in Their Mother's Host Choice. *Entomologia Experimentalis et Applicata* **2015,** *155*, 206–217.
23. Moya-Raygoza, G.; Early Development of Leaf Trichomes Is Associated with Decreased Damage in Teosinte, Compared with Maize, by *Spodoptera frugiperda* (Lepidoptera: Noctuidae). *Ann. Entomol. Soc. Am.* **2016,** *109*, 737–743.
24. Sarwar, M. Management of Rice Stem Borers (Lepidoptera: Pyralidae) Through Host Plant Resistance in Early and Medium and Late Plantings of Rice (*Oryza sativa* L.). *J. Cereals Oilseeds* **2012,** *3*, 10–14.
25. Rao, C. N.; Panwar, V. P. S. Morphological Plant Characters Affecting Resistance to Chilopartellusin Maize. *Annu. Plant Protection Sci.* **2002,** *8*, 145–149.
26. Dhillon, M. K.; Chaudhary, D. P. Biochemical Interactions for Antibiosis Mechanism of Resistance to *Chilo partellus* (Swinhoe) in Different Maize Types. *Arthropod-Plant Interactions* **2015,** *9*, 373–382.
27. Sharma, V. K.; Chatterji, S. M. Studies on Some Chemical Constituents Concerning Differential Susceptibility of Some Maize Germplasms to *Chilo zonellus* (Swinhoe). *Indian J. Ent.* **1971,** *33*, 419–424.
28. Kabre, G. B.; Ghorpade, S. A. Susceptibility to Maize Stem Borer, *Chilo partellus* (Swinhoe) Concerning Sugars, Proteins, Free Amino Acids Content of Maize Germplasm, F1 Hybrids. *J. Insect Sci.* **1999,** *12*, 37–40.
29. Jalali, S. K.; Lalitha, Y.; Kamath, S. P.; Mohan, K. S.; Head, G. P. Baseline Sensitivity of Lepidopteran Corn Pests in India to Cry1Ab Insecticidal Protein of *Bacillus thuringiensis*. *Pest Manage. Sci.* **2010,** *66*, 809–815.
30. Lokesh; Mehla, J. C.; Study on Morphological, Biochemical Features Associated with Maize (*Zea mays* L.) Resistance Against *Chilo partellus*. *Int. J. Pure Appl. Biosci.* **2017,** *5*, 1011–1021.
31. Chen, Y.; Ni, X.; Buntin, G. D. Physiological, Nutritional, Biochemical Bases of Corn Resistance to Foliage-Feeding Fall Armyworm. *J. Chem. Ecol.* **2009,** *35*, 297–306.
32. Rasool, I.; Wani, A. R.; Nisar, M.; Dar, Z. A.; Nehru, R. K.; Hussain, B. Antixenosis, Antibiosis as a Resistance Mechanism to *Chilo partellus* (Swinhoe) (Lepidoptera: Pyralidae) in Some Maize Genotypes. *J. Entomol. Zool. Stud.* **2017,** *5*, 22–27.
33. Hedin, P. A.; Williams, W. P.; Davis, F. M.; Buckley, P. M. Roles of Amino Acids, Protein and Fibre in Leaf-Feeding Resistance of Corn to the Fall Armyworm. *J. Chem. Ecol.* **1990,** *16*, 1977–1995.
34. Niemeyer, H. M. Hydroxamic Acids (4-hydroxy-1, 4-benzoxazin-3-ones), Defense Chemicals in the Gramineae. *Phytochemistry.* **1988,** *27*, 3349–3358.
35. Wiseman, B. R.; Snook, M. E.; Isenhour, D. J.; Mihm, J. A.; Widstrom, N. W. Relationship Between the Growth of Corn Earworm, Fall Armyworm Larvae (Lepidoptera: Noctuidae) and Maysin Concentration in Corn Silks. *J. Econ. Entomol.* **1992,** *85*, 2473–2477.
36. Hatfield, R. D.; Marita, J. M. Enzymatic Processes Involved in the Incorporation of hydroxycinnamates Into Grass Cell Walls. *Phytochem. Rev.* **2010,** *9*, 35–45.
37. Qi, J.; Malook, S. U.; Shen, G.; Gao, L.; Zhang, C.; Li, J.; Zhang, J.; Wang, L.; Wu, J. Current Understanding of Maize, Rice Defense Against Insect Herbivores. *Plant Divers.* **2018,** *40*, 189–195.

38. Snook, M. E.; Wiseman, B. R.; Widstrom, N. W.; Wilson, R. L. Chemicals Associated with Maize Resistance to Corn Earworm, Fall Armyworm. In *Insect Resistant Maize: Recent Advances, Utilization; Proceedings of an International Symposium Held at the International Maize, heat Improvement Center (CIMMYT)* 27 November–3 December 1994; Mihm, J. A. Ed.; CIMMYT: Mexico, D.F., 1997.
39. Santiago, R.; Malvar, R. A.; Baamonde, M. D.; Revilla, P. Free Phenols in Maize Pith, Their Relationship with Resistance to *Sesamia nonagrioides* (Lepidoptera: Noctuidae) Attack. *J. Econ. Entomol.* **2005,** *98*, 1349–1356.
40. Lakshmi Soujanya, P.; Sekhar, J. C.; Ratnavathi, C. V.; Shobha, E.; Chikkappa, G.; Karjagi, Suby, S. B,. Sunil, N.; Sujay Rakshit. Role of Soluble, Cell Wall-Bound Phenolics, Tannin and Flavonoid Contents in Maize Resistance to Pink Stem Borer *Sesamia inferens* Walker. *Maydica* **2020,** *65*-M8.
41. Barry, D.; Alfaro, D.; Darrah, L. L. Relation of European Corn Borer (Lepidoptera: Pyralidae) Leaf-Feeding Resistance, DIMBOA Content in Maize. *Environ. Entomol.* **1994,** *23*, 177–182.
42. Gutierrez, C.; Castanera, P.; Torres, V.; Wound-Induced Changes in DIMBOA (2, 4 dihydroxy7methoxy2H1, 4 benzoxazin3 (4H) One) Concentration in Maize Plants Caused by *Sesamia nonagrioides* (Lepidoptera: Noctuidae). *Ann. Appl. Biol.* **1998,** *113*, 447–454.
43. Nicollier, G. F.; Hedin, P. A.; Davis, F. M. 5-, 6-, 7-Methoxybenzoxazolinone: Carbon-13 NMR Spectra, Biological Activity. *J. Agric. Food Chem.* **1982,** *30*, 1133–1135.
44. Hedin, P. A.; Davis, F. M.; Williams, W. P. 2-hydroxy-4, 7-dimethoxy-1, 4-benzoxazin-3-one (N-O-Me-DIMBOA), a Possible Toxic Factor in Corn to the Southwestern Corn Borer. *J. Chem. Ecol.* **1993,** *19*, 531–542.
45. Oikawa, A.; Ishihara, A.; Hasegawa, M.; Kodama, O. Iwamura, H. Induced Accumulation of 2-hydroxy-4, 7-dimethoxy-1, 4-benzoxazin-3-one glucoside (HDMBOA-Glc) in Maize Leaves. *Phytochemistry.* **2001,** *56*, 669–675.
46. Niemeyer, H. M. Hydroxamic Acids Derived from 2-hydroxy-2*H*-1, 4-benzoxazin-3(4*H*)-one: Key Defense Chemicals Of Cereals. *J. Agric. Food Chem.* **2009,** *57*, 1677–1696.
47. Glauser, G.; Marti, G.; Villard, N.; Doyen, G. A.; Wolfender, J. L.; Turlings, T. C.; Erb, M. Induction, Detoxification of Maize 1, 4-benzoxazin-3-ones by Insect Herbivores. *Plant J.* **2011,** *68*, 901–911.
48. Schmelz, E. A.; Kaplan, F.; Huffaker, A.; Dafoe, N. J.; Vaughan, M. M.; Ni, X.; Rocca, J. R.; Alborn, H. T.; Teal, P. E. Identity, Regulation and Activity of Inducible Diterpenoid Phytoalexins in Maize. *Proc. Natl. Acad. Sci.* **2011,** *108*, 5455–5460.
49. Praveen, H. D.; Ugalat, J.; Singh, H. Biochemical Changes During the Crop Growth Period of Resistance, Susceptible Varieties of Maize Against Stem Borer. *Environ. Ecol.* **2013,** *31*, 1621–1626.
50. Campbell, M. M, Sederoff, R. R. Variation in Lignin Content, Composition (Mechanisms of Control, Implications for the Genetic Improvement of Plants). *Plant Physiol.* **1996,** *110*, 3–13.
51. Santiago, R.; Barros-Rios. J.; Malvar, R. A. Impact of Cell Wall Composition on Maize Resistance to Pests, Diseases. *Int. J. Mol. Sci.* 2013. *14*, 6960–6980.
52. Shivaji, R.; Camas, A.; Ankala, A.; Engelberth, J.; Tumlinson, J. H.; Williams, W. P.; Wilkinson, J. R.; Luthe, D. S. Plants on Constant Alert: Elevated Levels of Jasmonic Acid and Jasmonate-Induced Transcripts in Caterpillar-Resistant Maize. *J. Chem. Ecol.* **2010,** *36*, 179–191.

53. Smith, W. E.; Shivaji, R.; Williams, W. P.; Luthe, D. S.; Sandoya, G. V.; Smith, C. L.; Sparks, D. L.; Brown, A. E. A Maize Line Resistant to Herbivory Constitutively Releases (E) -Beta-Caryophyllene. *J. Econ. Entomol.* **2012,** *105*, 120–128.
54. Dunse, K. M.; Stevens, J. A.; Lay, F. T.; Gaspar, Y. M.; Heath, R. L. Anderson, M. A. Coexpression of Potato Type I, II Proteinase Inhibitors Gives Cotton Plants Protection Against Insect Damage in the Field. *Proc. Natl. Acad. Sci.* **2010,** *107*, 15011–15015.
55. Pechan, T.; Cohen, A.; Williams, W. P.; Luthe, D. S. Insect Feeding Mobilizes a Unique Plant Defense Protease That Disrupts the Peritrophic Matrix of Caterpillars. *Proc. Natl. Acad. Sci.* **2002,** *99*, 13319–13323.
56. Brioschi, D.; Nadalini, L. D.; Bengtson, M. H.; Sogayar, M. C.; Moura, D. S.; Silva-Filho, M. C. General Up-Regulation of *Spodoptera frugiperda* Trypsins, Chymotrypsins Allows Its Adaptation to Soybean Proteinase Inhibitor. *Insect Biochem. Mol. Biol.* **2007,** *37*, 1283–1290.
57. Louis, J.; Basu, S.; Varsani, S.; Castano-Duque, L.; Jiang, V.; Williams, W. P.; Felton, G. W.; Luthe, D. S. Ethylene Contributes to mir1-Mediated Maize Defense Against the Phloem-Sap Sucking Insect *Rhopalosiphum maidis*. *Plant Physiol.* **2015,** *169*, 313–324.
58. Jadhav, A. R.; War, A. R.; Nikam, A. N.; Adhav, A. S.; Gupta, V. S.; Sharma, H. C.; Giri, A. P.; Tamhane, V. A. *Capsicum annuum* Proteinase Inhibitor Ingestion Negatively Impacts the Growth of Sorghum Pest *Chilo partellus*, Promotes Differential Protease Expression. *Biochem. Biophys. Rep.* **2016,** *8*, 302–309.
59. Harfouche, A. L.; Shivaji, R.; Stocker, R.; Williams, P. W.; Luthe, D. S.; Ethylene Signalling Mediates a Maize Defense Response to Insect Herbivory. *Mol. Plant-Microbe Interactions* **2006,** *19*, 189–199.
60. Privalle, L.; Estruch, J.; Wright, M.; Hill, M. B.; Dowd, P. F.; Lagrimini, L. M.; Novartis Finance Corp, Ohio State University, US Department of Agriculture, Methods for Conferring Insect Resistance to a Monocot Using a Peroxidase Coding Sequence, 1999. U.S. Patent 6,002,068.
61. Rodriguez, V. M.; Santiago, R.; Malvar, R. A.; Butrón, A. Inducible Maize Defense Mechanisms Against the Corn Borer *Sesamia nonagrioides*: A Transcriptome, Biochemical Approach. *Mol. Plant-Microbe Interactions* **2012,** *25*, 61–68.
62. Marti, G.; Erb, M.; Boccard, J.; Glauser, G.; Doyen, G. R.; Villard, N.; Robert, C. A. M.; Turlings, T. C.; Rudaz, S.; Wolfender, J. L. Metabolomics Reveals Herbivore-Induced Metabolites of Resistance, Susceptibility in Maize Leaves, Roots. *Plant, Cell Environ.* **2013,** *36*, 621–639.
63. Qi, J.; Sun, G.; Wang, L.; Zhao, C.; Hettenhausen, C.; Schuman, M. C.; Baldwin, I. T.; Li, J.; Song, J.; Liu, Z.; Xu, G. Oral Secretions from *Mythimna separata* Insects Specifically Induce Defense Responses in Maize as Revealed by High-Dimensional Biological Data. *Plant, Cell, Environ.* **2016,** *39*, 1749–1766.
64. Arimura, G. I.; Matsui, K.; Takabayashi, J. Chemical, Molecular Ecology of Herbivore-Induced Plant Volatiles: Proximate Factors, Their Ultimate Functions. *Plant Cell Physiol.* **2009,** *50*, 911–923.
65. War, A. R.; Paulraj, M. G.; War, M. Y.; Ignacimuthu, S.; Jasmonic Acid-Mediated-Induced Resistance in Groundnut (*Arachis hypogaea* L.) Against *Helicoverpa armigera* (Hubner) (Lepidoptera: Noctuidae). *J. Plant Growth Regulation* **2011,** *30*, 512–523.
66. Howe, G. A.; Jander, G. Plant Immunity to Insect Herbivores. *Annu Rev Plant Biol.* **2008,** *59*, 41–66.

67. He, J.; Chen, F.; Chen, S.; Lv, G.; Deng, Y.; Fang, W.; Liu, Z.; Guan, Z.; He, C. Chrysanthemum Leaf Epidermal Surface Morphology, Antioxidant, Defense Enzyme Activity in Response to Aphid Infestation. *J. Plant Physiol.* **2011,** *168*, 687–693.
68. Alborn, H. T.; Turlings, T. C. J.; Jones, T. H.; Stenhagen, G.; Loughrin, J. H.; Tumlinson, J. H. An Elicitor of Plant Volatiles from Beet Armyworm Oral Secretion. *Science.* **1997,** *276*, 945–949.
69. Tamiru, A.; Bruce, T. J.; Woodcock, C. M.; Caulfield, J. C.; Midega, C. A.; Ogol, C. K.; Mayon, P.; Birkett, M. A.; Pickett, J. A.; Khan, Z. R. Maize Landraces Recruit Egg, Larval Parasitoids in Response to Egg Deposition by a Herbivore. *Ecol. Lett.* **2011,** *14*, 1075–1083.
70. Tamiru, A.; Bruce, T. J.; Richter, A.; Woodcock, C. M.; Midega, C. A.; Degenhardt, J.; Kelemu, S.; Pickett, J. A.; Khan, Z. R. A Maize Landrace That Emits Defense Volatiles in Response to Herbivore Eggs Possess a Strongly Inducible Terpene Synthase Gene. *Ecol. Evol.* **2017,** *7*, 2835–2845.
71. Turlings, T. C. J.; Bernasconi, M.; Bertossa, R.; Bigler, F.; Caloz, G.; Dorn, S. The Induction of Volatile Emissions in Maize by Three Herbivore Species with Different Feeding Habits: Possible Consequences for Their Natural Enemies. *Biol. Control.* **1998,** *11*, 122–129.
72. Rasmann, S.; Köllner, T. G.; Degenhardt, J.; Hiltpold, I.; Toepfer, S.; Kuhlmann, U.; Gershenzon, J.; Turlings, T. C. Recruitment of Entomopathogenic Nematodes by Insect-Damaged Maize Roots. *Nature.* **2005,** *434*, 732–737.
73. Köllner, T. G.; Held, M.; Lenk, C.; Hiltpold, I.; Turlings, T. C. J.; Gershenzon, J.; et al. A Maize (E)-β-caryophyllene Synthase Implicated in Indirect Defense Responses Against Herbivores Is Not Expressed in Most American Maize Varieties. *Plant Cell.* **2008,** *20*, 482–494.
74. Mutyambai, D. M.; Bruce, T. J.; van den Berg, J.; Midega, C. A.; Pickett, J. A.; Khan, Z. R. An Indirect Defense Trait-Mediated Through Egg-Induced Maize Volatiles from Neighbouring Plants. *PloS One* **2016,** *11*, e0158744.
75. Tamiru, A.; Paliwal, R.; Manthi, S. J.; Odeny, D. A.; Midega, C. A.; Khan, Z. R.; Pickett, J. A. Bruce, T. J. Genome-Wide Association Analysis of a Stem Borer Egg Induced "call-for-help" Defense Trait in Maize. *Sci. Rep.* **2020,** *10*, 1–12.
76. Chen, H.; Wilkerson, C. G.; Kuchar, J. A.; Phinney, B. S.; Howe, G. A. Jasmonate-Inducible Plant Enzymes Degrade Essential Amino Acids in the Herbivore Midgut. *Proc. Natl. Acad. Sci.* **2005,** *102*, 19237–19242.
77. Ray, S.; Gaffor, I.; Acevedo, F. E.; Helms, A.; Chuang, W. P.; Tooker, J.; Felton, G. W.; Luthe, D. S. Maize Plants Recognize Herbivore-Associated Cues from Caterpillar Grass. *J. Chem. Ecol.* **2015,** *41*, 781–792.
78. Acevedo, F. E.; Peiffer, M.; Tan, C.-W.; Stanley, B. A.; Stanley, A.; Wang, J.; Jones, A. G.; Hoover, K.; Rosa, C.; Luthe, D.; Felton, G. Fall Armyworm-Associated Gut Bacteria Modulate Plant Defense Responses. *Mol. Plant-Microbe Interactions* **2017,** *30*, 127–137.
79. Dafoe, N. J.; Thomas, J. D.; Shirk, P. D.; Legaspi, M. E.; Vaughan, M. M.; Huffaker, A.; Teal, P. E.; Schmelz, E. A. European Corn Borer (*Ostrinia nubilalis*) Induced Responses Enhance Susceptibility in Maize. *PLoS One.* **2013,** *8*, e73394.
80. Felton, G. W.; Chung, S. H.; Hernandez, M. G. E.; Louis, J.; Peiffer, M.; Tian, D. Herbivore Oral Secretions Are the First Line of Protection Against Plant-Induced Defenses. *Annu. Plant Rev.* **2014,** *47*, 37–76.

81. Tian, D.; Peiffer, M.; Shoemaker, E.; Tooker, J.; Haubruge, E.; Francis, F.; Luthe, D. S.; Felton, G. W. Salivary Glucose Oxidase from Caterpillars Mediates the Induction of Rapid, Delayed-Induced Defenses in the Tomato Plant. *PLoS One* **2012,** *7*, e36168.
82. Erb, M.; Flors, V.; Karlen, D.; De Lange, E.; Planchamp, C.; D'Alessandro, M.; Turlings, T. C.; Ton, J. Signal Signature of Aboveground-Induced Resistance Upon Belowground Herbivory in Maize. *Plant J.* **2009,** *59*, 292–302.
83. Zhao, L. Y.; Chen, J. L.; Cheng, D. F.; Sun, J. R.; Liu, Y.; Tian, Z.; Biochemical, Molecular Characterizations of *Sitobion avenae*-Induced Wheat Defense Responses. *Crop Protection* **2009,** *28*, 435–442.
84. Scott, I. M.; Thaler, J. S.; Scott, J. G. Response of a Generalist Herbivore *Trichoplusia ni* to Jasmonate-Mediated Induced Defense in Tomato. *J. Chem. Ecol.* **2010,** *36*, 490–499.
85. He, J.; Chen, F.; Chen, S.; Lv, G.; Deng, Y.; Fang, W.; Liu, Z.; Guan, Z.; He, C. Chrysanthemum Leaf Epidermal Surface Morphology, Antioxidant, Defense Enzyme Activity in Response to Aphid Infestation. *J. Plant Physiol.* **2011,** *168*, 687–693.
86. Erb, M.; Veyrat, N.; Robert, C. A.; Xu, H.; Frey, M.; Ton, J.; Turlings, T. C. Indole Is an Essential Herbivore-Induced Volatile Priming Signal in Maize. *Nat. Commun.* **2015,** *6*, 1–10.
87. Yang, F.; Zhang, Y.; Huang, Q.; Yin, G.; Pennerman, K. K.; Yu, J.; Liu, Z.; Li, D.; Guo, A. Analysis of Key Genes of a Jasmonic Acid-Mediated Signal Pathway for Defense Against Insect Damages by Comparative Transcriptome Sequencing. *Sci. Rep.* **2015,** *5*, 16500.
88. Varsani, S.; Grover, S.; Zhou, S.; Kyle, G. K.; Huang, P-C.; Michael, V. K.; Williams, W. P.; Heng-Moss, T.; Sarath, G.; Luthe, D. S.; Jander, G.; Louis, J. 12-Oxo-Phytodienoic Acid Acts as a Regulator of Maize Defense Against Corn Leaf Aphid. *Plant Physiol.* **2019,** *179*, 1402–1415.
89. Anna, K. B.; Hunter, C. T.; Sattler, S. E.; Rering, C.; McDonald, S.; Basset, G. J.; Christensen, S. A. Fighting on Two Fronts: Elevated Insect Resistance in Flooded Maize. *Plant, Cell Environ.* **2020,** *43*, 223–234.
90. Wang, J.; Peiffer, M.; Hoover, K.; Rosa, C.; Zeng, R.; Felton, G. W.; *Helicoverpa zea* Gut-Associated Bacteria Indirectly Induce Defenses in Tomato by Triggering a Salivary Elicitor (s). *New Phytol.* **2017,** *214*, 1294–1306.
91. Brarr, K. L.; Hearne, L. B.; Briesacher, S.; Clark, T. L.; Davis, G. E. Microbial Symbionts in Insects Influence Down-Regulation of Defense Genes in Maize. *PLoS One* **2010, 5**, e11339.
92. Wang, J.; Yang, M.; Song, Y.; Acevedo, F. E.; Hoover, K.; Zeng, R.; Felton, G. W. Gut-Associated Bacteria of Helicoverpa Zea Indirectly Trigger Plant Defenses in Maize. *J. Chem. Ecol.* **2018,** *44*, 690–699.
93. Lakshmi Soujanya, P.; Suby, S. B.; Sekhar, J. C.; Vasudeva Rao, V.; Rakshit, S. *Major Insect Pests of Maize, Their Management*; ICAR-Indian Institute of Maize Research: Ludhiana, Punjab 2020, 141004 pp. 62.
94. Szczepaniec, A.; Widney, S. E.; Bernal, J. S.; Eubanks, M. D. Higher Expression of Induced Defenses in Teosintes (*Zea* spp.) Is Correlated with Greater Resistance to Fall Armyworm and *Spodoptera frugiperda*. *Entomol. Exp.et Applicata.* **2013,** *146*, 242–251.

CHAPTER 9

Insect Pest Management in Millet Cropping Systems

J. STANLEY, K. SRINIVASA BABU, G. SHYAM PRASAD*, A. KALAISEKAR, B. SUBBARAYUDU, and B. GANGAIAH

ICAR- Indian Institute of Millets Research, Hyderabad, Telangana, India

Corresponding author. E-mail: shyam@millets.res.in

ABSTRACT

Millets are usually grown under adverse agroecological conditions with minimum management. Different kinds of pests affect millets and they are of pests affecting the seedling stage, the foliage, the panicle, stem borers, and the soil pests. Pest management in millet crops is mostly attributed to host plant resistance and cultural management. Many species of shoot flies affect millet crop and they are managed by growing resistant or tolerant varieties and using cultural management like sowing seeds at a high seed rate and removal of infected seedlings. Stem borers are suppressed by natural enemies like parasitoids and predators; however, when the infestation crosses economic threshold levels (ELTs), insecticides is used. Of late, fall armyworm is also found to attack the sorghum especially in the early crop stages. Pest management in millets is to be aimed as a system approach rather than managing individual pests in millets. In this view, pest management using strategic crops in crop sequence, intercropping with nonhost plants to reduce the pest management in both the companion crops, selection and growing of millets in alley cropping aiming in pest management through an ecosystem approach are aimed. In this chapter, pests and their management in millet-based cropping sequence, intercropping, mixed cropping, and alley cropping are given in detail.

Integrated Pest Management in Diverse Cropping Systems. Devendra Pal Singh, PhD (Ed.)

9.1 INTRODUCTION

Millets also regarded as wonder crops are the major food for the people in arid and semiarid regions of the world that are also featured in their traditional cuisines. They are usually grown in conditions where limited availability of water resources and with minimum or no fertilizers and other agricultural inputs. They are regarded as harder crops and can sustain even the most adverse climatic conditions. However, millets did not get its importance at the time of green revolution in India. In view of changing climate, revisiting the forgotten millets has become inevitable apart from its gaining importance due to its health benefits partially because of the increased awareness in health by the public. Millets once called as "coarse cereals" are now being called as "nutri-cereals" as they are rich in essential nutrients. The protein content in millets is close to that of wheat, that is, 11% by weight and with high fiber content than rice and wheat. Millets are also rich in B vitamins, calcium, iron, magnesium, potassium, and zinc. Besides a nutritious crop to the humans, it also gives good quality fodder.

As stated above, millets are being grown in resource-poor conditions without assured irrigation water, and fertilizers with miniscule care and management. Pest management in millets is not exceptional and get very minimal attention by the grower. So, pest management in millets is aimed at having resistant varieties and natural management measures rather than measures like chemical or classical biological control. Extensive cultivation of millets caused the pests that damage them also to rise. However, host plant resistance and cultural pest management measures are practiced with insecticides at the last resort. Fully defined integrated pest management is seldom practiced in millet crops. Thus, pest management through cropping system approach gains importance as the colonization and infestation of pest also depend upon the preceding crop and soil management practices. If an entirely different, nonhost plant or a strategic plant is introduced in the cropping sequence, it helps to break the burgeoning population of the pest.

In general, the incidence of pest is reportedly low in intercropping system as compared with a monocrop grown extensively over a larger area. This is mainly because of the interruption caused during searching of host plant, alightment, and colonization of the pest by the nonhost crop near the host plant. Monocropping allows the pest to locate the host quickly and easily aiding in faster feeding and achieving higher reproduction rates. On the other hand, intercropping interrupts the insect pest to feed on host plant and colonize in large numbers. Moreover, some reports reveal intercropping may attract more predators and prasitoids than monocropping. This is also true in the case of mixed cropping and alley cropping systems to a certain extend. Thus studies on pest management as a system approach will result in practically meaningful and effective pest management strategy.

9.2 INSECT PESTS OF MILLETS

9.2.1 INSECT PESTS INFESTING SEEDLINGS

Shoot flies are the most important pests of seedling stage of millet crops. Many species of shoot flies (Diptera: Muscidae) attack different kinds of millets in India[1] and Africa.[2] Shoot flies cause a yield loss of about 20–50% in pearl millet,[3] 39% in little millet,[4] and 36% in proso millet.[5] In sorghum grown in Asia, Africa, and Mediterranean Europe, the sorghum shoot fly, *Atherigona soccata* is an economic pest.[6] The pearl millet shoot fly, *Atherigona approximata* is a common pest of pearl millet in India.[7] *Atherigona simplex* was found affecting the kodo millet naturally in Madhya Pradesh, India.[8] The tef shoot fly, *Atherigona hyalinipennis* causes damage to seedling and panicles of tef to an extent of 9–20% in Ethiopia.[9] Apart from shoot flies, seedlings of millets are also infested by leaf beetles, grey weevils, and stem flies.[10]

9.2.2 FOLIAGE PESTS

Leaf rollers (*Marasmia trapezalis*) sugarcane leaf hopper (*Pyrilla perpusilla),* shoot bug (*Peregrinus maidis),* and hairy caterpillars (*Amsacta* spp.) are also found damaging pearl millet.[3] Some important insect pests of pearl millet grown in Africa are the leaf miner (*Heliocheilus albipunctella)*, armyworm (*Spodoptera exempta)*, and grasshoppers.[11] Foliage pests like hairy caterpillars (*Amsacta moloneyi*), armyworms (*Spodoptera* spp.; *Mythimna* spp.), and grasshoppers also attack millet crops.[12] Outbreaks of the noctuids (*Marasmia separata*, *Marasmia albistigma*, and *Mocis frugalis*) were reported in Karnataka and found attacking and defoliating finger millet and various other grasses.[13]

Shoot bugs, *P. maidis* though occur sporadically especially in rabi season,[14] is reportedly cause damage to the crop up to 11.2% and 21.1% losses in grain and fodder yields, respectively, with 51% increase in the incidence of sorghum stripe virus disease.[15] Sugarcane aphids, *Melanaphis sacchari,* is reportedly attack sorghum and millets.[16] In South Africa, grain yield losses by *P. maidis* reached 60%,[17] and 46–78%, without insecticide control.[18,19].

9.2.3 STEM BORERS

Stem borers are the major pest complex contributing to yield reduction in millets. In Indian conditions, *Chilo partellus* also known as the spotted stalk borer is predominant among the borers.[20] It inflicts a loss of 14–36% in different agroclimatic regions of the country[21] even leading to crop failures.[22] The other important pest is the pink borer, *Sesamia inferens,* which cause serious yield loss to the crop especially during rabi season in the peninsular India.[23] In Africa, south of Sahara, stem borers such as *Busseola fusca, C. partellus, Sesamia calamistis,* and *Eldana saccharina* are major pests of sorghum and pearl millets.[24,25] In Ethiopia, *B. fusca* was found in higher altitudes (>1500 m), whereas *C. partellus* dominated in the lower and warmer areas and *S. calamistis* was cosmopolitan.[26] Pearl millet appears to be immune to borer attack at initial stages of crop growth but at later stages it is prone to internode injury.

9.2.4 PANICLE PESTS

The millet head miner, *H. albipunctella* is the most important pest of pearl millet in Sahel region, that is, between the Sahara to the north and the Sudanian savanna to the south.[27] Earhead bug, *Calocoris angustatus* is a key pest in the southern part of India, that is, in the states of Andhra Pradesh, Karnataka, and Tamil Nadu. The bug attacks from the head emergence to hard-dough stage of grain development. Hundreds of adults and nymphs can be observed on a single earhead and a yield loss of 5.8–84.3% is recorded.[28] The earhead caterpillars, *Eublemma silicula* and *Cryptoblabes gnidiella,* cause yield loss in millets especially pearl millet.[3] Head caterpillars (*Helicoverpa armigera, E. silicula, C. gnidiella, Euproctis subnotata, Nola analis, Celama* spp.) feed on the developing grains of sorghum. In cultivars with compact panicles, the inside of the panicle may be completely damaged while the panicle may look healthy externally.[28] Grain midge *(Geromyia penniseti*) attacks late millets and causes a considerable loss in yield.[12] Sorghum midge, *Stenodiplosis sorghicola,* is reported to be one of the major pest affecting grain sorghum worldwide.[29] In Mali, blister beetles, *Psalydolytta* spp., and *Mylabris* spp. were found to infest pearl millet even to a range of 12.3 beetles/panicle.[30]

9.2.5 SOIL PESTS

Among soil pests of millets, termites and cutworms that feed and damage the roots of millets are relatively less important.[31] However, soil pests like white grubs (Coleoptera: Scarabaeidae) can impact millet growth severely and attain the status of economic pest especially in the arid and semiarid regions.[32,33]

9.3 INSECT PEST MANAGEMENT IN MILLETS

9.3.1 CULTURAL MANAGEMENT

Crop husbandry practices that directly or indirectly help to reduce pest damage have become an integral component of farming systems. Intercropping of pearl millet with sunflower and pigeon pea is reportedly suppressed the white grub incidence and thus the damage. The adult beetles congregate in neem or acacia trees especially after first monsoon showers and collection/trapping of those beetles is considered as a good management practice.[34] In south Indian conditions, sorghum is recommended to be sown immediately after the first monsoon shower as a management of shoot fly. Staggered sowing is also recommended to contain the build-up of shoot fly population. A higher seed rate of 4–10 kg/ha is recommended and the infected seedlings are thinned within 15 days of sowing. In case of late crop, transplantation is recommended. Deep summer plowing to expose egg pods in soil and scrapping of bunds and weed-free cultivation are some of the cultural practices for the management of grasshoppers. In case of termites, collection and burning of stubbles/plant refuge and use of well decomposed FYM and timely irrigation subdues termite damage in millet fields.

Altering the planting time in order to create a mismatch between insect larvae stages and critical plant developmental stages effectively reduce the insect damage. Normally, sorghum is sown with the monsoon showers. Synchronous sowing of cultivars in similar maturity groups reportedly reduce yield losses caused by shoot fly, head bug, and midge. Effective midge control requires the successful integration of several cultural practices that adversely affect its population density and potential to cause crop damage. Early planting in kharif avoids the peak midge activity during the flowering period. Planting of uniform flowering varieties and hybrids and sowing of early flowering types under delayed sowing so as to synchronize in a given region avoids and minimizes the risk of heavy population build-up. Prevention of carryover of diapausing larvae by destruction of chaffy panicles and earhead residues as well as the removal of Johnson grass and other grassy types, which serve as an alternate hosts are highly beneficial. Plowing before planting and after harvest greatly reduces the carryover of pests like stem borer, white grubs, grasshoppers, and hairy caterpillars as they are not only exposed to birds and other predators and parasites but also to the adverse weather factors such as high temperatures and low relative humidity.[28]

High seed rate and destroying the dead hearts after removal to maintain an optimal plant stand were effective against shoot fly. Weeding suppress many pests as many common weeds act as hosts for oviposition by shoot fly, stem borers, armyworms, etc. Weed-free sorghum crops harbor less armyworms as compared with weed-infested crops. Burning of stubbles and infected/chaffy earheads is reportedly disrupts the carryover of stem borer. Stalks from the previous season should be fed to cattle or burnt before the onset of monsoon rains to reduce the carryover of stem borer. Destroying volunteer eliminates shoot fly, stem borer, midge, sugarcane aphid, shoot bug, and other panicle pests.

Fertilizing crops can increase infestation and survival of borers through an increase in the nitrogen content of plants. Sorghum plants grown without fertilizers were less preferred for oviposition by *C. partellus*.[35] It was also observed in another report that nitrogen fertilization enhances borer development as well as the plant's tolerance to borer attack.

9.3.2 BIOLOGICAL CONTROL

9.3.2.1 PARASITOIDS

The parasitoid, *Trichogrammatoidea simmondsi* caused 8.8–12.3% egg parasitism in sorghum + cowpea intercrop in Burkina Faso.[36] *Aprostocetus* sp., *Callitula bipartitus*, *Neotrichoporoides* sp., and *Neotrichoporoides nyemitawus* are recorded in many places and especially in the month of August.[37] *N. nyemitawus* significantly reduced the population build-up of shoot fly.[36] Intercropping of sorghum with cowpea has a beneficial effect in increasing the levels of parasitism.[36] An endo-parasite, *Trichoplasta* sp. is reported to parasitise shoot fly pupae. *Anagrus virlai* was reported to be a good egg parasitoid of *Dalbulus maidis* and *P. maidis*.[38] *Aprostocetus gala*, *Eupelmus popa*, and *Orius maxidentex* were reported on sorghum midge; and *Cotesia ruficrus* on armyworm.

Biological control of pearl millet leaf miner in fields of West Africa is achieved by the release of the larval parasitic wasp, *Habrobracon hebstor*.[39]

The parasitic wasp, *N. nyemitawus,* was found to parasitize 7–19% of larvae of *A. hyalinipennis* in eastern Ethiopia.[40] Parasitoids, *Cotesia flavipes* and *Xanthopimpla stemmator* were recorded in on stem borer, *C. partellus.* About 9–20% parasitism of *Cotesia sesamiae* was observed in all the stem borer populations surveyed in sorghum and pearl millet in Eritrea along with pupal parasitoids, *Psilochalsis soudanensis, Pediobius furvus,* and *Procerochasimas nigromaculaus.*[41]

9.3.2.2 PREDATORS

Spiders play a major role in suppressing the shoot fly population.[42] *Abrolophus* spp. (Acari: Erythraeidae) was reported to feed on the larvae and eggs of *A. soccata.*[43] Several species of spiders predates on shoot fly eggs.[44] It was reported that salticids, aranaeids, and thomisids were predominant in the sorghum monocrop, but the population of *Araneus* sp., *Misumenops* sp., *Neoscona* sp., *Latrodectus geometricus, Meioneta prosectes, Pardosa injuncta,* and *Steatoda badia* were more in sorghum-cowpea intercropping system.[36] *Coccinella septempunctata* and *Chrysoperla rufilabris* predates on the sugarcane aphid in sorghum.[45] Many predators such as *Aphelinus nigritus, Lysiphlebus testaceipes*, coccinellids, syrphids, green lacewings, brown lacewings, and pirate bugs are found suppressing sugarcane aphid in sorghum field.[46] Black ants (*Pheidole megacephala* Fabricius), earwigs (*Diaperasticus erythrocephala* Olivier), and spiders were found to feed on the larva of *B. fusca* and *C. partellus.*[41]

9.3.2.3 MICROBES

Beauveria bassiana was reportedly effective against shoot bug, *P. maidis* in sorghum.[47] Deepthi et al.[48] evaluated biorational pesticides and Bt treatment gave the best leaf protection and less deadheart. Three sprays of Bt (var. *kurstaki*) were reported to effectively manage sorghum stem borer (*C. partellus*) and enhanced grain yield.[49]

9.3.3 MANAGEMENT THROUGH BOTANICALS

Tobacco, *Nicotiana* sp. leaf powder (3 kg in 500 L water/ha) sprayed on tef exhibited a pest mortality of 80–83% and revealed as the best treatment followed by neem seed powder extract (3 kg in 500 L water/ha).[9] Neem spray was found effective in reducing blister beetles in pearl millet even to a tune of 75–76% in Mali.[30] Neem seed kernel extract 5% was found effective against stem borer, *C. partellus.*[50]

9.3.4 HOST PLANT RESISTANCE

Host plant resistance forms the backbone of pest management in millets. For some insect species, it is the only way of effective pest control. The salient feature of using resistant cultivars is that virtually no skill in pest control application techniques or cash investment is involved. Over the last 50 years, many world millet germplasm collections are being assessed for resistance to insect pests resulting in the identification of many resistance lines to major insect pests. Identification of resistant lines on one hand and underpinning the mechanism of resistance to insects pests on the other is needed to have a successful host plant resistance lines in shoot fly (*A. soccata*), stem borer (*C. partellus*), shoot bug (*P. maidis*), midge (*S. sorghicola*), head bug (*C. angustatus*), and aphid (*M. sacchari*).[51] Many marker traits are identified such as glossiness for resistance to shoot fly, short and tight glumes for midge, long glumes for head bug, etc.

The susceptible sorghum genotypes to shoot bug, *P. maidis* is associated with biophysical characters such as green color leaves, nonglossiness, leaves with more trichome, and waxiness.[52] However, Anaji and Balikai[53] found no correlation between any of the morphological characters and shoot bug infestation. Sorghum hybrid HG35W was reportedly resistant to sugarcane aphids, *M. sacchari.*[16] Many wild relatives of sorghum were found to be less attractive for sorghum midge especially for oviposition except *Sorghum halepense.*[29]

9.3.5 CHEMICAL CONTROL

For soil pests like white grubs, imidacloprid 600FS seed treatment at 10–12 mL/kg was reported to reduce seedling mortality from 28.6 to 2.1% and doubling grain yield from 12.30 to 27.52 q/ha resulting in the highest cost: benefit ratio (1:24.8).[33] Clothianidin 50WDG at 10 g/kg or mixing of chlorpyriphos 20 EC or quinalphos 25 EC at 4.0 L/ha along with irrigation water 3 weeks after seedling emergence is found effective against soil pests.[1] Chlorpyriphos or quinalphos dust can be mixed with FYM and applied in furrows at the time of planting is also effective against soil pests.[10] Chlorpyriphos 20 EC @ 1.25 L is applied along with irrigation water for termite management in millets. Dusting of quinalphos 1.5 DP or malathion 5 DP @ 25 kg/ha for grey weevil is recommended.

Seed treatment with thiamethoxam 70 WS @ 3 g/kg,[54,55], thiamethoxam 30 FS @10 mL/kg,[56] and imidacloprid 600 FS @ 7 mL/kg seed[57] gave good control against shoot fly in millets. For sprays, imidacloprid 17.8 SL @ 75 mL/ha, indoxacarb 15.8 SC @ 125 mL/ha, chlorantraniliprole 18.5 SC @ 100 mL/ha, and fipronil 5SC @ 175 mL/ha at 3–4 leaf stage was reported to reduce the shoot fly incidence to a greater extent.[57] Whorl application of carbofuran 3G @ 8–10 kg/ha at 20 days after emergence is found effective against stem borer, *C. partellus*.[58] In case of shoot bug and aphids, acephate 75 SP @ 1 g/L, imidacloprid 17.8 SL @ 0.3 mL/L, fipronil 5 SC @ 0.5 mL/L were found effective.[45] Poison baits comprising 10 kg rice bran + 1 kg jaggery + 1 L quinolphos was found effective against cutworms.[58] Imidacloprid 70 WS @ 2 g/kg seed treatment was reported effective in reducing the population of shoot bug.[42] Thiamethoxam 70WS @ 3 g/kg seed or imidacloprid 70WS @ 5 g/kg seed are the most effective treatment against shoot bugs.[59]

9.3.6 INTEGRATED PEST MANAGEMENT

One method cannot solve all insect pest problems in a crop and different methods should be used in integration to attain successful pest management. The use of insect-resistant millet germplasm is also a tactic of IPM that functions to reduce the pest abundance.[60] A combination of techniques like host plant resistance, proper fertilizer application followed by need-based insecticide application was effective against *A. soccata* resulting in enhanced grain yield in sorghum.[61] Application of carbofuran 3G @ 2 g/m row in soil + use of high seed rate (~10 kg/ha) and thinning at 20 days of germination + egg parasitoid, *T. chilonis* three times release @ 5 lakhs at 7, 14, and 21 days old crop as a full package of IPM resulted in low shoot fly incidence.[62] Seed treatment with thiamethoxam 70 WS @ 3 g/kg seed and intercropping of sorghum with chickpea (2:2) and spray of NSKE 5% at 45 DAE of crop were effective in reducing the shoot fly population.[15] It was also reported that only seed treatment with thiamethoxam followed by spray of NSKE 5% at 35 DAE as effective against shoot fly.[63] Imidacloprid 70 WS @ 3 g/kg seed treatment and spraying of NSKE 5% at 45 days DAE as the best module, which recorded lowest deadhearts caused by shoot flies (6.74%), aphids (13.85/sq cm) and shoot bugs (8.63/plant) resulting in highest yield of 26.6 q/ha.[64] An approach of proper planting time of a resistant variety and need-based use of insecticide effectively manage sugarcane aphid on grain sorghum.[65] Lama et al.[66] concluded that the use of resistant variety along with nitrogen management itself can effectively manage sugarcane aphids in sorghum.

9.4 PEST MANAGEMENT IN SEQUENTIAL CROPPING SYSTEMS

Crop sequence is a pattern applied within a single field to succeeding crops over time. It could be the same crop or different crops in sequence.[67] Use of a strategic crop in sequence that improves soil fertility, break the cycle of crop associated weeds, insect pest, and disease by crop rotation is beneficial.[68] The remnant insects and pathogens harboring the crop stubbles and other residues may infect subsequent crop immediately.[69] A number of experiments concluded that growing of legume enrich soil nitrogen content and thus beneficial to the succeeding crop.

Pearl millet–Green gram/sesame/cowpea is usually practiced in the states of Rajasthan, Gujarat, Haryana, Uttar Pradesh, and Madhya Pradesh, whereas in the southern states of Tamil Nadu, Karnataka, Andhra Pradesh, and Telangana, pearl millet–pigeon pea/green gram/sunflower/soybean is grown. Pearl millet–wheat is one of the most important cropping systems in the Indo-Gangetic Plains of India, which is just next to the famous rice–wheat cropping system in terms of importance.[70] The pearl millet–wheat cropping system spreads over an area of about 2.26 million ha.[71] However, continuous practice of pearl millet–wheat cropping led to reduced productivity and a legume crop is introduced. Thus, pearl millet–wheat–cluster bean–barley was recommended for Rajasthan.[72]

Winter infestations of Hessian fly, *Mayetiola destructor* were significantly greater in wheat–pearl millet rotation than the wheat–soybean.[73] Infestation of false chinch bugs, *Nysius raphanus* in pearl millet seedlings was high in pearl millet grown after canola than that of wheat or rye as preceding crop.[73] The severity of root rot was low when winter wheat was planted after either legumes or sorghum[74] than continuous winter wheat; however, it is unaffected with proso millet as a sequential crop.[75] The incidence of stalk and neck rot (*Fusarium graminearum*) on pearl millet was less when grown after wheat than having canola as a preceding crop.[76]

Sorghum is grown in peanut fields as crop rotation to manage root-knot nematode, *Meloidogyne arenaria* in peanut. This crop sequence was found to reduce the infestation of *M. arenaria* to an extent of 44.3%. Though sorghum reduced the juvenile nematode population in peanut, it was found to recover quickly in the peanut crop even after two consecutive cultivation of sorghum. The value of using sorghum as a sequence crop in *M. areanaria* management in peanut is demonstrated when ethylene dibromide (EDB) is used as a control measure. Higher yields of peanut were realized in plots cultivated with sorghum for the past two season and use of EDB in peanuts. The yield of sorghum was also found high because of preceding peanut crop.[77] Thus growing of sorghum–peanut is found beneficial to both the crops. Like sorghum, pearl millet–peanut was also found to reduce galling of *M. arenaria* in peanut. However, the stem rot or Rhizoctonia limb rot are not found affected by this crop sequence.[78]

Rice fallow sorghum is usually under cultivation in coastal regions of Andhra Pradesh. Due to high humidity in coastal regions, heavy infestation of insect–pests and diseases was observed. Among the major pests attacking rice fallow sorghum, shootfly and aphids, and stem borer are found dominant.[79] In rice fallow sorghum, percent deadhearts were found to be in a range of 11–36 and the variety CSV 14R was found to be resistant.[80] Stem borer. *C. partellus* was found attacking sorghum grown in rice fallow under zero tillage conditions. The genotype CSH 16(C) was found to be resistant and NTJ-2 (C), NLCW-6, and N-14 were highly susceptible.[81]

9.5 PEST MANAGEMENT IN INTERCROPPING SYSTEMS

Intercropping is the concept of cultivating two or more crops simultaneously in a field. Ideally, intercropping offers many benefits, however, competition for space, light, or nutrients may result in adverse effect, which needs proper planning. Such competitions can be minimized by following patterns like paired row and paired-wider row or skip-row.[82] The crop chosen for intercropping system should act as a companion crop rather than a competitive crop. Castor intercropped with sorghum, though registered very low pest population, it was reported that the sorghum suppressed the growth of castor because of its faster growth and canopy formation. In fact, the small and lean castor plants are not suitable for pest infestation.[83,84] The yield advantage apart from pest suppression is to be achieved by reducing the competition and selecting a companion crop and paired row pattern is the best.[85,86] In terms of pest reduction, intercropping not only affects the establishment of insect pests right from alighting of pest on a crop plant by interacting the olfactory, gustatory, and visual clues but also influences their survival and reproduction. Further, intercropping influences the availability of natural enemies, and intercrops are demonstrated to have more number of generalized natural enemies than monocropping systems.[87] Karabo et al.[88] summarized four hypotheses to explain the effectiveness of intercropping in insect pest management.

i) ***THE NATURAL ENEMY HYPOTHESIS:*** This suggests intercropping patterns attract more predators and parasitoids than monocropping.[89]

ii) ***RESOURCE CONCENTRATION HYPOTHESIS:*** This suggests that herbivores will be more in monocrop than intercropping system due to the ability of the pest to locate the host plant quickly and easily, faster feeding, and ability to achieve higher reproduction rates.[90]

iii) ***APPROPRIATE–INAPPROPRIATE HYPOTHESIS:*** This suggests that unrelated plants closer to the host plants interrupt the behavior of the pest resulting in difficulty in host finding and host acceptance.[91]

iv) ***HOST–PLANT QUALITY HYPOTHESIS:*** This suggests that the intercropping negatively affects the quality and chemical suitability of the hosts to the pest.[92,93]

In one way or other, the intercropping by default reduces the pest build-up and thus aid in pest management.

9.5.1 INTERCROPPING OF SORGHUM

Sorghum intercropped with legume is reported to be the most economical and beneficial in terms of insect pest management. The stem borer larva collected from sorghum intercropped with legumes were found infested with parasitoids such as *Stenobracon rufus*, *Temelucha* sp., *P. furvus*, *Psilochalis sondanensis*, and *C. sesamiae* but field parasitization was not found in the borer larva collected from pure crop of sorghum.[88]

Sorghum + Black Gram

Black gram intercropped with sorghum was found to have less infestation of whiteflies and leaf hoppers. The presence of whitefly, *Bemesia tabacci* per leaf was 1.3 and 2.8 in intercrop and sole crop, respectively similarly leaf hopper, *Empoasca kerri* was 0.57 and 2.0 per leaf. In an intercropping system of black gram with sorghum, the occurrence of bean podborer, *M. vitrata* was 3.1 larva/plant as compared with 9.6 larvae/plant in sole crop of black gram. The damage in harvested pods of black gram due to *M. vitrata* was reported to be as low as 8.7% in intercrop whereas it was 18.4% in pure crop. The blue butterfly, *Lampides boeticus* was reportedly infesting the black gram pods more when grown as pure crop than when it is grown along with sorghum. The damage caused by *L. boeticus* in sole crop was 6.1% as compared with 4.7% in intercrop. The reduction of pod borer, *H. armigera* in black gram pods grown in intercropping system was 24.1% as compared with sole crop.[94] The leaf damage of beetle, *Madurasia obscurella* in black gram was found to be 69.5% in sole crop of black gram, whereas it was only 44.5% in black gram intercropped with sorghum. The predatory coccinellids also were reportedly high up to 4.7/plant in intercrop of sorghum + black gram whereas it was only 1.3/plant in sole crop of black gram.[94]

Sorghum + Cowpea

Pure crop of sorghum was reported to be infected by stem borer, *C. partellus* to a tune of 32.6%, while the incidence was very low to a tune of 9.2 and 16.4% in sorghum intercropped with lablab and cowpea, respectively.[95] Ampong-Nyarko et al.[96] reported an increased incidence of stem borer, *C. partellus* in sorghum monocrop (9.5 insects/10 plants) as compared with sorghum + cowpea (5.3 insects/10 plants) in the first season. In the second year also, 69.8% increased incidence of *C. partellus* was observed in monocrop of sorghum as compared with sorghum+ cowpea grown as single alternating rows using a partial replacement series with a sorghum and cowpea at inter-row spacing of 90 cm each and monocrop at 60 cm spacing. In cowpea, about 422 flower thrips, *Megalurothrips sjostedti* per 20 flowers were noticed in monocrop as compared with 287 thrips/20 flowers, registering a reduction of 31.9% in intercropping system. Further, intercropping was reportedly increase the effectiveness of insecticides used for managing both stem borer in sorghum and thrips in cowpea.[96] This experiment with sorghum and cowpea intercrop at alternate rows shows an advantage in terms of pest management in both the crops viz., sorghum and cowpea. It was also found that even the phytochemicals extracted from lablab or cowpea when sprayed on sorghum prevented the oviposition of stem borer, *C. partellus* for up to six days.[95] However, cowpea intercropped in maize was reported to favor foliage-feeding beetles, *Ootheca* spp.[97] and pod sucking bugs, *Clavigralla tomentosicollis* and *C. shabadi.*[98] So, selection of intercrop is to be done based on the main crop and the pests to be managed.

Sorghum + Pigeon Pea

Sorghum intercropped with pigeon pea reduced the incidence of leaf hopper, *E. kerri* compared with untreated control. An increased activity of predatory spiders and coccinellids is also found in the intercropping system.[99] Sorghum as intercrop in pigeon pea was reported to reduce the leaf hopper incidence in pigeon pea.[100]

Sorghum + Peanut

Sorghum as intercrop reduced the leaf hopper incidence in peanut.[101] The sorghum stem borer larva was parasitized by braconids significantly higher numbers in sorghum intercropped with peanut, than that found in monocrop. The population of coccinellids was found similar in monocrop of peanut and also when intercropped with sorghum.[88]

9.5.2 INTERCROPPING OF PEARL MILLET

Pearl millet offers a unique and ideal plant structure because of its excessive tillering for exploitation to manipulate different cropping ecosystems with different possible intercrops for the reduction of insect pests.[102,103] The productivity of pearl millet + cowpea and pearl millet + peanut intercrop system can be increased by choosing appropriate cultivars,[104,105] adjusting plating date, populations, spacing,[106,107] and application of fertilizer.[108]

Pearl Millet + Peanut

Pearl millet intercropped with peanut at a ratio of 1:2 or 1:1 was found to register 3.8 and 5.7 stem borer larva/plant, as compared with 9.2 stem borer larva/plant in the sole crop of pearl millet. The reduced incidence of stem borer larva in intercrop may be due to the difficulty in finding the host for oviposition by adult female stem borers in pearl millet plots intercropped with peanut[25] thus limiting the population build up. The percent infestation of stem borer in pearl millet was found more in sole crop (10.1%) than pearl millet intercropped with peanut in a ratio of 1:2 (6.6%). A reduction of 34.6% of stem borer infestation in pearl millet when intercropped with peanut was realized at a ratio of 1:2 in Sudan.[103] A corresponding increase in grain yield in pearl millet to a tune of 1249.3 kg/ha was realized in pearl millet + peanut intercrop (1:2) as compared with 1209.3 kg/ha in sole crop.[103]

Peanut when grown with pearl millet at a ratio of 8:1 was not only found to harbor very low insect pests but also economical as land equivalent ratio (LER) is high.[109] Low incidence of insect pests was found in peanut + pearl millet intercropping than peanut + soybean, peanut + red gram and peanut + cowpea. Peanut intercropped with pearl millet harbored more number of predators and parasitoids.[109] Particularly, this intercropping system was reported to increase the population of *Goniozus* sp., an important parasitoid of peanut leaf miner, thus aids in pest management in peanut.[110]

Pearl Millet + Soybean

Pearl millet + soybean (1:1) intercropping system was reported to have less defoliation and pod damage by pests. Grain yields were significantly higher in the pearl millet + soybean intercrop system than in the pearl millet + soybean + ground nut + cowpea system.[111]

Pearl Millet + Cowpea

Traditionally pearl millet intercropping systems in Africa involve hill planting of tall, late-maturing varieties and planting photosensitive, indeterminate cowpea varieties after 2–6 weeks of millet planting.[104] Low incidence of thrips, *M. sjostedti* and *Frankliniella schultzei* in cowpea was reported in cowpea pearl millet intercropping system. The crop configuration of a row of pearl millet with 10 rows of cowpea was reported to significantly reduce the larval population of millet stem borer, *Acigona ignefusalis* and also millet head miner, *H. albipunctella* infestation.[31] In Nigera, an intercropping system of cowpea + pearl millet (1:1) a reduction of 4.7% and 13.8% of aphids, *Aphis craccivora* and thrips, *M. sjostedti* were noticed in intercrop with respect to pure crop of cowpea. However, a significant high population of thrips was noticed in intercrop of millet: cowpea ratio of 2:4.[112] Care should be taken to finalize the crop with proper row patterns for effective pest Management.

Pearl Millet + Black Gram

In the intercropping system of pearl millet and black gram, a reduction in leaf beetle, *M. obscurella* damage in black gram to a tune of 23.7% was reported than the sole crop. The occurrence of whiteflies, *B. tabacci* and hopper, *E. kerri* were found to be 1.75 and 1.16/leaf in intercrop as against 2.8 and 2.0/leaf in sole crop of black gram. The incidence of bean pod borer, *M. vitrata* was found to be low in black gram intercropped with pearl millet (4.8 larva/plant) than sole crop of black gram (9.5 larva/plant). The damage caused by *M. vitrata* was 24.8% lower in black gram intercropped with pearl millet than sole crop. The reduction in damage caused by blue butterfly, *L. boeticus* and pod borer, *H. armigera* in black gram intercrop with pearl millet with respect to pure crop was 26.2% and 68.2%, respectively.[94] Thus black gram intercropped with pearl millet was reported to have pest management benefits in all the pests studied, *viz.*, *M. obsurella*, *B. tabacci*, *E. kerri*, *M. vitrata*, *L. boeticus*, and *H. armigera*. The predatory coccinellids were found 3.1 times more in intercrop than sole crop.[94]

9.5.3 INTERCROPPING OF FINGER MILLET

Finger Millet + Black Gram

Finger millet intercropped with black gram (1:1) was found to have more incidence of leaf folder, *Anticarsia irrorata* than sole crop of black gram. The incidence of the rice leaf roller, *Cnaphalocrocis medinalis* was at a rate of 1.3% in finger millet and 2% in black gram and 2% in finger millet + black gram intercrop. Thus intercropping of finger millet + black gram is not found beneficial for pest management aspects especially for *A. irrorata* and *C. medinalis*. However, a reduction of 12.5% in the infestation of legume pod borer, *Maruca testulalis* was observed in black gram grown as intercrop with finger millet than that of sole black gram crop. The sole crop of black gram was severely affected and caused yield reduction.[113]

Finger Millet + Horse Gram

The incidence of leaf folder, *A. irrorata* was observed in horse gram grown as sole crop and also in intercrop (finger millet + horse gram) (1:1) during 6 weeks stage and the pest was not observed afterward in the intercrop whereas the sole crop was continued to get infested. The initial infestation was also low in intercrop compared with the sole crop of horse gram. Further, *C. medinalis* was found only in the sole crop of horse gram and not in the intercrop.[113]

9.5.4 INTERCROPPING OF OTHER SMALL MILLETS

Intercropping of soybean and foxtail millet at row proportion of 2:1 and 4:2 registered the most advantageous LER and higher B:C ratio. The LER was found good in intercropping system of peanut and foxtail millet in a ratio of 6:1.[116] In the Northern zone of Karnataka, little millet and foxtail millet with pigeon pea in 4:2 row proportions is recommended.[117] In alfisols of Karnataka, little millet + pigeon pea intercropping system at a ratio of 5:1 recorded more dry weight, long ears, and more grain weight.[118]

TABLE 9.1 Pest Reactions in Some Millet based Intercropping Systems.

Millet crop	Intercrop	Pest Reaction
Sorghum	Black gram	Low whitefly, *B. tabacci* & leafhopper, *E. kerri*[94]
	Black gram	Very low podborer, *M. vitrata, H. armigera*[94]
	Black gram	Low blue butterfly, *L. boeticus*[94]
	Cowpea	Very low stem borer, *C. partellus*[95]
	Cowpea	Low thrips, *M. sjostedti*[96]
	Pigeon pea	Low leaf hopper, *E. kerri*[100]
	Lab lab	Low stem borer, *C. partellus*[95]
	Peanut	High braconid parasitization[88]
	Peanut	Similar population of coccinellids[88]
	Black gram	Increased population of predatory coccinellids[94]
	Brinjal	Low little leaf of brinjal[69]
Pearl millet	Peanut	Low *C. partellus*[25]
	Peanut	Very low *C. partellus*[103]
	Soybean	Less defoliation and podborer damage[111]
	Cowpea	Low thrips, *M. sjostedti* and *F. schultzei*[31]
	Cowpea	Low stem borer, *A. ignefusalis* and head miner, *H. albipunctella*[31]
	Cowpea	Very low aphids, *A. craccivora* and thrips *M. sjostedti*[112]
	Black gram	Very low *M. obscurella, B. tabacci, M. vitrata, H. armigera*[94]
	Black gram	Low *E. kerri, L. boeticus*[94]
	Peanut	High parasitoids and predators[109]
	Peanut	High peanut leaf miner parasitoid, *Goniozus* sp.[110]
Finger millet	Black gram	More leaf folder, *A. irrorata*[113]
	Black gram	Similar rice leaf folder, *C. medinalis*[113]
	Black gram	Low *M. testulalis*[113]
	Horse gram	Very low leaf folder, *C. medinalis*[113]
	Sesame	Low incidence of gall midge but no difference in gall midge damage in capsule[114]
	Desmodium	Low *C. partellus*[115]

In Bihar and Uttar Pradesh, proso millet + green gram (2:1) was observed as a common combination of intercropping.[34] Intercropping of pigeon pea with proso millet (1:2) resulted in higher net return and B:C ratio over sole cropping.[119] In China, proso millet is intercropped with mung bean, which had a yield advantage of 37.3% compared with sole crop.[120] Milenkovic et al.[121] reported proso millet + soybean (1:1) resulted in high biomass yield in Serbia. Midega et al.[115] reported that intercropping of finger millet with desmodium significantly reduce stem bore damage on finger millet.

9.5.5 INTERCROPPING WITH VEGETABLES

When sorghum is grown in brinjal fields and little leaf of brinjal was found to occur in low levels. This may be because of breaking the continuity of feeding of vector in brinjal. The nonhost plant in mixtures may emit chemicals or odor that affects the insect pests, thereby protecting the host plants.[69] Egg laying and dead heart caused by shoot fly, *A. pulla* was found very low in little millet + onion/garlic system. About 0.27 and 0.43 eggs/plant was observed in little millet + onion and little millet + garlic, whereas it was 1.57 eggs/plant in sole crop of little millet. The reduction in deadheart to a tune of 74.7 and 72.8% was obtained in little millet onion and garlic intercropping, respectively as compared with sole crop.[122] Karibasavaraja et al.[123] reported a low incidence of shoot flies in sorghum intercropped with onion or garlic.

9.6 PEST MANAGEMENT IN MIXED CROPPING SYSTEMS

The tolerance of millets to high temperatures, less rainfall, and unfertile soils leads to its cultivation in such areas also. However, farmers think that the sole cultivation of millets in such fragile environment is risky and often lead to crop failure, they grow black gram, green gram, cluster bean cowpea, moth bean etc in mixed cropping system apart from intercropping.[124] Conventionally, pearl millet and cluster bean are broadcasted as crop mixtures and grown together in Rajasthan.[125]

Mixed cropping of finger millet with pulses or oilseeds is reportedly reduces the weed and pest attack.[117] Mixed cropping of pearl millet and cowpea had very less effect on the important insect pests and natural enemies. However, cowpea aphids, *A. craccivora* were found low in mixed cropping system than sole crop of cowpea, but the predatory coccinellids (*Cheilomenes vicina, Exochomus flavipes, Adonia* sp. and an unidentified species) were found more in sole crop. Though adults of the bean podborer, *M. vitrata* was high in mixed cropping system (14.1 insects per 100 m row) than sole crop (3.5 per 100 m), the infestation was found similar in mixed and sole crop. Incidence of African pod bug, *Clavigralla tomentosicollis* was reported similar in both the cropping system. The infestation of beanfly, *Ophiomyia phaseoli* was found similar in both sole and intercrop, but the parasitoid, *Opius* sp. was found 2.17 times more in mixed crop, that is, pearl millet and cowpea.[126]

9.7 PEST MANAGEMENT IN ALLEY CROPPING SYSTEMS

Alley cropping is planting of trees in wide-spaced rows with companion crops grown between their rows. The concept of alley cropping itself is enhancement of production and providing protection to the crop. In India, especially in the semiarid regions, alley cropping of pearl millet with wild tamarind tree (*Leucaena leucocephala*) is practiced. Though the yields of alley-cropped finger millet is less than the sole crop, a spacing of 2.8 or 3.3 m planting of tree and six rows of finger millet in the alley space allow the crop to evade shading effect resulting in good yield.[127] Reduction in the infestation of stem borers in maize when grown in alley with *L. leucocephala* than in a maize monocrop was reported in Kenya.[128]

Growing of pearl millet under the apple-ring acacia trees, *Faidherbia albida* is common in West Africa.[129] It is a deep-rooted, leguminous tree that drops all the leaves during the growing season[130] and do not have shade effect on the crop. Pearl millet grown under *F. albida* are reported to have significantly higher yield of 36–169%,[131,132] which is attributed to the soil nutrients, availability of water, conducive microclimate, and better soil physical properties.[133]

In Senegal, pearl millet is grown along with apple-ring acacias (*F. albida*), soapberry trees (*Balanites aegyptica*), baobabs (*Adansonia digitata*), tamarinds (*Tamarindus indica*), and red acacias (*Acacia seyal*). Apart from giving shade to the crop, the trees harbor insectivorous birds and bats thus aid in pest management. The Mauritian tomb bat, *Taphozous mauritianus* was found to feed on the moth and larva of millet head miner, *H. albipunctella* (Lepidoptera, Noctuidae) whereas the weaver bird (*Ploceus cucullatus*) feed only on the late instar larva, thus revealing bat as an effective predator. However, the emergence of pearl millet panicle coincides with the nest building of weaver birds, thus aid in collection of large number of late instar larva to feed the chicks.[134] This study demonstrates pest management through natural habitat conservation. The trees played a role in the natural regulation of the millet head borer indirectly by providing habitat to the insectivorous birds (neem tree for village weavers) and bats (palms for tomb bats).[134] However, pearl millet grown under baobass (*A. digitata*) was found to have an increased incidence of millet head borer, which could be due to the food and refuge given by the tree to the pest.[135]

9.8 MILLETS AS TRAP CROP, BARRIER CROP, AND COVER CROP

Trap cropping in pest management is managing the pest with the most preferred plant over the cash crop.[136] In United States, sorghum is reported to be an effective trap crop for southern green stink bug, *Nezara viridula* in cotton. Most of the *N. viridula* adults are found in sorghum and remained there instead of moving into cotton. Sorghum is also reported to be an attractant and trap crop to the brown marmorated stink bug, *Halyomorpha halys* followed by sunflower, okra, and pearl millet.[137] Overall, percent parasitism of *N. viridula* adults by tachinid fly, *Trichopoda pennipes* was also more in sorghum.[138] The tachinid fly, *Trichopoda pennis,* was reported as the most important parasitoid of *N. viridula* reducing the pest population.[139] Further, 85.7% of reduction in insecticide sprays for stink bug was reported in cotton crop with trap crop of sorghum than sole crop of cotton. The ETLs were not reached in cotton or corn fields with sorghum plots.[138] Pearl millet and sorghum besides acting as trap crop for stink bugs and leaf-footed bugs also serve as pollen source for bees and parasitic Tachinidae (*Trissolcus* spp.) of stink bug and other general beneficial insects like ladybeetle and lacewings.[140,141]

Sorghum or pearl millet can be recommended as excellent trap crops since they are amenable for ratooning to increase their temporal efficacy. Planting of several commercially available cultivars with a range in maturity dates to make it available and attractive for the insects pests for a longer duration is also effective. Ratooning of sorghum works best if the plants are cut at a height of 0.5 m, as mature seed heads are not attractive to stink bugs. The plants will tiller and set new panicles in 3–4 weeks.[142]

In orchards and vineyards, cover crops can alter the pest infestations by favoring the population of beneficial arthropod, keeping the pest population well below ETL.[69] Millet hedges when grown with watermelons are reported to act as a barrier and significantly reduced watermelon mosaic virus infection.[143] However, sorghum-sudan grass and pearl millet grown usually as summer cover crops in vegetable fields reportedly increased root-knot (*Meloidogyne incognita*), ring (*Mesocriconema* spp.) and lesion (*Pratylenchus* spp.) nematodes.[144] So selection of cover crops is to be done carefully keeping in view of the targeted pest.

9.9 CONCLUSION

Millets are hardy crops grown under resource poor conditions and practically integrated pest management in millets is challenging. However, pest management tactics like host plant resistance and cultural management are being practiced. Pest management in millets is to be aimed as a system approach rather than managing an individual pest, as that of an important pest in a cash crop. Proper selection of crops in millet based cropping sequence can lead to effective pest management by suppression the pest build-up and keeping it well below levels that cause economic damage, that is, ELTs. Intercropping besides having advantages of crop diversification and assured income generation to farmers, aids in effective pest management also. Strategic companion crops interrupt pest colonisation and build-up by influencing the host selection and colonization by the pest. Millets grown under specific trees in alley cropping were reported to have less pest incidence, as the trees act as an attractant to insectivorous birds and bats by giving food and shelter, thus aiming in an ecosystem approach of pest management in millets.

KEYWORDS

- **millets**
- **pest management**
- **cropping sequence**
- **intercropping**
- **mixed cropping**
- **alley cropping**
- **trap crop**

REFERENCES

1. Kalaisekar, A.; Padmaja, P. G.; Bhagwat, V. R.; Patil, J. V. *Insect Pests of Millets: Systematics, Bionomics and Management*; Elsevier: New York, 2017.
2. Nigus, C.; Damte, T. Identification of the Tef Shoot Fly Species from Tef, *Eragrostis tef* (Zucc.), Trotter Growing Areas of Ethiopia. *Afr. J. Insects* **2018,** *5*, 181–184.
3. Kishore, P. Evolving Management Strategies for Pests of Millets in India. *J. Entomol. Res.* **1996,** *20* (4), 287–297.
4. Selvaraj, S.; Natarajan, V. S.; Raghupathy A. On the Occurrence of Shoot Fly and Its Damage in Some Varieties of Little Millet. *Indian J. Entomol.* **1974,** *44*, 556–557.
5. Natarajan, V. S.; Selvaraj, S.; Reghupathy A. Assessment of Loss in Grain Yield Caused by Shoot Fly, *Atherigona destructor* M. (Anthomyiidae: Diptera) in Certain Varieties of Panivarugu, *Panicum miliaceum. Sci. Culture* **1974,** *40*, 502–504.
6. Dhillon, M. K.; Sharma H. C.; Singh, R.; Naresh, J. S. Mechanisms of Resistance to Shoot Fly, *Atherigona soccata* in Sorghum. *Euphytica* **2005,** *144*, 301–312.
7. Biradar, A.; Sajjan, S. Management of Shoot Fly in Major Cereal Crops. *Int. J. Pure Appl. Biosci.* **2018,** *6*, 971–975.
8. Singh, O.; Sharma, R. P.; Singh, R. P. Effect of Sowing Dates on the Incidence of Shootfly (*Atherigona simplex*) and Yield of Some Improved Kodo (*Paspalum scrobiculatum*) Strains. *Bull. Entomol.* **1984,** *25* (1), 22–26.
9. Mideksa, A.; Negeri, M.; Shiberu, T. Management of Tef Shoot Fly, *Atherigona hyalinipennis* (Reg.) (Diptera: Muscidae) on Tef at Ambo, West Showa of Ethiopia. *J. Entomol. Nematol.* **2014,** *6*, 134–139.
10. Gahukar, R.T.; Reddy, G. V. P. Management of Economically Important Insect Pests of Millet. *J. Integrated Pest Manage.* **2019,** *10* (1), 28.
11. Mason, S.; Maman, N.; Pale, S. 2015. Pearl Millet Production Practices in Semi-Arid West Africa: A Review. *Exp. Agric.* **2015,** *51* (4), 501–521.
12. Gahukar, R. T. Insect Pests of Pearl Millet in West Africa: A Review. *Trop. Pest Manage,* **1984,** *30* (2), 142–147.
13. Prasad, G. S.; Babu, K. S. Insect Pest Resistance in Pearl Millet and Small Millets. In *Biotic stress resistance in millets*; Das, I. K., Padmaja, P. G., Eds.; Academic Press, Elsevier Publications, 2016; pp 221–233.
14. Akshatha, S.; Karabhantanal, S. S.; Jagginavar, S. B. Shoot Bug, *Peregrinus maidis* (Ashmead): A Potential Threat to Rabi sorghum. *J. Entomol. Zool Stud.* **2020,** *8* (6), 1387–1391.
15. Balikai, R. A.; Bhagwat V. R. Evaluation of Integrated Pest Management Components for the Management of Shoot Fly, Shoot Bug and Aphid in Rabi Sorghum. *Karnataka J. Agric. Sci.* **2009,** *22*, 532–534.
16. Serba, D. D.; Michaud, J.P. Sugarcane Aphid Resistance in Pearl Millet. *Kansas Field Res.* **2019,** *5* (6), 146–151.
17. Matthee, J. J. Waak Teen Plantluise op kafferkoring. *Boerdin Soc. Afr.* **1962,** *38* (10), 27–29.
18. Van Rensburg, N. J.; Van Hamburg, H. Grain Sorghum Pests: An Integrated Control Approach. In *Proceedings of the First Congress of Entomological Society of Southern Africa*; 1975; pp 151–162.
19. Van den Berg, J. Resistance of Sorghum Hybrids of the Sorghum Aphid, *Melanaphis sacchari* (Zehntner) (Homoptera: Aphididae). *SA J. Plant Soil.* **2002,** *19*, 151–155.

20. Swathy, K. J. Manimegalai, S. Evaluation of Grain Sorghum Entries Against Spotted Stem Borer, *Chilo partellus* (Swinhoe). *Madras Agric. J.* **2017,** *104*, 363–367.
21. Ganguli, R. N.; Chaudhary, R. N.; Ganguli, J. Effect of Time of Application of Chemicals on Management of Maize Stem Borer, *Chilo partellus* (Swinhoe). *Int. J. Pest Manage.* **1997,** *43*, 253–259.
22. Sekhon, S. S. Kanta U. Effect of Seed Treatment and Other Insecticidal Formulations on the Control of Maize Borer, *Chilo partellus* (Swinhoe). *J. Insect Sci* **1992,** *5*, 45–47.
23. Pavani, T.; UmaMaheswari, T.; Sekhar, J. C. Evaluation of Efficacy of Different Insecticides and Bioagents Against *Sesamia inferens* Walker in Maize. *Eur. J. Zool. Res.* **2013,** *2* (4), 98–102.
24. Seshu-Reddy, K. V. Studies on the Stem Borer Complex of Sorghum in Kenya. *Insect Sci. App.* **1983,** *4*, 3–10.
25. Harris, K. M. Recent Advances in Sorghum and Pearl Millet Stem Borer Research. In *International Workshop on Sorghum Stem Borers*, 17–20 November **1987,** ICRISAT Centre, Patanchueru, A.P.; 1989, India; pp 9–16.
26. Gebre-Amlak, A. Survey of Lepidopterous Stem Borers Attacking Maize and Sorghum in Ethiopia. *Ethiopian J. Agric. Sci.* **1985,** *7*, 15–26.
27. Gahukar, R. T.; Guevremont, H.; Bhatnagar, V. S.; Doumbia, Y. O.; Ndoye, M.; Pierrard, G. A Review of the Pest Status of the Millet Spike Worm, *Raghuva albipunctella* De Joannis (Noctuidae: Lepidoptera) and Its Management in the Sahel. *Int. J. Trop. Insect Sci.* **1986,** *7*, 457–463.
28. Prasad, G. S.; Stanley, J.; Babu, K. S.; Subbarayudu, B.; Kalaisekar, A. Major Pests: Status, Approaches, and Strategies for Management. In *Sorghum in the 21st Century: Food – Fodder – Feed – Fuel for a Rapidly Changing World*; Tonapi, V. A., Talwar H. S., Are, A. K., Bhat, B. V., Reddy, C. R., Dalton, T. J., Eds.; Springer: Singapore, 2020.
29. Sharma, H. C.; Franzmann, B. A. Host-Plant Preference and Oviposition Responses of the Sorghum Midge, *Stenodiplosis sorghicola* (Coquillett) (Dipt.; Cecidomyiidae) Towards Wild Relatives of Sorghum. *J. Appl. Entomol.* **2001,** *125* (3), 109–114.
30. Toure, G. K.; Edwards, R. C.; Traore, S. H. Approaches to Blister Beetle Control on Millet: Botanical and Biological Agents, Associational Resistance, Varietal Resistance, and Light and Pheromone Traps, 1999, Virgina Tech accessed through https://vtechworks.lib.vt.edu/handle/10919/66501
31. Gahukar, R. T. Pest and Disease Incidence in Pearl Millet Under Different Plant Density and Intercropping Patterns. *Agric. Ecosyst. Environ.* **1989,** *26*, 69–74.
32. Stanley, J.; Supyal, V. Insect Pests of Millets and Their Management. In Food Security Through Millet Crop Production; Sood, S., Khulbe, R., Pandey, B. M., Gupta, A., Bhatt, J. C., Eds.; Uttarakhand State Department of Agriculture: Almora, 2013; 74 p.
33. Choudhary S. K.; Tandi, B. L.; Singh, S. Management of White Grub *Holotrichia consanguinea* Blanchard in Pearl Millet by Seed Treatment. *Indian J. Entomol.* **2018,** *80* (3), 619–622.
34. Chapke, R. R.; Prabhakar, Prasad, G. S.; Das I. K.; Tonapi, V. A. Improved Millet Production Technologies and Their Impact. Technology Bulletin, ICAR-Indian Institute of Millets Research: Hyderabad, 2018, India; 84p.
35. Van den Berg, J.; Van Rensburg, J. B. J. Unavoidable Losses in Insecticidal Control of *Chilo partellus* (Swinhoe) (Lepidoptera: Pyralidae) in Maize and Grain Sorghum. *SA J. Plant Soil.* **1991,** *8*, 12–16.

36. Zongo, J.; Vincent, O. C.; Stewart, R. K. Effects of Intercropping Sorghum-Cowpea on Natural Enemies of the Sorghum Shoot Fly, *Atherigona soccata* Rondani (Diptera: Muscidae) in Burkina Faso. *Biol. Agric. Hort.* **1993,** *9*, 201–213.
37. Jotwani M. G. Integrated Approach to the Control of the Sorghum Shoot Fly. *Insect Sci. App*, **1981,** *2*, 123–127.
38. Hill, J. G.; Albarracin, E. L.; Araoz, M. V. C.; Virla, E. G. Effects of Host Species and Host Age on Biological Parameters of *Anagrus virlai* (Hymenoptera: Mymaridae), an Egg Parasitoid of *Dalbulus maidis* (Hemiptera: Cicadellidae) and *Peregrinus maidis* (Hemiptera: Delphacidae). *Biol. Control.* **2019,** *131*, 74–80.
39. Payne, W.; Tapsoba, H.; Baoua, I. B.; Malick, B. N.; N'Diaye, M.; Dabire-Binso, C. On-Farm Biological Control of the Pearl Millet Head Miner: Realization of 35 Years of Unsteady Progress in Mali, Burkina Faso and Niger. *Int. J. Agric. Sustainabil.* **2011,** *9*, 186–193.
40. Sileshi, G. Biology of Shoot Fly, *Atherigona hyalinipennis* van Emden in Eastern Ethiopia. *Int. J. Trop. Insect Sci.* **1997,** *17*, 349–355.
41. Haile, A.; Hofsvang, T. Survey of Lepidopterous Stem Borer Pests of Sorghum, Maize and Pearl Millet in Eritrea. *Crop Protection* **2001,** *20* (2), 151–157.
42. Bailey, C. L.; Chade, H. L. Spider Population in Grain Sorghum. *Entomol. Soc. Am.* **1968,** *61*, 567–571.
43. Reddy, K. V. S.; Davies, J. C. A Predaceous Mite on the Eggs of Sorghum Shoot Fly, *Atherigona soccata* (Diptera: Muscidae) at Hyderabad. *Acarol. Newslett.* **1978,** *6*, 9.
44. Delobel, A. G. L.; Lubega, M. C. Rainfall as a Mortality Factor in the Sorghum Shoot Fly, Atherigona soccata Rondani (Diptera: Muscidae) in Kenya. *Insect Sci. App.* **1984,** *2*, 67–71.
45. Hewlett, J. A.; Szczepaniec A.; Eubanks, M. D. The Effects of Sugarcane Aphid Density in Sorghum on Predation by Lady Beetles and Lacewings. *Biol. Control.* **2019,** *129*, 171–177.
46. Maxson, E. L.; Brewer, M. J.; Rooney, W. L.; Woolley, J. B. Species Composition and Abundance of the Natural Enemies of Sugarcane Aphid, *Melanaphis sacchari* (Zehnter) (Hemiptera: Aphididae), on Sorghum in Texas, 2019.
47. Balikai, R. A.; Kambrekar, D. N.; Natikar, P. K.; Anaji. R. Bio-Ecology and Management of Shoot Bug, *Peregrinus maidis* (Ashmead) on Sorghum and Maize—A Review. *Biochem. Cell Arch.* **2017,** *17* (1), 27–40.
48. Deepthi, J.; Shekharappa, R.; Patil, R. K. Evaluation of Biorational Pesticides for the Management of Stem Borer, *Chilo partellus* Swinhoe in Sweet Sorghum. *Karnataka J. Agric. Sci.* **2008,** *21*, 293–294.
49. Kandalkar, H. G.; Men, U. B.; Efficacy of *Bacillus thuringiensis* var. Kurstaki Against Sorghum Stem Borer, *Chilo parellus* (Swinhoe). *J. Biol. Control* **2006,** *20* (1), 101–104.
50. Ambarish, S.; Biradar, A. P.; Jagginavar, S. B.; Karbhanatanal, S. S.; Sajjanar, G. M. Management of Insect Pests in Rabi Sorghum*Sorghum bicolor* (L.) Moench. *Int. J. Chem. Stud.* **2017,** *5* (6), 642–646.
51. Sharma, H. C.; Bhagwat, V. R.; Padmaja, P. G. Techniques to Screen Sorghums for Resistance to Insect Pests. In *Sorghum Improvement in the New Millennium Patancheru*; Reddy, B. V. S., Ramesh, S., Ashok, K. A., Gowda, C. L. L., Eds.; International Crops Research Institute for the Semi-Arid Tropics: Andhra Pradesh, India, 2008; pp 31–49.

52. Rajasekhar, P. Assessment of Crop Loss and EIL Due to Shoot Bug, *Peregrinus maidis* Ashmead on Sorghum. *Ann. Agric. Res.* **1996,** *17*, 333–334.
53. Anaji, R.; Balikai, R. A. Correlations of Biophysical and Biochemical Characters with Shoot Bug Incidence in Rabi Sorghum. *Int. J. Agric. Sci.* **2006,** *2* (2), 450–451.
54. Balikai, R. A. Seed Treatment, an Eco-Friendly Management Tactic for the Suppression of Insect Pests in Sorghum. *Int. J. Plant Protection* **2011,** *4* (2), 381–384.
55. Daware, D. G.; Bhagwat, V. R.; Ambilwade, P.P.; Kamble, J.R. Evaluation of Integrated Pest Management Components for the Control of Sorghum Shoot Pests in Rabi Season. *Indian J. Entomol.* **2012,** *74*, 58–61.
56. Kumar, R.; Tiwana, U. S. Control Efficacy of Different Seed Dressing Insecticides Against Sorghum Shoot Fly, *Atherigona soccata* (Rondani) in Forage Sorghum, *Sorghum bicolor* (L.) Moench. *J. Entomol. Zool. Stud.* **2018,** *6* (2), 795–799.
57. Sandhu, G. S. Evaluation of Management Components Against Shoot Fly in Sorghum. *Ann. Plant Protection Sci.* **2016,** *24*, 67–70.
58. Prasad, G. S.; Bhagwat, V. R.; Babu, K. S.; Tonapi, V. A. *Pests of Millets-a Ready Reckoner*; Indian Institute of Millets Research, Rajendranagar, Hyderabad, Telangana, 2016; 61p.
59. Anaji, R.; Balikai, R. A. Management of Shoot Bug, *Peregrinus maidis* (Ashmead) Through Seed Treatment in Rabi Sorghum. *Ann. Plant Protection Sci.* **2012,** *20* (1), 37–41.
60. Teetes, G. L.; Peterson, G. C.; Nwanze, K. N.; Pendleton, B. B. Genetic Diversity of Sorghum: A Source of Insect Resistant Germplasm. In *Global Plant Genetic Resources for Insect-Resistant Crops*; Clement, S. L., Ed.; CRC Press, 2019; 320 p.
61. Obonyo, D. N.; Ogola, J. B. O.; Kamau, A. W. Effects of Time of Planting, Fertilizer and Insecticide on Sorghum Shoot Fly, *Atherigona soccata* (Diptera: Muscidae), on Two Varieties of Sorghum. *Afr. Entomol.* **2008,** *16*, 185–195.
62. Balikai, R. A. Integrated Pest Management for Shoot Fly (*Atherigona soccata* Rondani) in Rabi Sorghum. *Agric. Sci. Digest.* **2003,** *23*, 291–293.
63. Daware, D. G.; Ambilwade, P. P. Comparative Study Between IPM and non IPM Module (Farmer's Practice) for the Control of Sorghum Shoot Pests in Kharif Season. *J. Entomol. Res.* **2014,** *38*, 17–21.
64. Karabhantanal, S. S.; Prasad, G. S.; Biradar, A. P.; Boranayaka, M. Evaluation of Integrated Pest Management Components in the Management of Insect Pests in Rabi Sorghum. *J. Exp. Zool.* **2018,** *21* (2), 1245–1249.
65. Haar, P. J.; Buntin, G. D.; Jacobson, A.; Pekarcik, A.; Way, M. O.; Zarrabi, A. Evaluation of Tactics for Management of Sugarcane Aphid (Hemiptera: Aphididae) in Grain Sorghum. *J. Econ. Entomol.* **2019,** *112*, 2719–2730.
66. Lama, L.; Wilson, B. E.; Davis, J. A.; Reagan, T. E. Influence of Sorghum Cultivar, Phenological Stage, and Fertilization on Development and Reproduction of *Melanaphis sacchari* (Hemiptera: Aphididae). *Florida Entomol.* **2019,** *102* (1), 194–201.
67. Steinmann, H.; Dobers, E. S. Spatio-Temporal Analysis of Crop Rotations and Crop Sequence Patterns in Northern Germany: Potential Implications on Plant Health and Crop Protection. *J. Plant Dis Protection* **2013,** *120*, 85–94.
68. Bullock, D. G. Crop Rotation. *Crit. Rev. Plant Sci.* **1992,** *11*, 4309–4326.
69. Rana, S. S.; Rana, M. C. *Cropping System*; Department of Agronomy, College of Agriculture, CSK Himachal Pradesh Krishi Vishvavidyalaya: Palampur, 2011; 80p,
70. Kumar, P.; Nanwal, R. K.; Yadav, S. K. Integrated Nutrient Management in Pearl Millet (*Pennisetum glaucum*)–Wheat (*Triticum aestivum*) Cropping System. *Indian J. Agric. Sci.* **2005,** *75*, 640–643.

71. Yadav, R. L.; Subba-Rao, A. V. M. *Atlas of Cropping Systems in India*; PDCSR Bulletin No. 2001–02, Project Directorate for Cropping Systems Research. Modipuram, Meerut, Uttar Pradesh, 2002.
72. Sammauria, R.; Meena, O. P.; Yadav, M. R.; Gupta, K. C.; Gupta, A. K.; Yadav H. L. Crop Diversification of Predominant Pearl Millet Based Cropping System for Higher Productivity, Resource-Use Efficiency and Profitability in Semi-Arid Rajasthan. *Legume Res. Int. J.* **2020,** *43*, 568–572.
73. Buntin, G. D.; Cunfer, B. M.; Phillips, D. V.; Wilson, J. P. Sequence and Rotation Effects on Pest Incidence and Grain Yield of Double-Cropped Soybean and Pearl Millet After Wheat and Canola. *Crop Manage.* **2007.** DOI:10.1094/CM-2007-1023-01-RS.
74. Wildermuth, G.; McNamara, R. B. Effects of Cropping History on Soil Populations of *Bioolaris sorokiniana* and Çommon Root Rot of Wheat. *Austr.J. Agric. Res.* **1991,** *41*, 1–10.
75. Anderson, R. L. Diversity and No-Till: Keys for Pest Management in the U.S. Great Plains. *Weed Sci.* **2008,** *56*, 141–145.
76. Wilson, J. P.; Cunfer, B. M.; Phillips, D. V. Double-Cropping and Crop Rotation Effects on Diseases and Grain Yield of Pearl Millet. *J. Prod. Agric.* **1999,** *12*, 198–202.
77. Rodriguez-Kabana, R.; Touchton, J. T. Corn and Sorghum as Rotational Crops for Management of *Meloidogyne arenaria* in Peanut. *Nematropica* **1984,** *14* (1), 26–36.
78. Timper, P.; Brenneman, T. B. Wilson, J. P. Pearl Millet as a Rotation Crop for Peanut. *Plant Health Progress*. 2007. https://apsjournals.apsnet.org/doi/abs/10.1094/PHP-2007-0202-02-RS
79. Chapke, R. R.; Babu, S.; Subbarayudu B.; Tonapi, V. A. Growing Popularity of Sorghum in Rice Fallows—An IIMR Case Study, Technology Bulletin, ICAR-Indian Institute of Millets Research, Hyderabad, 2017; 40p.
80. Yogeswari, P.; Rani, C. S.; Rao, G. R.; Kumar, V. M. Influence of Bio-Physical Factors on Shoot Fly Infestation in Different Sorghum Genotypes Cultivated Under Zero Tillage in Rice Fallows. *J. Entomol Zool. Stud.* **2019a,** *7* (1), 808–814.
81. Yogeswari, P.; Rani, C. S.; Rao, G. R. Performance of Sorghum Genotypes Under Zero Tillage Conditions in Rice Fallows with Reference to Stem Borer *Chilo partellus*. *Curr. Sci.* **2019b,** *117* (4), 692–699.
82. Adesiyun, A. A. Some Effects of Intercropping of Sorghum, Millet and Maize on Infestation by Lepidopterous Stalkborers, Particularly *Busseola fusca*. *Insect Sci. App.* **1983,** *4*, 387–391.
83. Trenbath, B. R. Intercropping for the Management of Pests and Diseases. *Field Crops Res.* **1993,** *34*, 381–405.
84. Rao, M. S.; Rao, C. A. R.; Srinivas, K.; Pratibha, G.; Vidya-Sekhar, S. M.; SreeVani G.; Venkatswarlu, B. Intercropping for Management of Insect Pests of Castor, *Ricinus communis*, in the Semi–Arid Tropics of India. *J. Insect Sci.* **2012,** *12*, 14.
85. Ofori, F.; Stern, W. R. Cereal-Legume Intercropping Systems. *Adv. Agron.* **1987,** *41*, 41–90.
86. Chatterjee, B. N. Mandal, B. K. Present Trends in Research on Intercropping. *Indian J. Agric. Sci.* **1992,** *62*, 507–518.
87. Langer, V.; Kinane, J.; Lyngkjaer, M. Intercropping for Pest Management: The Ecological Concept. In *Ecologically Based Integrated Pest Management*; Koul, O., Cuperus, G. W., Eds.; CAB International, 2007; pp 74–110.
88. Karabo, O.; Obopile, M.; Tiroesele, B. Insect Diversity and Population Dynamics of Natural Enemies Under Sorghum–Legume Intercrops. *Trans. R. Soc. SA* 2019. https://www.tandfonline.com/doi/abs/10.1080/0035919X.2019.1658654

89. Vandermeer, J. *The Ecology of Intercropping*; Cambridge University Press: Cambridge, 1989; 237p.
90. Root, R. B. Organisation of Plant–Arthropod Association in Simple and Diverse Habitats: The Fauna of Collards (*Brassica oleraceae*). *Ecol. Monographs* **1973,** *43*, 95–124.
91. Finch, S.; Collier, R. H. Host-Plant Selection by Insects—A Theory Based on 'Appropriate/ Inappropriate Landings' by Pest Insects of Cruciferous Plants. *Entomologia Experimentalis et Applicata* **2000,** *96*, 91–102.
92. Bach, C. E. Host Plant Growth form and Diversity: Effects on Abundance and Feeding Preference of a Specialist Herbivore, *Acalymma vittata* (Coleoptera: Chrysomelidae). *Oecologica* **1981,** *50*, 370–375.
93. Van-Lenteren, J. C. Sustainable and Safe Crop Protection: A Reality? In Oomen, P. A., Forster, R., Lewis, G. G., Eds.; *Proc. 50th Int. Symp. Crop Protection 63* (2b). Gent. **1998,** 409–413 pp.
94. Soundararajan, R. P.; Chitra, N. 2012. Impact of Intercrops on Insect Pests of Blackgram, *Vigna mungo* L. *J. Entomol. 9* (4), 208–219.
95. Mahadevan, N. R.; Chelliah, S. Influence of Intercropping Legumes with Sorghum on the Infestation of the Stem Borer, *Chilo partellus* (Swinhoe) in Tamil Nadu, India. *Trop. Pest Manage.* **1986,** *32* (2), 162–163.
96. Ampong-Nyarko, K.; Seshu Reddy, K. V.; Nyangor, R. A. and Saxena, K. N. Reduction of Insect Pest Attack on Sorghum and Cowpea by Intercropping. *Entomol. Experimentalis et Applicata* **1994,** *70*, 179–184.
97. Karel, A. K.; Lakham, D. A.; Nduguru, B. S. Intercropping of Maize and Cowpea. Effect of Plant Population on Insect Pest and Seed Yield. In *Proceedings of the 2nd Symposium on Intercropping in Semi-Arid Areas*; Keswani, C. L., Ndunguru, B. J., Eds.; International Development Research Centre: Canada, 1980; pp 102–109.
98. Gethi, M.; Khaemba, B. M. Damage by Pod-Sucking Bugs on Cowpea When Intercropped with Maize. *Trop. Pest Manage.* **1991,** *37*, 236–238.
99. Rao, M. S.; Reddy, K. D.; Singh, T. V. K. Impact of Intercropping on *Empoasca kerri* of Pigeonpea in Rainy and Post Rainy Season. *Indian J. Entomol.* **2003,** *65*, 506–512.
100. Sekhar, J. C.; Singh, K. M.; Singh, R. N.; Yeshbir, S.; Malik, K. S. Impact of Intercropping on the Incidence of Green Jassid, *Empoasca kerri* on Pigeon Pea. *Indian J. Entomol.* **1997,** *59*, 119–123.
101. Nath, P. Singh, A. K. Effect of Intercropping on Ground Nut with Millets and Pigeon Pea on the Relative Incidence of Insect Pests. *Ann. Plant Protection Sci.* **1998,** *6*, 151–154.
102. Willey, R. W. Intercropping Its Importance and Research Needs 1. Competition and Yield Advantage and 2. Agronomy and Research Approaches. *Field Crop Abstracts.* **1979,** *32*, 73–85.
103. Degri, M. M.; Mailafiya, D. M.; Mshelia J. S. Effect of Intercropping Pattern on Stem Borer Infestation in Pearl Millet (*Pennisetum glaucum* L.) Grown in the Nigerian Sudan Savannah. *Adv. Entomol.* **2014,** *2* (2), 45429.
104. Reddy, K. C.; van der Ploeg, J.; Maga, I. Genotype Effects in Millet/Cowpea Intercropping in the Semi-Arid Tropics of Niger. *Exp. Agric.* **1990,** *26*, 387–396.
105. Ntare, B. R. Intercropping Morphologically Different Cowpea with Pearl Millet in a Short Season Environment in the Sahel. *Exp. Agric.* **1990,** *26*, 41–47.
106. Ntare, B. R.; Williams, J. H. Response of Cowpea Cultivars to Planting Pattern and Date of Sowing in Intercrops with Pearl Millet in Niger. *Exp. Agric.* **1992,** *28*, 41–48.

107. Reddy, K. C.; Visser, P.; Buckner, P. Pearl Millet and Cowpea Yields in Sole and Intercrop Systems, and Their After-Effects on Soil and Crop Productivity. *Field Crops Res.* **1992,** *28*, 315–326.
108. Ntare, B. R.; Bationo, A. Effects of Phosphorus on Yield of Cowpea Cultivars Intercropped with Pearl Millet in Niger. *Fertiliz. Res.* **1992,** *32*, 143–147.
109. Kennedy, J. S.; Balaguranathan. R.; Christopher, A.; Rajamanickam, K. Insect Pest Management in Peanut: A Cropping System Approach. *Trop. Agric.* **1994,** *71* (2), 116–118.
110. Dhaliwal, G. S.; Arora, R. *Principles of Insect Pest Management*; National Agricultural Technology Information Centre: Ludhana, 1996.
111. Sastawa, B. M.; Lawan, M.; Maina, Y. T. Management of Insect Pests of Soybean: Effects of Sowing Date and Intercropping on Damage and Grain Yield in the Nigerian Sudan Savannah. *Crop Protection* **2004,** *23* (2), 155–161.
112. Muhammed, I. B.; Yusuf A. U.; Wudil, B. S. Influence of Cowpea Genotype and Millet-Cropping System on Cowpea Damage by Thrip and Aphid Pests in the Sudan Savannah of Nigeria. *Bayero J. Pure Appl. Sci.* **2010,** *3* (2), 110–114.
113. Pan, R. S.; Sarkar, P. K.; Shinde, R.; Kumar, R.; Mishra, J. S.; Singh, A. K.; Bhatt, B. P. Effect of Diversified Cropping System on Insect and Weed Pest Diversity. *J. Entomol. Zool. Stud.* **2020,** *8* (1), 387–391.
114. Egonyu, J. P.; Kyamanywa, S.; Anyanga1, W.; Ssekabembe, C. K. Review of Pests and Diseases of Sesame in Uganda. *Proc. Afr. Crop Sci. Conf. Afr. Crop Sci. Soc.* **2005,** *7*, 1411–1416.
115. Midega, C. A. O.; Zeyaur, R. K.; David, M. A.; Jimmy, P.; John, A. P. Integrated Management of *Striga hermonthica* and Cereal Stemborers in Finger Millet (*Eleusine coracana* (L.) Through Intercropping with *desmodium intortum*. *Int. J. Pest Manage.* **2010,** *56* (2), 145–151.
116. Shwethanjali, K. V.; Naik, A. H. K.; Basavaraj Naik, T.; Kumar, M. D. Effect of Groundnut + Millets Intercropping System on Yield and Economic Advantage in Central Dry Zone of Karnataka Under Rainfed Condition. *Int. J. Curr. Microbiol. Appl. Sci.* **2018,** *7*, 2921–2926.
117. Meena, D. S.; Gautam, C.; Patidar, O. P.; Singh, R.; Meena, H. M.; Vishwajith and Prakash, G. Management of Finger Millet Based Cropping Systems for Sustainable Production. *Int. J. Curr. Microbiol. Appl. Sci.* **2017,** *6* (3), 676–686.
118. Patil, N. B.; Halikatti, S. I.; Sujay, Y. H.; Kumar, P. B. H.; Topagi, S. C.; Pushpa, V. Influence of Intercropping on the Growth and Yield of Little Millet and Pigeon Pea. *Int. J. Agric. Sci.* **2010,** *6*, 573–577.
119. Manjunath, M.; Vajjaramatti and Kalaghatagi, S. B. Performance of Pigeon Pea and Millets in Intercropping Systems Under Rainfed Conditions. *J. Farm Sci* **2018,** *31*, 199–201.
120. Gong, X.; Ferdinand, U.; Dang, K.; Li, J.; Chen, G.; Luo, Y.; Yang, P. Feng, B. Boosting Proso Millet Yield by Altering Canopy Light Distribution in Proso Millet/Mung Bean Intercropping Systems. *Crop J.* **2019,** *8*, 365–377.
121. Milenkovic, M.; Simic, M.; Brankov, M.; Milojkovic–Opsenica, D.; Kresovic, B.; Dragicevic, V. Intercropping of Soybean and Proso Millet for Biomass Production. *J. Process. Energy Agric.* **2019,** *23*, 38–40.
122. Sathish, R.; Manjunatha M, Rajashekarappa, K. Effect of Intercropping on Incidence of Shoot Fly, *Atherigona pulla* (Wiedemann) in Little Millet. *Int. J. Pure Appl. Biosci.* **2017,** *5* (4), 1845–1849.

123. Karibasavaraja, L. R.; Balikai, R. A.; Deshpande, V. P. Intercroppings for Management of Sorghum Shoot Fly. *Ann. Plant Protection Sci.* **2005,** *13* (1), 237–238.
124. Rani, S.; Goyat, R.; Soni, J. K. Pearl Millet (*Pennisetum glaucum* L.) Intercropping with Pulses Step Towards Increasing Farmer's Income Under Rainfed Farming: A Review. *Pharma Innov. J.* **2017,** *6* (10), 385–390.
125. Singh, R. P. Studies on Intercropping in Guar. *Annual Report Kharif 1977*; Dryland Agriculture Main Centre: Jodhpur, India, 1977.
126. Bottenberg, H.; Tamo, M.; Singh, B. B. Occurrence of Phytophagous Insects on Wild *Vigna* sp. and Cultivated Cowpea: Comparing the Relative Importance of Host-Plant Resistance and Millet Intercropping. *Agric. Ecosyst. Environ.* **1998,** *70* (2–3), 217–229.
127. Corlett, J. E.; Black, C. R.; Ong, C. K.; Monteith, J. L. Above and Belowground Interactions in a Leucaena/Millet Alley Cropping System. II. Light Interception and Dry Matter Production. *Agric. Forest Meteorol.* **1992,** *60* (1–2), 73–91.
128. Ogol, K. P. O.; Spence, J. R.; Keddie, A. Maize Stem Borer Colonization, Establishment and Crop Damage Levels in a Maize-Leucaena Agroforestry System in Kenya. *Agric. Ecosyst.*. **1999,** *76* (1), 1–15.
129. Garrity, D. P.; Akinnifesi, F. K.; Ajayi, O. C. et al. Evergreen Agriculture: A Robust Approach to Sustainable Food Security in Africa. *Food Security* **2010,** *2*, 197–214.
130. Roupsard, O.; Ferhi, A.; Granier, A.; Pallo, F.; Depommier, D.; Mallet, B.; Joly, H. I. and Dreyer, D. Reverse Phenology and Dry-Season Water Uptake by *Faidherbia albida* (Del.) A. Chev. in an Agroforestry Parkland of Sudanese West Africa. *Funct. Ecol.* **1999,** *13*, 460–472.
131. Mokgolodi, N. C.; Setshogo, M. P.; Ling-ling, S.; Yu-jun, L.; Chao, M. A. Achieving Food and Nutritional Security Through Agroforestry: A Case of *Faidherbia albida* in Sub-Saharan Africa. *Forestry Studies China.* **2011,** *13*, 123–131.
132. Reij, C.; Tappan, G.; Smale, M. Agroenvironmental Transformation in the Sahel: Another Kind of 'Green Revolution'. In Discussion Paper 00914; International Food Policy Research Institute: Washington, DC, 2009.
133. Kho, R. M.; Yacouba, B.; Yaye, M.; Katkore, B.; Moussa, A.; Iktam, A.; Mayaki, A. Separating the Effects of Trees on Crops: The Case of *Faidherbia albida* and Millet in Niger. *Agroforestry Syst.* **2001,** *52*, 219–238.
134. Sow, A.; Seye, D.; Faye, E.; Benoite, L.; Galan, M.; Haran J.; Brevault, T. Birds and Bats Contribute to Natural Regulation of the Millet Head Miner in Tree-Crop Agroforestry Systems. *Crop Protection.* **2020,** *132*, 105127.
135. Schroth, G.; Krauss, U.; Gasparotto, L.; Aguilar, J. A. D.; Vohland, K. Pests and Diseases in Agroforestry Systems of the Humid Tropics. *Agrofor. Syst.* **2000,** *50*, 199–241.
136. Hokkanen, H. M. T. Trap Cropping in Pest Management. *Annu. Rev. Entomol.* **1991,** *36*, 119–138.
137. Nielsen, A. L.; Dively, G.; Pote, J. M.; Zinati, G.; Mathews, C. Identifying a Potential Trap Crop for a Novel Insect Pest, *Halyomorpha halys* (Hemiptera: Pentatomidae), in Organic Farms. *Environ. Entomol.* **2016,** *45* (2), 472–478.
138. Tillman, P. G. Sorghum as a Trap Crop for *Nezara viridula* L. (Heteroptera: Pentatomidae) in Cotton in the Southern United States. *Environ. Entomol.* **2006,** *35*, 771–783.
139. Temerak, S. A.; Whitcomb, W. H. Parasitoids of Predaceous and Phytophagous Pentatomid Bugs in soybean Fields at Two Sites of Alachua County, Florida. *Zeitschr. Angew. Entomol.* **1984,** *97*, 279–282.

140. Bugg, R. L.; Dutcher, J. D. *Sesbania exaltata* (Rafinesque-Schmaltz) Cory (Fabaceae) as a Warm-Season Cover Crop in Pecan Orchards: Effects on Aphidophagous Coccinellidae and Pecan Aphids. *Biol. Agric. Hort.* **1993,** *9*, 215–229.
141. Tedders, W. L. Insect Management in Deciduous Orchard Ecosystems: Habitat Manipulation. *Environ. Manage.* **1983,** *7*, 29–34.
142. Mizell, R. F.; Riddle, T. C.; Blount, A. S. Trap Cropping System to Suppress Stink Bugs in the Southern Coastal Plain. *Proc. State Hort. Soc.* **2008,** *121*, 377–382.
143. Chalfant, R. B.; Jaworski, C. A.; Johnson, A. W.; Sumner, D. R. Reflective Film Mulches, Millet Barriers and Pesticides: Effects on Watermelon Mosaic Virus, Insects, Nematodes, Soil-Borne Fungi, and Yield of Yellow Summer Squash. *J. Am. Soc. Hort. Sci.* **1977,** *102*, 11–15.
144. Bhan, M.; McSorley R.; Chase, C. A. Effect of Cropping System Complexity on Plant-Parasitic Nematodes Associated with Organically Grown Vegetables in Florida. *Nematropica* **2010,** *40*, 53–70.

CHAPTER 10

Advances in Pest Management in Pulses-Based Cropping Systems under Changing Climate

N. P. SINGH, R. K. MISHRA*, SANJAY M. BANDI, and NARENDRA KUMAR

ICAR-Indian Institute of Pulses Research, Kanpur, India

Corresponding author. E-mail: rajpathologist@yahoo.com

ABSTRACT

Pulses are the principal sources of dietary protein among the vegetarian populations and are an integral part of the daily diet in several forms worldwide. These crops are mainly grown under rainfed conditions and the productivity levels of these crops are quite low mainly because of several biotic and abiotic stresses affect at various stages of crops. Recently, climate variability is also leading to changes in crop pest and diseases scenario which may also increase the losses of crop plants. Management of these biotic stresses of pulses and chemical pesticides is the most effective but due to several adverse effects of these pesticides in human, animal, and plants, integrated pest management (IPM) strategies are the most adoptive and effective integrative approaches worldwide. Crop rotation and cultural practices may reduce the intensity of diseases and insect pests are another important component of IPM. Biological control agents/biopesticides are another important area of IPM which are widely accepted and using worldwide for the management of plant diseases and insect pest and also for the promotion of plant growth. Hence, there is a urgent need to explore and develop an economically viable and cost-effective strategy for diseases and insect pest management in pulses in relation to climate change.

Integrated Pest Management in Diverse Cropping Systems. Devendra Pal Singh, PhD (Ed.)

10.1 INTRODUCTION

The production of food grain in India has continuously increased in the past few decades. It thus provided self-sufficiency in food. A significant change was observed through popularizing the new advances technologies including the high-yielding varieties, innovation in integrated pest management (IPM) strategies, use of balanced fertilizers, safer pesticides, and expansion of cropped area. However, the growth in agricultural commodities needs to be continued feeding the increasing population in the country. The available newer potential technology will be a key component to the future growth of agriculture commodities. In India, a quantum jump in the production of pulse has been witnessed since 2016–2017. This record pulse production is attributed to 19% increase in area (29.29 M ha) resulting in 36% increase in production (23.98 MT) during 2017–2019 as compared to those during 2014–2016. Pulse crops are highly susceptible against various biotic and abiotic stresses at different crop stages. These crops are mainly grown by resource-poor farmers in rainfed conditions and biotic factors are major constraints in realizing the potential yield of pulses. The insect pests and diseases cause major yield losses of approximately 30% every year in pulses.[1]

The increased heat stress, changes in monsoon patterns are significantly affecting the production of pulses mainly in the tropics than the temperate regions.[2] Due to the uncertain and unpredictable nature of climate phenomenon and their associated consequences on pulse production, the development of mitigation strategies is a difficult task.[3] In this chapter, we reviewed the pulse inclusive cropping systems, distribution, and symptomatology of key pests of and diseases of major pulses, namely, chickpea, pigeonpea, lentil, fieldpea, rajmash, mungbean, urdbean, cowpea, and their management.

10.2 PULSES PRODUCTION SCENARIO OF INDIA

India ranked first both in the production and consumption of pulses. Due to the high level of grain protein content, the pulses are predominantly eaten by vegetarians in India. Therefore, pulses hold a great promise in view of its role in alleviation of protein hunger and malnutrition that is prevalent amongst the poor in the society. Pulse production in India has witnessed a quantum jump since 2016–2017. There is more than 19% increase in area (29.29 M ha) and 36% increase in production (23.98 MT) of pulses during 2017–2019 as compared to those during 2014–2016. The average pulse availability was 28.70 MT during 2017–2019 and significantly (30%)

increased during 2014–2016 (22 MT). This has also put a break on increasing import basket of pulses which has reduced from the highest level of 6.61 MT during 2016–2017 to nearly 2.53 MT during 2018–2019. Thus, it can loudly be said that a pulse revolution has been ushered in India making the country self-sufficient in Pulses.

During 2018, the global grain legumes production was 94.99 million tonnes from an area of 96.70 million hectares with an average yield of 982 kg/ha.[109] India not only contributes maximum (around 25–28%) to the total global grain pulses production but also largest consumer in the world. Since 2010, the country has witnessed upward trend in pulses production and reached to all time highest of 25.42 mt during 2017–2018 (Figure 10.1). The availability of pulses per capita in India has also shown increasing in last two decades although it is still below the recommendation of ICMR of 52 g/capita/day. With several other options available for meeting protein requirements and also their changing food habits, about 32 million tonnes of pulses are needed by the year 2030 and 39 million tonnes by the year 2050 with an annual growth rate of 2.2%.[4]

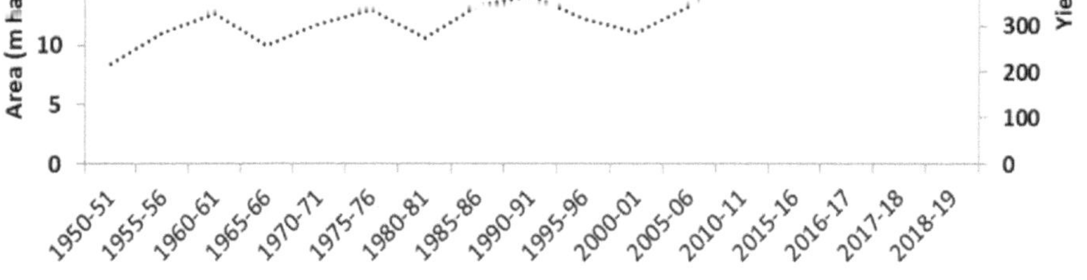

FIGURE 10.1 Trends of pulses production in India.

The pulse production during 2018–2019 was 23.40 million tonnes with 29.03 million ha area and 806 kg/ha yield.[5] On the share of pulses (2018–2019), chickpea accounted for 43.3% of total pulses production, while pigeonpea 15.3%, urdbean 13.1%, mungbean 10.5%, and lentil 5.3%. State-wise, the highest share of pulses (2018–2019) comes from Madhya Pradesh (33.39%), followed by Rajasthan (15.71%), Uttar Pradesh (10.24%), Maharashtra

(9.23%), Karnataka (7.94%), and which together share about 76.5% of the total pulse production while the remaining 23.5% is contributed by Andhra Pradesh, Jharkhand, Tamil Nadu, Chhattisgarh, Gujarat, West Bengal, and Bihar. Pulses are mainly grown under rainfed (81% area) and formed an integral part of the different cropping systems in the country.

10.3 MAJOR PULSE-BASED CROPPING SYSTEMS IN THE COUNTRY

After the Green Revolution, cereal-based systems, namely, rice–wheat, rice–rice, and maize–wheat becomes dominant cropping systems in the country. The following of monoculture resulted in several soil, environment, and pests in different agroecologies of the country. In increased areas under high inputs, agriculture has aggravated the second generation problem. Hence, to reverse the negative processes associated with continuous following of cereal–cereal cropping systems, the diversification/intensification of existing cereal-based cropping systems could be a strategic option.

Among 250 double-cropping systems, about 30 are of primary importance.[6] Pulses, being low inputs requirement and ability to fix atmospheric nitrogen in association with *Rhizobium,* are candidate crops for diversification/intensification of cereal-based cropping systems. The positive effects of the inclusion of pulses on soil health improvements and agricultural sustainability have been documented by many researchers.[7]

10.4 CLIMATE CHANGE AND CROP PRODUCTION

In spite of evidence on effect of climate change on status and intensity pests and diseases, the accurate prediction of their influence is very difficult. Changes in temperature, moisture, and atmospheric gases are altering growth and generation rates of plants, pathogens, and insects. The weather parameters particularly temperature, rainfall, relative humidity, hours of sunshine, and speed of wind are impacting the disease incidence. The deforestation or desertification leads to reduced land cover that can make the remaining plants and animals increasingly vulnerable to pests and diseases. The population dynamics of insect pests and incidence of disease and distribution is reported to be under the influence of weather parameters. The impact of increased heat stress, shift in monsoons, and drier soils could be a much greater threat to the production of pulses in the tropics than the temperate countries.[2] The rise in heat stress is reported to impact pest populations in several complex ways.[9]

There are unequivocal evidence on global climate change and its gravest challenge to agriculture and allied activities.[10] Based on the projections of Intergovernmental Panel on Climate Change, the increase in temperature up to 1.7°C in *kharif* and 3.2°C during *rabi* season are made. The mean rainfall is expected to increase by 10% by 2070.[11] Probability of 10–40% loss in crop productivity in India is expected by 2080–2100.[110] The pulses are mostly cultivated in rainfed areas and may be affected more.[12,13]

10.5 INSECT-PEST PROBLEMS AND CROP LOSSES IN PULSES

The grain legumes are damaged by different species of insect pests in the field and storage.[14] The high-yielding varieties and change in cropping pattern affected the pest complex of pulses. The minor pests of past have emerged as major pests in pulses in recent years.[15]

The pod borer (*Helicoverpa armigera* Hubner) is major pests in chickpea and pigeonpea in Asia, Africa, and Australia.[16] Likewise, the spotted pod borer effects the cowpea and pigeonpea adversely. Pod fly is the major threat to pigeonpea in India. The defoliators (*Spodoptera* spp.,) also affect pulses. Bihar hairy caterpillar (*Spilosoma obliqua* Walk) is a major pest in mungbean and urdbean in North India and red hairy caterpillar (*Amsacta* sp.) is found on rainy season pulses in south and central India.[17] The sucking pests of the order hemipteran and thysanoptera are also causing extensive damage in pulses and often have long-lasting physiological effect on the plant growth.[18] The major sucking pests in pulses are aphids, leafhoppers, pod sucking bugs, thrips, and whitefly. Apart from direct feeding, these are of several viral diseases. The bruchids the principal postharvest pests of pulses causing extensive damage worldwide.

Many of the insect pests are sporadic in the distribution or have restricted distribution to certain pockets or they seldom present at high densities causing economic losses. However, few of them can be devastating to pulse crops. The loss due to biotic stresses is estimated from 8 to 20%.[19] In general, the yield losses estimates in pulses were found to vary from 5 to 10% in the temperate regions and 50 to 100% in the tropics.[20] In chickpea such losses are about 15%.[21] The *H. armigera* alone causes losses of US$330 and US$310 million, respectively, in chickpea and pigeonpea annually in the world.[22] The bruchids are the major field-carry-over pest of stored pulses and reported to cause upto 90% grain damage in storage.[23] Deshmukh et al. (2010)[24] reported 41% avoidable loss of grain yield in chickpea due to *H. armigera.*

10.6 MAJOR INSECT PESTS OF PULSES

10.6.1 PIGEONPEA

The insect pests are adapted to feed on all the parts of pigeonpea crop.[25] However, the pests which occur continuously are economically important[26] (Table 10.1).

TABLE 10.1 The Insect Pests of Pigeonpea.

S. No.	Common name	Scientific name	Distribution
1	Pod borer	*Helicoverpa armigera*	Asia, Africa, and Australia
2	Spotted pod borer	*Maruca vitrata*	Asia, Africa, Australia, and Americas
3	Spiny pod borer	*Etiella zinckenella* Treitschke	Asia, Africa, and Americas
4	Leaf webber	*Grapholita critica* Meyr	Asia and Africa
5	Pod fly	*Melanagromyza obtusa* Malloch	Widespread in Asia
6	Pod sucking bugs	*Clavigralla gibbosa* Spinola, *Anoplocnemis* spp. *Riptortus* spp. *Nezaraviridula* (L.)	Asia, Africa, and Australia
7	Blister beetles	*Mylabris* spp.	Asia and Africa
8	Leaf hoppers	*Empoascakerri* Pruthi *Empoascafabae* Harris	Asia, Americas
9	Aphids	*Aphiscraccivora* Koch *Aphisfabae scopoli, Myzus persicae* (Sulzer)	Worldwide
10	Thrips	*Megalurothrips usitatus* (Bagnall) *Scirtothrips dorsalis* Hood	Worldwide
11	Cowbugs	*Otinotus oneratus* W. *Oxyrhachis tarandus* F.	Asia, Africa, and Caribbean
12	Scale insects	*Ceroplastodes cajani* Maskell; *Icerya purchasi* Maskell	Asia and Africa
13	Mites	Eriophyid mite, *Aceriacajani* Channabasavanna; *Tetranychus* spp.	Asia and Africa
14	Bruchids	*Callosobruchus* spp.	Worldwide

Pod Borer: H. Armigera (Hubner), Noctuidae: Lepidoptera

Podborer (*H. armigera*) is one of the major polyphagous pests distributed throughout the tropics and subtropics in Asia.[27] This pest is considered as a major constraint to pigeonpea production in South Asia. In addition to its ability to feed to the large number of host plants and wider distribution, the pest has high resistance toward insecticides.

Several hundred eggs on all plant parts are produced by female but have a preference for young shoots and florets. The eggs are deposited on flower buds and young pods in pigeonpea, whereas these are deposited on foliage in chickpea.[17] The young larva eats tender leaves, buds, and flowers, and subsequently bore the pods and feeds on seeds. A single larva is reported to destroy about 30–40 pods before it reaches maturity. Larvae make conspicuous entry holes on the pods and these feed on developing seeds. The foliages are commonly affected in chickpea.

In the case of different maturity groups of pigeonpea, and causes 29.70% pod damage in short-duration pigeonpea in North-West India, 13.10% in long-duration crop of North India, 24.30% in medium- and long-duration crop of central India, and 36.40% in south India.[28] The larva passes through five to six instars.[29] This pest completes one generation in about 4 weeks under favorable conditions.[30] It can complete seven or more generations per year in the southern parts of India and three to four generations in the northern states.[17]

Pod Sucking Bugs

Several species of pod-feeding bugs are commonly referred to as pod-sucking bugs. Bugs belonging to genus, *Anoplecnemis*, *Clavigralla,* and *Riptortus*. These are considered as serious pests of pulses.[31] Green stink bug (*Nezara viridula)* is present in most areas in India.[32,33] The other hosts of sucking bugs are beans, mungbean, urdbean, and cowpea.

These pod-sucking bugs deposit the eggs in clusters on leaf and pod surface. *C. scuttelaris* lays eggs in clusters of 18–20 eggs whereas; *C. gibbosa* deposits 10–12 sculptured eggs. Most of these species complete one generation in about 4 to 5 weeks. Different life stages of the bugs can be simultaneously observed on the pulse crops. Grains in damaged pods had dark patches, and turn shriveled and became unsuitable for food and seed purposes.

Pod Fly: Melanagromyza Obtusa Malloch, Agromyzidae: Diptera

Pod fly is widely distributed and causes major economic losses in northern parts of India, especially in long-duration pigeonpea varieties. It results 70–80% damage in pigeonpea.[35]

The maggots of the pod fly upon hatching mine into soft seeds and starts feeding. The second and third instar larvae drill deep into the seeds consuming the starchy food as well as the embryo and deposit excreta outside. Normally the maggot completes its development in a single seed. Infested seeds show characteristic mining galleries and lose more than two-thirds of their weight. The pest infestation is more on medium- and late-maturing varieties of the crop and causes 60–80% grain damage. But in south India, pod fly was more damaging to early maturing pigeonpea cultivars (56–59%) than medium- and late-maturing varieties (27–58%). In Uttar Pradesh, the pod fly damage to pigeonpea is huge (24.65%).[36]

The pod fly has been observed to breed on early and late pigeonpea varieties from September to April. During the off-season (May–August), it probably survives on perennial pigeonpea and alternative wild hosts, for example, *Rhynchosia minima*, *Flemingia congesta*, *Atylosia* spp., and *Tephrosia purpurea*. The crop maturing in February–March has low pod fly damage because its multiplication rate is low during winter (December–February). Therefore, advancing the sowing of long-duration varieties helps in minimizing the pod fly damage.[37] The adult female fly inserts a single egg into the developing pods. A single fly is found to lay 80–100 eggs.[38] Partly matured pods were preferred as compared to very tender or fully matured pods. The freshly hatched larva feeds upon developing seeds. It comes out of the seed and cuts a circular window on the pod wall, leaving a thin papery outer layer before pupation, which becomes the exit hole for the adult emergence. The larva pupates either in the mines of the seed or outside the seed in the pod locule. The lifecycle is normally between 25 and 40 days depending on weather factors. The adult fly is small shining and metallic blue or black in color, and its emergence takes place in the morning hours.[37]

Bruchids

Pulses in the tropical and subtropical countries are prone to infestation by bruchids of the subfamily *Bruchinae* in the field as well as storage. They are cosmopolitan pests with worldwide distribution. The major bruchid species, namely, *Callosobruchus chinensis* (L.) and *C. maculatus* (Fab.) are widespread across the globe while *C. analis* (F.) is found only to southeast Asia.

The female bruchids glue the eggs to the testa of the seeds.[38] The larvae are internal feeders and more than one larvae are reported to develop within a single seed. The reproductively mature and emerging adults do not require either food or water resources to reproduce.[39] Infested seed has round exit holes with the "flap" of the seed coat. The presence of life stages of the bruchids and completely hollowed seeds and are not suitable to human consumption. The heavy infestation results in moldiness, poor germination, and reduced weight.[40] Some bruchid species can infest the crop in the field and preharvest infestation is about 1–2%[18] and this field-carry-over inoculums causes up to 100% loss of stored seeds within 6–8 months of storage.[41,42] One generation is completed in about 4–5 weeks under optimum conditions. In the case of pigeonpea, infestations begin in the field.

10.6.2 CHICKPEA

A very few insect pests are found to infest chickpea as compared to other pulse crops. The fewer pest load of chickpea is probably because it is a cool-season crop and also because of presence of the dense glandular trichomes on all green tissues which excrete the acidic substance that deter the herbivores. The *kabuli* genotypes of chickpea are susceptible to insect pests.[17] Nearly 60 insect species are known to infest chickpea under field and storage conditions[29] (Table 10.2).

TABLE 10.2 The Major Insect Pests of Chickpea.

S. No.	Common name	Scientific name	Distribution
1	Pod borer	*Helicoverpa armigera* (Hübner)	Asia, Africa, Australia, and Americas
2	Aphids	*Aphis craccivora* Koch *Acrythosiphon pisum* (Harris)	Worldwide
3	Cutworms	*Agrotis ipsilon* (Hfn.)	Worldwide
4	Semiloopers	*Autographa nigrisigna* Walker	Asia
5	Tobacco caterpillar	*Spodoptera litura* F	Asia
6	Termites	*Microtermes obesi* (Holm.), *Odontotermes obesus* Ramb.	Asia
7	White grub	*Holotrichia consanguinea* Blanch	Asia

APHIDS

Two major species of aphids, *A. craccivora* and *A. pisum,* attack cultivated pulses in Africa and India,[43] America continent, and Australia. However, *A. craccivora* is reported as one of the most severe pest of major and minor pulses.[44]

The adults and nymphs of *A. craccivora* are shiny black in color with brown to yellow legs and nymphs will be covered with light waxy growth on their body. Females are apterous and viviparous.[45] Nymphs and adults congregate all along with young parts of the plant and cause direct damage by sucking the plant sap. Infestation is manifested by twisting of leaves followed by stunting of affected parts. In the case of severe infestation, wilting of plants, poor root or nodule development, and death of plants are observed.[34,26] Apart from direct feeding damage, *A. craccivora* transmits number of viral diseases in chickpea[46] and pea leaf roll virus causing chickpea stunt disease has assumed economic importance.[47]

Aphids remain active throughout the year but high incidence evident under drought conditions. They reproduce through parthenogenesis in the tropics and are viviparous. A gravid female can produce up to 100 nymphs in a span of 15 days. Nymphs undergo four molts and life cycle from nymph to adult stage is completed in 5–8 days.[48] Adult lives for about 6–15 days.[49] The generation completes in two weeks under favorable conditions.

10.6.3 MUNGBEAN AND URDBEAN

Nearly 64 species of insect pests are reported on mungbean.[50] Mungbean and urdbean are mostly infested by common species of insect pests[34] and sucking pests are considered to be important biotic constraints in mungbean and urdbean (Table 10.3).

Whitefly: Bemisia Tabaci Genn. Hemiptera: aleyrodidae

Apart from mungbean and urdbean, whiteflies infest several minor pulses and survive on several wild hosts during off-season. In India, whiteflies are distributed along southern and northern parts.[51]

Whiteflies are very small insects with whitish body and wax-coated wings. They lay eggs about 300 eggs during its lifespan on under the surface of leaves. Egg period lasts for about 3–5 days during summer and takes more time (5–33 days) in winter. Nymphal develops in 9–14 days during summer time and 17–73 days in the winter season. The adult lives for about 1–3 weeks upon hatching, nymphs find suitable site for feeding and stay intact until they pupate in flat, oval-shaped, scale-like puparium.[26] One generation can be completed in less than 30 days.

They suck the sap from leaves and reduce the plant vitality. In the case of severe infestation, plants show downward cupping of leaves, plants become sick and die. The sooty mold develops on honeydew excreted by whiteflies results in blackening of leaves and affect the normal photosynthesis of the host plant. It also acts as vector for Gemini viruses causing yellow mosaic disease. Among the pulses, whiteflies prefer urdbean followed by mungbean, ricebean, and pigeonpea.[52,53]

TABLE 10.3 The Major Insect Pests of Mungbean and Urdbean.

S. No.	Common name	Scientific name	Crop damaged
1	Pod borer	*Helicoverpa armigera* (Hubner)	Mungbean and urdbean
2	Spotted pod borer	*Maruca vitrata* (Geyer)	Mungbean and urdbean
3	Pod sucking bugs	*Riptortus pedestris* Fb., *Clavigralla gibbosa* (Spinosa), *Nezara viridula* L.	Mungbean
4	Defoliators	*Spodoptera litura* F., *Spilosoma obliqua* Walk., *Amsacta* sp.	Mungbean and urdbean
5	Aphids	*Aphis craccivora* Koch,	Mungbean and urdbean
		Acyrthosiphon pisum Harris *Myzus persicae* Sulz	Mungbean
6	Whitefly	*Bemisia tabaci* Genn.	Mungbean and urdbean
7	Thrips	*Megalurothrips distalis* (Karny), *M. usitatus* (Bagnall)	Mungbean and urdbean
8	Leafhoppers	*Empoasca kerri* Pruthi	Mungbean and urdbean
9	Blister beetles	*Mylabris* spp.	Mungbean and urdbean
10	Bruchids	*Callosobruchus* spp.	Mungbean and urdbean

10.6.4 LENTIL

Among the several insect pests infesting the lentil under field and storage conditions, only a few are economically important requiring control measures.[31] Lal and coworkers[54] reported 21 species of insect pests infesting lentil crops in India[54] (Table 10.4).

Aphids

Aphid species, namely, *A. craccivora* and *A. pisum*, are the major pests of lentil in America, Europe, Africa, Australia, and Asia.[55] These species infests and feed on the leaves, young shoots, inflorescence, fruits, and even tender stem portion. The nymphs and adults suck the sap of leaves, stem, and pods. They colonize young leaves and growing points resulting in characteristic deformation which affects the normal plant growth and crop yield. Apart from direct feeding damage, *A. craccivora*is reported to transmit tobacco streak virus in lentil.[54]

TABLE 10.4 Major Insect Pests of Lentil.

S. No.	Common name	Scientific name	Pest status
1	Pod borer	*Helicoverpa armigera* (Hubner)	Occasional pest
2	Lima-bean pod borer	*Etiella zinckenella* Triet	Moderate/minor pest
3	Cowpea aphid	*Aphis craccivora* Koch *Acyrthosiphon pisum* Harris	Major pests
4	Thrips	*Megalurothrips distalis* (Karny)	Minor pest
5	Armyworm	*Spodoptera exigua* Hubner	Minor pest
6	Cutworms	*Agrotis ipsilon* Hubnagel	Major pests
7	Bud weevil	*Apion arrogans* Wenck	Minor pest
8	Leaf weevils	*Sitona crinitus* Herbst (*S. macularius* Marsham)	Major pest
9	Bruchids	*Callosobruchus* spp., *Bruchus* sp.	Major

Leaf Weevils

Leaf weevils are the major pests of lentil in southern Europe, North Africa, and West Asia.[55] The adult weevils are gray-brown and a pronotum with three longitudinal lines and elytra have three rows of dark and white spots. The females deposit the spherical yellow eggs on the soil near the lentil plants. The adult weevils feed on foliage making a semicircular notch on the leaf edges while the larvae feed on root nodules affecting the N_2 fixing capacity of the plant. In the case of severe infestation, foliage becomes yellow giving the appearance of N_2 deficiency.

Cutworms

This is the polyphagous pest and females deposit eggs in batches of dozen or so on the leaves and even on the soils. The first two larval stages feed on foliage of the plant. Later instars become solitary in habits.[56] The late-stage larvae cut the several plants above the root crown which wither and dry. Mature larvae pupate deep into the soils.

10.6.5 COWPEA

Cowpea is infested by several pest species which are adopted to feed on every part of cowpea plant.[57] Over 85 insect species are reported to attack cowpea.[58] Out of these, 20 species regularly occurrence on cowpea and are of economic importance worldwide.[59] The important sucking pests of cowpea are described hereunder (Table 10.5).

TABLE 10.5 The Major Insect Pests of Cowpea.

S. No.	Common name	Scientific name	Occurrence
1	Pod borer	*Helicoverpa armigera* (Hubner)	Postflowering and podding stage
2	Legume pod borer	*Maruca vitrata* (Geyer)	Postflowering and podding stage
3	Pod sucking bugs	*Riptortus dentipes* F., *Clavigralla tomentosicollis* Stal., *Anoplocnemis ccurvipes* F., *Nezaraviridula* (L.)	Postflowering and podding stage
4	Cowpea aphid	*Aphis craccivora* Koch	Seedling and vegetative stage
5	Leafhoppers	*Empoasca kerri* Pruthi	Seedling and vegetative stage
6	Flower thrips	*Megalurothrips distalis* (Karny), *Megalurothrips usitatus* (Bagnall), *Megalurothrips sjostedti* Trybom	Flowering stage
7	Blister beetles	*Mylabris* spp.	Flowering stage
8	Whitefly	*Bemisia tabaci* Genn.	Seedling and vegetative stage
9	Bruchids	*Callosobruchus* spp.	Postharvest storage

Legume Pod Borer

Legume pod borer is found in tropics and subtropics causing extensive damage. It is reported to be major pest of cowpea in Africa and southeast Asia.[111] Larvae cause damage to flowers and flower buds in cowpea. The crop losses up to 80% are reported due to this pest in cowpea.[61]

The adult moth is nocturnal in habit and deposits about 150 eggs on leaf buds, flower buds, and flowers. The early generation cause damage to peduncles and tender stem. The characteristic larval feeding is by webbing together the plant parts with frass on pods and shoot tips. The pupae develop in soil and moth emergence is favored by rainfall or high soil moisture content.[44]

Cowpea Aphids

Aphids are the key foliage pests of cowpea. The infestation causes distortion of leaves and premature defoliation, stunting, and death of cowpea seedlings. The sooty mold reduces the photosynthetic efficiency of plants.[62] Apart from direct feeding damage, an aphid acts as a vector of cowpea mosaic virus.[44,63]

The damage is more in cowpea in the dry season.[57] In case of severe infestation, yield loss can reach 50% or more.[64]

Flower Thrips

This is the early major insect pest of cowpea at the reproductive stage of the cowpea.[65] The population is generally higher during the dry season.[57,66] The adult thrips are shiny black insects found feeding in flower buds and flowers. Under severe infestation, the plants do not produce any flowers and open flowers become distorted and discolored. The damage also results in shortening of peduncle. Affected flowers fall off in the early stage and as a result pods are not formed.

Pod Sucking Bug Complex

Pod bugs are serious pests of cowpea at postflowering stage. Both nymphs and adults thrust their stylets through the pod wall and suck nutrients from the developing seeds.[67] The infestation under uncontrolled conditions causes losses in the magnitude of 80 and 95% of pod and seed yield, respectively.[68] Adults and nymphs suck the sap from green pods which shrivel and dry prematurely resulting in loss of seed.[69]

The bugs of the species, deposit the eggs on leguminous trees and weeds but few are found on cowpeas. There are five nymphal instars and nymphal period lasts for 2 to 3 weeks. They suck the sap from green pods.[70] The species of *C. tomentosicollis* deposit brown color eggs on the parts of cowpea. Both adults and nymphs feed on a single pod.[70] Green stink bugs (*N. viridula*) are distributed in the tropics and subtropics. Both adults and nymphs suck the sap from the developing pods causing damage.

10.7 MAJOR DISEASES OF PULSES AND YIELD LOSSES

The pulse crops are highly susceptible to several phytopathogens which subsequently results in economic losses[71] (Table 10.6). The major diseases are dry root rot (DRR), wilt, and phytophthora stem blight (PSB) in the pulses (Table 10.7 and Figure 10.2). Fusarium wilt decreases the both seed yield and seed weight[73] is found to cause annual yield losses of about 10–15% in India;[74] however, 70% to total crop loss can occur in years of severe outbreaks.[75] Fusarium wilts in chickpea, pigeonpea, and lentil causing over 20–25% yield losses all over the country.[112] Similarly

DRR of chickpea caused 10 to 100% yield loss annually.[76,77] Stem rot of pulses found to cause almost 10% losses annually around the world.[113] The Fusarium wilt disease can occur at all the stages of pigeonpea and reported 30 to 100% yield losses in susceptible cultivars.[78–81] Chickpea is found to be most vulnerable to Botrytis gray mold (BGM) infection during flowering and podding stages[82] and infection at these stages may lead to a complete failure of the crop. Yield losses up to 50% were reported under endemic situations and if favorable conditions of BGM infections prevail during vegetative and reproductive growth of plants, the losses may reach up to 100%.[83]

TABLE 10.6 The Major Diseases of Pulses and Their Causal Agents.

S. No.	Crops	Diseases	Causative organisms
1	Pigeonpea	Phytophthora stem blight	*Phytophthora drechsleri* f. sp. *cajani.*
		Wilt	*Fusarium oxysporum* f. sp. *udum* (Butler) Snyd. and Hans.
		Leaf spot	*Cercospora* spp.
		Powdery mildew	*Erysiphe polygoni* DC
2	Chickpea	Wilt	*F. oxysporum* f. sp. *ciceri* Matuo and Sato
		Dry root rot	*Rhizoctonia bataticola*
		Stem rot	*Sclerotinia sclerotiorum*
		Rust	*Uromyces ciceris-arietinii (Grogn.)* Jacz. & Boy.
		Wet root rot	*Rhizoctinia solani*
		Botrytis gray mold	*Botrytis cinerea*
3	Lentil	Wilt	*F. oxysporum* f. sp. *lentis*
		Root rot	*Macrophomina phaseolina*
		Rust	*Uromyces fabae*
		Ascochyta blight	*Ascochyta fabae* f. sp. *lentis*
4	Pea	Rust	*U. fabae*
		Powdery mildew	*E. polygoni*
		Root rot	*R. solani*
5	Mungbean and urdbean	Root rot	*M. phaseolina*
		Leaf spot	*Cercospora* spp.

FIGURE 10.2 Diseases and pests of pulses; (a) dry root rot in chickpea; (b) chickpea rust; (c) chickpea stem rot; (d) chickpea wilt; (e) aschochyta blight of chickpea; (f) *Alternaria* blight on pigeonpea; (g) pigeonpea wilt affected field; (h) *Phytophthora* blight of pigeonpea; (i) pigeonpea stem rot; (j) pea rust; (k) lentil rust; (l) *Anthracnose* affected bean; (m) rajmash powdery mildew; (n) *H. armigera* infestation on pigeonpea (o) and chickpea; (p) *Maruca* infestation on pigeonpea.

Fusarium Wilt of Pulses

Distribution and Symptomatology

Fusarium wilt is a major soil-borne pathogen of chickpea, pigeonpea, and lentil causing over 20–25% yield losses every year.[71,73,85,86] Wilt-infected plants show drooping of petioles, rachis, leaflets, and later dark brown discoloration of internal xylem.[47] Ultimate drooping, yellowing, and gradual killing of plant, leaf yellowing, xylem necrosis, darkening of xylem vessels, and pith were also produced in infected plants.[84] Infected seeds are also a source of primary inoculum of the disease.

Phytophthora Stem Blight (PSB)

Distribution and Symptomatology

PSB (*Phytophthora drechsleri* f. sp. *Cajani)* is a soil-borne disease of pigeonpea. It is favored by excessive rains fall within a short span of time, hot and humid weather. It is reported in India in 1966 by Williams and coworker[89] and now present in Asia,[90] Africa, America,[88] Australia, Dominican Republic, Kenya, Panama, and Puerto Rico.[71] The severity of PSB on grain yield, depends on the onset of the disease in relation to crop growth and disease incidence.

The characteristic disease symptoms are appeared on $\geq$ 1-month-old seedlings. The water-soaked lesions are developed on the primary and triplicate leaves that become necrotic within a week when humidity $\geq$ 80% and temperature 20–30°C persist. The lesions on stems may be extending up to 20 cm and form girdles and cracks on the stem. The stems are swollen and turn to cankerous structures. Infected plant parts break easily in the wind. Phloem vessels show brown to gray colored discoloration although xylem vessels remain healthy.

Dry Root Rot

Distribution and Symptomatology

Rhizoctonia bataticola (synonym—*Macrophomina phaseolina*) causes DRR in chickpea and pycnidial stage causes disease in green gram and blackgram. It is one of the most important destructive constraints to chickpea productivity and production, causing 10–20% yield loss annually. It is distributed in the Indian subcontinent, Australia, Ethiopia, Iran, and several countries.[47] In India, it is most serious in central and southern parts where the crop is grown under rainfed conditions. High day temperature (~30°C) with moisture stress conditions at podding and flowering stage rapidly aggravate disease incidence.[115] The disease is more severe in sandy soils and appears during podding and flowering stage. Affected plant petiole and leaf lets droops only at top of the plant. Affected root system devoid of lateral roots and tap root become black. When we uproot the plant usually lower portions of the tap root remain in the soil. Black sclerotial bodies can be seen on the exposed roots or inside the base of the stem and when we split color region vertically, mycelium, or minute sclerotia. Pathogen has a wide host range particularly belonging to *Leguminosae* family. High day temperature (~30°C) with moisture stress conditions at podding and flowering stage rapidly aggravate disease incidence.[115,116]

Powdery Mildew

Distribution and Symptomatology

Powdery mildew is a most important disease of pulses including, fieldpea, rajmash, cowpea, mungbean, and urdbean. The disease occurs worldwide. The situation worsens in dry weather. Early varieties of crops are known to show less damage. It appears in severe form almost every year.

Rust Diseases

Distribution and Symptomatology

Rust is caused by *Uromyces* sp. which is an autoecious fungus; completing its entire life cycle on the same host. The weather conditions with high humidity, cloudy weather with temperatures of 20–22°C favor the development of disease. The plants turn dark brown or blackish patches in the field.[92] For the management of rust, volunteer plants are removed. Host plant resistance is the best option of effective management of this disease.

Ascochyta Blight

Distribution and Symptomatology

Ascochyta blight (*Ascochyta rabiei)* is one of the important foliar diseases of chickpea. The major incidence of this disease occurs in the area where growing season of chickpea coincides with cool and moist weather. The substantial decline in chickpea area was observed in western Canada. The yield loss can be up to 100% when weather conditions are favorable for the development of this disease. This disease is a seed and stubble-borne producing both airborne ascospores and water-splashed conidia during the cropping season.[93] Some strategies like crop rotation may contribute for the management of this disease to become epidemic. It causes damage in Australia, Canada,[117] Latin America, southern Europe, the USA,[118] and West

Asia.[96] Infection on petioles, leaflets, and on young branches of chickpea includes early symptoms such as epinasty and loss of turgor, followed by water soaking. This pathogen secretes a toxin that maintains the symptom. This disease spreads mainly by the infected seeds produced from infected crops. The introduction of this disease to Australia, Canada, Iran, and the USA[118] is primarily due to the use of infected seeds for cropping. The chickpea seeds with light infection by *A. rabiei* appear usually discolored with very low seed weight. However, in severe infection, the pathogen appears on the seed coat, cotyledons, and embryo axes. On the infected chickpea seed, the mycelium of *A. rabiei* grows on the seed coat on cotyledons and on the emerging seedlings.[119]

Pycnidium is a structure for the survival of *A. rabiei* on the debris lying on the surface of soil. It can also survive as teleomorph having ability to colonize debris and form viable pseudopothesia on it. Pycnidium produces conidia in favorable conditions and acts as primary inoculums for the initiation of disease in the new crop. However, the teleomorph develops on infected chickpea debris that remains in the field and in favorable conditions produces ascospores for long-distance spread of the disease.

Botrytis Gray Mold

Distribution and Symptomatology

BGM causing *Botrytis cinerea* pers is one of the important fungal foliar diseases of chickpea. It occurs in cool, cloudy, and humid weather. In the off-season, this pathogen survives on the infected crop debris and seeds. The asexual stage of this necrotrophic fungus is dominant in crop chickpea. The mycelial growth of *B. cenerea* appears as white, cottony, which turns light gray after few days' incubation on PDA. The young hyphae of *B. cinerea* are thin, hyaline, and 8–16 μm wide, which become septate with age. Conidiophores are light brown, septate, and erect ramified pseudo dichotomically. They have slightly enlarged tips, bearing small pointed sterigmata which bear 1–2-celled, hyaline, oval, or globose conidia forming clusters. The size of conidia is $4–25 \times 4^{-18}$ μm and $4–16 \times 4^{-10}$ μm on infected chickpea plants and PDA, respectively.[85] Joshi and Singh[120] have suggested that sporodochia formed by the pathogen on the host surface measure 0.5–5.0 μm in diameter and may turn into hard sclerotial masses. Sporulation of *B. cinerea* largely depends upon a range of factors including temperature, nutrients, and ecological factors. BGM disease has been reported on chickpea from India.[121,131] However, the first epidemic of this disease with about 95% crop loss was reported in Argentina also.[122] Since then, a number of epidemics caused by BGM were reported showing almost complete loss from many chickpea growing countries. The disease has caused serious concerns in India, Bangladesh, Nepal, Pakistan, and Australia. BGM is reported on chickpea globally. Numbers of reports from Argentina have shown almost complete yield loss (about 100%) under conducive environmental conditions. However, this disease is also reported from Canada, Chile, Colombia, Hungary, Mexico, Myanmar, Spain, Turkey, the USA, and Vietnam.[94] During sexual stage, *B. cinerea* fertilized sclerotia germinate by the emergence of apothecia that release sexually produced ascospores. The natural occurrence of the sexual structure was not known, however, its occurrence is reported in India under laboratory conditions.[97]

An antifungal compound produced by *Botrytis* is reported to kill yeast and is most often used to stop fermentation, prior to sufficient accumulation of alcohol in the fermenter. During 1978–1979, this disease had become epidemic in India and destroyed about 20,000 hectares of chickpea.[123–125] Similarly, this disease occurs almost every year in Nepal with an average yield loss of about 15% (Joshi, 1992). In Bangladesh, the disease was first documented during 1981.[126] Pod yield depends upon the onset of disease, its growth, disease severity, weather conditions, and level of inoculums of the pathogen.

Stem Rot of Pulses

Distribution and Symptomatology

Stem/white rot is the most devastating soil-borne pathogen causing *Sclerotinia sclerotiorum* (Lib.) de Bary, with a very wide host range of different agricultural crops including grain legumes.[112] It survives as a mass of fungal hyphae covered by an environmentally resistant black rind called sclerotia. When conditions are favorable, such as at canopy closure when humidity levels in the canopy are high and the crop begins flowering, an inverted mushroom structure called an apothecium produces ascospores which then infect the flowers. Hyphae can also grow directly into plant tissues, especially when plant–soil contact is established. The fungus kills tissue and fills the stem with white hyphae and sclerotia. In this way, it survives in the soil for many years. This fungus is a soil-borne fungus that attacks plants in all stages of growth including young seedlings, mature plants, and their harvested products distributed worldwide.[108] This fungus may cause almost 100% losses to the crop.[108,113] The survival of the off-season structure of this disease is sclerotia, which can survive for several years. After maturity, the apothecia release a large amount of ascospores that spread over a long distance and serve as the primary source of inoculum for infection. Due to these features, the management of this disease is difficult. In India, the regular occurrence of the disease has been reported in some districts of Himachal Pradesh, where it causes considerable yield losses. The yield loss ranges from 21 to 47%.[124] Limited variety resistance may be available in dry pea and lentil cultivars. No resistance has been observed among chickpea cultivars. White mold is more common in winter-sown lentils than in spring lentils due to their exposure to prolonged wet and cool spring weather.

Yellow Vein Mosaic

Distribution and Symptomatology

The mungbean yellow mosaic virus is transmitted by the whitefly (*B. tabaci*). This viral disease is found on several alternate and collateral hosts. The tender leaves show yellow mosaic spots which subsequently enhances with progression of disease, thus ultimately leading to complete yellowing. Yellowing results in less flowering and pod development. The infected plants are dwarf and occur in patches.

10.8 CONTROL MEASURES OF PESTS IN PULSES

Considerable research findings have been made toward the development of IPM system in pulse crops, particularly pigeonpea, and chickpea. The information on the various components of pest management has been documented for insect pest control.[17,34,127] The regular monitoring of pest and diseases is the key to adopt the timely management practices and thus avoiding the major crop losses. Monitoring of *H. armigera* populations through light or pheromone traps has been followed in Asia.

10.8.1 CULTURAL PRACTICES

The different life stages of the insect pests and pathogen propagules are found to persist. These survive in the soil during the absence of host crop or adverse conditions of weather. Hence deep summer ploughing of fields after crop harvest to expose the life stages of the insect pests present in the soil to biotic and abiotic mortality factors and to destroy the over-wintering stages of *H. armigera* and other noctuid pests should be followed after crop harvest. The predatory birds play significant role in checking the population of several pests in pulses. During intercultural operations, the predatory birds, namely, common myna (*Acridotheres tristis*), egrets (*Egretta* spp.), bulbul (*Bubulcus ibis*), and drongo (*Dicrurus adsimilis*) feed on the insect pests. The provision of tall crops (sorghum) or artificial bird perches should be made which serve as resting sites and attract the insectivorous birds in the field. The maintenance of a high standard of weed control during early crop growth stage helps reduce the alternate hosts to these insect pests and pathogens. The growing of barrier crops, namely, maize, sorghum, pearl millet in mungbean significantly reduced the incidence of whitefly and leafhoppers in cowpea. The growing of marigold is served as a trap crops for *H. armigera*. Intercropping of mungbean and urdbean with sorghum, pearl millet, sesame reduced the incidence of sucking pests compared to sole crops. Sorghum or millets mixed or intercropping also helps in conserving the beneficial insects. Uprooting and destruction of sterility mosaic affected pigeonpea plants and infected stubbles of previous crop are practiced to discourage the survival and carryover of the vector mite. Weed plant, *Atylosia* spp. is reported to harbor the mite vectors, hence removal of weed helps in minimizing the mite population.

The use of healthy seeds and 3–4 years crop rotation have been recommended for effective management of diseases and insect pests. Best sowing time of different pulses and legume crops in *rabi, kharif*, and *spring* season for maximum yield and minimum diseases incidence. The control of volunteer plants over the summer and removal of infected plant debris is recommended for the control of rust disease. Crop rotations have more impact in reducing populations of some soil-borne pathogens.

10.8.2 CHEMICAL APPROACHES

The systemic insecticides recommended for use in pulse crops can be employed for presowing seed treatment or spraying whenever the pest population crosses the threshold level. The chemical control of soilborne disease is poor in the case of soil-borne pathogen since these survives in form of chlamydospores up to 6 years.[73] The deployment of resistant cultivars is the most practical and cost-effective method for the management of soil-borne pathogens of pulse crops. The systemic and nonsystemic fungicides are used in reducing such diseases.[98] Carbendazim or thiram as seed treatment and two or three sprays of mancozeb were effective in managing the diseases.[128] The seed treatment with tebuconazole (0.1%) and foliar sprays of tricyclazole (0.03%) were effective in reducing diseases on leaves and pods.[99] Seed treatment with diclobutrazole is recommended for control of rust disease. The sprays of mancozeb (0.25%), propiconazole (0.1%), and hexaconazole (0.1%) at the early development stage of rust are effective.[100]

10.8.3 MECHANICAL CONTROL

The installation of yellow sticky traps at 1–2 traps per 50–100 m^2 helps to bring down the active flying adult populations of sucking pests like white-flies, leafhoppers. The use of pheromone traps (20–25/ha) is recommended for monitoring and trapping of *H. armigera* moths.

10.8.4 BIOLOGICAL APPROACHES

The conservation of common natural enemies, namely, ladybird beetles (*Menochiluss exmaculatus* (F.), *Brumussuturalis* (F.), and *Coccinella septempunctata* L., *C. transversalis* (F.)), Syrphids (*Ischiodanjavana* Weid.), and green lacewings (*Chrysoper lacarnea* Stephens) helps in reducing the pest population. The egg parasitoid, *Gryon clavigrallae* (Hymenoptera: Scelionidae) reported to parasitizes up to 69% of eggs and up to 100% of *Clavigralla* egg clusters in India.[127] The crude extract of neem seeds (5%) should be obtained at farm level and used against lepidepterous pests as and when economic threshold level is attained. Spraying with *nuclear poly hedrovirus* (HearNPV) @ 450 LE /ha (9 × 10^{11} POBs) along with 0.1% tinopol for the control of *H. armigera*. The ichneumonid, *Campoletis chlorideae* Uchida is reported as an important larval parasitoid of *H. armigera* in chickpea and pigeonpea.

Considerable efforts were made to manage different diseases of pulses by incorporating bioagents colonized natural substrates into the soil. Among the antagonistic fungi, *Trichoderma* species are the most studied bioagents in the pulse ecosystem.[129] They are successfully used to control wilts, root rots, collar rot, and stem rot diseases incited by different *Fusarium* spp., *Rhizoctonia solani, R. bataticola, Sclerotium rolfsii, S. sclerotiorum, Ascochyta, Cercospora, Alternaria* spp., *Phytophthora* spp., and *Pythium* spp. in different pulse crops. *Trichoderma* is reported to suppress the fungal diseases in pulses are *Aspergillus candidus, Aspergillus niger, Pacilomyces lilacinus, Glomus mosseae, Verticillium chlamydosporum, Absidia cylindrospora,*

Coniothyrium minitans, *Talaromyces flavus*, and *Trichothecium roseum*. The different biological control agents effectively employed against different phytopathogens of pulses are listed hereunder (Table 10.7).

TABLE 10.7 Biological Control Agents Employed Against Major Diseases of Pulses.

Crops	Diseases	Pathogens	Effective biocontrol agents
Pigeonpea	Wilt	*Fusarium udum*	*Trichoderma harzianum*, *T. hamatum*, *T. viride*, *T. konningi*, *B. subtilus*
	Phytophthora stem blight	*Phytophthora drechsleri* f. sp. *cajani*	*T. harzianum*, *T. hamatum*, *Glomus mosseae*, *Pseudomonas fluorescens*, *Bacillus subtilis*
	Seed borne diseases	*Pseudomonas compestris* pv. *vinae*	*T. viride*, *T. harzianum*
Chickpea	Wilt	*Fusarium oxysporum* f. sp. *ciceri*	*Trichoderma harzianum*, *T. viride*, *T. virens*, *B. subtilus*, *A. niger AN 27*
	Root rot	*Rhizoctonia bataticola*	*Trichoderma harzianum*, *T. viride*
	Collar rot	*Sclerotium rolfsii*	*T. viride*, *T. harzianum*, *P. fluorescens*
	Gray mold	*Botrytis cenerea*	*Trichoderma* spp.
	Stem rot	*Sclerotinia sclerotiorum*	*T. harzianum*
Cowpea	Wilt	*Fusarium oxysporum*	*T. harzianum*
	Charcoal rot	*M. phaseolina*, *F. oxysporum f.* sp. *tracheiphilum*	*T. viride*, *T. harzianum*, *T. koningii*, *T. pseudokonnigii*
Lentil	Wilt	*Fusarium oxysporum* f. sp. *lentis*	*T. viride*, *T. harzianum*, *G. virens*, *Pseudomonas fluorescens*
	Root rot	*Macrophomina phaseolina*	*T. viride*, *T. harzianum*, *G. virens*, *P. fluorescens*
Mungbean	Root rot	*M. phaseolina*	*T. harzianum*, *T. viride*
Soyabean	Dry root rot	*M. phaseolina*	*T. harzianum*, *T. viride*
Fieldpea	Rust	*Uromyces fabae*	*Pseudomonas fluorescens*, *P. auriginosa*, *Bacillus subtilis*
	Powdery mildew	*Erysiphe polygoni* DC	*T. harzianum*, *T. konninghii*, *T. longibrachiatum*, *P. fluorescens*, *Bacillus subtilis*
	Root rot	*Rhizoctonia solani*	*T. harzianum*, *T. longibrachiatum*, *P. fluorescens*, *B. subtilis*,

10.8.5 MOLECULAR APPROACHES

In recent years, myriad developments were seen in the area of pulses genomics which has resulted in access to a wide variety of modern genomic tools and technologies for pulses improvement. From the point of resistance breeding, DNA markers have emerged as the promising molecular tools to effectively and precisely address the problem of biotic stresses, in particular disease resistance that severely constrains productivity of these crops. Diverse molecular marker systems have been employed across various pulse crops for molecular mapping of resistance to various diseases. Consequently, DNA markers tightly linked with the disease resistance genes/QTLs were made available for the scientific community.[101–105] Once genomic location of a trait linked gene/QTL/marker is known, molecular breeding techniques like marker assisted backcrossing (MABC) can be implemented to transfer the genomic region of interest to different genotypes. The MABC is particular suitable for transferring the genomic segments that govern substantial amount of phenotypic variation for the trait.[106] In the case of transfer of disease resistance from donor species to elite lines, MABC has been remarkably successful.

Importantly, with the advent of next generation sequencing (NGS) technologies, the development of breeder-friendly DNA markers could be performed in a time-saving and high throughput manner. In the context to disease resistance, NGS-based protocols like restriction-site associated DNA (RAD) sequencing led to the generation of over 30 markers associated with anthracnose disease resistance in lupin.[107]

10.9 CONCLUSION

The pulse crops suffer from a range of biotic and abiotic constraints, of which, pests and diseases are the most significant impediments to pulse production. For the management of pest and diseases, chemical pesticides are generally used across the world. It is well established that the injudicious use agricultural chemicals cause not only the pest resistance and resurgence but also the environmental and health hazards. The majority of the pathogens of pulses are soil-borne in nature hence their management by chemicals has not yielded the anticipated results. The pest status, distribution, and intensity of many pests and diseases have changed over the years in India. The IPM modules have been successfully refined and validated. Despite the development of various IPM modules to tackle economically important diseases of food legumes, a considerable knowledge gap persists between scientists and farmers. The future activities should focus on the expansion of IPM technology by increasing the awareness among the farmers. The cultural practice which reduces the intensity of insect pests and diseases is one of the important components of the pest management program. The emphasis must be on popularizing the cropping systems that encourage the activity and abundance of natural enemies and beneficial microbes. The eco-friendly formulations, namely, neem seed kernel extract, *Bacillus thuringiensis*, and *nucleopolyhedrovirus, Trichoderma* have been found effective in many cases, but issues limiting their regular application for pest management are timely supply, availability, quality control, and economic feasibility. The climate change is recently impacting the early emergence and incidence of pests, hence, efforts should be made to understand the weather effects on pest biology, distribution, and natural enemy abundance in order to develop the mitigation strategies. The major emphasis should be toward the development of location-specific pest management strategies to suit prevailing farming practices in that region.

KEYWORDS

- **pulse crops**
- **diseases**
- **pests**
- **cropping pattern**
- **climate change**
- **IPM**

REFERENCES

1. Dhaliwal, G. S.; Arora, R. An Estimate of Yield Losses Due to Insect Pests in Indian Agriculture. *Indian J. Ecol.* **1996,** *23*, 70–73.
2. Rosenzweig, C.; Liverman, D. Predicted Effects of Climate Change on Agriculture: A Comparison of Temperate and Tropical Regions. In *Global Climate Change: Implications, Challenges, and Mitigation Measures*; Majumdar, S. K., Ed.; The Pennsylvania Academy of Sciences, USDA: Easton, PA, 1992; pp. 342–361.
3. Sharma, H. C.; Srivastava, C. P.; Durairaj, C.; Gowda C. L. L. Pest Management in Grain Legumes and Climate Change. In *Climate Change and Management of Cool*

Season Grain Legume Crops; Yadav, S. S. et al., Eds.; Springer Science+Business Media, 2010; pp 115–139.

4. Ahlawat, I. P. S.; Sharma Purusottam; Singh, Ummed. Production, Demand and Import of Pulses in India. *Indian J. Agron.* **2016,** *61*, 33–41.
5. Pareek A.; Meena B. M.; Sharma, S.; Teterwal, M. L. Impact of Climate Change on Insect Pests and Their Management Strategies. In *Climate Change and Sustainable Agriculture*; New India Publishing Agency, 2017; pp 253–286.
6. IPCC (Intergovernmental Panel on Climate Change). *Climate Change 2007: Impacts, Adaptation and Vulnerability*; Cambridge University Press: Cambridge, 2007.
7. Shetty, P. K.; Ayyappan, S.; Swaminathan, M. S. *Climate Change and Sustainable Food Security*; National Institute of Advanced Studies, Bangalore and Indian Council of Agricultural Research: New Delhi, 2013. ISBN: 978-81-87663-76-8 27.
8. Swaminathan, M. S.; Kesavan, P. C. Agricultural Research in an Era of Climate Change. *Agric. Res.* **2012,** *1*, 3–11.
9. Clement, S. L.; Wightman, J. A.; Hardie, D. C.; Bailey, P.; Baker, G.; McDonald, G. Opportunities for Integrated Management of Insect Pests of Grain Legumes. In *Linking Research and Marketing Opportunities for Pulses in the 21st Century*; Knight, R., Ed.; Kluwer Academic: Dordrecht, 2000; pp 467–480.
10. Saxena, H.; Bandi, S. Devindrappa, M. Pests of Pulses. In *Pests and Their Management*; Omkar, Ed.; Springer Nature Singapore Pte Ltd. 2018; pp 99–136.
11. Sharma, O. P.; Bantewad, S. D.; Patange, N. R.; Bhede, B. V.; Badgujar. A. G.; Ghante. P. H.; Kadam, M.; Bhagat, S.; Kumari, A. Implementation of Integrated Pest Management in Pigeonpea and Chickpea Pests in Major Pulse-Growing Areas of Maharashtra. *J. Integr. Pest Manage.* **2015,** *15* (1), 12.
12. Sharma, H. C, Clement, S. L.; Ridsdill-Smith, T. J.; Ranga Rao, G. V.; El Bouhssini, M.; Ujagir, R.; Srivastava, C. P.; Miles, M. Insect Pests Management in Food Legumes: The Future Srategies. In *Food Legumes for Nutritional Security and Sustainable Agriculture*, Vol. 1; Kharkwal, M. C., Ed.; ISGPB, 2008; pp 522–544.
13. Saxena, H.; Bandi, S. M.; Revanasidda. Sucking Pests of Pulse Crops. In *Sucking Pests of Crops*; Omkar, Eds.; Springer: Singapore, 2020; pp 107–133.
14. Dhar, V.; Ahmad, R. Integrated Pest management in Chickpea and Pigeonpea. In *Integrated Pest Management in Indian Agriculture, Proceedings-11*; Birthal, P. S., Sharma, O. P., Eds.; NCAP and NCIPM: New Delhi, 2004; p 279.
15. Van Emden, H. F.; Ball, S. L.; Rao M. R. Pest and Disease Problems in Pea, Lentil, Fababean, and Chickpea. In *World Crops: Cool Season Food Legumes*; Summerfield, R. J., Ed.; Kluwer Academic: Dordrecht, 1988; pp 519–534.
16. Chandrashekar, K.; Gupta, O.; Yelshetty, S.; Sharma, O. P.; Bhagat, S.; Chattopadhyay, C.; Sehgal, M.; Kumari, A.; Amaresan, N.; Sushil, S. N.; Sinha, A. K.; Asre, R.; Kapoor, K. S.; Satyagopal, K.; Jeyakumar, P. Integrated Pest Management for Chickpea, **2014,** p. 43.
17. Ranga Rao, G. V.; Shanower, T. G. Identification and Management of Pigeonpea and Chickpea Insect Pests in Asia. Information Bulletin no. 57. (In En. Summaries in En, Fr.) Patancheru, 502 324, A. P.; India: International Crops Research Institute for the Semi-Arid Tropics. 1999; p 96.
18. Southgate, B. J. Biology of Bruchidae. *Ann. Rev. Entomol.* **1979,** *24*, 449–473.
19. Deshmukh, S. G.; Sureja, B. V.; Jethva, D. M.; Chatar, V. P. Estimation of Yield Losses by Pod Borer, Helicoverpa Armigera (Hubner) on Chickpea. *Legume Res.* **2010,** *33* (1), 67–69.

20. Lateef, S. S.; Reed, W. Insect Pests of Pigeonpea. In: *Insect Pests of Tropical Food Legumes*; Singh, S. R., Ed.; John Wiley & Sons: Chichester, 1990; pp 193–242.
21. DES. Agricultural Statistics at a Glance 2019. Directorate of Economics & Statistics, Department of Agriculture and Cooperation and Farmers Welfare, Ministry of Agriculture and Farmers Welfare, Government of India, 2020; p 315.
22. Yadav, R. L.; Prasad, K. Identifications of Predominant Cropping Systems in Different Sistricts of India. Annual Report, Project Directorate of Cropping Systems Research, Modipuram, 1997; pp 30–49.
23. Kumar, N.; Nath, C. P.; Hazra, K. K.; Das, K.; Venkatesh, M. S.; Singh, M. K.; Singh, S. S.; Prahara, J. C. S.; Singh, N. P. Impact of Zero-Till Residue Management and Crop Diversification with Legumes on Soil Aggregation and Carbon Sequestration. *Soil Tillage Res.* **2019,** *189* (6), 158–167.
24. Singh, K. K.; Ali, M.; Venkatesh, M. S. Pulses in Cropping Systems. Technical Bulletin, IIPR, Kanpur, 2009.
25. Shanower, T. G.; Romeia, J.; Minja, E. M. Insect Pests of Pigeon Pea (*Cajanus cajan*) and Their Management. *Annu. Rev. Entomol.* **1999,** *44*, 77–96.
26. Reed, W.; Lateef, S. S.; Sithanantham, S.; Pawar, C. S. Pigeonpea and Chickpea Insect Identification Hand-Book. Information Bulletin No. 26. ISBN 92-9066-171-2. ICRISAT, Patancheru, India, 1989; p 123.
27. Sachan, J. N.; Lal, S. S. Integrated Pest Management of Pod Borers Complex of Chickpea and Pigeonpea in India. In *Recent Advances in Pulses Research. Indian Society of Pulses Research and Development*; Asthana, A. N., Ali, M., Eds.; IIPR: Kanpur, 1997; pp 349–376.
28. Ahmad Rai, A. B. *25 Year of Research on Helicoverpa at IIPR*; Indian Institute of Pulses Research: Kanpur, 2005; p 54.
29. Reed, W.; Cardona, C.; Sithanantham, S.; Lateef, S. S. The Chickpea Insect Pests and their Control. In *The Chickpea*; Saxena, M. C., Singh, K. B., Eds.; CAB International: Wallingford, UK, 1987; pp 283–318.
30. Shanower, T. G.; Anita, V.; Bhagwat, V. R.; Dreyer, H. Parasitism of *Clavigralla* spp. (Hemiptera: Coreidae) Eggs by *Gryon clavigrallae* Mineo (Hymenoptera: Scelionidae). *J. Biol. Contr.* **1996,** *10*, 1–7.
31. Yadav, D. K.; Sachan, S. K.; Singh, G.; Singh, D. V. Insect Pests Associated with Pigeonpea Variety UPAS 120 in Western Uttar Pradesh, India. *Plant Arch.*, **2016,** *16* (1), 140–142.
32. Singh, S. R.; Van Emden, H. F. Insect Pests of Grain Legumes. *Annu. Rev. Entomol.* **1979,** *24*, 255–278.
33. Sharma, O. P.; Gopali, J. B.; Yelshetty, S.; Bambawale, O. M.; Garg, D. K.; Bhosle, B. B. *Pests of Pigeonpea and Their Management*; ICAR-NCIPM: New Delhi, 2010a; p 100.
34. IIPR. *25 Years of Pulses Research at IIPR*; Indian Institute of Pulses Research (IIPR): Kanpur, 2009, 219p.
35. Sharma, H. C.; Srivastava, C. P.; Durairaj, C.; Gowda C. L. L. Pest Management in Grain Legumes and Climate Change In *Climate Change and Management of Cool Season Grain Legume Crops*; Yadav, S. S. et al., Eds.; Springer Science+Business Media, 2010b; pp 115–139.
36. Eker, T.; Felai, E.; Alper, A. Screening of Chickpea Accessions Far Resistance Against the Pulses Beetle *Callasobruchus chinesis* L. (Coleoptera bruchidae). *J. Stored Prod. Res.* **2018,** *76*, 51–57.

37. Olubayo, F. M.; Port, G. R. The Efficacy of Harvest Time Modification and Intercropping as Methods of Reducing the Field Infestation of Cowpeas by Storage Bruchid in Kenya. *J. Stored Prod. Res.* **1997,** *33* (4), 271–276.
38. Singh, R. R.; Luse, R. A.; Leuschuner, K. Nangju, D. Groundnut Oil Treatment for the Control of *Callosobruchus maculatus* (F.) During Cowpea Storage. *J. Stored Prod. Res.* **1978,** *14*, 77–80.
39. Caswell, G. H. The Infestation of Cowpeas in Western Region of Nigeria. *Trop. Sci.* **1961,** *3*, 154–158.
40. Saxena, H. P. Pests of Grain Legumes and Their Control in India. In: *The Pests of Grain Legumes: Ecology and Control*; Singh, S. R., van Emden, H. F., Taylor, T. A.; Academic Press: New York, 1978; pp 15–23.
41. Singh, S. R.; van Emden, H. F. *Ann. Review Entom.* **1978,** *24*, 255–278.
42. Patel, M. B.; Srivastava, K. P. Biology of Groundnut Aphid, *Aphis craccivora* Koch on Cowpea, *Vigna unguiculata* (Linnaeus) Walpers. *Bull. Entomol.* **1989,** *30* (1), 65–73.
43. Kaiser, W. J.; Ghanekar, A. M.; Nene, Y. L.; Rao, B. S.; Anjaiah, V. Viral Diseases of Chickpea. In *Chickpea in the Nineties: Proceedings of the Second International Workshop on Chickpea Improvement.* Dec 4–8, 1989; ICRISAT: Hyderabad, India, 1990; pp 139–142.
44. Nene, Y. L.; Reddy, M. V.; Haware, M. P.; Ghanekar, A. M.; Amin, K. S. Field Diagnosis of Chickpea Diseases and Their Control. In Information Bulletin No. 28. ed. by Int. Crops Res. Inst. for the Semi Arid Tropics, Patancheru, India,1991.
45. Sharma, H. C.; Gowda, C. C. L.; Stevenson, P. C.; Ridsdill-Smith, T. J.; Clement, S. L.; Rao, G. V. R.; Romies, J.; Miles, M.; Bouhssini, M. E. Host Plant Resistance and Insect Pest Management in Chickpea. In *Chickpea Breeding and Management*; Yadav, S., Redden, R., Chen, W., Sharma, B., Eds.; CAB International: Wallingford, 2006; pp 527–544.
46. Singh, S. R.; Allen, D. J. Cowpea Pests and Diseases (Manual Series No. 2). International Institute of Tropical Agriculture: Ibadan, Nigeria, 1979; pp 3–27.
47. Lal, S. S. Insect Pests of Mung, Urd, Cowpea and Pea and Their Management. In: Plant Protection in Field Crops; Rao, M, V., Sithanantham, S., Eds.; Plant Protection Association of India: Hyderabad, India, 1987, pp 185–201.
48. Srivastava, K. M.; Singh, L. N. A Review of the Pest Complex of Kharif Pulses in Uttar Pradesh. *Int. J. of Pest Mgt.* **2009,** *22* (3), 333–335.
49. Nene, Y. L. *A Survey of Viral Disease of Pulse Crops in Uttar Pradesh. Research Bulletin. 4, G. B*; Pant University of Agriculture and Technology: Pantnagar, India, 1972; pp 88–95.
50. Murugesan, S.; Chelliah, S. Effect of Yellow Mosaic Infection of the Host Green Gram on the Biology of *Bemisia tabaci* (Genn.) Entomon. **1978,** *3* (1), 41–43.
51. Lal, S. S. Insect Pests of Lentil and their Management—A Review. *Agric. Rev.* **1992,** *13* (4), 225–232.
52. Stevenson, P. C.; Dhillon, M. K.; Sharma, H. C.; Bouhssini, M. E. Insect Pests of Lentil and Their Management. In *Lentil*; Yadav, S. S., McNeil, D. L., Stevenson, P. C., Eds.; Springer: Dordrecht. 2007; pp 331–348.
53. Hill, D. S. *Agrotis ipsilon* (Hfn) In *Agricultural Insect Pests of the Tropics and Their Control*, 2nd ed.; Cambridge University Press: Cambridge, London, New York, New Rochelle, Melbourne, Sydney, 1983; pp. 357–358.
54. Jackai, L. E. N.; Daoust, R. A. Insect Pests of Cowpeas. *Annu. Rev. Entomol.* **1986,** *31*, 95–119.

55. Booker, R. H. List of Insect Species Found in Association with Cowpea at Sumaru. Samaru M. Sc. Paper No 9, IAR, Abuja, Zaria. In FAO, 1982, Breeding for Durable Disease and Pest Resistance, FAO Plant Production and Protection Paper, **1965,** *55*, 139.
56. Oyewale, R. O.; Bamaiyi, L. J. Management of Cowpea Insect Pests. *Sch. Acad. J. Biosci.* (SAJB). **2013,** *1* (5), 217–226.
57. Lal, S. S.; Srivastava, C. P.; Rai, A. B. Insect Pests of Pigeonpea and their Management. In: *Advances in Pigeonpea Research*; Ali, M., Kumar, S., Eds.; Indian Institute of Pulses Research, 2005, pp 262–283.
58. Dannon, E. A.; Tamo, M.; van Huis, A.; Dicke, M. Functional Response and Life History Parameters of Apantelestaragamae, a Larval Parasitoid of Marucavitrata. *Biocontrol* **2010,** *55* (3), 363–378.
59. Singh, S. R. Insects Damaging Cowpeas in Asia. In *Cowpea Research, Production and Utilization*; Singh, S. R., Rachie, K. O., Eds.; Wiley: Chichester, 1985; pp 247–264.
60. Ofuya, T. I. Control of the Cowpea Aphid, *Aphis craccivora* Koch (Homoptera: Aphididae), in Cowpea, *Vigna unguiculata* (L.) Walp. *Integr Pest Manag. Rev.* **1997,** *2* (4), 199–207.
61. Obopile, M.; Ositile, B. Life Table and Population Parameters of Cowpea Aphid, *Aphis craccivora* Koch (Homoptera: Aphididae) on Five Cowpea *Vigna unguiculata* (L. Walp.) Varieties. *J. Pest. Sci.* **2010,** *83* (1), 9–14.
62. Togola, A.; Boukar, O.; Belko, N.; Chamarthi, S. K.; Fatokun, C. Host Plant Resistance to Iinsect Pests of Cowpea (*Vigna unguiculata* L. Walp.): Achievements and Future Prospects. *Euphytica* **2017,** 213–239.
63. Abudulai. M.; Salifu, A. B.; Haruna, M. Screening of Cowpeas for Resistance to the Flower Bud Thrips, *Megalurothrips sjostedti* Trybom (Thysanoptera: Thripidae). *J. Appl. Sci.* **2006,** *6*, 1621–1624.
64. Jackai, L. E. N.; Nokoe, S.; Tayo, B. O.; Koona, P. Inferences on Pod Wall and Seed Defences Against the Brown Cowpea Coreid Bug, *Clavigralla tomentosicollis* Stal. (Hem.; Coreidae) in Wild and Cultivated *Vigna* Species. *J. Appl. Entomol.* **2001,** *125* (5), 277–286.
65. Egho, E. O. Effects of Two Agro-Ecological Zones on Insect Species of Cowpea (*Vigna unguiculata* L.) Walp During the Late Cropping Season, Delta State, Southern Nigeria. *Agric. Biol. J. North. Am.* **2011,** *2* (3), 448–453.
66. Singh, S. R. Grain Legume Entomology, IITA, Ibadan, Nigeria, **1977,** 55 pp.
67. Taylor, T. A. Nigeria Grower Producer. **1964,** *3*, 1–4.
68. Nene, Y. L.; Sheila, Y. K.; Sharma, S. B. *A World List of Chickpea and Pigeonpea Pathogens*, 5th ed.; Patancheru, Andhra Pradesh, India: International Crops Research Institute for the Semi-Arid Tropics, 1996.
69. Kumar, N.; Hazra. K. K.; Singh, M. K.; Venkatesh. M. S.; Kumar, L.; Singh, J.; Nadarajan, N. Weed management Techniques in Pulse Crops. Technical Bulletin 10/2013. Indian Institute of Pulses Research, Kanpur (Uttar Pradesh), India, 2013, p 47.
70. Haware, M. P.; Nene, Y. L. Influence of Wilt at Different Stages on the Yield Loss in Chickpea. *Trop. Grain Legume Bull.* **1980,** *19*, 38–40.
71. Riaz A. K.; Bhat, T. A.; Kumar, K. Management of Chickpea Dry Root Rot Caused by Rhizoctonia bataticola Inter. *J. Res. Phar. Biomed. Sci.* **2012,** *3*, 15–39.
72. Munjal, R. L.; Chenulu, V. V.; Hora, I. S. Assessment of Losses due to Powdery Mildew *Erysiphe polygoni* on Pea. *Indian Phytopath.* **1963,** *16*, 260–267.

73. Kannaiyan, J.; Nene, Y. L. Influence of Wilt at Different Growth Stages on Yield Loss in Pigeonpea. *Trop. Pest Manage.* **1981,** *27*, 141.
74. Reddy, M. V.; Nene, Y. L.; Kannaiyan, J.; Raju, T. N.; Saka, V. W.; Daudi, A. T.; Songa, W. P.; Omanga, P. Pigeonpea Lines Resistant to Wilt in Kenya and Malawi. *Int. Pigeonpea Newsl.* **1990,** *12*, 25–26.
75. Okiror, M. A. Evaluation of Pigeonpea (*Cajanus cajan*) Germplasm for Resistance to Fusarium Wilt. *Indian J. Agric. Sci.* **1999,** *69*, 600–601.
76. Dhar, V.; Reddy, M. V.; Chaudhary, R. G. Major Diseases of Pigeonpea and Their Management In *Advances in Pigeonpea Research*; Ali, M., Kumar S., Eds.; ICAR-IIPR: Kanpur (India), 2005, pp 229–261.
77. Khan, M. R.; Anwer, M. A.; Shahid, S. Management of Gray Mold of Chickpea, *Botrytis cinerea* with Bacterial and Fungal Biopesticides Using Different Modes of Inoculation and Application. *Biol. Control.* **2011,** *57* (1), 13–23.
78. Pande, S.; Galloway, J.; Gaur, P. M.; Siddique, K. H. M.; Tripathi, H. S.; Taylor, P.; MacLeod, M. W. J.; Basandrai, A. K.; Bakr, A.; Joshi, S.; Kishore, K. G.; Isenegger, D. A.; Rao, N. J.; Sharma, M. Botrytis Grey Mould of Chickpea: a Review of Biology, Epidemiology, and Disease Management. *Austr. J. Agric. Res.* **2006,** *57* (11), 1137–1150.
79. Erwin, D. C. Fusarium and Verticillium Wilt Diseases of *Ciceris arietinum* L. *Phytopathology* **1957,** *47*, 10.
80. Pande, S.; Sharma, M. Rao, J. N. Etiology, Biology and Management of Diseases of Food Legumes. In *Food Legumes for Nutritional Security and Sustainable Agriculture*; Kharkwal, M. C., Ed.; Proceedings of the Fourth International Food Legumes Research Conference (IFLRC-IV), October 18–22, 2005; New Delhi, India, 2007; pp 1–15.
81. Singh, G.; Chen, W.; Rubiales, D.; Moore, K.; Sharma, Y. R.; Gan, Y. Diseases and their Management. Pages 520–537. In *Chickpea Breeding and Management*; Yadav, S. S.; Redden, R.; Chen W.; Sharma, B.; CAB International, 2007.
82. Williams, F. J.; Amin, K. S.; Baldev, B. Phytophthora Stem Blight of *Cajanus cajan*. Phytopathology. **1975,** 65, 1029–1030.
83. Kannaiyan, J.; Nene, Y. L.; Reddy, M. V.; Ryan, J. G.; Raju, T. N. Prevalence of Pigeonpea Diseases and Associated Crop Losses in Asia, Africa and the Americas. *Trop. Pest Manage.* **1984,** *30*, 62–71.
84. Williams, F. J.; Grewal, J. S.; Amin, K. S. Serious and New Diseases of Pulse Crops in India in 1966. *Plant Dis. Reptr.* **1968,** *52*, 300–304.
85. Pal, M.; Grewal, J. S.; Sarbhoy, A. K. A New Stem Rot of Arhar Caused by Phytophthora. *Indian Phytopath.* **1970,** *23*, 583–587.
86. Pande, S.; Pathak, M.; Sharma, M.; Narayana Rao, J.; Tomar, O. S. Resistance to Phytophthora Blight in the Improved Pigeonpea Lines at ICRISAT, Patancheru, India. *Int. Chickpea Pigeonpea Newslett.* **2006,** *13*, 42–44.
87. Nene. Y. L. Reddy, M. V. Preliminary Information on Chickpea Stunt. *Trop. Grain Legume Bull.* **1976,** *5*, 31–32.
88. Akem, C. Ascochyta Blight of Chickpea: Present Status and Future Priorities. *Int. J Pest Manage.* **1999,** *45*, 131–137.
89. Akem, C.; Malhotra, R. S.; Bayaa, B.; Mouselli, M. N.; Kabbabeh, S. *Taking Blight off Chickpea*; ICARDA-CARAVAN2, 2000; p 18.
90. Singh, G. Epidemiology of Botrytis Graymold of Chickpea. In: *Recent Advances in Research on Botrytis Gray Mold of Chickpea*; Haware, M. P., Lenne, J. M., Gowda, C. L. L., Eds.; ICRISAT: Patancheru, AP, India, 1997; pp 47–50.

91. Gupta, H. S. *Climate Change and Indian Agriculture: Impacts, Mitigation and Adaptation. X Agricultural Science Congress on Soil, Plant and Animal Health for Enhanced and Sustained Agricultural Productivity;* ICAR: New Delhi, 2011; pp 73–81.
92. Bohra, A.; Jha, U. C.; Kavi Kishor, P. B.; Pandey, S.; Singh, N. P. Genomics and Molecular Breeding in Lesser Explored Pulse Crops: Current Trends and Future Opportunities. *Biotechnol. Adv.* **2014a,** *32*, 1410–1428.
93. Bohra, A.; Pandey, M. K.; Jha, U. C.; Singh, B.; Singh, I. P.; Datta, D.; Chaturvedi, S. K.; Nadarajan, N. Varshney, R. K. Genomics-Assisted Breeding in the Four Major Pulse Crops of Developing Countries: Present Status and Prospects. *Theor. Appl. Genet.* **2014,** *127* (6), 1263–1291.
94. Varshney, R. K.; Murali Mohan, S.; Gaur, P. M.; Gangarao, N. V. P. R.; Pandey, M. K. Bohra, A. Achievements and Prospects of Genomics-Assisted Breeding in Three Legume Crops of the Semi-Arid Tropics. *Biotechnol. Adv*. **2013,** *31*, 1120–1134.
95. Bohra, A. Emerging Paradigms in Genomics-Based Crop Improvement. *Sci. World J.* Article ID 585467. **2013,** 17.
96. Yang, H.; Tao, Y.; Zheng, Z.; Shao, D.; Li, Z.; Sweetingham, M. W. Rapid Development of Molecular Markers by Next-Generation Aequencing Linked to a Gene Conferring Phomopsis Stem Blight Disease Resistance for Marker-Assisted Selection in Lupin (*Lupinus angustifolius* L.) Breeding. *Theor. Appl. Genet.* **2013b,** *126*, 511–522.
97. Purdy, L. H. *Sclerotinia sclerotiorum*: History, Diseases and Symptomatology, Host Range, Geographic Distribution, and Impact. *Phytopathology* **1979,** *69* (8), 875–880.
98. Food and Agriculture Organization (FAO). *The State of Food Insecurity in the World*; FAO: Rome, 2020.
99. Aggarwal, P. K.; Singh, A. K.; Samra, J. S.; Singh, G.; Gogoi, A. K.; Rao, G. G. S. N.; Ramakrishna, Y. S. Introduction. In *Global Climate Change and Indian Agriculture*; Aggarwal, P. K., Ed.; ICAR: New Delhi, 2009; pp 1–5.
100. Singh, S. R.; Allen, D. J. *Cowpea Pest and Diseases. Manual Series No. 2*; International Institute of Tropical Agriculture: Ibadan, 1979.
101. Nene, V.; Musoke, A.; Morzaria, S. Conservation of the Sporozoite p67 Vaccine Antigen in Cattle-Derived *Theileria parva* Stocks with Different Cross-Immunity Profiles Infect. *Immun.* **1996,** *64*, 2056–206.
102. Tahtamuni, M. E. W.; Hameed, K. M.; Saadoun, I. M. Biological Control of *Sclerotinia sclerotium* Using Indigenous Chitinolytic Actinomycetes in Jordan. *Plant Pathol. J.* **2006,** *22* (2), 107–114.
103. Patil, B. L. Pigeonpea Sterility Mosaic Virus: a Legume-Infecting Emaravirus from South Asia. *Mol. Plant Pathol.* **2015,** *16* (8), 775–786.
104. Gurha, S. N.; Singh, G.; Sharma, Y. R. Diseases of Chickpea and their Management. In: Chickpea Research in India; Ali, M., Kumar, S., Singh, N. B., Eds.; Indian Institute of Pulses Research: Kanpur, India, 2003; pp 195–227.
105. Singh. A.; Sharma, S. Composition of a Crop Residue Through Treatment with Micro-organisms and Subsequently Vermicomposting. *Bioresour. Technol.*, **2002,** *85* (2), 107–111.
106. Muehlbauer, F. J.; Kaiser, W. J.; Kutlu, Z. International Plant Genetic Resources Newsletter, Issue 78/79, **1989,** 33–34.
107. Luthra, J. C.; Bedi, K. S. Some Preliminary Studies on Gram Blight with Reference to Its Cause and Perenation. *Indian J. Agric. Sci.* **1932,** *2*, 499–515.

108. Joshi, M. M.; Singh, R. S. A Botrytis Gray Mould of Gram. *Indian Phytopathol.* **1969,** *22*, 125–128.
109. Carranza, J. M. Wilt of Chickpea (*Cicer arietinum* L.) Caused by *B. cinerea* (In Spanish). *Revista de la Facultad de Agronomia, Universidad Nacional de la Plata*, **1965,** *41*, 135–138.
110. Grewal, J. S.; Laha, S. K. Chemical Control of Botrytis Blight of Chickpea. *Indian J. Phytopathol.* **1983,** *36*, 516–520.
111. Grewal, J. S.; Pal, M.; Rewal, N. Botrytis Gray Mold of Chickpea in India. In *Botrytis Gray Mold of Chickpea*; Haware, M. P., Faris, D. G., Gowda, C. L. L., Eds.; ICRISAT: Patancheru, AP, India, 1992; pp 6–8.
112. Haware, M. P. Diseases of Chickpea. In *The Pathology of Food and Pasture Legumes*; Allen, D. J., LenneJ. M., Eds.; ICARDA, CAB International: Wallingford, 1998; pp 473–516.
113. Bakr, M. A.; Ahmed, F. Botrytis Gray Mold of Chickpea in Bangladesh. In *Botrytis Gray Mold of Chickpea. Summary Proceedings of the BARI/ICRISAT Working Group Meeting*; Haware, M. P., Faris, D. G., Gowda, C. L. L.; ICRISAT: Patancheru, AP, India, 1992; pp 10–12.
114. Shanower, T.; Romeis, J.; Minja, E. M. Insect pests of pigeonpea and Their Management. *Annu. Rev. Entomol.* **1999,** *44* (1), 77–96.
115. Sharma, P. N.; Sugha, S. K.; Panwar, K. S.; Sagwal, J. C. Reactions of Land Races and Exotic Collections of Kidney Bean to Anthracnose (*Colletotrichum lindemuthianum*). *Indian J. Agric. Sci.* **1993,** *63*, 456–457.
116. Mishra, R. K.; Pandey, S.; Mishra, M.; Rathore U. S.; Naimuddin, Kumar K.; Singh, B. Assessment of Biocontrol Potential of *Trichoderma* Isolates Against Wilt in Pulses. *J. Food Legumes* **2020b,** *33* (1), 48–52.

CHAPTER 11

Integrated Management of Diseases in Pigeonpea for Ensuring Food Security under Changing Climate in India

SANJEEV KUMAR* and U. K. KHARE

Department of Plant Pathology, College of Agriculture, Jawaharlal Nehru Krishi Vishwa Vidyalaya-Jabalpur, M.P.

**Corresponding author. E-mail: sanjeevcoa@gmail.com*

ABSTRACT

The pigeonpea varieties suffer due to diseases and pests; therefore, management of these is done in an integrated manner. Integrated disease management combines a wide array of all the methods, and practices for the control of diseases, at the same time protecting the environment in the long run as natural enemies are conserved and pesticide use is minimized. The wisdom to adopt this technology is that no single method of disease control was useful in containing the losses. Various cultural practices like field sanitation and soil solarization, summer ploughing, and green manuring are done in the case of *Fusarium* wilt. The amendment of soil using neem or castor oil cake is also recommended. The sterility mosaic is managed by planting crop early in pigeonpea. The planting of the crop on ridges helps to reduce the incidence of phytophthora blight. Three to four years of rotation with tobacco/sorghum, cereal crops reduce the intensity of wilt, *Phytophthora* blight diseases. Seed dressing with *Trichoderma* @ 4 g/kg of seed and soil application @ 10 g in 1 kg of farm yard manure (FYM) manages the wilt disease. Seed treatment with carbendazim+ thiram (1:2) @ 2.5 g/kg of seed is effective against wilt and reduces the incidence of *Alternaria* blight immensely. In *Phytophthora* prone area, metalaxyl should be used for seed treatment @ 2.5 g/kg of seed.

Integrated Pest Management in Diverse Cropping Systems. Devendra Pal Singh, PhD (Ed.)

Control of vector mites by foliar spray of neem oil @ 2% or neem seed kernel extract (NSKE) @5% or chemical insecticides, for example, dicofol 18.5 EC (2.5 mL/L) reduces the spread of the disease considerably. Foliar spray of Carbendazim+Mancozeb 0.25% reduces the spread of *Alternaria* blight considerably.

11.1 INTRODUCTION

The total area coverage and production of pigeonpea (arhar) have been about 45 lakh ha and 42 lakh tons, respectively, in India. The states like Maharashtra, Madhya Pradesh, Jharkhand, Karnataka, Andhra Pradesh, Gujarat, Uttar Pradesh, Telangana, Odisha, and Tamil Nadu are major producers of pigeonpea.[1] The biotic factors like phytophthora blight, sterility mosaic, and foliar diseases lead to 10–30% yield losses in pigeonpea.[7]

Diseases of pigeonpea and their regions of importance in India

Wilt (*Fusarium udum*),[3,4] stem blight caused by *Phytophthora drechleri* f.sp. *cajani, Pigeonpea* sterility mosaic due to pigeonpea sterility mosaic virus,[12] Cyst nematode (*Heterodera cajani*)

Tamil Nadu, Uttar Pradesh, Gujarat, Odisha

Yellow mosaic disease is caused by the mung bean yellow mosaic virus.

Uttar Pradesh, Delhi, Madhya Pradesh, Tamil Nadu, Karnataka, Maharashtra

Dry root rot (*Macrophomina phaseolina)*

Uttar Pradesh, Bihar

Cercospora leaf spots (*Cercospora cajani*)

Uttar Pradesh, Maharashtra, Tamil Nadu

Powdery mildew (*Leveillula taurica*)

Uttar Pradesh, Gujarat, Karnataka, Andhra Pradesh

Lance nematode (*Hoplolaimus* spp.)

Gujarat, Uttar Pradesh

Root-knot nematode (*Meloidogyne* spp.)

Madhya Pradesh, Maharashtra Andhra Pradesh, Gujarat, Uttar Pradesh, Karnataka

Reniform nematode (*Rotylenchulus reniformis* Linford, Oliveira)[2]

The farmers often resort to indiscriminate the use of pesticides to manage diseases and insect pests which results pest resurgence, resistance, pesticide residues, and associated environmental problems. It has dictated the need to look for other alternatives like the use of botanical pesticides, bioagents, and manipulation of cultural practices, and lastly the use of eco-friendly pesticides. Integrated disease management (IDM) combines a wide array of methods and practices. It also protects the environment in the long run as natural enemies are conserved and pesticide use is minimized. The major diseases of pigeonpea are as follows.

11.2 FUSARIUM WILT (*FUSARIUM UDUM* BUTLER)

Diagnostic Symptoms

The wilting initiates furom seedlings and may go up to the adult stage. It may be partial and the branches on only one side may show wilting.[3] The yellowing and drying of leaves are initial symptoms and later whole plant is infected. Severe wilting occurs at the flowering and pod development stages. The infected stem turns brown or dark purple at the base. The affected plants may dry partial or completely. The infected branches show the wilting symptoms first.

Survival and Spread

The pathogen survives in the soils. It can spread up to 3 m with soil in one season. The pathogen can survive in stubbles of the crop for 2.5–3 years.

Favorable Conditions

The disease is favored by low soil temperature and appears more on mature plants. About 30% soil water-holding capacity along with a temperature range of 20–30°C are the most favorable conditions for pathogen.

11.3 PHYTOPHTHORA BLIGHT (*PHYTOPHTORA DRECHSLERI* F. sp. *CAJANI*)

Diagnostic Symptoms

The infected plants show wilting, rolling, and drying of leaflets with withered petioles and small stems. On leaves brown, water-soaked spots with circular or irregular shape develop which later turn necrotic. The lesions result girdles on the stems and such plants break at the infection point. The stem looks swollen and later turns in cankerous structures. The infected seedlings die. The fungus develops white-pink mycelium on the blighted area under favorable weather conditions.

Survival and Spread

It is a soil-borne disease. The pathogen survives in soil and in plant debris.

Favorable Conditions

The disease incidence is more in soils with poor drainage, under heavy rains, and a temperature range of 28–30°C.

11.4 POWDERY MILDEW (*LEVEILLULA TAURICA*)

Diagnostic Symptoms

White powdery patches develop on both surfaces of leaves and infected leaves fall prematurely. Severe infection may cause yellowing and crinkling of leaves. The fungus is airborne.

11.5 ALTERNARIA BLIGHT (*ALTERNARIA ALTERNATA*)

Diagnostic Symptoms

The light to dark prominent brown colored small necrotic spots develops on the leaves and pods. Severe infection may cause complete defoliation. Disease affects older leaves more during the postrainy season.

11.6 CERCOSPORA LEAF SPOT (*CERCOSPORA INDICA*)

Diagnostic Symptoms

The spots are circular to irregular brown in color (1–2 mm in diameter) and become necrotic later. These develop on lower surface of leaves in most of the cases. These lesions coalesce and thus cause leaf blight. These may develop on petiole and on young branches.

11.7 STERILITY MOSAIC (PIGEONPEA STERILITY MOSAIC VIRUS (PPSMV) TRANSMITTED BY: ERIOPHID MITE ACERIA CAJANI

Diagnostic Symptoms

Early stage infection may cause 95–100% losses and older plants suffer 26–97% losses. The malformed, crinkled leaves, mosaic, reduced size with ring spots are common symptoms of the disease on leaves. The sterility of plants may be full or partial. The seeds are not developed, and if developed then they are discolored and shriveled.

Survival and Spread

The eryophid mites (*A. cajani*) transmit the disease. The virus survives on perennial and volunteer pigeonpea.

Favorable Conditions

The incidence of the virus increases under shade and humidity.

11.8 MUNG BEAN MOSAIC

The disease symptoms appear as yellow, diffused irregularly distributed spots on the leaf lamina. The affected leaflets show board and yellow patches. The entire lamina may turn yellow. The size of the leaf is reduced. The higher incidence of the virus in late-sown pigeonpea is found in peninsular India. Whitefly (*Bemisia tabaci*) acts as a vector.

Survival and Spread

Whitefly vector transmits the disease. The virus overwinters on perennial and volunteer pigeonpea.

Favorable Conditions

The disease proliferates under shade and high humidity.

11.9 CYST NEMATODE (*HETERODERA CAJANI*)

The affected plants show small cysts on roots and retarded growth. During off-season, the nematode survives in a cyst stage that contains eggs and larvae.

11.10 INTEGRATED DISEASE MANAGEMENT

The IDM is currently defined as: "a sustainable approach to managing the disease by combining all the available methods of control, which are easily accessible to the resource-poor farmers."[5] The IDM components are as follows:[10]

1. Use of resistant variety.
2. Farm management practices reduce the disease incidence.
3. Facilitation of activity of biocontrol agents.
4. Judicious use of pesticides

Following IDM modules are developed in pigeonpea.

Summer Ploughing

The spores and propagules of pathogens get killed at a temperature above 40°C for several days. It is effective against species of *Fusarium*, *Rhizoctonia*, *Phytophthora*, and nematodes.[5]

Soil Solarization

A thin polythene sheet cover is put on infested soil for 6–7 weeks during April and May. It raises soil temperature up to 60°C. It is quite effective to manage *H. cajani* and *F. udum*.

Plant Nutrients

Pigeonpea is a legume crop; therefore, nitrogen is required in lesser quantity. Therefore, diammonium phosphate @ 100 kg per ha is recommended. The incidence of *Phytophtora* blight is minimized with the application of 20 kg of K per ha. The wilt is also reduced in fields where manures and oil cakes are applied in the soil since it increases beneficial microorganisms on the one hand, as well as reduces the fungal and bacterial propagules on the other hand

Varieties

Table 11.1 shows resistant and tolerant, as well as high-yielding pigeonpea varieties.

TABLE 11.1 Disease Resistant/Tolerant Varieties of Pigeonpea.

Disease	Resistant/tolerant varieties
Fusarium wilt	ICP 9145, H 76–65, ICPL 267, Mukta, and ICP 8863 (Maruti)
Phytophthora blight	ICPL 304, KPBR Hy 4, ICPL 288, KPBR 80-1-4, and ICPL 150[8]
Sterility mosaic	Bahar, MA166, PDA 10, Rampur Rahar, Sharda, TT 6, Bageshwari, DA 11, ICPL 87051, PDA 2, Prabhat, PDA 10, DA 13, TT 5, MA 165
Alternaria blight	MA128-2, 20-105, DA 2, MA1 28-1
Wilt, Sterility mosaic	Asha, Narendra Arhar-1, BSMR 736, MA 3

Seed Treatment

Seed treatment with *Trichoderma* formulations @ 5 g per kg of seed reduces the incidence of wilt considerably. In *Phytophthora* prone area, Mancozeb + Metalaxyl should be used for seed treatment @ 2.5 g/kg of seed. Seed treatment with Carbendazim+ Thiram (1:2) @ 2.5 g/kg of seed reduces the incidence of wilt, *Alternaria* blight immensely.

Sowing Time and Season

Phytophthora blight incidence is reduced in premonsoon sown crops as compared to late sown crops of which early plant stage is caught by cloudy weather, continuous rains. The disease occurs more on plants at the age of 15–20 days, the susceptibility decreases with advancing age and being minimum at 120 days. The pre rabi pigeonpea crop escapes the attack of *Phytophthora* blight as well as wilt. On the contrary, *Alternaria* blight is serious on prerabi crops as compared to rainy season crops.

Method of Sowing

Ridge sowing helps in the better crop establishment than the flatbed sowing method. Ridges not only provide proper drainage but also avoid the water to come in contact of plants for the longer duration. Otherwise, water stagnated in the field facilitates congenial conditions for sporangium formation, germination, liberation of zoospores. It also helps in free-swimming of zoospore and infection of seedling to cause *Phytophthora* blight disease.

Crop Gometry

The crop geometry affects the growth of crops and microclimate also. The dense plant population favors diseases like *Phytophthora* blight and stem rot.

Inter/Mixed Cropping

The mixed- or inter-cropping of pigeonpea with sorghum, sunhemp, urdbean, sesame, and mung bean reduces soil-borne pathogens in pigeonpea.

Crop Rotation

The incidence of *Phytophthora* blight and wilt reduces under crop rotation with tobacco, sorghum, and cereal crops

Sanitation

The pathogens, *Phytophthora, Fusarium,* and *Sclerotium* survive on the infected debris of plant. The mite vector as well as the virus of SDM survives on the perennial plants of pigeonpea growing in the field bund or near the houses. Such plants should be uprooted and destroyed to eliminate the inoculums. In the standing crop also, the diseased plants may be destroyed as soon as they appear in the field.

Chemical Management

Drench the soil with Carboxin + Thiram @ 1.0 g/L of water in the case of severe disease incidence of wilt in the field. The diseases like *Alternaria* blight, *Cercospora* blight, powdery mildew, etc., are controlled by foliar sprays with fungicides like Carbendazim 12% + Mancozeb (0.2%). The soil application of phorate@1 kg a.i/ha at the time of sowing can protect the pigeonpea plant from infection of sterility mosaic diseases for the period of 75 days after sowing.

Control of vector mites by foliar spray of neem oil @2% or NSKE @5% or chemical insecticides, for example, dicofol 18.5 EC (2.5 mL/L) is done.[13]

Integrated Disease Management Packages[14]

- Selection of disease-free fields.
- Use of tolerant varieties like Pusa-9, NP 15, NP 38, and AL 1430.
- Rotate pigeonpea with tobacco, pearl millet, sorghum, and cotton up to 3 years.
- Wide row interspacing, sowing on ridges, and mixed cropping with sorghum.
- Use of oil cakes, boron, zinc, and manganese.
- Apply Trichoderma @ 1 kg in 100 kg FYM.[15]
- Seed treatment using hexaconazole + captan @ 2.5 g per kg or metalaxyl @ 2.0 g per kg or carbendazim or thiram @ 2 g per kg of seed 24 h before sowing and later with *Trichoderma viride* powder @ 5 g per kg.[16]
- Foliar sprays of Carbendazim 12% + Mancozeb @ 0.25% Mancozeb +Metalaxyl @ 0.2% at the interval of 120 days.[9]
- Eradication of self-sown pigeonpea plant in fields to manage sterility mosaic.
- Control of vector mites by foliar spray of neem oil @ 2% or NSKE @ 5% or chemical insecticides, for example, dicofol 18.5 EC (2.5 mL/L) reduces the spread of the disease considerably.

Ecological Engineering for Disease Management

The disease management in pigeonpea has gone a paradigm shift and now rely cultural techniques which manipulate habitat of pathogens and also enhance biological control.[18,19] The approach includes:

- Keep soils covered intact throughout the year by allowing vegetation and crop residue.
- Soil amendments using FYM, vermicompost, and crop residue to enhance biodiversity in soil.
- Reduced tillage practices to help natural enemies of pathogens.
- Use of balanced doses of fertilizers including biofertilizers.
- Application of mycorrhiza and PGPR.
- Enhanced use of *Trichoderma viride, T. harzianum,* and *Pseudomonas fluorescens* for treatment of seeds, seedlings, and planting materials. Also, these are used in nursery treatment and soil application.[11,16]

11.10.1 DO'S AND DON'TS IN IDM

S. No.	DO'S	DON'TS
1	Deep ploughing (May–June) and exposure of field to sunlight for 2–3 weeks	Allow 2–3 weeks sunlight exposure of bulbs or rhizomes of perennial weeds
2	Apply NPK fertilizers according to the soil test report	Do not use imbalanced fertilizers
3	Row sowing at optimum depths under proper moisture	Keep the seed depth up to 5–7 cm
4	Early sowing	Avoid late sowing
5	Use recommended varieties	Avoid old varieties and those not recommended for the season or the region
6	Seed treatment with biopesticides and chemicals	Do not sow untreated seeds
7	Follow crop rotation	Do not go for monocropping
8	Use safe pesticides	Avoid pesticides, 7 days before harvest

11.10.2 IMPACT OF THE TECHNOLOGY

- Reduction in the quantity of pesticides used, improvement in the environmental safety.
- Conservation of natural fauna, enhancement in the activity of the natural enemies.
- Utilization of locally available need seeds for preparation of NSKE and creating employment for rural youths.
- Effective but absence of low-cost mass production, storage technology, need for timely availability of bioagents has opened opportunities for small entreprepreneurs to take up mass production through cottage industry.
- The socioeconomic status of farmers is raised.
- Reduction in the use of chemical pesticides.

11.11 CONCLUSION

In India, no specific IDM modules except for some fruit crops for disease management are available. In fact, IDM should be now the backbone of the entire disease management program but still awaiting the due attention in the field that it deserves. To increase the reach of IDM amongst the farmers, it is also essential to invest in developing forewarning technologies for epidemiologically potential diseases. Similarly strengthening the surveillance mechanisms and educating the farmers through the transfer of technologies shall be of great help in the adoption of IDM and in the future for disease management.

KEYWORDS

- **fusarium wilt**
- ***Fusarium udum***
- **phytophthora blight**
- ***Phytophthora drechsleri* sp.**
- ***cajani.***
- **powdery mildew**
- ***Leveillula taurica***

REFERENCES

1. Annonymous. Pulses Revolution—From Food to Nutritional Security Min. of Agri. & FW (DAC&FW), GOI: New Delhi, 2018; p 115.
2. Satyagopal, K.; Sushil, S. N.; Jeyakumar, P.; Shankar, G.; Sharma, O. P.; Boina, D. R.; Sain, S. K.; Lavanya, N.; Sunanda, B. S.; Asre, R.; Kapoor, K. S.; Arya Sanjay, Kumar, S.; Patni, C. S.; Jacob, T.; Santhosh, J.; Eapen, C. N.; Biju, K.; Dhanapal, H.; Ravindra, B. C.; Hanumanthaswamy, Linga Raju, S.; Ramesh Babu, Sathyanarayana, N.; Latha, S. AESA Based IPM Package for Redgram, **2014,** p 42.
3. Butler, E. J. The Wilt Disease of Pigeonpea and Pepper. *Agric. J. India* **1906,** *1*, 25–36.
4. Butler, E. J. The Wilt Disease of Pigeonpea, the Parasitism of *Neocosmospora vasinfecta* Smith. *Memoirs Depart. Agric. India, Botanical Series* **1910,** *2*, 1–64.
5. Chauhan, Y. S. Pigeonpea: Optimum Agronomic Management. In *The Pigeonpea*; Nene, Y. L., Hall, S. D., Sheil, V. K., Eds.; CAB International: Wallingford, 1990; pp 257–278.
6. Gouder, S. B.; Kulkarni, S. Compatibility Effect of Antagonists, Seed Dressers Against *Fusarium udum*-The Causal Agent of Pigeonpea Wilt. *Karnataka J. Agric. Sci.* **1999,** *12* (1–4),197–199.
7. Kannaiyan, J.; Nene, Y. L.; Reddy, M. V.; Ryan, J. G.; Raju, T. N. Prevalence of Pigeonpea Diseases and Associated Crop Losses in Asia, Africa and the Americas. *Trop. Pest Manage.***1984,** *30*, 62–71.
8. Kannaiyan, J.; Nene, Y. L.; Raju, T. N.; Sheila, V. K. Screening for Resistance to Phytophthora Blight of Pigeonpea. *Plant Dis.* **1981,** *65*, 61–62.

9. Kannaiyan, J.; Nene, Y. L. Efficacy of Metalaxyl for Control of Phytophthora Blight of Pigeonpea. *Indian Phytopathol.* **1984,** *37*, 506–510.
10. Karimi, R.; Owuoche, J. O.; Silim, S. N. Importance and Management of Fusarium Wilt (*Fusarium udum* Butler) of pigeonpea. *Int. J. Agron. Agric. Res.* **2012,** *2* (1), 1–14.
11. Khan, J.; Ooka, J. J.; Miller, S. A.; Madden, L.V.; Hoitink, H. A. J. Systemic Resistance Induced by *Trichoderma hamatum* 382 in Cucumber Against Phytophthora Crown Rot, Leaf Blight. *Plant Dis.* **2004,** *88*, 280–286.
12. Khare, D.; Satpute, R. G.; Tiwari, A. S. Present State of the Wilt and Sterility Mosaic Disease of Pigeonpea. *Indian J. Genet.* **1975,** *54*, 331–346.
13. Kulka, M.; Von Schmeling, B. In *Modern Selective Fungicides-Properties and Mechanism of Action*; Longman Scientific & Technical, 1987; p 119.
14. Sanjeev, K.; Singh, S. N.; Khare, U. K.; Bhale, U.; Bhatt, J.; Bhale, M. S. Integrated Disease Management in Pigeon Pea. *Dis Field Crops: Diagnosis Manage.* 2020; pp 2133–2143.
15. Sanjeev, K.; Upadhyay, J. P.; Rani, Archana. Evaluation of *Trichoderma* Species Against *Fusarium udum* Butler Causing Wilt of Pigeonpea. *J. Biol. Control* **2009,** *23* (3), 329–332.
16. Mandhare, V. K.; Suryawanshi, A. V. Application of *Trichoderma* Species Against Pigeonpea Wilt. *JNKVV Res. J.* **2004,** *38*, 99–100.
17. Sharma, M. Climate Change and Emerging Diseases in Chick Pea, Pigeonpea: Occurrence, Distribution, Epidemiology. *6th International Conference Plant Pathogens*, People February 23–27, **2016,** New Delhi, India. **2016,** P. No.53.
18. Gurr, G. M.; Wratten, S. D.; Altieri, M. A. *Ecological Engineering for Pest Management Advances in Habitat Manipulation for Arthropods*; CSIRO Publishing: Collingwood, Australia, 2004a.
19. Gurr, G. M.; Wratten, S. D.; Altieri, M. A. Ecological Engineering: A New Direction for Pest Management. *AFBM J.* **2004b,** *1*, 28–35.

CHAPTER 12

IPM in Groundnut: Current Scenario and Future Prospects

NATARAJA V. MAHESHALA, ANANTH KURELLA, RUPAK JENA, HARISH GANGAIAH, and RADHAKRISHNAN THANKAPPAN*

ICAR-Directorate of Groundnut Research, Junagadh, Gujarat, 362001, India

**Corresponding author. E-mail: radhakrishnan.nrcg@gmail.com*

ABSTRACT

Integrated pest management (IPM) has evolved and reinvented itself over the years. New tools have been added after thorough research, which are tailor-made for specific agro-climatic regions, seasons of crop cultivation, stages of crop, economics, and feasibility in adaptation of tools and for either individual or multiple pests. The IPM for groundnut is dynamic, as the crop is grown in several states experiencing vagaries of edapho-climatic factors. The overdependence on synthetic pesticides in pest control has resulted in resistance development in major pests, minor pests becoming major, and increased cost of crop protection. Subsistence farming of groundnut in resource poor soils has made IPM even more important for Indian agrarian community. IPM begins even before planting of crop and wherein selection of crop varieties and removal of crop residues/voluntary plants and preparation of land, addition of soil amendments/bioagents/pesticides, planting time, space, etc., are crucial. This effectively reduces the primary inoculums of pathogens, insect-pests, and nematodes. Emphasis should be given for monitoring of pest activity during the crop growth period. Trap crops and border rows, pheromones/sticky traps, light traps, etc., comes in handy to gauge the levels of infestation and time the application of bio-control agents. In-situ conservation of natural enemies of pests is both economical and

Integrated Pest Management in Diverse Cropping Systems. Devendra Pal Singh, PhD (Ed.)

ecologically safe. Host-plant resistance is exploited for managing several pests of groundnut; however, availability of quality seeds is a concern. Chemical pesticides should always be the last resort whenever pest populations cross economic threshold levels (ETLs). Choice of pesticides solely rests upon the users and, in this case, the growers; but, educating them about pesticide label claims is necessary. Some pests, like white grubs in Saurashtra region of Gujarat and Fusarium leaf blight/wilt in Rajasthan, are gaining importance in recent years, necessitating development of IPM tools. Post-harvest concerns due to bruchids and aflatoxin contamination remain unaltered; the gap between lab and land (including storage) should be bridged. In the future prospects of IPM section, several research thrust areas have been brought forward, such as quick, correct detection methods for pests, especially plant viruses; development of nano-formulations of pesticides and pheromones; novel biotechnological applications for imparting biotic stress resistance; and development of good agricultural practices for groundnut.

12.1 INTRODUCTION

Groundnut (*Arachis hypogaea* L.) is a self-pollinating intermediate herbaceous legume. It belongs to family Leguminosae, is a hybrid of *A. duranensis* and *A. ipaensis,* wild groundnut species. This crop ranks fourth and third position in contribution toward world's edible oil and protein, respectively. Groundnut is considered as the cheapest food for the poor, is highly rich in calories, and has dietary fibers, like vitamins and minerals. Nuts contain 26–28% protein, 48–50% oil, and are rich in omega-3 fatty acids. A South American native of nuts is now grown largely in the tropical and the sub-tropical regions of the world.[1] China is the largest producer as well as consumer of groundnut globally with 16.6 million tonnes followed by India and Nigeria.[2] As per the fourth advanced estimates of the Government of India, the average productivity of groundnut is 1868 kg/ha with a production of nearly 9.2 million tonnes from an area of 4.9 million hectare.[3]

In India, traditional groundnut growing states are Andhra Pradesh, Karnataka, Tamil Nadu, Maharashtra, and Gujarat; however, significant increase in acreage under groundnut is observed in states like Rajasthan, West Bengal, Odisha, and Uttar Pradesh.[4] More than 80% of cropped area is under rainfed condition, and is constantly under threat by erratic rainfall, prolonged dry spells of monsoon, soil salinity, and alarming biotic constraints like plethora of diseases and pest attacks.[5]

Diseases caused by fungal and viral infections and nematodes are devastating in groundnut. Fungal diseases like leaf spots (*Cercospora arachidicola* Hori, *Phaeoisariopsis personata* Berke and Curt), rust (*Puccinia arachidis* Speg.), stem rot (*Sclerotium rolfsii* Sacc.), and collar rot (*Aspergillus niger* van Tieghem) are economically important. Recently, Fusarium leaf blight/wilt caused by a complex of *Fusarium incarnatum–F. equiseti* has affected groundnut production in Rajasthan.[6] Viral diseases like groundnut bud necrosis (*groundnut bud necrosis virus*), groundnut stem necrosis (*tobacco streak virus*), peanut mottle (*peanut mottle virus*), and peanut clump (*Indian peanut clump virus*) are important. The loss caused by the diseases, including both at the seedling and foliar stages, ranges from 10 to 70% all over the world.

Insect-pests of national importance are tobacco caterpillar (*Spodoptera litura* Fab.), leaf miner (*Aproaerema modicella* Deventer), aphids (*Aphids craccivora* Koch), leafhoppers/jassids (*Empoasca kerri* Pruthi and *Balclutha hortensis* Lindberg), and thrips (*Scirtothrips dorsalis* Hood, *Frankliniella schultzei* Trybom, *Thrips palmi* Karny, *Caliothrips indicus* Bagnall, and *Megalurothrips usitatus* Bagnall). Insect-pests of regional importance are white grubs, termites, red-headed hairy caterpillar, etc. Losses caused by insect-pests in groundnut are reported to be 47.3%.[7] Groundnut bruchids, *Caryedon serratus* (Olivier), are primary storage pests, which occur throughout India and cause losses to stored groundnut, both in quality and quantity.[8,9] Apart from bruchids, aflatoxin contamination in groundnut also affects the trade prospects of India.[10]

Root knot nematodes (*Meloidogyne hapla* Chitwood, *M. arenaria* Chitwood, *M. javanica* Chitwood and *M. incognita* Kofoid and White) are major plant pathogenic nematodes infesting groundnut. In India, an annual yield loss of 21.6% due to infestation by *M. arenaria* and *M. javanica* in groundnut have been reported.[11] Similarly, pod scab nematode, *Tylenchorynchus brevilineatus* Williams, endemic to Chittoor and Nellore districts of Andhra Pradesh, is known to cause *Kalahasti Malady* in groundnut.

The congenial and erratic biotic conditions are supplementing the new and resurgent pests, causing diseases to explode and cause extensive crop losses in groundnut. Adoption of integrated pest management is necessary to reduce crop losses and also the cost of pest management, thereby increasing the crop productivity and economic sustainability. In this chapter, an earnest effort has been made by the authors to brief on the major biotic constraints and their distribution and host range and nature of damage and biology/life cycle and their integrated management. Future prospects and emerging research areas were also discussed for knowledge on constraints and opportunities in groundnut pest management.

12.2 MAJOR GROUNDNUT DISEASES

12.2.1 SOIL AND SEED-BORNE DISEASES

12.2.1.1 COLLAR ROT (ASPERGILLUS NIGER)

Collar rot is most prevalent in Bikaner of Rajasthan and Ludhiana of Punjab. The maximum disease incidence (25–50%) was reported in Rajasthan.[12] Spores or hyphae of fungi available in the soil or on seed surface germinate and attack seeds causing pre-emergent rotting of seeds. The other characteristic diagnostic symptoms are hypocotyl rotting, seedling blight, and rapid wilting of plant. Collar region of the affected plant shred is covered by black mycelia and spores (Figure 12.1a). Soil-borne inoculums are the primary source of infection. Disease development is favored in sandy loam, medium black soils with soil moistures ranging from 13 to 16% in temperatures between 31°C and 35°C.[13]

FIGURE 12.1 Soil-borne diseases of groundnut: (a) collar region covered by spores, mycelia of *A. niger*; (b) white mycelial growth of *S. rolfsii* on stems; (c) blackened root system with micro sclerotia of *M. phaseolina*. Photo credit: Thirumalaisamy PP, ICAR-NBPGR (RS), Thrissur.

12.2.1.2 STEM ROT (SCLEROTIUM ROLFSII)

It occurs throughout the groundnut-producing regions of India. Hanumangarh in Rajasthan, Latur in Maharashtra, Junagadh in Gujarat, and Raichur and Dharwad in Karnataka have been identified as "hot spots" for the disease. Stem rot causes yield losses up to 80%; however, losses less than 27% are more common.[13]

Initial symptoms of stem rot are wherein stems and branches that are in direct contact with the infected soils show either partial or complete wilting. Both on infected plant parts and adjacent soil, white mycelial growth (Figure 12.1b) and brown sclerotia are visible. Wilted leaves turn brown and remain attached to the plant. Other symptoms caused by stem rot are peg infection, pod rot, and leaf blight. Being necrotrophic fungi, *S. rolfsii*, can colonize both living and dead plant tissues. Sclerotia present near soil surface remain viable for many years; whereas, sclerotia buried deep in soil can survive about a year or less.[14] Defoliated leaves facilitate plant-to-plant spread. Infected soil, wind splashed rain, and sclerotia are the primary sources of inoculums. Crop residues in soil influence sclerotial germination, infection by *S. rolfsii*. Disease development is favored by soil moisture (40 to 50%) and optimal temperatures of 29°C to 32°C (day) and 25°C (night).[13]

12.2.1.3 DRY ROOT ROT (MACROPHOMINA PHASEOLINA)

Disease is sporadic but severe in Rajasthan, Maharashtra Andhra Pradesh, and Tamil Nadu.[14] Dry root rot is also known as charcoal rot. It causes severe mortality in groundnut seedlings resulting in reduced crop stand. At first, water-soaked necrotic lesions appear on the stem, above collar region, and they encircle the stem and cause wilting. Blackening, rotting, and shredding of roots are characteristic symptoms of this disease (Figure 12.1c). Micro-sclerotia produced by fungi in excess cause blackening of the inner surfaces of the pods.

Macrophomina phaseolina is a facultative saprophyte living in soil, and hence has a wide host range. Primary inoculums for the disease are infected soil, plant debris, and infected pods. Dry root rot fungi require temperatures from 29°C to 35°C for seedling infection and from 26°C to 32°C for pod invasion.[13]

12.2.1.4 AFLAROOT (ASPERGILLUS FLAVUS)

The aflaroot or yellow mold fungi can be found in both healthy and infected groundnut pods. Pathogen first infects cotyledons upon their emergence from soil. Stunted plant growth and reduced leaf lamina with a pointed tip are the characteristic symptoms.[14] In addition, leaves also show veinal clearing and chlorosis. Development of secondary root system in infected seedlings is lacking, such plants become reproductively inactive. Spores and hyphae in the soil act as primary source of inoculum. Disease development is favored by low soil moisture levels and temperatures between 25°C and 35°C.[13]

12.2.2 FOLIAR FUNGAL DISEASES

12.2.2.1 LEAF SPOT DISEASES: EARLY LEAF SPOT (CERCOSPORA ARACHIDICOLA) AND LATE LEAF SPOT (PHAEOISARIOPSIS PERSONATA)

Leaf spot diseases are the major foliar diseases in groundnut with an estimated yield loss ranging from 15 to 59%.[14] They are most commonly occurring diseases throughout the country. Severe loss in photosynthetic area reduces pod yields in groundnut coupled with adverse effect on the fodder quality.

Early leaf spot (ELS) typically infects crop four-weeks after sowing; whereas, late leaf spot (LLS) is observed from the eighth week till harvest. Lesions caused by ELS are 1–10 mm in diameter, sub-circular, and appear dark brown on upper leaf surface due to sporulation. ELS lesions are usually surrounded by yellow halo (Figure 12.2a). Lesions caused by LLS are minute, circular, appear black (Figure 12.2b), and rough on the lower leaf surface due to sporulation.[15] In both diseases, lesions coalesce and infected leaves drop prematurely. Lesions may also occur on the stem, and in branches in severe cases.

The pathogens survive by producing conidia, which remain in crop residue or on pod surface. Self-sown groundnut plants act as bridge between one season to another for the survival of pathogens. Disease development is favored by prolonged leaf wetness, temperatures ranging from 25°C to 30°C, and high RH (>80%).[13] Air-borne conidia causes secondary infection.

12.2.2.2 RUST (PUCCINIA ARACHIDIS)

Rust causes an estimated yield loss ranging from 10 to 52% and reduction in seed size and oil content.[14] This disease is prevalent throughout the country and can infect six-weeks-old groundnut plants. Except flowers, all aerial parts of the plant are vulnerable to pathogen. Initially, chlorotic flecks appear on the upper surface of the leaves, while correspondingly on the lower surface, orange-collared pustules appear, which later turn to brown (Figure 12.2c). In contrast to leaf spot diseases, infected leaves remain attached to plant, turn necrotic and dry in severe conditions. Kernels obtained from infected plants are small and shriveled.

Uredinio spores help in disease perpetuation. Volunteer groundnut plants during the off-season enable year-round survival of pathogen. For the development of this disease, high humidity, leaf wetness, and temperatures around 20°C are favorable.[14] Secondary infection is caused through wind, insect-activity, and rain splashes.

FIGURE 12.2 Foliar fungal diseases of groundnut: (a) brown lesions with yellow halo caused by *C. arachidicola*; (b) lower leaf surface showing black lesions caused by *P. personata*; (c) lower leaf surface showing orange-colored pustules caused by *P. arachidis*; (d) "V" shaped lesions at the tip of the leaves due to *A. Alternate*; and (e) withering and drooping of aerial parts caused by *F. Incarnatum* and *F. equiseti*. Photo credit: Thirumalaisamy PP, ICAR-NBPGR (RS), Thrissur.

12.2.2.3 ALTERNARIA LEAF BLIGHT (ALTERNARIA ALTERNATA)

Alternaria leaf blight a minor disease, but severe in some parts of Gujarat (Junagadh, Rajkot, and Kutch districts), specially on summer groundnut.[14] Depending on disease severity, it reduces pod and fodder yield.

Blightening of leaf starts from tip downward in characteristic V-shape (Figure 12.2d), that later turns to dark brown. Lesions coalesce to impart ragged, blighted appearance to leaves, which later curl inward and turn brittle.

Volunteer plants serve as primary source while conidia serve as secondary source of infection. Aquatic plants like water hyacinth (*Eichhornia crassipes*) and pistia (*Pistia stratiotes*) are weed hosts. During epidemics, inoculum load comes from crop plants like sesame and sunflower, apart from weed hosts and groundnut debris. For disease development, high relative humidity (>85%) and optimum temperature (25°C–30°C) are favorable.[16]

12.2.2.4 FUSARIUM LEAF BLIGHT/WILT (FUSARIUM INCARNATUM–F. EQUISETI SPECIES COMPLEX)

Since 2012, Fusarium leaf blight/wilt have regularly occurred infecting groundnut in Rajasthan, with the highest incidence (52%) reported from Asarlai of Jodhpur.[6,17] Water-soaked lesions appear initially on leaves. Tips and margins of leaves show chlorosis, which turn brown and give blighted appearance to leaves. The symptoms progress toward the midrib and petiole, and water soaking, withering, drooping appearances of aerial parts, and wilting of flaccid stems are the characteristic symptoms of the disease (Figure 12.2e). The roots are devoid of any symptoms and signs. The entire plant gets dried up during advanced stages. The disease appears in patches and reappears in the same area with high intensity in the subsequent years. Even though the primary site of infection is either leaf or stem, pathogen survives in the soil.[6]

12.2.3 VIRAL DISEASES

12.2.3.1 BUD NECROSIS (GROUNDNUT BUD NECROSIS VIRUS, GBNV)

Bud necrosis in groundnut is a major viral disease in India with an estimated annual yield loss ranging from 30 to 90% depending upon plant growth stage at the time of infection.[14] Several hotspot locations have been identified for screening of groundnut germplasm, namely, Jagtial and Hyderabad in Telangana, Raichur in Karnataka, Latur in Maharashtra, Tikamgarh in Madhya Pradesh, and Mainpuri in Uttar Pradesh.[18]

Necrosis of the terminal buds is the characteristic symptom of the disease (Figure 12.3a). Death of terminal bud results in stunting of main shoot and proliferation of axillary shoots. Under severe conditions, necrosis spreads to the petiole and stem, causing death of the plant. Late infection causes reduction in pod size and shrivelling of kernels.

The GBNV is placed in the genus *Tospovirus* of the family *Bunyaviridae*.[18] Thrips are vectors of GBNV, which transmit virions in circulative and propagative manner. Larvae (first and second larval stages) acquire the virus by feeding on infected plants. Thrips are viruliferous throughout their adult life time, but transovarial transmission is absent. Both vector and virus have wide host range; they can survive on crop plants (tomato, brinjal, green gram, black gram, beans, and pea), ornamentals (*Zinnia*, cosmos, and sunflower), as well as weeds (*Ageratum conyzoides*, *Cassia tora*, *Acanthospermum hispidum*, and *Desmodium triflorum*).[14] Temperatures above 30°C and wind speed of 10 km/h favor thrips migration.[14]

FIGURE 12.3 Viral diseases of groundnut: (a) terminal bud necrosis due to *PBNV*; (b) large necrotic lesions on young leaflets caused by *TSV*; (c) severe stunting of plants due to *IPCV*. Photo credit: (a and b) Vemana K, ARS, Kadiri, (c) Thirumalaisamy PP, ICAR-NBPGR (RS), Thrissur.

12.2.3.2 STEM NECROSIS (TOBACCO STREAK VIRUS, TSV)

Stem necrosis in groundnut, a regular phenomenon in Rayalseema region of Andhra Pradesh (Anantapur, Kadapa, Kurnool, and Chittoor districts) as well as in Raichur district of Karnataka.[14] It is caused by TSV of the genus *Ilarvirus* (Family: *Bromoviridae*). Large necrotic lesions appear on young leaves (Figure 12.3b), which coalesce at the later stages. Finally, necrosis spreads to stems resulting in eventual death of the plant.

Though TSV is pollen-borne, thrips carry pollen on their bodies from infected to healthy plants. Carrot grass or congress grass, *Parthenium hysterophorus*, is an obnoxoius weed, a symptomless carrier for TSV.[19] Also, crop plants like sunflower and tobacco serve as hosts for TSV.

12.2.3.3 PEANUT MOTTLE (PEANUT MOTTLE VIRUS, PMV)

Peanut mottle is reported in summer groundnut from states like Gujarat, Maharashtra, and Andhra Pradesh. Mottling and veinal clearing appear on freshly opened leaves. Matured leaves curl upward showing inter-veinal depressions and formation of dark green islands. Pods harvested from PMV infected plants appear small with irregular green to brown patches and have discolored kernels.[18]

Transmission of virus is through seed and aphid vectors.[14] Virus survives on legume crops like groundnut and soybean as well as on weeds like *Cassia obtusifolia, C. leptocarpa, C. occidentalis,* and *Desmodium canum.*

12.2.3.4 PEANUT CLUMP (INDIAN PEANUT CLUMP VIRUS, IPCV)

Peanut clump in late-infected plants can cause yield loss up to 60%.[14] The virus has a wide host range; and several weeds that commonly occur in groundnut also get infected by the virus.

The characteristic symptom of IPCV include severe stunting of infected plant (Figure 12.3c) and peeling of root epidermis. Mosaic mottling of leaves vary from young leaves (Chlorotic rings) to that of older leaves (dark green). The IPCV is transmitted through seed[20] and a soil-borne fungus (*Polymyxa graminis*).[21] The latter helps in virus perpetuation over seasons due to availability of different hosts.

12.3 MAJOR INSECT-PESTS OF GROUNDNUT

12.3.1 SOIL-INHABITING INSECT-PESTS

12.3.1.1 WHITE GRUBS (HOLOTRICHIA SP., APOGONIA SP., SCHIZONYCHA SP., ADORETUS SP.)

White grubs are soil-inhabiting polyphagous pests and are also known as root grubs. They feed on organic matter in soil as well as roots of many economic crops. They belong to Scarabidae family of Coleopteran insect order. Species of white grubs, which infest groundnut belong to three subfamilies; Melolonthinae, Rutelinae, and Dynastinae. The genus *Holotrichia* includes the major species of white grubs inflicting serious damage to groundnut cultivation in rainy season. *H. consanguinea* is pre-dominant in Rajasthan, Gujarat, Haryana, Punjab, Bihar, and Uttar Pradesh, whereas, *H. serrata* is most destructive in Maharashtra, Karnataka, Tamil Nadu, and Andhra Pradesh, but also is reported from western and peninsular India.[22] The species of white grub endemic to Saurashtra region of Gujarat are *Apogonia rauca, Schizonycha ruficollis*, and *Adoretus* sp.[23,24]

Apart from groundnut, they are known to attack roots of guava, sugarcane, coconut, areca nut, tobacco, potato, many other oilseeds, pulses, and vegetable crops. White grubs are present throughout the year; however, their activity can be seen only during the rainy season (May/June–December). Adults emerge in large numbers for mating after the first monsoon rains in mid-May or June months; again they return to soil for rest of its life cycle. White grubs prefer light clay soils rich in organic matter, aeration, moisture, and soil temperature around 25°C.

The adult beetles (Figure 12.4a) emerge from the soil during dusky hours (7– 10 PM) after good rains in mid-May or later. Once active, the beetles follow the daily rhythm of emergence, congregation on host trees at dusk, and return to soil at dawn. The beetles feed on the foliage of variety of host trees such as ber, drumstick, tamarind, banyan, acacia, neem, mahagoni, jamun, guava, sapota, banana, mango, fig, etc., found in and around the farm. The mated females return to soil in the early morning, start egg-laying. *H. consanguinea* takes 76–96 days for one life cycle whereas; *H. serrata* takes 141–228 days. The grubs are active in their second, third, and fourth instar stages during rainy seasons (July–October) and feed on organic matter and plant roots. The young grubs are white translucent whereas mature grubs are pale in color, "C- shaped" (Figure 12.4b). Before pupation, grub ceases to feed and descends down in the soil to the depth of 40–70 cm. A short, pre-pupal period is spent preparing an earthen chamber for pupation. Adult beetles remain in the soil in an inactive state at a depth of about one meter till the next monsoon rains. All the species of white grubs attacking groundnut are having single generation per year with an active larval diapause stage.

Damage is caused by grubs where, young grubs feed on fine rootlets while mature grubs feed on both roots and pods. Varying levels of wilting showed by the affected plants. Wilted plants eventually die, can easily be pulled out. The grubs cause infestation in patches leading to "patchy appearance" of field (Figure12.4c). In endemic areas, the damage to groundnut crop ranges from 20 to 80%.

12.3.1.2 TERMITES (ODONTOTERMES OBESUS & MICROTERMES OBESI)

Termites of economic significance to groundnut are *O. Obesus* and *M. obesi*.[25] They are distributed in all groundnut growing states.[25,26] They are highly polyphagous in nature and their occurrence mainly depends on the soil characteristics (light, well-aerated clay, sandy loam soils) and favorable environmental conditions (dry weather and higher temperature). Termites infest groundnut in severe form during dry seasons (summer) or when rainfall is sparse or scanty in rainy (September–October) seasons.[27]

FIGURE 12.4 White grubs infesting groundnut: (a) adult beetle; (b) C-shaped mature grub; (c) patchy appearance in the infested field due to wilting and death of plants.

They are eusocial and found in subterranean tunnels or in above-ground termitaria. Worker termites damage the groundnut plants from seedling stage to maturity. They feed on soft tissues of roots and make them hallow (Figure 12.5a). Infested plants wilt and die prematurely, especially in sandy and red soils (Figure 12.5b). They also attack pods by scrapping tissues between the veins causing scarification. Scarified pods are prone to invasion by soil fungi *A. flavus* that produce aflatoxin. The scarified pods are liable to shatter or crack during harvest. Breakage results in loss of kernels especially during mechanical harvesting. In Odisha, the loss in pod yield is 6 to 11% in rainy, 2.5 to 28% in post-rainy seasons.[28]

FIGURE 12.5 Termites affecting groundnut: (a) termite workers hallowed tap root system; (b) dead plants in the field.

12.3.1.3 MINOR SOIL-INHABITING INSECT-PESTS OF GROUNDNUT

Earwigs belonging to family Anisolabididae of Dermaptera, wireworms belonging to family Elateridae of Coleoptera are inflicting damage to groundnut pods. They are particularly severe in Tamil Nadu, Maharashtra, Manipur, and Gujarat states.[29,30] Common earwigs that infest groundnut are *Euborellia (=Anisolabis) anulipes* (Lucas) and *E. stali* (Dohrn). Apart from groundnut, they are known to attack cabbage, cotton, sorghum, and onion. They are known to infest groundnut throughout the year in Tamil Nadu; however, in Manipur, their incidence is high during May–July months.[29] Both grubs, adults make holes in the pods and feed on developing kernels.[30] The infested pods are either empty or partly filled with excreta, soil particles. In endemic areas (Tamil Nadu and Manipur), they may inflict 26 to 65% yield loss.

12.3.2 DEFOLIATING INSECT-PESTS

12.3.2.1 THE GROUNDNUT LEAF MINER (APROAERERMA MODICELLA)

Leaf miner belonging to Gelechiidae of Lepidoptera has been reported from Andhra Pradesh, Gujarat, Karnataka, Maharashtra, and Tamil Nadu. It is an oligophagous pest and found mainly attacking groundnut, soybean, pigeon pea and alfalfa. They occur in severe form from March to October, coinciding with high temperature. The amount of rainfall (mm) and minimum temperature are negatively correlated to leaf miner population.

Leaf miner moths (Figure 12.6a) lay individual eggs either on underside of the leaves, sometimes on petiole or stem. Upon hatching, first instar larvae cut open epidermis and reach the leaf mesophyll. Initially, they mine the leaflets in between the epidermal layers for a week time. Once big enough, they leave their mines to fold leaves together, continue feeding (Figure 12.6b). Due to the mining and defoliation the infested leaves dry and wither away. A severely infested field gives a burnt appearance (Figure 12.6d). Full grown larvae pupate in leaf webs (Figure 12.6c). Life cycle completes in 13–28 days in South India or 37–45 days in North India. Depending on the pest incidence, the yield losses are up to 50% in different parts of India.[31]

FIGURE 12.6 The groundnut leaf miner: (a) adult moth; (b) larvae in leaf web; (c) pupae in leaf web; and (d) heavily infested plant.

12.3.2.2 THE TOBACCO CATERPILLAR (SPODOPTERA LITURA)

The tobacco caterpillar is a cosmopolitan pest, recorded on more than 74 species of plants. Tobacco, cotton, and several pulse and vegetable crops act as alternate hosts.[32] It occurs in severe form twice during April to October and November to December months in South India, and once during October to February in North India.[33] Usually population starts after the receipt of rainfall; however, positive correlation was observed with maximum temperature and bright sunshine hours.[34]

A female moth lay about 300–700 eggs in clusters covered by scales (Figure 12.7a). Neonate larvae gregariously feed on leaves by scrapping epidermis resulting in skeletonization of leaves. Later, instar larvae (Figure 12.7b) feed solitarily, voraciously on leaf lamina leaving only the mid ribs (Figure 12.7c); also known to feed on tender shoots, even exposed pods. A severely attacked field gives a cattle grazed appearance. The yield losses are up to 40% depending on the level of infestation. Mature larvae drops down to soil or plant debris and pupates in an earthen cell.

FIGURE 12.7 Tobacco caterpillar on groundnut: (a) egg mass on leaf; (b) full grown larvae; (c) defoliated plants.

12.3.2.3 GRAM POD BORER (HELICOVERPA ARMIGERA)

Similar to tobacco caterpillar, it is a cosmopolitan pest with a wide host range (>96 species crops, 61 species of wild plants). Cotton, and several pulse and vegetable crops serve as alternate hosts.[35] The pest activity is seen throughout the year due to availability of host plants. Its population starts after the receipt of rainfall but positive correlation is with maximum temperature and duration of sunshine.[34]

Female moths (Figure 12.8a) lay individual eggs on tender parts of the plant. Upon egg hatching, larvae feed on leaf (Figure 12.8b), pegs, flowers, and exposed pods. They defoliate leaves resulting in reduced photosynthetic area. The yield loss due to gram pod borer is highly variable depending on level of incidence. Mature larvae pupate in earthen cell or fallen leaves at the base of the plant.

12.3.2.4 THE HAIRY CATERPILLARS

The common hairy caterpillars which infest groundnut are: (a) red-headed hairy caterpillar, *Amsacta albistriga* and *A. moorei*; and (b) Bihar hairy caterpillar, *Spilosoma obliqua. A. albistriga* is confined to southern India whereas; *A. moorei* is widely distributed in central and northern India.[36,37] Whereas, *S. obliqua* is found in Gujarat, Maharashtra, Uttar Pradesh, West Bengal, Punjab, Bihar, Assam, and Madhya Pradesh. They are highly polyphagous in nature; apart from groundnut they also feed on cotton, cowpea, finger millet, sunflower, castor, sorghum, pearl millet, soybean, black gram, pigeon pea, cluster bean, green gram, sun hemp, jute, etc.[38] *A. albistriga* moth activity is seen just after the receipt of monsoon showers in southern India. They are known to have two generations per year[39] whereas; *A. moorei* has one generation, *S. obliqua* has eight generations per annum at moderate temperatures (32°C–34°C), high relative humidity.

FIGURE 12.8 Gram pod borer: (a) adult moth resting on leaf; (b) larvae feeding on foliage.

In all three species of hairy caterpillars, female moths lay eggs in clusters on leaves. The neonate larvae gregariously scrape the green portion from leaves resulting in papery appearance of leaves. Later, stage larvae cause complete defoliation leaving mid-ribs, main stems of plant (Figure 12. 9a). The severely infested crop gives a cattle grazed appearance (Figure 12.9b). Larvae form marching bands, migrate from one field to another in the late hours of the day. Full grown larvae pupate in soil, diapauses till the onset of next monsoon rains. When the pest out break occurs, cent per cent yield losses may be observed in endemic areas of Andhra Pradesh, Karnataka, and Tamil Nadu.

12.3.3 SAP-FEEDING INSECT-PESTS

12.3.3.1 APHIDS (APHIS CRACCIVORA)

Aphids are severe, widely distributed in regions with warm, dry climate. They are polyphagous in nature, apart from groundnut they also feed on cotton, other pulses, and vegetables.[40] Intermittent droughts during the rainy season (July–September) and summer seasons (February–May) are better suited for aphid multiplication; however, heavy rainfall wipes out population. In Gujarat, the aphid intensity is greater during *rabi*. It is positively correlating to morning and evening relative humidity.[41]

FIGURE 12.9 Red-headed hairy caterpillar, *A. albistriga*: (a) larvae congregated on leaves; (b) severely infested plot giving cattle-grazed appearance in comparison to insecticide-sprayed plot. Photo credit: Prabhu Ganiger, ARS, Pavagada.

Aphids occur in both alate and apterous forms, where alate forms can only be seen at the beginning of colonization and during the migration stages. Alate forms sexually reproduce to give rise apterous forms, which lack wings. Apterous forms reproduce mostly through parthenogenesis and rarely by sexual means. Aphids congregate on young leaves and leaf buds sometimes even on petiole (Figure 12.10a), flowers, and pegs. They insert their stylets into phloem vessels and desap photosynthates resulting in chlorosis of plants and curling of leaves. Aphid colonies on plants are guarded by ants and in exchange for honey dew secretions of aphids. Pod losses up to 40% reported due to direct feeding damage[42] but they also cause indirect damage by acting as vectors of about 23 viral diseases including the Peanut Mottle and Groundnut Rosette.[43]

12.3.3.2 LEAFHOPPERS/JASSIDS (EMPOASCA KERRI, BALCLUTHA HORTENSIS)

Leafhoppers have gained their importance as pests of groundnut in parts of Gujarat, Maharashtra, Karnataka, and Tamil Nadu and are known to infest castor, soybean, cluster bean, maize, green gram, black gram, cowpea, and pigeon pea. The peak population density of leafhoppers was governed primarily by abiotic factors. *B. hortensis* populations occur throughout the year on groundnut; however, their density is high during *rabi* (January–March) followed by *kharif* (June–October) and summer (April–May).[44]

Both the leafhopper species lay around 30 eggs, individually in leaf tissue. Nymphs pass through four stages to reach adulthood. Both nymphs, adults

(Figure 12.10b) suck plant sap from the under surface of leaf, midribs. Under prolonged leafhopper infestation, "V"-shaped yellowing (Figure 12.10c) can be observed on tip of leaflets and they later spread to entire lamina resulting in yellowing of leaves. The pod yield losses due to leafhoppers are as high as 22%.

12.3.3.3 THRIPS (SCIRTOTHRIPS DORSALIS, FRANKLINIELLA SCHULTZEI, THRIPS PALMI, CALIOTHRIPS INDICUS AND MEGALUROTHRIPS USITATUS)

The thrips species viz., *S. dorsalis*, *F. schultzei*, *C. indicus*, *M. usitatus* are predominant in western India whereas, *T. palmi* is severe on groundnut in southern India. Apart from groundnut they infest, soybean, green gram, black gram, peas, brinjal, tobacco, onion, cotton, mango, chillies, etc..[45] In general, thrips appear at both vegetative stage and flowering stages of the crop where their population is governed primarily by daily mean temperature. Their population densities are greater during summer (April–May) followed by *rabi* (February–March), and *kharif* (June–October).[46] The abundance of thrips species inhabiting terminal leaf buds is in the order of *S. dorsalis* > *T. palmi* > *F. schultzei*, whereas, in case of flowers, it is *F. schultzei* > *T. palmi* > *S. dorsalis*.[47]

Adult thrips (Figure 12.10d) lay more than 30 eggs, individually on either sides of leaf midrib. Young larvae pierce leaf tissues, suck oozing sap. Emptied cellular spaces in leaf are filled with air, giving a silvery appearance (Figure 10e). A characteristic symptom produced by thrips feeding activity is known as "pouts" wherein, fully opened leaves look distorted. Thrips also act as vectors of GBNV and TSV in groundnut (as described under viral diseases section). As a pest, it causes pod yield loss up to 28%, but as a vector of viral diseases, cent per cent yield losses are common in endemic areas. *S. dorsalis* is also known to successfully transmit *peanut yellow spot virus* in groundnut.[48]

12.3.3.4 MEALYBUGS

The geographical distribution of mealybug species attacking groundnut in India are given in Table 12.1. Apart from groundnut, they are known to infest vegetables, fruit trees, and ornamental shrubs in warm humid regions. In Andhra Pradesh, *D. breviceps* on groundnut was reported infesting from September to October months whereas, *M. hirsutus* occurred during

February to March months.[49,50] In Gujarat, *F. virgata* was found on harvested groundnut plants in the month of May whereas, *P. solenopsis* observed in June to October months.[51,52] Mealybugs may inflict losses to pod yield to the tune of 25%.[53]

FIGURE 12.10 Sap-feeding insect-pests of groundnut: (a) aphid colony on leaf petiole; (b) adult leafhopper on leaf; (c) "V" shaped yellowing of leaf from tip downward; (d) adult thrips observed under microscope; (e) silvery white patches on groundnut seedling; (f) mealybugs on stem.

TABLE 12.1 Mealybug Species Infesting Groundnut in India.

Mealybug species	Common name	Region	References
Dysmicoccus brevipes	Pineapple mealybug	Andhra Pradesh, Karnataka, Tripura	[49, 53–54]
Ferrisia virgata	Striped mealybug	Gujarat	[51]
Maconellicoccus hirsutus	Pink hibiscus mealybug	Andhra Pradesh	[50]
Paracoccus marginatus	Papaya mealybug	Tamil Nadu	[55]
Phenacoccus solenopsis	Solenopsis mealybug	Gujarat	[52, 56]

They live in colonies on under surface of leaves and shoots, and sometimes even on pegs as well as pods. Females lay eggs in ovisacs, containing up to 600 eggs. Upon egg hatching, nymphs crawl toward growing plant parts and settle down. In about six weeks' time, they grow fully to reach adulthood. The females are wormiform while the males are alate, an example for sexual dimorphism. Nymphs and female adults suck sap from almost all parts of the plants like, leaves, shoots (Figure 10f), petiole, flowers, roots, nodules, pegs, and pods. Due to continuous feeding on sap, the plants appear pale, yellow, and wilt. Due to feeding on roots, the supply of nutrients to the plant is cut off. Hence, plants show stunted growth and reduced pod yields. Ants actively protect and help in dispersal of mealybugs in return for honeydew.

12.4 POST-HARVEST CHALLENGES

After the harvest of groundnut, insect-pests like bruchid (*Caryedon serratus*), *pod sucking bug (Elasmolomus sordidus),* rust red flour beetle (*Tribololium castaneum*), rice moth (*Corcyra cephalonica*), merchant grain beetle (*Oryza-ephilus mercator),* and almond moth *(Ephestia cautella)* inflict damage. The later four species are secondary pests, which prefer broken or processed kernels, while bruchid is the primary pest of groundnut as it attacks whole pod/kernel. Due to mechanical injuries caused either by drought during crop production or insect-activity during storage, mold fungi (*Aspergillus* sp.) infects groundnut. Mycotoxins produced by *Aspergillus* fungi and presence of insect-pests in the groundnut consignments pose great threat export.

12.4.1 BRUCHID (CARYEDON SERRATUS)

Buchid, *C. serratus* infesting groundnut and tamarind are respectively referred to as groundnut or tamarind bruchid (Chrysomelidae: Coleoptera). Bruchids are known to infest stored groundnut in almost all the groundnut growing states; however, they are severe in Gujarat.[8,9,57] They are highly specific, attack only plants belonging four genera of Fabaceae (*Bauhinia, Cassia, Piliostigma, Tamarindus*) apart from *Arachis*.[58] Tamarind is most suitable host for bruchid multiplication in laboratories.[59]

Bruchids observed in warm humid regions from June to December months. High temperature (28±2°C) and high humidity (70±5%) are most conducive for bruchid growth and development. The increase in temperature from 25°C to 40°C shortens generation doubling duration by three times.[60]

The adult beetles (Figure 12.11a) lay individual eggs on pod surface. Grubs upon hatching scrape the pod surface; they penetrate, feed on kernel (Figure 12.11b). Fully grown grubs emerge out by making an exit hole, construct a tough silken cocoon on the surface of pods (Figure 12.11c) or kernels. On an average, 1.9 adults only can emerge from a kernel irrespective of egg density on it.[61]

FIGURE 12.11 Groundnut bruchid: (a) bruchid adult; (b) infested kernels; (c) infested pods with silken cocoons; (d) exit holes.

Bruchid can cause both quantitative and qualitative losses to the stored groundnut wherein, it causes extensive damage to kernels than pods. First visual symptom is the exit hole of final instar larva before pupation (Figure 12.11d). Grubs feed internally on pods and kernels. The quantitative loss is by the weight loss wherein, depending on the storage period, the level of infestation, the losses may go up to 73%.[62] Active insect population generates heat and moisture, which is conducive to mold growth. Aflatoxin contaminated produce is not fit for consumption contributing to qualitative losses. The favorable factors like temperature (28°C ± 2°C), relative humidity (70±5%), along with seed moisture status (>9%), the previously infested/broken seeds are conducive for the growth and development of bruchid population.[10,57] The bruchid infestation and the availability of the broken seeds will result in the damage by the secondary insect-pests.

12.4.2 AFLATOXIN CONTAMINATION

Mold fungi, *Aspergillus flavus* (Figure 12.12a) and *A. parasiticus* cause aflatoxin contamination in groundnut. Aflatoxins, B1, B2, G1, and G2 are major among 18 reported types of aflatoxins.[10,14] They are carcinogenic, mutagenic, and immunosuppressive agents.

Levels of contamination depend on density of *A. flavus*, moisture, and temperatures in soil during the pod development to maturity. A fungus invades, colonizes, and produces toxins in groundnut pods kernels (Figure 12.12b, 12.12c). In majority of the samples tested, a positive correlation witnessed between population densities of *A. flavus,* levels of aflatoxin contamination.[14] Cropping system (i.e., *kharif* vs. summer) also influenced the *A. flavus* population density. In India, groundnuts and groundnut products are classified under high-risk commodities for aflatoxin contamination.[63]

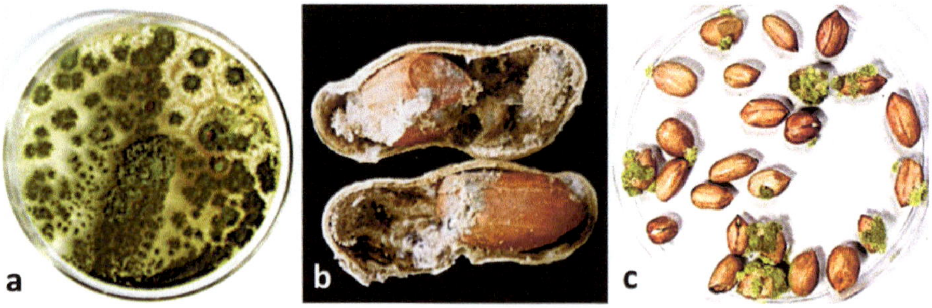

FIGURE 12.12 Mold fungi, *A. flavus* infecting groundnut: (a) *A. flavus* colony; (b) fungal colony inside pod; (c) fungi colonizing kernels. Photo credit: Thirumalaisamy PP, ICAR-NBPGR (RS), Thrissur.

12.5 MAJOR NEMATODE-PESTS OF GROUNDNUT

The plant parasitic nematode pests of groundnut include root knot nematodes (*Meloidogyne arenaria, M. javanica, M. hapla*), root lesion nematode (*Pratylenchus brachyurus, P. coffeae*), pod lesion nematode (*Tylenchorhynchus brevilineatus*, *T. brevicadatus*), groundnut testa nematode (*Aphelenchoides arachidis*), and seed and pod nematode (*Ditylenchus destructor*).

12.5.1 ROOT KNOT NEMATODE (MELOIDOGYNE SP.)

Root knot nematodes are most important nematode pests of groundnut crop in India. Peanut root knot nematode, *Meloidogyne arenaria* and Javanese root knot nematode, *M. javanica* are reported to cause an annual yield loss of 21.6%.[11] Northern root knot nematode (*M. hapla*) is of minor importance in groundnut. Crop affected by root knot nematodes tend to be susceptible to the pod rot, white mold, other soil-borne diseases. They not only act as act as primary pathogens but also exposes the plant to secondary organisms aggravating the crop loss.[64] They have a wide host range (>3000 plant species), are most destructive against the crops grown under protected cultivation.[65] In India, they are reported from several groundnut growing states.[66]

Root knot nematodes are generally endo-parasitic in nature. Juvenile-2 (J_2) is the infective stage (Figure 12.13a) which attacks, penetrates the roots, pegs, or pods, moves intracellular to a region near the differentiating vascular tissue forming giant cells around its feeding zone. The groundnut plants often infected with *M. arenaria* develop enlarged roots (Figure 12.13b), pegs caused by internal swelling of roots, pegs called as galls. Noticeable symptoms can be observed at 45–55 days while severe symptoms are detected between 90 and 120 days. Symptoms occur in patches of varying sizes exhibiting rusty, yellowish, mottled appearances. The slow dying, browning of plants presents a mottled effect among the greener plants, but even when such plants show conspicuous galling their neighboring plants in apparent vigour are usually also infected. Nematode infected fields give characteristic appearance of yellowing, stunting of plants in patches (Figure 12.13c). Symptoms due to nematodes tend to camouflage themselves with symptoms of nutrient deficiency, buy themselves time and ecological benefit to aggravate the disease development.

12.5.2 THE LESION NEMATODE (PRATYLENCHUS SP.)

It is a major nematode parasite, with a distribution mainly in the warmer groundnut production regions of the world. Apart from the root lesion nematode (*P. penetrans*), smooth-headed lesion Nematode (*P. brachyurus*) have also been identified as the predominant plant-parasitic nematodes of groundnut.[67] They have a wide host range (> 350 plant species), reported from several states of India.[66]

FIGURE 12.13 Root knot nematode, *M. arenaria*: (a) infective J2 stage; (b) swollen root system, (c) field showing patchy yellowing, stunted plants. Photos credit: (a and b) Prasanna Holajjer, ICAR-NBPGR (RS), Hyderabad.

The nematodes are migratory endo-parasites moving intercellular, feeding on parenchymatous cells. The endo-parasites attack groundnut roots, pods, and pegs, and weaken them. They feed within the parenchymatous tissues. Heavy infection damage is mostly seen on pods that show distinct light brown lesions, and cause extensive discoloration of below-ground plant parts and reduced root systems and pod weights. Above-ground symptoms may include slight stunting with unthriftiness and yellow-green foliage. On pod surface, the most obvious symptom is appearance of small lesions with coloration varying from purplish-brown to black (Figure 12.14), that is, speckled appearance.[68] Secondary infections by soil-borne pathogens cause lesions to swell in size and pegs and pods to rot. The lesions on mature pods are "purplish-brown," have a somewhat darker color with distinct boundaries as compared with those induced by a soil–microbial complex. Symptoms may be less conspicuous on Virginia-type groundnut than on Spanish, Runner types.

12.5.3 THE STUNT NEMATODE (TYLENCHORYNCHUS BREVILINIATUS)

The stunt nematode causes *Kalahasti malady* disease of groundnut grown at Srikalahasti, Andhra Pradesh.[69] It is one of the serious disease causing 20–50% yield losses in parts of Chittoor and Nellore districts of Andhra

Pradesh. They are prolific in sandy, sandy loam soil. Nematode is a migratory ectoparasite that feeds along the root surface penetrating epidermal cells of root and root hairs. Disease appears in patches, where infected plants show stunted growth with dark green foliage. Small brownish yellow lesions appear on pegs causing shortening of peg length. On young developing pods, brownish-yellow lesions coalesce, turn black giving scab-like appearance, in advanced stages of infection. Discoloration is also evident on roots. Diseased pods (Figure 12.15a) are smaller than healthy pods (Figure 12.15b).

FIGURE 12.14 Brown- to black-colored lesions on pod surface (pod speckling). Photo credit: Prasanna Holajjer, ICAR-NBPGR (RS), Hyderabad.

12.6 MINOR NEMATODE PESTS OF GROUNDNUT

The parasitic nematode, *Ditylenchus africanus*, infects the hulls, the seeds resulting in lower quality grade and reduced groundnut yield.[70] The seeds exhibited blemishes, premature germination before harvest with dark brown and a corky appearance. The most distinct symptom of advanced disease is dark brown to black discoloration of veins that extended longitudinally in the exocarp just beneath the pod surface. Infected pods lacked shine and appear dead. Infected seed appear shrunken with darker micropyle.[71] The nematodes also survive in the planting seeds and in the soil without the host. The peanut seed testa nematode (*Aphelenchoides arachidis)* was reported in Nigeria for the first time, and more recently in Egypt and South Africa.[72] They are found in the parenchymatous tissue of the groundnut testa, root cortex, and hypocotyl. All stages of the nematode are found in testa's of young seeds; however, in mature seeds mainly juveniles are observed along with very few adults. Nematode activity paves the way for secondary infection by *Fusarium* spp., *M. phaseolina*, *Rhizoctonia solani*, and *S. rolfsii*. Infected pods, kernels are unmarketable. The seed coat becomes discolored when large number of nematodes (>2000/testa) are present.[73] The Yam nematode *Scutellonema cavenessi* are ectoparasites of groundnut crop that feeds on root tissue and affects nodulation. Necrotic lesions are also observed on the roots during infestation. This nematode causes "yellow patch" of groundnut in some parts, chlorosis, and the characteristics yellow patches in the field with a reduction in nitrogen fixing ability. When affected with sting nematode (*Belonolaimus longicaudatus*), taproot becomes stubby with small dark necrotic lesions on them. Affected plants become chlorotic and stubby.

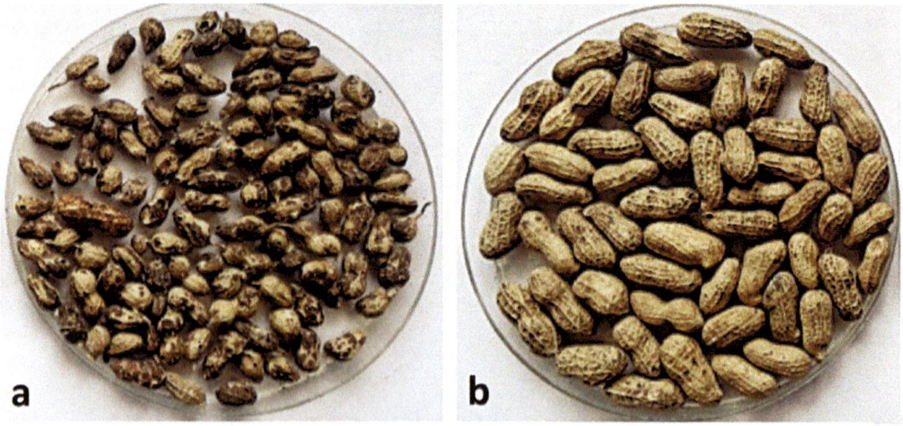

FIGURE 12.15 (a) The stunt nematode infected pods are small but discolored; (b) healthy pods. Photo credit: Prasanna Holajjer, ICAR-NBPGR (RS), Hyderabad.

12.7 INTEGRATED PEST MANAGEMENT (IPM)

Integrated pest management (IPM) is an optimum blend of feasible, economically viable options of pest management for different agro-climatic regions depending on the occurrence, importance of the pest.[15] It encompasses use of cultural practices, resistant/tolerant varieties, botanicals, biocontrol agents and synthetic chemical pesticides

12.7.1 CURRENT SCENARIO OF PEST MANAGEMENT IN GROUNDNUT

12.7.1.1 CULTURAL PRACTICES

The basic principle that underlies in any cultural practice is either to reduce the inoculums of pest/pathogen available for establishment or to strike an imbalance between availability of susceptible stage of crop with that of virulent stage of pest/pathogen. Most common cultural practice is deep summer ploughing to a depth of 8–10 inches recommended for management of defoliator, soil-inhabiting insect-pests as well as soil-borne fungi and nematodes. Safe disposal of previous season plants, their residues, volunteer plants, and weeds effectively reduces the inoculum load of all kinds of insect-pests, diseases, and nematodes. Removal of *Parthenium* weeds from groundnut field and surrounding areas is necessary to control the spread of *TSV*.[19] Incorporation of organic amendments like castor cake, neem cake, karanj cake, or well-fermented farm yard manures into the soil before sowing of crop reduces load of soil pathogens, nematodes, and soil-inhabiting insect-pests.[74]

Crop rotations with cotton, wheat, maize, sorghum, onion, and garlic are recommended for defoliator insect-pests and soil-borne disease management.[74] Before the onset of monsoon rains, trimming the trees, shrubs in, around the groundnut field will reduce white grub incidence. Setting up bonfires on community basis on field bunds from 1900 to 2200 hours starting from the next day of the rain is advised for management of hairy caterpillars in the endemic areas. Growing four rows pearl millet on all four-sides of groundnut crop restricts the movement of airborne sap-feeding insect pests (aphids, thrips), fungal spores, and virus particles. Early planting following wider row spacing (40–45 cm) is recommended for management of thrips, leaf miner, white grubs, and leaf spots and rust.[15]

Intercropping with soybean is followed for managing leaf miner[75]; with finger millet (7:1), pearl millet (2:1), and pigeon pea (2:1) for tobacco caterpillar and gram pod borer management.[76] Intercropping with either castor or pigeon pea (8:2) is practiced in areas endemic to hairy caterpillars.[77] Similarly, intercropping with either cluster bean (3:1) or sunflower (3:1), with either castor (3:1) or hybrid cotton (3:1) is recommended for leafhopper and thrips management, respectively.[78] For viral, nematode disease management, follow intercropping with either pearl millet or sorghum or pigeon pea or maize at 3:1.[64,79] Growing of trap plants within the groundnut crop is also recommended for early detection of insect-pests, trapping the initial colonies and destruction either through uprooting/burning or spray of chemical pesticides. For example, marigold is a trap for gram pod borer; castor is for tobacco caterpillar; cowpea and jatropha, calotropis are for hairy caterpillars; and soybean is for leaf miner.

Installing pheromone traps, light traps, sticky traps provide decision-making advantage to the grower. Setting up of light traps (one petro max light trap/ha; active during 1900 to 2200 hours) would be beneficial to monitor, trap adults of defoliator, soil-inhabiting insect-pests of groundnut. Similarly, planting pheromone traps helps in both monitoring and mass trapping of insect-pests depending on the density of traps per hectare. For leaf miner mass trapping, 25 traps/ha planted, lure [(Z)-7, 9-decadienyl acetate: (E)-7-decanyl acetate: (Z)-7- decanyl acetate in 100:20:14] must be changed once in 15 days.[80] For monitoring tobacco caterpillar, gram pod borer moth activity, two traps/ha are recommended with spodolure [(Z, E)-9, 11-tetra decadienyl acetate: (Z, E)-9, 12-tetra decadienyl acetate in 10:1 ratio], helilure [(Z, E)-9, 11-tetra decadienyl acetate: (Z, E)-9, 12-tetra decadienyl acetate in 10:1 ratio], respectively, with a fortnightly change of lure.[76,81] Similarly, installing either yellow or blue sticky traps (5/ha) just above the crop canopy help in monitoring aphid, leafhopper, and thrips populations. Placing three pheromone (synthetic pheromone methoxy benzene or anisole) dispensers/tree at 15-meter radius for three consecutive evenings after the first monsoon rains, employing chemical pesticides followed by collection of fallen adults are found to be effective.[82] Egg masses, gregarious larvae of defoliators, and adults of white grubs may be collected and destroyed to reduce primary inoculum, thereby reduce further increase in pest density.

Severity of post-harvest problems can be minimized in the field itself. Regular earthing up of soil around the plant base before peg initiation, harvesting the produce at right maturity stage, avoiding heaping the produce in the field itself, sun-drying the pods to reduce the moisture content to safe level (<9%), carryover of pest infestation from field to store, sorting, safe disposal of broken, damaged pods from the main produce reduces the attack by bruchids, *A. flavus*.[10,57] Use of new gunny bags or triple-layered gunny bags or Purdue Improved Crop Storage (PICS) bags protects produce from bruchid infestation.[10,83] Produce stored for seed purpose can be done safely in super grain bags or galvanised metallic/Polyvinyl chloride seed bins.[57]

12.7.1.2 HOST PLANT RESISTANCE (HPR)

Crop losses due to pests and cost of their management may be effectively brought down by opting for resistant cultivars. *Host plant resistance* (HPR) operates through tolerance, antixenosis, and antibiosis for insect-pests. Sources of resistance have been identified both in the primary gene pool, in the wild relatives of *A. hypogea*. Germplasm accessions, breeding lines, wild species, and land races have been screened for resistance. Table 12.2 gives a brief list of resistant varieties.

12.7.1.3 BIOLOGICAL CONTROL

Biological control of pest can be achieved either by conserving resident/introduced natural enemies or by application of their products for reducing pest populations immediately or in the near future. Both preventive and curative modes of biological control of pest are available. Preventive or inoculative control can be achieved by incorporating or treating with bio-control agents. Soil incorporated with *Metarrhizium anisoplae* along farm-yard manure kills all soil-inhabiting insect-pests, pupal stages of defoliator insect-pests. Conserving natural enemies by growing pollen, nectar-producing plants (marigold) and shade-giving plants (cowpea, pearl millet, lucerne, berseem, alfalfa, etc.) in the field are beneficial to predators like coccinellids, syrphids, lycanids, spiders, and hymenopeteran parasitoids.[93]

Augmentative release of *Telenomus remus* at 50,000 adults/ha thrice at 15 days interval after seeing the tobacco caterpillar moth activity in field are recommended.[74] Similarly, releasing adults of *Trichogramma chilonis* at 100,000/ha and *Chrysoperla carnea* at 50,000/ha coinciding with gram pod borer moth activity, oviposition.[94] Augmentative release of *Rhynocoris marginatus* adults at 5000/hectare, three times (30, 50, and 70 DAS) effectively controlled aphids.[95] Releasing adult coccinellids, syrphid larvae, and earwigs at 3:2:1 ratio per 1000 aphids gives good control of aphids.[96] Spraying entomopathogenic fungal formulations of *Beauveria bassiana, Verticillium lecanii* at 1 kg/ha effectively controlled sap-feeding insect-pests of groundnut.[97–100]

Foliar spray application of *Beauveria bassiana* B2 strain at $1X10^8$ cfu/g along with *Pseudomonas fluorescens* TDK1 + Pf1 strains at $3X10^8$ cfu/g effectively managed leaf miner populations in *kharif* season.[101] Similarly, commercial formulations of nuclear polyhedrosis viruses at 250 larval equivalents are effective for management of *S. litura* and *H. armigera.*

TABLE 12.2 Groundnut Varieties Resistant/Tolerant to Biotic Stresses.[84–92]

Name of pest/disease	Name of groundnut variety
Insect-pests	
Leafminer	BG-2, Co (Gn)-5, Durga, Girnar-3, GG-14, GG-16, ICGV 87160, ICGV 86031, ICGV 86699, Pratap Mungphali-1, Pratap Mungphali-2, Pratap Raj Mungphali, Phule Unap, VRI (Gn)-7
Tobacco caterpillar	Abhaya, Co (Gn)-5, GG-14, GG-16, HNG-123, ICGV 86699, ICGV 87264, ICGV 86590, Kadiri Harithandhra, Phule Unap, Pratap Mungphali-1, Pratap Mungphali-2, Pratap Raj Mungphali, Vasundhara
Gram pod borer	ICGV 86699, ICGV 86031, ICG 2271, ICG 1697, Kadiri Harithandhra
Thrips	Girnar-2, Girnar-3, GG-16, Durga, Kadiri Harithandhra, Pratap Mungphali-1, Pratap Mungphali-2, Pratap Raj Mungphali, Abhaya, Jawahar Groundnut-3, Phule Unap, Prutha, Vasundhara, Kadiri-7, Kadiri-8, GG-14.
Aphids	BG 2, Girnar 1, Girnar-2, Jawahar Groundnut-3, Kadiri-7, Kadiri-8, Prutha
Leafhoppers/Jassids	Abhaya, AK-159, ALR-2, CO-1, CO-3, Dh-3-30, Durga, GG-3, GG-5, GG-7, Girnar-1, Girnar-2, Kadiri-7, Kadiri-8, Kadiri Harithandhra, Kalahasti, Jawahar Groundnut-3, Pratap Raj Mungphali, Prutha
Thrips	AK-159, CO-1, CO-3, GG-2, GG-3, GG-5, Girnar-1, ICG-FDRS10, ICGV-86590, OG-52-1, SG-84
Bruchids	Chandra, Dh 8, DRG 12, DRG 17, GG-3, GG-4, GG-6, ICGS 44, ICGS 5, R 8808, RS 1, TAG 24, TMV 10, VRI 3
Diseases	
Collor rot	BAU-13, GG-8, GJG-22, HNG-69, M-548, Mallika, Smruti, TG-37A
Stem rot	GG-8, GJG-31, GJG-17, HNG-69, ICGV-00350, Kadiri Harithandhra, Phule Unap, Ratneshwar, Smruti, TG-38B, TG-51, Vasundhara
Dry root rot	TG-51, VL-Moongphali-1.
Early leaf spot	Ambar, BAU-13, Chitra, HNG-69, ICGV-00348, Jawahar Groundnut-3, Jawahar Groundnut-23, Kadiri-5, Kadiri-6, Kadiri-8, Kadiri Harithandhra, Kaushal, M-522, M-548, Pratap Mungphali-1, Pratap Mungphali-2, Pratap Raj Mungphali, Smruti, Somnath, Vemana, VRI-2

TABLE 12.2 *(Continued)*

Name of pest/disease	Name of groundnut variety
Late leaf spot	Abhaya, AK-265, ALR-3, Ambar, Apoorva, BAU-13, Co 6, Co (Gn)-4, GG-7, Girnar-2, GPBD-5, ICGV-00348, ICGV-00350, ICGV-91114, Jawahar Groundnut-3, Jawahar Groundnut-23, Kadiri-4, Kadiri-5, Kadiri-6, Kadiri-7, Kadiri-8, Kadiri Harithandhra, M-548, Phule Unap, Pratap Mungphali-1, Pratap Mungphali-2, Pratap Raj Mungphali, Frutha, Ratneshwar, Somnath, Smruti, TG-26, TG-37A, TMV (Gn)-13, Vemana, Vikas, VL-Moongphali-1, VRI-2, VRI (Gn)-5, VRI (Gn)-6, VRI (Gn)-7
Rust	AK-265, ALR-3, Ambar, Chitra, Co 6, Co (Gn)-4, Co (Gn)-5, Girnar-2, GPBD-5, ICGV-00348, ICGV-00350, ICGV-91114, Kadiri Harithandhra, Phule Unap, Ratneshwar, Smruti, TG-26, TG-37A, TMV (Gn)-13, Vikas, VRI-2, VRI (Gn)-5, VRI (Gn)-6 and VRI (Gn)-7
LLS, rust (Multiple disease resistance)	GG 20, GJG 9 and GJGHPS 1
GBNV	Ajeya, Apoorva, BAU-13, Co (Gn)-5, Divya, GG-8, GG-16, Kadiri-4, Kadiri Harithandhra, Kalahasti, Mallika, Pratap Mungphali-1, Pratap Mungphali-2, Pratap Raj Mungphali, SG-99, TG-26, TMV (Gn)-13, R-9251, Ratneshwar, Vasundhara, Vemana Vijetha and VRI (Gn)-6
TSV	Girnar-2
Aflatoxin (Seed colonization)	Greeshma, TKG-19A
Nematodes	
M. arenaria	Abuarbaa, Ah 25, Ah 3328, Ah 4515, Khargaon, Kigung
M. javanica	Apexy, Dangi, C166, Jh223, No 923, PI 268594
Pratylenchus sp.	PI 365553, PI 390606, PI 395233,
Tylenchorynchus sp.	Kalahasti, Prasuna, TCG 1518, Tirupati-2, Tirupati-3

For managing *A. albistriga*, spray the crop with granular virus formulations at 500 larval equivalents. Similarly, spray application of commercial formulations of *Bacillus thuringensis* or *Nomuraea rileyi* or *B. bassiana* at 1 kg/ha during the evening hours of the day is recommended for defoliator insect-pests.[74]

For stem rot management, mahua cake with *T. viride* each at 5 g/kg of soil reduced stem rot incidence effectively.[102] Seed treatment with commercial formulation of *T. harzianum* or *T. viride* @ 10 g/kg seeds, soil application of *Trichoderma* spp. @ 4 kg enriched in 250 kg FYM or 200 kg castor cake prevents seed-, soil-borne diseases.[74,103] Effective suppression of stem rot and leaf spots by *Pseudomonas* spp. were reported in groundnut.[104,105] *Paecilomyces lilacinus, B. subtilis, Trichoderma* sp. reduced *M. arenaria, M. Javanica*, and *Pratylenchus* sp. damage.[106,107]

Serratia marcescens GPS-5, *B. circulans* GRS-242 have effectively inhibited in vitro conidial germination of *P. personata* in controlled environment.[108] Biocontrol agents against late leaf spot, rust such as *P. islandicum, V. lecanii* can be mass multiplied, formulated using cheap and simple substrates.[109] Another antagonist *F. chlamydosporum* has been found to rapidly colonize the rust pustules.[110]

12.7.1.4 BOTANICALS

Plants are reservoirs of effective therapeutants and act as renewable sources of bio-pesticides. For example, spray of neem (*Azadirachta indica*) formulations like oil at 2% or seed kernel extract at 5% with soap at 1g/L are recommended when moth activity is observed in the crop.[75] Azadirachtin in neem has both antifeedant and ovicidal properties.[111] Likewise, foliar spray application of aqueous leaf extract of sorghum at 10% helps in bringing down thrips population. Groundnut pods treated with either neem oil at 5% (v/w) or pongamia/eucalyptus/castor oil at 10% (v/w) is useful for bruchid control.[8]

Spray application of neem and mehandi (*Lawsonia inermis*) leaf extracts at 2% or *Hemionitis arifolia* chloroform extracts effectively controls leaf spot and rust diseases.[112,113] Likewise, growth of *S. rolfsii* inhibited by all the tested concentrations of the extracts of *Agave americana, Cassia occidentalis, A. indica* and the maximum inhibition reported at 10% concentration.[114] Soil incorporation of fresh leaves of *P. hysterophorus*, neem, *Pongamia glabra,* wild sorghum at 500 kg/ha effectively controlled stem rot.[115]

12.7.1.5 CHEMICAL METHODS

Synthetic chemical pesticides are measures of last resort in IPM, when the pest population crosses thresholds or when economic damage starts. Economic threshold levels (ETLs) have been worked out for insect-pests (Table 12.3). Pesticides with label claims for crop are recommended for use in groundnut pest management.

TABLE 12.3 ETLs for Major Insect-Pests of Groundnut.

Name of insect-pest	ETLs
Leaf miner	Five mines per plant at 30 days of crop age
Tobacco caterpillar	20–25% defoliation at 40 days of crop age
Gram pod borer	20–25% defoliation at 40 days of crop age
Hairy caterpillars	20–25% defoliation at 40 days of crop age
Thrips	Five thrips/terminal bud at seedling stage
Leafhopper/Jassids	5–10 adults per plant at 30 days after crop emergence
Aphids	5–10 aphids per terminal at seedling stage
White grub/Root grub	One grub/square meter

Treating seeds with imidacloprid at 2 ml/kg protects groundnut crop from sap-feeding insect-pests like thrips and leafhoppers for a period of 25 days.[116] Spreading poison bait (rice bran 12.5 kg + molasses/jaggery 2.5 kg+ carbaryl 50 WP 1.25 kg) in the field during the evening hours will help in managing the grown-up larvae of hairy caterpillars and tobacco caterpillars.[74] Dusting furrow trenches around the field with either methyl parathion 2 DP or carbaryl 50 WP is effective for managing migrating red-headed hairy caterpillars in endemic areas.[74]

For effective defoliator pest management, spray application of chlorpyriphos 20 EC at 1250 ml, quinalphos 25 EC at 1000 ml, fenvalerate 20 EC at 250 ml, novaluron 10 EC at 500, emamectin benzoate 5 SG at 100 g, phosalone 35 EC at 750 ml, or spinosad 45 SC at 200 ml/ha are recommended.[76,117–120]

Two sprays (25, 40 days after sowing) of dimethoate 30 EC at 1200 ml, imidacloprid 17.8 SL at 150 ml, thiacloprid 480 SC at 150 ml, or thiamethoxam 25 WG at 100 g/L were found effective in managing sap-feeding insect-pests.[74,121] For white grub adult management, spraying of imidacloprid 17.8 SL at 1.5 ml/L on trees and shrubs during the evening hours is recommended. Whereas, furrow application of insecticides such as thiamethoxam 25 WS at 1900 ml/ha or fipronil 5 FS at 2000 ml/ha were recommended for white grub control.[74] In airtight godowns, placing aluminium phosphide 56% at 1 pouch (10 g)/ton of pods is advised for bruchid management.[57] At time of earthing up, incorporate carbofuran 3 G at 33 kg/ha and gypsum at 200 kg/ha is recommended for effective management of nematodes.[74]

Treating seeds with tebuconazole 2 DS at 1.5 g/kg followed by two sprays of tebuconazole 25.9 EC at 1 ml/L at 15-days interval beginning with the occurrence of disease symptoms has reduced leaf spots and rust.[122] Seed dressing chemicals like tebuconazole at 1.5 g/kg seed, mancozeb at 3g/kg seed, and carbendazim + mancozeb at 3g/kg seed are recommended for managing soil-borne diseases.[123] For leaf spots and rust management, need-based spray application of propiconazole 25 EC at 1 ml/L or hexaconazole 5 EC at 1 ml/L or tebuconazole 25.9 EC at 1.5 ml/L were recommended.[74]

12.7.1.6 BIOTECHNOLOGICAL APPROACHES

Potential biotechnology approaches through engineering, expression of novel genes can improve stress tolerance/resistance that help in crop adaptation, and ultimately lead to yield enhancement. *Agrobacterium* mediated and balistic method of genetic engineering are widely employed for transforming groundnut varieties.[124] To address the biotic stresses, genes like glucanase, chitinase, chloro peroxidase, coat proteins, crystal proteins, and pathogen-related genes have been targeted in groundnut.[124]

12.8 FUTURE PROSPECTS OF IPM IN GROUNDNUT

12.8.1 NANOTECHNOLOGY—A POTENTIAL TOOL IN PEST MANAGEMENT

Nanotechnology may help in reducing toxicity, improving shelf-life, and increasing the solubility of pesticides. Thus, nanotechnology has role to play in reducing the impacts of pesticides on environment. Nano-materials find their usage in two ways: (1) nanoparticles as protectants and (2) nano carriers of either pesticides or RNAi molecules (Figure 12.16). Though advantageous, not many nanoparticle-based formulations are available for agricultural application. This may be attributed to inadequacy in field testing and poor utilization of available host-pest/pathogen interactive systems. Fresh imputes are required in research and development of nanoproducts and their applications in agriculture. For example, to increase efficacy of insect attraction and to enhance lure duration, a nanogel impregnated pheromones dispensers or formulations was developed recently.[125]

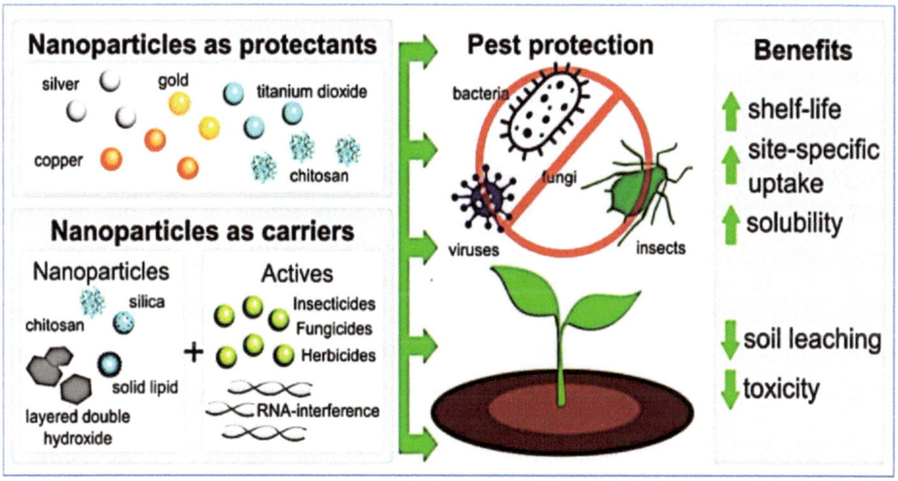

FIGURE 12.16 Nanomaterials in pest management either as protectants or carriers of pesticides. Source: Ref. [126].

12.8.2 DETECTION AND DIAGNOSIS OF PLANT PATHOGENS AND INSECT-PESTS

Timely, correct identification or diagnosis is crucial for pest management. Hence, quick, sensitive, and accurate detection methods help in decision-making process. Improved detection methods have become essential strategies for disease control.[127] The early, accurate detection of plant diseases, especially viruses, is an essential component to minimize the yield losses, and, therefore, plant disease management. Diseased plants express different kinds of symptoms such as mild mosaic, mottling, chlorotic, necrotic ring spots, etc.; often these symptoms are confused with mixed infection and abiotic stresses; hence, symptomatic diagnosis may not be appropriate under field conditions. Application of on-site virus detection methods like loop-mediated isothermal amplification (LAMP) and lateral flow devices (LFDs) are limited by lack of robustness and high-throughput.[128–130] Therefore, there is a scope in research and development of rapid, sensitive on-site detection techniques for plant diseases. Zero tolerance is the need of hour in storage and use of improved monitoring techniques like electronic nose technology for accurate monitoring and detection of insect pests, especially bruchids, is required.

12.8.3 NOVEL BIOTECHNOLOGICAL APPROACHES

The strengths of conventional, technological advancements are integrated to effectively deal with emerging pathogens like viruses. Managing GBNV in groundnut has become difficult due to the availability of alternate hosts, favorable climate. Some germplasm show field resistance to bud necrosis only due to their resistance to vector. Sources with genetic resistance to GBNV are lacking necessitating transgenic research. Expressing genes responsible for production of viral glycoproteins in plants restricts virus acquisition by thrips. Similarly, expression of genes governing production of modified forms of movement proteins of viruses and virus-specific anti-bodies may help in developing transgenic plant resistant to viruses. Virus disease management must include components like virus resistance, cultural practices, and vector management.[18]

12.8.4 INCREASED BIO-CONTROL RESEARCH

It has been known fact that biocontrol agents differ in their requirements, modes of action against crop pests. Hence, advanced research required understanding their requirements, modes of action for optimizing pest control. Development of broad-spectrum, pesticide-, thermo-tolerant biocontrol agents can be a useful for IPM. Future challenges in bio-control research: (1) Distribution of pathogens, their natural enemies in the environment, (2) Favorable conditions for improving efficacy of biocontrol agents, (3) Influence of different management practices on native, introduced populations, and d) Studies on colonization, biocontrol traits, components and dynamics of plant host defenses.

12.8.5 DEVELOPMENT OF GOOD AGRICULTURAL PRACTICES

Proper drying and storage of groundnuts in the post-harvest stage is required for use/export of insect-, mold-free produce. Hence, there is a need to develop groundnut harvesters that can make the harvested plants upside-down in field while harvesting, which will minimize probability of aflatoxin contamination of groundnut kernels in the field. Building modern storage structures like silos, which are having sensors to detect infestation, automatically pump hot air/carbon dioxide to manage insect infestation. Hermetic storage need to be practiced like use of PICS triple-layered bags and ICAR-DGR technology of using airtight 160-micron High Density Polyethylene bags for safe storage of groundnuts.[10,57,83]

12.9 CONCLUSION

The integrated pest management is useful for effective and economical pest management. It encompasses judicious use of practices like cultural, HPR, biological control, botanicals, chemical pesticides, and biotechnological approaches. IPM reduces the frequent, injudicious use of pesticides, thereby helping to conserve natural enemies of pests and pollinators. It also delays the onset of pesticide resistance and outbreaks of minor pests. It becomes far more necessary in the times of climatic change, which hastens emergence of new virulent strains of pests. The right amount of blending of conventional, advanced technologies in IPM would help achieve effective management of groundnut pests.

KEYWORDS

- **groundnut**
- **diseases**
- **insect-pests**
- **post-harvest challenges**
- **nematode-pests**

REFERENCES

1. Nigam, S. N. Groundnut at a Glance. **2014,** 121.
2. https://www.worldatlas.com/articles/top-peanut-groundnut-producing-countries (accessed on Dec 08, 2020).

3. https://eands.dacnet.nic.in/Advance_Estimate/4th_Adv_Estimates2019-20_Eng.pdf (accessed on Dec 08, 2020).
4. Ahir, K. C.; Saini, A.; Rana, B. S. Estimation of Yield Losses Due to Major Insect Pests of Groundnut (*Arachis hypogaea* L.). *J. Entomol. Zool. Stud.* **2018,** *6* (2), 312–314.
5. Ngele, K. K.; Kalu, U. N. Studies on Different Species of Plant Parasitic Nematodes Attacking Vegetable Crops Grown in Afikpo North L. G. A, Nigeria. *Direct Res. J. Agric. Food Sci.* **2015,** *3* (4), 88–92.
6. Thirumalaisamy, P. P.; Ram Dutta, Jadon, Kuldeep Singh, Nataraja, M. V.; Ramesh D. Padvi, Riddhi Rajyaguru; Sana Yusufzai. Association, Characterization of the *Fusarium incarnatum–F. equiseti* Species Complex with Leaf Blight, and Wilt of Peanut in India. *J. Gen. Plant Pathol.* **2019,** *85*, 83–89.
7. Baskaran, R. K. M.; Rajavel, D. S. Yield Loss by Major Insect Pests in Groundnut. *Ann. Plant Protection Sci.* **2013,** *21*, 176–223.
8. Harish, G.; Nataraja, M. V.; Prasanna Holajjer, Thirumalaisamy, P. P.; Jadon, K. S.; Savaliya, S. D.; Padavi, R. D.; Koradia, V. G.; Gedia, M. V. Efficacy, Insecticidal Properties of Some Essential Oils Against *Caryedon serratus* (Oliver)-a Storage Pest of Groundnut. *J. Food Sci. Technol.* **2014,** *51* (11), 3505–3509.
9. Harish, G.; Nataraja, M. V.; Ajay, B. C.; Prasanna Holajjer, Savaliya, S. D.; Gedia, M. V. Comparative Efficacy of Storage Bags, Storability, Damage Potential of Bruchid Beetle. *J. Food Sci. Technol.* **2014,** *51*, 4047–4053.
10. Thirumalaisamy, P. P.; Nataraja, M. V.; Radhakrishnan, T. Good Storage Practices for Groundnut Pods, Kernels: Protection from Stored Pests and Aflatoxin Contamination. *ICAR-Directorate Groundnut Res.* **2016,** 2.
11. Jain, R. K.; Mathur, K. N.; Singh, R. V. Estimation of Losses Due to Plant Parasitic Nematodes on Different Crops in India. *Indian J. Nematol.* **2007,** *2*, 219–221.
12. Kumari, M.; Singh, M.; Godika, S.; Choudhary, S.; Sharma. J. Effect of Different Fungicides, Plant Extracts on Incidence and Varietal Screening Against Collar Rot of Groundnut (*Arachis hypogea* L.) Caused by *Aspergillus niger* Van Tiegham. **2016,** *11* (4), 2835–2839.
13. Pal, K. K.; Dey, R.; Tilak, K. V. B. R. Fungal Diseases of Groundnut: Control, Future Challenges. In *Future Challenges in Crop Protection Against Fungal Pathogens*; Goyal, A.; Manoharachary, C., Eds.; Springer Publ: New York, 2014; pp 1–29.
14. Ghewande, M. P.; Desai, S.; Basu, M. S. Diagnosis and Management of Major Diseases of Groundnut. *NRCG Bull.* **2002,** 8–9.
15. Kumar, V.; Thirumalaisamy, P. P. Diseases of Groundnut. In *Diseases of Field Crops*; Dubey S. C.; Rashmi Aggarwal, Patro, T. S. S. K.; Sharma, P., Eds.; Today, Tomorrows Printers and Publishers Publ: New Delhi, 2016; p 50.
16. Reis, R. F.; de Goes, A.; Mondal, S. N.; Shilts, T.; Brentu, F. C.; Timmer, L. W. Effect of Lesion Age, Humidity and Fungicide Application on Sporulation of *Alternaria alternata*, the Cause of Brown Spot of Tangerine. *Plant Dis.* **2006,** *90*, 1051–1054.
17. Nataraja, M. V.; Singh, R.; Kumar, Y.; Garwal, R. S. A Hitherto Unknown Disease of Groundnut Noticed in Jodhpur, Rajasthan. *DGR Newslett.* **2013,** *12* (3–4), P 8.
18. Radhakrishnan, T.; Thirumalaisamy, P. P.; Vemana, K.; Kumar, A.; Rathnakumar, A. L. Major Virus Diseases of Groundnut in India, and Their Management. In *Plant Viruses: Evolution, Management*; Gaur, R. K.; Petrov, N. M.; Patil, B. L.; Stoyanova, M. I., Eds.; Springer Science+Business Media Pvt. Ltd. Publ: Singapore, 2016; pp 253–272.

19. Vemana, K.; Venkateswarlu, N. C.; Rajesh, A. P.; Naik, K. S. S. Field Screening Technique for Peanut Stem Necrosis Disease Using *Parthenium hysterophorus* Infector Border, Impact of Disease on Yield. *Indian J. Plant Protection* **2016,** *44* (2), 239–245.
20. Nolt, B. L.; Reddy, D. V. R. Peanut Clump. In *Compendium of Peanut Diseases*; Porter, D. M., Smith, D. H., Rodriguez-Kabana, R., Eds., American Phytopathological Society: St. Paul, Minn, 1984; pp 50–51.
21. Ratna, A. S.; Rao, A. S.; Nolt, B. L.; Reddy, D. V. R.; Vijayalakshmi, M.; McDonald, D. Studies on the Transmission of Indian Peanut Clump Virus Disease by *Polymyxa graminis*. *Ann. Appl. Biol.* **1991,** *118*, 71–78.
22. http://ainpwhitegrubs.com/major-species.pdf (accessed on Dec 08, 2020).
23. Kapadia, M. N.; Butani, P. G.; Beria, N. N. White Grub Species Attacking Groundnut in the Saurashtra Region in Gujarat, India. *Int. Arachis Newslett.* **2006,** *26*, 28–29.
24. https://krishi.icar.gov.in/jspui/bitstream/123456789/4520/1/WHITE%20GRUBS%20AND%20THEIR%20MANAGEMENT%20IN%20GROUNDNUT.pdf (accessed on Dec 08, 2020).
25. Logan, W. M.; Rajagopal, D.; Wightman, A.; Pearce, M. J. Control of Termites, Other Soil Pest of Groundnuts with Special Reference to Controlled Release Formulations of Non-Persistent Insecticides in India, Sudan. *Bull. Entomol. Res.* **1992,** *82*, 57–66.
26. Ranga Rao, D. V.; Singh, K. N.; Wightman, J. A.; Ranga Rao G. V. Economic Status of Neem Cake Mulch for Termite Control in Groundnut. *Int. Arachis Newslett.* **1991,** *9*, 12–13.
27. Wightman, J. A. W.; Dick, M.; Ranga Rao, G. V.; Shanower, T. G.; Gold, C. G. Pests of Groundnut in the Semi-Arid Tropics. In *Insect Pests of Tropical Food Legumes*; Singh, S. R., Ed.; John Wiley, Sons Publ: Chichester, 1990; pp 243–322.
28. Mohapatra, H. K.; Padhi, J.; Samalo, A. P.; Patra, G. J. Screening Promising Groundnut Varieties Against Termite Damage at Bhubaneswar, Orissa, India. *Int. Arachis Newslett.* **1995,** *15*, 59–60.
29. Barwal, R. N.; Gupta, D. K. Seasonal Incidence, Control of Pod Borer (*Euborellia annulipes*), Cercospora Leaf Spots in Groundnut. *Indian J. Plant Protection* **1991,** *19*, 156–160.
30. Senguttuvan, T.; Dhanakodi, C. V. Management of Pod Borer in Groundnut Through Manipulation of Cultural Practices in Alfisols. *J. Oilseeds Res.* **1997,** *14*, 269–273.
31. http://archive.deccanherald.com/Deccanherald/aug162005/snt1041182005815.asp (accessed on Dec 08, 2020).
32. Singh, S. P.; Jalali, S. K. Management of *Spodoptera litura* (Fabricius) (Lepidoptera: Noctuidae), In *Spodoptera litura* in India: Proceedings of the National Scientist's Forum on *Spodoptera litura* (F.); Wightman, J. A.; Ranga Rao G. V., Eds.; ICRISAT: Hyderabad, 1997; pp 27–65.
33. Ranga Rao, G. V.; Wightman, J. A.; Ranga Rao, D. V. World Review of the Natural Enemies and Diseases of *Spodoptera litura* (F.) (Lepidoptera: Noctuidae). *Insect Sci.* **1993,** *14*, 273–284.
34. Harish, G.; Nataraja, M. V.; Jasrotia, P.; Holajjer, P.; Savalia, S. D.; Gajera, M. Population Dynamics of *Spodoptera litura, Helicoverpa armigera* on Groundnut. In *International Conference on Plant Health Management*; Hyderabad, 2012, p 103.
35. Sharma, H. C.; Pampathy, G.; Dhillon, M. K.; Ridsdill-Smith, J. T. Detached Leaf Assay to Screen for Host Plant Resistance to *Helicoverpa armigera*. *J. Econ. Entomol.* **2005,** *98* (2), 568–576.

36. Qayyum, M. A.; Lakshminarayana, M.; Sanghi, N. K. Management of Red Hairy Caterpillar, *Amsacts* spp. Through Light Traps. In *National Seminar on Oilseeds, Oils-Research, and Development Needs in the Millennium*; Hyderabad, 2000.
37. Arora, R.; Bhattu, G. S.; Bath, D. S.; Singh, M.; Arora, R.; Singh, M. Management of Bihar Hairy Caterpillar, *Spilosoma oblique* (Walker) with Nuclear Polyhedrosis Virus. *Indian J. Ecol.* **1996,** *23* (2), 138–140.
38. Ahlawat, D. S.; Saini, R. K.; Singh, H. Biology of the Red Hairy Caterpillar, *Amsacta moorei* Butler (Arctiidae: Lepidoptera). *Ann. Biol.* **1993,** *9*, 118–122.
39. Ganiger, P. C.; Sannaveerappanavar, V. T. Field Biology of Red-Headed Hairy Caterpillar, *Amsacta albistriga* (Walker) on Groundnut. *Int. J. Trop. Agric.* **2007,** *25*, 271–278.
40. Jaba, J.; Haseena, B.; Tripathy, S.; Hosamani, A. C.; Amaresh, Y. S. Olfactory Response of Cowpea Aphid, *Aphis craccivora* Koch, to Host Odours, Population of Conspecifics. *J. Biopesticides*. **2010,** *3* (1), 405–407.
41. Prasad, T. V.; Nandagopal, V, Gedia, M. V. Effect of abiotic factors on the population dynamics of Aphis craccivora Koch in groundnut in Saurashtra region of Gujarat. *Indian J. Entomol.* **2008,** *70* (4), 309–313.
42. Jena, B. C.; Kuila, B. Effect of Insecticides on Population of Aphids and Their Natural Enemies in Groundnut. *Madras Agric. J.* **1997,** *84* (2), 101–102.
43. Reddy, D. V. R. Groundnut Viruses, Virus Diseases: Distribution, Identification and Control. *Rev. Plant Pathol.* **1991,** *70*, 665–678.
44. Prasad, T. V.; Nandagopal, V.; Gedia, M. V. Effect of Weather Parameters on the Seasonal Abundance of Leafhopper, *Balclutha hortensis* in Groundnut Ecosystem. *Indian J. Plant Protection.* **2008,** *36* (1), 43–47.
45. Vijayalakshmi, K. Transmission, Ecology of *Thrips palmi* Karny, the Vector of Peanut Bud Necrosis Virus. Hyderabad and Andhra Pradesh Agricultural University Ph.D. thesis, **1994,** P 99.
46. Prasad, T. V.; Nandagopal, V.; Gedia, M. V. Seasonal Abundance of Sesbania Thrips, *Caliothrips indicus* Bagnall in Groundnut. *J. Agrometeorol.* **2008,** *10*, 211–214.
47. Ranga Rao, G. V.; Wightman, J. A. *First Annual Rabi/Summer Groundnut Research Workers Group Meeting*; Kharagpur, 1994; p 36.
48. Gopal, K.; Krishna Reddy, M.; Reddy, D. V. R.; Muniyappa, V. Transmission of Peanut Yellow Spot Virus (PYSV) by Thrips, *Scirtothrips dorsalis* Hood in Groundnut. *Arch. Phytopathol. Plant Protection.* **2010,** *43* (5), 421–429.
49. Singh, T. V. K.; Goud, T. R.; Azam, K. M. Attack of Mealybug, *Dysmicoccus breviceps* on Groundnut. *Indian J. Entomol.* **1986,** *48* (3), 358.
50. Rao, D. V. S.; Srinivasan, S. *Maconellicoccus hirsutus* (Green), a New Mealy Bug Pest of Groundnut in Andhra Pradesh. *Entomon* **1987,** *12* (2), 115.
51. Anonymous. A new pest on groundnut. *NRCG Newslett.* **2003,** *2* (2), 1.
52. Nataraja, M. V.; Poonam Jasrotia, Harish, G. Mealybug: An Emerging Threat to Groundnut Crop. *DGR Newslett.* **2013,** *12* (1–2), 12.
53. Das, B. B.; Ray, S. Insect Pest Incidence of Groundnut in Tripura. *Indian J. Entomol.* **1988,** *50* (3), 387–388.
54. Rajagopal, D.; Siddaramegowda, T. K.; Rajagopal, B. K. Incidence of Pineapple Mealy Bug, *Dysmicoccus brevipes* (Cockerell) on Rhizobium Nodules of Red Gram and Groundnut. *J. Soil Biol. Ecol.* **1982,** *2* (2), 97–98.

55. Selvaraju, N. G, Sakthivel, N. Host Plants of Papaya Mealybug (*Paracoccus marginatus* Williams, Granara de Willink.) in Tamil Nadu. *Karnataka J. Agric. Sci.* **2011,** *24* (4), 567–569.
56. Harish, G.; Nataraja, M. V. Groundnut. In *Mealybugs, Their Management in Agricultural and Horticultural Crop*; Mani, M.; Shivaraju, C., Eds.; Springer: New Delhi, 2016; pp 251–256.
57. Nataraja, M. V.; Harish, G.; Jasrotia, P.; Dutta, R. *Bruchid: A Major Storage Pest of Groundnut*; Directorate of Groundnut Research: Junagadh, 2014; p 8.
58. Nilsson, J. A.; Johnson, C. D. New Host, *Bauhinia variegata* L.; New Locality Records for *Caryedon serratus* (Olivier) in the New World. *Pan-Pacific Entomol.* **1992,** *68*, 62–63.
59. Jasrotia, P.; Nataraja, M. V.; Harish, G.; Mahatma, M. K.; Holajjer, P.; Savalia, S. D.; Gajera, M. Biochemical Basis of Host Suitability for Groundnut Bruchid. In *International Conference on Changing Scenario of Pest Problems in Agri-Horti Ecosystem, Their Management*; Udaipur, 2014, pp 147–148.
60. Prasad, T. V.; N,agopal, V.; Gedia, M. V.; Savaliya, S. D. Life-Table, Intrinsic Rate of Increase of Seed Beetle (*Caryedon serratus*) on Groundnut at Different Temperatures. *Indian J. Agric. Sci.* **2011,** *81* (6), 550–554.
61. Prasad, T. V.; N,agopal, V.; Gedia, M. V. Effect of Egg Density on Weight Loss and Adult Emergence of *Caryedon serratus* in Groundnut Kernels. *Indian J. Agric. Sci.* **2008,** *78* (5), 479–481.
62. Harish, G.; Holajjer, P.; Savaliya, S. D.; Gedia, M. V. Population Density on Damage of Groundnut by *Caryedon serratus*. *Ann. Plant Protection Sci.* **2012,** *20*, 217–219.
63. Kumar, V.; Basu, M. S.; Rajendran, T. P. Mycotoxin Research, Mycoflora in Some Commercially Important Agricultural Commodities. *Crop Protection* **2008,** *27*, 891–905.
64. Grabau, Z. J.; Dickson, D. W. Management of Plant-Parasitic Nematodes in Florida Peanut Production. *EDIS*, **2018,** 1.
65. http://nemaplex.ucdavis.edu/Taxadata/G076.aspx (accessed on Dec 08, 2020).
66. Khan, M. R.; Jain, R. K.; Singh, R. V.; Pramanik, A.; Lin, T. *Economically Important Plant Parasitic Nematodes Distribution Atlas*; Directorate of Information and Publications of Agriculture, 2010.
67. Venter, C.; De Waele, D.; Van Eeden, C. F. Plant-Parasitic Nematodes on Field Crops in South Africa Groundnut. *Fundamentals App. Nematol.* **1992,** *15* (1), 7–14.
68. Bridge, J.; Starr, J. L. Plant Nematodes of Agricultural Importance: A Color Hand Book; Elsevier, 2007.
69. Reddy, D. D. R.; Subrahmanyam, P.; Reddy, G. H. S.; Reddy, C. R.; Rao, D. V. S. A Nematode Disease of Peanut Caused by *Tylenchorhynchus brevilineatus*. *Plant Dis.* **1984,** *68* (6), 526–529.
70. McDonald, A. H. Management of *Ditylenchus africanus* (Nematoda: Tylenchidae) on Groundnut (*Arachis hypogaea*). Potchefstroom University for Christian Higher Education, Ph.D. Thesis, 1998.
71. Steenkamp, S.; Jordaan, A.; McDonald, A. H.; De Waele, D. Differential Responses of Resistant, Susceptible Groundnut Genotypes at Cellular Level to *Ditylenchus africanus*. *Nematology* **2011,** *34* (2), 177–183.
72. Lesufi, M.; Swart, A.; McDonald, A. H.; Knoetze, R.; Tiedt, L. R.; Truter, M. Morphological and Molecular Studies on *Aphelenchoides arachidis* Bos, 1977 (Tylenchina: Aphelenchoididae) from Groundnuts in South Africa. *Nematology* **2015,** 433–445.

73. Minton, N. A.; Baujard, P. Nematode Parasites of Peanut. In *Plant Parasitic Nematodes in Subtropical and Tropical Agriculture*; Luc, M.; Sikora, R. A.; Bridge, J., Eds.; CABI Publ: Wallingford, 1990; pp 285–320.
74. Nataraja, M. V.; Jadon, K. S.; Holajjer, P.; Thirumalaisamy, P. P.; Jasrotia, P.; Dutta, R. *Integrated Pest and Disease Management in Groundnut*; Directorate of Groundnut Research, 2014, p 19.
75. Nandagopal, V. *Studies on Integrated Pest Management in Groundnut in Saurashtra*; Saurashtra University Ph.D. thesis: Rajkot, 1992; p 246.
76. Yambhatnal, R.; Hegde, M.; Patil, R. K.; Balikai, R. A. Evaluation of IPM Modules for Managing Defoliators in Groundnut. *Ann. Plant Protection Sci.* **2011,** *19* (2), 321–323.
77. Ganiger, P. C.; Sannaveerappanavar, V. T.; Reddy, V. C. Impact of Growing Intercrops on the Incidence of Red Headed Hairy Caterpillar, *Amsacta albistriga* (Walker) on Groundnut. *Karnataka J. Agric. Sci.* **2009,** *22* (3-Spl. Issue), 527–528.
78. Prasad, T. V.; Gedia, M. V. Intercropping Systems for Sacking Pest Management in Groundnut. *Indian J. Plant Protection.* **2011,** *39* (4), 278–282.
79. Sunkad, G.; Kenchanagoudar, P. V.; Naragund, V. B.; Naik, M. K. Management of Peanut Bud Necrosis Disease Through Intercropping. *Indian Phytopathol.* **2005,** *58* (2), 207–211.
80. Hall, D. R.; Peter, S. B.; Derek, G. C.; David J. C.; Alan Cork, Rosemary White, Aurelio Almetre, Thomas, J. H.; Nandagopal, V.; Wightman, J. A.; Ranga Rao, G. V. Identification, Synthesis of New Pheromones. *IOBC World Press Bull,* **1993,** *16,* 1–10.
81. Yashima, T.; Tamaki, T.; Kamano, S.; Oyama, M. Field Evaluation of a Synthetic Sex Pheromone "litlure" as an Attractant for Males of *Spodoptera litura* (F.) (Lepidoptera: Noctuidae). *Appl. Entomol. Zool.* **1974,** *9,* 147–152.
82. http://raridurgapura.org/Scheme-WhiteGrub.htm (accessed on Dec 08, 2020).
83. Sudini, H.; Ranga Rao, G. V.; Gowda, C. L. L.; Chandrika, R.; Margam, V.; Rathore, A.; Murdock, L. L. Purdue Improved Crop Storage (PICS) Bags for Safe Storage of groundnuts. *J. Stored Products Res.* **2015,** *64,* 133–138.
84. Rathnakumar, A. L.; Ranvir Singh, Parmar, D. L.; Misra, J. B. *Groundnut: A Crop Profile, Compendium of Varieties Notified in India*; Directorate of Groundnut Research: Junagadh, 2013; p 118.
85. Patel, B. A.; Patel, D. J.; Sharma, S. B.; Patel, H. V.; Patel, S. K. Nematode Problems of Groundnut and Their Management in Gujarat, India. *Int. Arachis Newslett.* **1996,** *16,* 38–39.
86. Yambhatnal, R.; Patil, R. K.; Mahabaleshwar Hegde, Balikai, R. A.; Kenchanagoudar, P. V.; Ninganur, B. T. Laboratory Evaluation of Groundnut Genotypes for Tobacco Cutworm (*Spodoptera litura* F.) Resistance. *Karnataka J. Agric. Sci.* **2011,** *24* (4), 461–463.
87. War, A. R.; Paulraj, M. G.; Ignacimuthu, S.; Sharma, H. C. Defensive Responses in Groundnut Against Chewing and Sap-Sucking Insects. *J. Plant Growth Regulation* **2013,** *32,* 259–272.
88. War, A. R.; Paulraj, M. G.; War, M. Y.; Ignacimuthu, S. Herbivore-, Elicitor-Induced Resistance in Groundnut to Asian Armyworm, *Spodoptera litura* (Fab.) (lepidoptera: noctuidae). *Plant Signal. Behav.* **2011,** *6* (11), 1769–1777.
89. Lingappa, S.; Patil, R. K.; Kulakarni, K. A. Management of Insect Pests of Oilseed Crops. In *Thematic Papers, National Seminar on Stress Management in Oilseeds for Attaining Self-Reliance in Vegetable Crops*; Rai, M.; Singh, H., Hegde, D. M., Eds.; ISOR Publ.: Hyderabad, 2003; pp 1–32.

90. Prasad, T. V.; Gedia, M. V.; Savaliya, S. D. Screening of Groundnut Varieties for Resistance to Sucking Insect Pests. *Indian J. Plant Protection Sci.* **2012,** *42* (2), 141–142.
91. Visalakshi, V. Non-preference of Mechanism of Resistance in Groundnut Cultures to the Leaf Miner *Aproaerema modicella* Deventer. *J. Oilseeds Res.* **1997,** *14*, 265–268.
92. Patil, R. K.; Gowda, M. V. C.; Nadaf, H. L. Screening of Spanish Bunch Breeding Lines of Groundnut Against *Spodoptera litura* (F.) Damage. *Int. Arachis Newslett.* **1991,** *10*, 26–27.
93. Shanower, T. G.; Wightman, J. A.; Guiterrez, A. P.; Ranga Rao, G. V. Larval Parasitoids, Pathogens of the Groundnut Leaf Miner, *Aproarema modicella* (Lep: Gelechiidae) in India. *Entomophaga* **1992,** *37* (3), 419–427.
94. Sundaram, M. K.; Dhandhapani, N.; Swamiapa, M.; Sundar Babu, P. L.; Jayaraj, S. A Study on the Management of Some Pests of Groundnut with Biocontrol Agents. *J. Biol. Control.* **1994,** *8* (1), 1–4.
95. Sahayaraj, K.; Martin, P. Assessment of *Rhynocoris marginatus* (Fab.) (Hemiptera: Reduviidae) as Augmented Control in Groundnut Pests. *J. Central Eur. Agric.* **2003,** *4* (2), 105–110.
96. Munyuli, T. Effects of Native Insect Predators on Population Densities of *Aphis craccivora,* Yields of *Vigna unguiculata, Arachis hypogaea* Grown Under Various Cropping Systems, in Kivu Province, Eastern Democratic Republic of Congo. *Tunisian J. Plant Protection* **2009,** *4*, 197–209.
97. Sahayaraj, K.; Namasivayam, S. K. R. Bioefficacy of Entomopathogenic Fungi Against *Aphis craccivora* in Groundnut. *Indian J. Plant Protection.* **2007,** *35* (2), 352–353.
98. Saranya, S, Ushakumari, R. Evaluation of Pure Cultures, Formulations of Entomopathogenic Fungi Against Cowpea Aphid, *Aphis craccivora* (Koch). *Ann. Plant Protection Sci.* **2011,** *19* (1), 88–92.
99. Tanwar, R. K.; Jeyakumar, P.; Monga, D. *Mealybugs and Their Management*; National Centre for Integrated Pest Management: New Delhi, 2007.
100. Suresh, B. C.; Khader Khan, H.; Prasanna, P. M. Efficacy of Different Entomopathogenic Fungi Against Cowpea Aphid, *Aphis craccivora* Koch Under Laboratory and Field Condition. *Int. J. Plant Protection.* **2012,** *5* (1), 68–71.
101. Senthilraja, G.; Anand, T.; Durairaj, C.; Raghuchander, T.; Samiyappan, R. Chitin-Based Bioformulation of *Beauveria bassiana* and *Pseudomonas flourescens* for Improved Control of Leaf Miner and Color Rot in Groundnut. *Crop Protection.* **2010,** *29*, 1003–1010.
102. Karthikeyan, V.; Sankaralingam, A.; Nakkeeran, S. Biological Control of Groundnut Stem Rot Caused by *Sclerotium rolfsii* (Sacc.). *Arch. Phytopathol. Plant Protection.* **2006,** *39* (3), 239–246.
103. Janila, P.; Mula, M. G. *Cultural Management Practices of Groundnut*; ICRISAT Publ: Hyderabad, 2015; p 16.
104. Rakh, R. R.; Raut, L. S.; Dalvi, S. M.; Manwar, A. V. Biological Control of *Sclerotium rolfsii*, Causing Stem Rot of Groundnut by *Pseudomonas* cf. *monteilii* 9. *Recent Res. Sci. Technol.* **2011,** *3* (3), 26–34.
105. Kishore, G. K.; Pandey, S.; Podile, A. R. Management of Late Leaf Spot of Groundnut (*Arachis hypogaea*) with Chlorothalonil-Tolerant Isolates of *Pseudomonas aeruginosa. Plant Pathol.* **2005,** *54* (3), 401–408.
106. Bridge, J. Nematode Management in Sustainable and Subsistence Agriculture. *Annu. Rev. Phytopathol.* **1996,** *34* (1), 201–225.

107. Dias-Arieira, C. R.; de Araújo, F. G.; Kaneko, L.; Santiago, D. C. Biological Control of *Pratylenchus brachyurus* in Soya Bean Crops. *J. Phytopathol.* **2018,** *166* (10), 722–728.
108. Kishore, G. K.; P,e, S.; Rao, J. N.; Podile, A. R. Evaluation of Chitinolytic Strains of *Serratia marcescens*, *Bacillus circulans* for Biological Control of Late Leaf Spot of Groundnut. In *Proceedings of the National Seminar on Stress Management in Oilseeds for Attaining Self-reliance in Vegetable Oils*; Directorate of Oil Seeds Research: Hyderabad, 2003; pp 19–20.
109. Desai, S.; Bagwan, N. B. *Mass Multiplication, Formulation of Fungal Biocontrol Agents Against Late Leaf Spot Pathogen of Groundnut in the Saurashtra region of Gujarat*; Asian Congress of Mycology and Plant Pathology: Mysore, 2002.
110. Mathivanan, N.; Murugesan, K. *Fusarium chlamydosporum*, a Potent Biocontrol Agent to Groundnut Rust, *Puccinia arachidis*. *J. Plant Dis. Protection* **2000,** *107* (3), 225–234.
111. Chaudhary, S.; Kanwar, R. K.; Sehgal, A.; Cahill, D. M.; Barrow, C. J.; Sehgal, R.; Kanwar, J. R. Progress on *Azadirachta indica* Based Biopesticides in Replacing Synthetic Toxic Pesticides. *Front. Plant Sci.* **2017,** *8*, 610.
112. Faisal, M.; Tiwari, S. Comparison of Bio-Agents, Botanicals with Fungicides Against Tikka and Anthracnose Diseases of Groundnut (*Arachis hypogea* L.). *Int. J. Plant Protection.* **2015,** *8*, 45–48.
113. Sahayaraj, K. Borgio, J. A. F.; Raju, G. Antifungal Activity of Three Fern Extracts on Causative Agents of Groundnut Early Leaf Spot and Rust Diseases. *J. Plant Protection Res.* **2009,** *49* (1), 53–56.
114. Sesha Kiran, K.; Lingaraju, S.; Adiver, S. S. Effect of Plant Extracts on *Sclerotium rolfsii*, the Incitant of Stem Rot of Groundnut. *J. Mycol. Plant Pathol.* **2006,** *36* (1), 77–79.
115. Ghewande, M. P.; Savaliya, S. D.; Hingrajia, H. M.; Padavi, R. D. Management of Stem Rot (*Sclerotium rolfsii* Sacc.) Through Organic Soil Amendments in Groundnut. In *Proceedings of the National Symposium on Recent Advances in Management of Plant Diseases*; Technology Development and Applications: Pune, 2003pp 9–10.
116. Nataraja, M. V.; Harish, G.; Holajjer, P.; Savaliya, S. D. Efficacy of Imidacloprid Seed Treatment for the Control of Leafhoppers and Thrips in Groundnut. *Legume Res.* **2015,** *39* (2), 305–309.
117. Gadhiya, H. A.; Borad, P. K.; Bhut J. B. Effectiveness of Synthetic Insecticides Against *Helicoverpa armigera* (Hubner) Hardwick, *Spodoptera litura* (Fabricius) Infesting Groundnut. *Bioscan* **2014,** *9* (1), 23–26.
118. Mutkule, D. S.; Patil, B. V.; Aglave, B. N.; Jagtap, P. K. Field Evaluation of New Molecules for the Control of *Spodoptera litura* (Fabricius) on Groundnut. *Indian J. Entomol.* **2009,** *71* (4), 292–295.
119. Sai Ram Kumar, D. V.; Raja Reddy, A. Management of Groundnut Leaf Miner, *Aproaerema modicella* with Chitin Synthesis Inhibitors and Insecticides. *Indian J. Plant Protection* **2012,** *40* (2), 105–108.
120. Praveena, Y. V.; Kotikal, Y. K.; Awaknavar, J. S.; Kenchanagoudar, P. V.; Somashekar. Evaluation of Newer Insecticides Against Groundnut Leaf Miner *Aproaerema modicella* Deventer. *Karnataka J. Agric. Sci.* **2010,** *24* (4), 542–545.
121. Kennedy, F. J. S.; Rajamanickam, K.; Balagurunathan, R.; Lourduraj, A. C. Yield Loss Due to Insect Pests and Their Control in Groundnut. *Groundnut News* **1992,** *4* (2), 7.
122. Kapadiya, H.; Moradiya A. M. Management of Groundnut Major Disease by Tebuconazole Alone and in Combination with Bio- Control Agent, Their Impact on Yield. *Int. J. Chem. Stud.* **2017,** *5* (6), 697–701.

123. Jadon, K. S.; Thirumalaisamy, P. P.; Kumar, V.; Koradia, V. G.; Padavi, R. D. Management of Soil Borne Diseases of Groundnut Through Seed Dressing Fungicides. *Crop Protection.* **2015,** *78*, 198–203.
124. Mallikarjuna, G.; Rao, T. S. R. B.; Kirti, P. B. Genetic Engineering for Peanut Improvement: Current Status and Prospects. *Plant Cell Tissue Organ Culture* **2016,** *125*, 399–416.
125. https://icar.org.in/content/icar-nbair-bengaluru-successfully-controls-holotrichia-consanguinea-slow-release-pheromone (accessed on Dec 08, 2020).
126. Worrall, E. A.; Hamid, A.; Mody, K. T.; Mitter, N.; Pappu, H. R. Nanotechnology for Plant Disease Management. *Agronomy* **2018,** *8*, 285.
127. Belete, T.; Boyraz, N. Biotechnological Tools for Detection, Identification and Management of Plant Diseases. *Afr. J. Biotechnol.* **2019,** *18* (29), 797–807.
128. https://iihr.res.in/arka-virus-kit-tomato-leaf-curl-bangalore-virus-tolcbv-lamp-detection-kit (accessed on Dec 08, 2020).
129. Ghosh, D. K.; Warghane, A.; Biswas, K. K. Rapid, Sensitive Detection of Citrus Tristeza Virus Using Reverse Transcription Loop-Mediated Isothermal Amplification (RT-LAMP) Assay. *Methods Mol. Biol.* **2019,** *2015*, 143–150.
130. Boonham, N.; Kreuze, J.; Winter, S.; van der Vlugt, R.; Bergervoet, J.; Tomlinson, J.; Mumford, R. Methods in Virus Diagnostics: From ELISA to Next Generation Sequencing. *Virus Res.* **2014,** *186*, 20–31.

CHAPTER 13

Integrated Disease Management in Rapeseed-Mustard-Based Cropping System

P. D. MEENA and P. K. RAI*

ICAR-Directorate of Rapeseed-Mustard Research, Bharatpur-321303, Rajasthan, India

Corresponding author. E-mail: director.drmr@gmail.com

ABSTRACT

Diseases caused by different fungi remain barriers to achieve higher productivity of oilseed Brassica. Monocropping with an application of higher doses of various inputs including fertilizers and irrigations as well as higher population per unit area has increased the incidence of diseases everywhere. The economic loss due to biotic stresses in oilseed Brassica crops is around 19.9%, at different growth stages of the crop. Several pathogens infect the crop but some are considered economically important based on emerging trends. Powdery mildews are becoming more aggressive and distressing disease of rapeseed mustard all over the world due to global warming. Alternaria blight disease severity decreased by 10% over different locations during the last two decades on *Brassica juncea,* may be due to defensive change between cultivars and species for exposing infection reaction. Breeding for disease resistance has not received enough attention so far. Resistance breeding has limitations due to narrow genetic base of the crop. The resistance level in cultivated varieties has been accomplished through multiple crosses, recurrent selection, diallele crossing, and selective mating system. The resistance has been introgressed with intra-specific pedigree,

Integrated Pest Management in Diverse Cropping Systems. Devendra Pal Singh, PhD (Ed.)

backcross, altered recurrent mass selection, inter-specific genome vis-a-vis chromosome substitution, and gene introgression against different pathogens. White rust resistance in mustard has been successfully transferred and is dominant and controlled by single or two genes. The molecular and genetic engineering techniques like genome manipulation, cytoplasmic genome manipulation, transformation, external gene expression, and embryo rescue for wide hybridization have been exploited in *Brassica*. However, the resistance or tolerance against other major pathogens is still lacking. Keeping in view the environmental consequences, now the effective biological and newer fungicides are required for disease control. There are options available for designing an effective and environment-compatible integrated disease management strategy for the crops. The conceptual models and primary and secondary decision-making tools need to be used effectively. Models with sensible prediction precision, once validated using an accurate data, will allow farmers to conscious around the status of disease to apply judicious and effective fungicidal sprays.

13.1 INTRODUCTION

Trends of rapeseed-mustard production in India show the erratic condition with a record high produce of 9.34 m t from 6.23 m ha with 1499 kg/ ha yield during 2018–2019.[1] The rapeseed-mustard has important role in the Indian oil economy with about 26.24% of the entire oilseed production. Production of 6.4 m t from 5.5 m ha of rapeseed-mustard during 2009–2010 by the fluctuating yield levels between 854 kg/ha (2002–2003) and 1499 kg/ha (2018–2019) during the preceding years and worldwide accounting for 17.5 and 10.8% of the entire acreage under cultivation and production, respectively.[2] However, the productivity of oilseed Brassica is lower in India than other countries of the world. Major causes of low productivity are higher acreage under rainfed cultivation, marginal holdings, unbalanced use of fertilizers, and damages due to biotic stresses. Though, a huge gap occurs among the potential yield and that expected at the farmers' field. It is mainly due to numerous insect–pest, pathogens, high temperature at seedling and terminal stage, draught and saline as well as alkaline stresses to which the oilseed Brassica crop is exposed. High soil temperature at sowing time and low, chilling temperatures, even frost during reproductive phase greatly affects the crop production and productivity. In general, the crop is grown under residual moisture situations with lower productivity. All these factors cumulatively declines the crop and oil yield. Diseases are the impediments on the way to achieving attainable production. The exhaustive sole cropping with higher inputs has further compounded the problem and the incidence of diseases has repeatedly developed extensively. Occurrence of disease epidemic drastically deteriorates the extent, worth of seed, and percent oil content in various rapeseed-mustard crops. The complete inherent genetic potential expression of a cultivar is governed by inputs that applied in to the cropping system. To reduce such damages, integrated disease management (IDM) in the oilseed Brassica cropping system is discussed here along with brief perspectives based on emerging trends for the future. Breeding for disease resistance is one of the methods of defending crops from damages due to biotic factors.[3] Inherited resistance is more valuable as it is economically and environmentally safe.

13.2 IMPORTANT DISEASES

The average yield loss around 19.9% is caused by diseases and pests at different growth stages of rapeseed-mustard crops. There are several pathogens responsible to distress the oilseed Brassica crop but about 18 are reflected to be economically important all over the world (Table 13.1). Among biotic stresses, some important diseases cause economic losses at different growth stages of the crop starting from initial sprouting stage (Downy mildew and white rust), early leaf development stage (white rust), stem elongation (Alternaria blight (AB)), flowering stage (Sclerotinia rot (SR)), and silique development and ripening stage (powdery mildew (PM)) are discussed under this chapter (Figure 13.1). The other diseases are minor, which occasionally causes considerable damage to the crop.

TABLE 13.1 Yield losses caused by major diseases in oilseed Brassica

Disease	Causal organism	Yield losses (%)
Downy mildew	*Hyaloperonospora parasitica* Constant. (Pers.:Fr) Fr.	37
White rust	*Albugo candida* (Lev.) Kuntze	47
Black spot	*Alternaria brassicae* (Berk.) Sacc. *A. brassicicola* (Schwein.) Wiltshire	70
White rot	*Sclerotinia sclerotiorum* (Lib.) de Bary	35
Powdery mildew	*Erysiphe cruciferarum* Opiz. L. Junell.	18
Clubroot	*Plasmodiophora brassicae* Woronin	100
Blackleg	*Leptosphaeria maculans* (Desm.) Ces.	86

A. Alternaria leaf blight B. Alternaria pods blight C. Club root

D. White rust staghead and pustule on leaves E. Staghead formation by *Albugo* F. Powdery mildew

G. Mycelial growth of Sclerotinia on stem H. Sclerotinia rot shot hole on leaf I. *Sclerotinia* proliferating on fallen petals; carpogenic germination of apothecia

J. Downy mildew affected cotyledons, leaves at different stage K. Staghead covered with *Hyaloperonospora* L. Downy mildew affected pods

FIGURE 13.1 (A-L) Symptoms of rapeseed-mustard diseases.

13.2.1 DAMPING OFF AND SEEDLING BLIGHT

Numerous complex fungal pathogens are accountable for causing the seed rot and seedling blight universally. But, *Rhizopus stolonifer* is stated to remain a most frequent.[4] Postemergence mortality with *Pythium aphanidermatum*,[5] *P. butleri*,[6] *Rhizoctonia solani*,[7] *Sclerotium rolfsii*,[8] *Macrophomina phaseolina*,[9] and *Fusarium* spp. is not common, being the associated fungi in India, causing 6–15% prevalence of the disease.[10] They frequently persist on plant debris and soil as diverse resting spores to contaminate the succeeding crop.

Symptomatically, a necrotic lesion of 1–2 cm length was observed at the base of the stem, with girdling occasionally appears marginally above soil surface. The taproot becomes discolored and occasionally wire-stem indications are visible. Salmon collared spore bunches of *Fusarium* are frequently detected on diseased plant tissues. Occasionally, the symptoms are restricted to roots comprising light brown wounds on the taproot and at the origins of bigger adjacent roots. Girdling of the primary root possibly occur and destruct the entire root system. Damping-off and seedling blight are typically challenged because farmers generally grow uncertified seeds contaminated with complex pathogens.

In order to avoid water stagnation, proper drainage system in field must exist to safeguard toward planting time of the crop. Sanitary cultivation and elimination of plant stubbles before planting are vital to succeed the pathogens. Collective impact of *Brassica napus* green manuring along with *Trichoderma* treated seed could be useful to reduce diverse soil-borne phytopathogenic fungi (*Pythium* and *Rhizoctonia*).[11] Improved plant stand with defense using seed treatment with thiophanate methyl 70 WP @ 2g/kg or Metalaxyl 35 SD 6 g/kg or by some new recommended seed protectant fungicide against *S. rolfsii, R. solani,* and *Fusarium oxysporum* was confirmed.[10]

13.2.2 DOWNY MILDEW

The downy mildew disease caused by an obligate pathogen *Hyaloperonospora parasitica* and is found normally in fluctuating proportions, particularly in northern humid foothill of India. The disease has been reported from Canada and several European countries, Japan, and South Asia.[12] It generally occurs on leaves or inflorescence either alone[13] or combined with white rust.[14–16]

Disease formed small angular transparent bright green lesions first on cotyledon or initial true leaves at the time of seedling stage. These lesions advanced to enlarge and converted into greyish-white and uneven necrotic spots on the leaves, show downy growth of the pathogen on its lower surface. Severe outbreak of the disease may cause dry up and shrivel leaves and its extent may depend on species of the crop.[17] Sometimes the disease is visible on silique very late in the season. Pith of the stem is affected more than the cortex, while thickening of the inflorescence is mainly caused by hypertrophy of the affected cells.[14] Oospore development may occur after the inflorescence dries up. Usually, the symptoms on leaves and inflorescence are found to be associated with white rust. Thick stunted growth of plant bearing profuse sporulation is resulted by systemic infection.[15,16]

Resistant genotypes including EC-129126 (*B. juncea*), PBN-9501, PBN-2002, GSL-1 (*B. napus*) are reported.[18,19] Though,[20] reported the risk of breakdown of resistance due to mixed infection of *H. parasitica* earlier to *Albugo candida*. Careful picking of infected hypertrophied racemes instantly after formation and their immediate destruction could also reduce the inoculum level of disease in the field. Location-specific appropriate sowing time need to be worked out.[15] Use of seed treatment with metalaxyl 35 SD @ 6 g/kg and foliar application of metalaxyl @ 0.01% a.i. is found effective in disease management.[21]

13.2.3 WHITE RUST

White rust or white blister disease caused by *A. candida* (Pers. Ex Fr.) Kuntz. is one of the most broadly familiar species, which is assumed to be the exclusive Brassicaceae pathogen on about 63 genera and 241 plant species.[22] There is an existence of enormous genetic diversity within *Albugo* infecting Brassicacae family,[23,24] and some of them detected lineages might found diverse species.[25] As a result of the recent lectotypification of *A. candid* and the taxonomic status of which was earlier not clear,[26] two specific *Albugo* species pathogenic to Brassicaceae have been defined within *Albugo*.[24,26] White blister recognized to *Albugo capparidis* infecting Capers (*Capparis spinosa*) or, spread over a wide range of species concept, to *A. candida*.[22]

White rust can result in yield loss up to 47% (Kolte, 1985) with each percent of disease severity and staghead formation causing reduction in seed yield of about 82 kg/ha and 22 kg/ha, respectively.[27] This disease has been observed from Brazil and many European and South Asian countries and Canada.[12] Total 241 Plant species in 63 genera of family cruciferae have been described to cause disease by *A. candida*.[28] Symptoms of the disease that appeared on leaves after 30–40 days of sowing are characterized by white or creamy yellow elevated pustules up to 2 mm in diameter and which later coalesce to form patches. The lower surface of leaves is covered with scattered pustules. While, the upper surface appeared with tan-yellow colored patches that enable to recognize the affected leaves. On maturity, pustule ruptures and releases a white powder of sporangia. Infected leaves become senescent once the necrosis around or in the pustule appear after maturity of pustules. Such pustules may also occur on the surface of a well-developed silique.[29] Thickening or hypertrophy of the infected foliage is generally not appearing in rapeseed-mustard like other crucifers.[30] Though, the development of deformed structure of inflorescence generally called staghead is witnessed in systemic infection or infection through stem or flower.[31] In such conditions, the affected flowers become malformed and petals develop green-like sepals, stamens may be distorted to leaf-like club-shaped sterile or carpelloid structures, which are established to continue on the flower rather than dropping premature as in normal plants. In general, typical symptoms shaped of downy mildew accompanying with white rust.[29] Spore production by the pathogen affects the consumption of accumulated nutrients vital to collapse and death of cells and drying of infected plants.[14]

Sources of resistance to WR in *B. juncea* including BIO YSR, NDUH-YJ-6, DRMR 2019, DRMR 2035, and DRMR 1–5, DRMR MJA 35, DRMRIJ 12 -48, IC313380, IC265495 has been identified.[12,32–35] Yellow-seeded mustard *(B. juncea)* variety "T4",[36] YRT-3184 and rapeseed (*B. Rapa* ssp. Yellow Sarson*)*, Type 6[16] have been described to be resistant to white rust. Incorporation of white rust resistance from *Brassica carinata* to *B. juncea* have been successful to some extent by developing immune plants under epiphytotic stress condition followed by their frequent backcrossing with *B. juncea* genotypes.[37] White rust resistance in mustard is dominant and governed by one or two genes with either dominant-recessive epistasis or broad dominance at both gene pair nevertheless one gene dominant epistatic to the other. Such genes might be well located on the same locus or diverse loci.[38] Resistance at true leaf stage and susceptibility at the cotyledonary leaf stage of *B. juncea* genotype EC 399301 seems to be governed by two independent genes. Therefore, white rust resistance screening at the cotyledonary leaf stage needs to be cautiously deliberated.[38]

The linkage map of genes governing white rust resistance in *B. napus* has been identified.[40] The resistance of *B. napus* cultivar "Regent" improved by independent dominant genes at three loci-designated as AC 7-1, AC 7-2, and AC 7-3.[41] Molecular mapping exposed the existence of two independent loci deliberating *A. candida* resistance in *B. juncea*.[42] Complete resistance to *A. candida* has been identified in some Australian germplasm.[43] However, there is scope to safeguard against the risk of breakdown of resistance by mixed infection with *H. parasitica.*[44] A closely linked marker for white rust resistance have been established by amplified fragment length polymorphism in combination with bulk segregant analysis.[45] Scheduling of defense-related gene expression is vital in defining chance of pathogenesis.[20]

There are several reports of fungicides to significantly reduce both foliar and staghead phase infection of disease. Fungicides namely, captafol, benomyl,[47] chlorothalonil,[47] mancozeb,[47–49] combination of metalaxyl 35 ES 6 ml/kg seed treatment + 0.2 g/l spray of combination of metalaxyl + mancozeb at 50, 65 DAS[36,41,47,50,51] have been reported. Recently, a cumulative awareness about environmental pollution caused by pesticides and development of fungicide resistant strains in plant pathogens have challenged the researchers to discover an ecofriendly approach for management of the disease. Aqueous bulb extract of *Allium sativum* 1% (w/v), an isolate of *Trichoderma viride* as seed treatment and in combination as individual foliar sprays were effective as that of mancozeb and combination of metalaxyl 35 ES 6 ml/kg seed treatment + 0.2 g/l spray of combination of metalaxyl + mancozeb in reducing the white rust severity on leaves and number of stagheads per plant.[51] Applying basal dose of potash @ 40 kg/ha reduced white rust severity and number of stagheads. Location-specific appropriate planting date and other epidemiological considerations are essential for decision-making to escape the disease. Chinese genotypes namely CBJ-001, CBJ-002, CBJ-003, CBJ-004, and Australian genotypes JR049, JM 06011 were found constantly resistant against *A. candida* pathotype prevailing in Australia during the different plant growth stages.[43,52,53]

13.2.4 ALTERNARIA BLIGHT (AB)

AB or black spot [*Alternaria brassicae* (Berk.) Sacc.] is a predominant and discouraging disease appears on the whole plant of oilseed Brassica. This is reported from all over the world but it is considered as a significant yield limiting disease in oilseed Brassica cropping system in India. Other species including *Alternaria brassicicola* and *Alternaria raphani* are also appeared rarely. Although, complete devastation of the crop due to the disease is not happen and considerable seed yield losses are 5–15%. Yield losses due to AB disease were reported up to 47% by influential photosynthetic metabolic activities and decline in seed quality namely, seed size, viability, etc..[54] AB severity fluctuates between seasons and geographical locations similarly among single crop contained by a zone in India.[55]

The symptoms of disease are described by development of lesions on all aerial parts of plant.[56] At the seedling stage, dark stem lesions appeared instantly after germination caused damping-off, or undersized seedlings. Normally, the disease commenced at 40–45 days of growth stage and most vulnerable at 75 days of plant growth has been reported.[57] Typically grey in color lesions formed by *A. brassicae* compared with black sooty velvety lesions produced by *A. brassicicola*. Lesions formed by *A. raphani* appear as diverse yellow halos surrounding them. Though, the appearance of leasion may fluctuate with prevalent host and environment. Disease symptoms initially appear on lower leaves with presence of dark black specks, which expand to progress into prominent, round, concentric rings of different dimensions spots. While disease progression, symptoms progresses on middle and upper leaves with minor size lesions, once the senescing and defoliation of lower leaves occur. Later, circular black visible spots formed on silique and stem, which may merge foremost to thorough darkening of silique or deteriorating of the stem with development of extended streaks. Decaying of the seed is noticeable just below the black spot on the silique of *Brassica rapa* ssp. Yellow Sarson or *Brassica rapa* ssp. Brown Sarson. Spots on silique of Indian mustard become visible as black or light brown having dissimilar grey center. As soon as the plants are infected, symptoms primarily appear on the adult leaves because they are nearer to the upper crust of soil. They are more freely infected as a result of rain splash or wind-blown rain. Secondary and primary branches and silique display the dark or blackened spots resulted in loss due to premature silique ripening and flaking of seeds. The AB infection on leaves and silique lessens the photosynthetic area substantially. The stage of disease on silique badly affects the usual seed setting, weight, color, percent oil content, and quality of the seed.[17]

Hot water treatment of seeds reduced the growth of Alternaria.[58] Though, conidia of fungi can persist on tissues of leaf for 8–12 weeks, whereas on tissues of stem up to 23 weeks and on seed over the year. Therefore, fields that are sown again after harvest often overlapped with an enormous volume of inoculum that may disturb the emergence of crop plants and early growth stages.[59,60] Accordingly, rotation with noncruciferous crops and eradication of cruciferous weed hosts can support to manage these pathogens and also fungicide spray in fields has to be done at the same time. Timely sowing[27] of clean certified seeds after deep plowing and timely weeding with the maintenance of optimum plant population and avoidance of irrigation at flowering and pod formation stages may help for management of the disease.

Many AB tolerant lines have been recognized by various researchers.[1,3,18] A dwarf *B. juncea* cultivar Divya reported for tolerance against AB.[61] *Brassica juncea* and *B. rapa* are highly vulnerable in comparison to *B. carinata* and *B. napus* among other diverse rapeseed-mustard crops.[62] Genotypes of *B. juncea* including DRMRM 163, DRMRM 177,[63] PAB 9511, PAB-9534, JMM-915, EC-399296, EC-399301, EC-399299, PHR-2, Divya and RN-490, *B. carinata* HC-1, PBC-9221 (Kiran), NRCDR-515 and DLSC-1, *B. napus* PBN-9501, PBN-9502, PBN-2001 and PBN-2002[1,64] were found tolerant to AB disease. While, resistance to AB have been prominent in wild relative Brassicas namely, *B. alba*,[65,66,68] *Camelina sativa, Capsella bursa-pastoris, Diplotaxis berthautii, D. catholica, D. cretacea, D. erucoides Eruca sativa, Neslia paniculata*,[68–70] *B. desnottesii, Coincya pseuderucastrum*, and *Erucastrum gallicum*.[17,71]

AB resistance is linked with factors like biochemical compounds including polyphenol oxidase, peroxidase, catalase, higher total soluble sugars content, and nitrogen.[72,73] The release of conidia is limited due to factors mainly elevated deposits of epicuticular wax to facilitate a physical obstacle as a hydrophobic varnish to lessen deposition of water-borne inoculums and reduce the speed of conidia germination and development of germ tube, etc..[74] *Brassica napus* (Tower, HNS-3), *B. carinata* (HC-2), *B. alba* have more wax on the leaf surface than *B. rapa* (BSH-1, YSPB-24), and *B. juncea* cultivar RH-30.[75,76]However, reports on mechanism of tolerance to AB indicated the outcome of additive genes or polygene or cluster of genes[77,78] with resistance being governed by nuclear genes of partial dominance.[79] There are indications of components of resistance being considerably linked to each other as regards to slow blighting.[80] Dominance (h) ensuring a leading role in genetic control on instance of emergence, additive × dominance predominant for other disease advancement elements namely, AUDPC, etc.[81,82] Wild relative Brassicas has been observed to obtain phytoalexins on challenge inoculation.[83] Actions of some compounds associated to Camalexin ($C_{11}H_8N_2S$) and 6-methoxycamalexin ($C_{12}H_{10}N_2SO$) are found to be maximum toxic to *A. brassicae*.[84] Phytotoxin dextruxin B elicits phytoalexin response in *B. alba*,[85] *B. rapa*, and *B. napus*.[86] *A. brassicae* resistance is observed to be covered and multicomponent with sensitivity to host-specific toxin dextruxin B, assessable and qualitative elicitation of phytoalexins, hypersensitive response and Ca-sequestration determining outcome of host–pathogen interaction.[68] Systemic acquired resistance has been induced mainly through inoculation by avirulent *A. brassicae*.[87]

Since, AB resistance is found to be governed by additive or polygenes and breeding for resistance might comprise pyramiding of minor genes manage the disease. Breeding methods suitable to deliver additive/poly-gene resistance against AB and introgression of genes from reported resistant sources are mutual recurrent selection or diallel selective mating,[78] wide hybridization (*B. alba*),[88] somatic hybridization from *Camelina sativa*,[89] transgenic expressing *Trichoderma harzianum* endo-chitinase gene,[90] pollen culture, and sensitivity investigation to dextruxin B,[91] etc. Transfer of resistance characters through extensive crossability has some complication, which is possible through speedy improvements in methods of tissue culture, protoplast fusion, embryo rescue, and genetic engineering. Transfer of a cDNA encoding having chitin-binding lectin from *Hevea brasiliensis* into *B. juncea* cv. RLM-198 showed resistance. Integration of the transgene has been confirmed in promising transgenics through southern analysis. Transgenics confirmed the parameters that are linked with resistance including longer incubation and latent period, smaller necrotic lesion size, lesser disease severity, and late senescence in entire plant bioassay under glasshouse conditions.[92] Genes being expressed during AB infection of *Brassica* has been confirmed recently.[93] Suppression subtractive hybridization used between RNA isolated from *A. brassicicola* conidia nurtured in water, and on the leaf surface of *A. thaliana* followed by cloning and sequencing of cDNA clones caused distinct expression. One gene (P3F2) was recognized that expressed during infection however its function remains to be defined. Such method with other pathogens could lead to developments in understanding of pathogenicity.

Spray of Metalaxyl+mancozeb 4% found effective for the management of AB disease in mustard. Disease infection on silique caused by *A. brassicae* has been effectively controlled by spraying Iprodione (Rovral).[94] Spray of aqueous bulb extract (1% w/v) of *Allium sativum* at 45 and 75 DAS is appropriate to minimize the disease severity on leaves and pods.[57,95,96] *Trichoderma viride* GR isolate was found suitable in checking blight severity on mustard leaves and silique. Use of potassium as basal reported to check AB disease in mustard.[97] Observations of mycelial growth, conidial germination, and germ-tube elongation reported to expose the presence resistance to both dicarboximides and phenylpyrroles against *A. brassicicola* isolates.[98] Keeping in view the environmental consequences now the effective biological and newer fungicides are required for control of the disease.

13.2.5 SCLEROTINIA ROT (SR)

SR or white rot caused by *Sclerotinia sclerotiorum* (Lib.) de Bary is a menace to rapeseed-mustard production with substantial yield losses world-wide including Australia, Europe, India and North America.[99–103] The first occurrence of disease on oilseed Brassica appears to have been made from India.[104]

This disease become important in India with up to 66% disease incidence and yield losses up to 39.9% leading to discourage the growers.[105] Yield losses differ with the percent disease incidence and the stage of growth of the crop at the time of infection in the similar region over the years. Disease incidence occur at the early flowering stage produce little or no seeds, and those infected at the late flowering stage set seed and may suffer less yield reduction. The pathogen is reported to have a wide host range, known to infect about 408 plant species[106] with no confirmed source of resistance described so far.

Based on the symptoms, the disease has been named white blight[107] and white rot[108] and stem blight[14] and stem canker, or rape canker. Commonly, the stem of the plant is seen affected more frequently but the entire aerial plant parts are reported to be infected by the disease under natural conditions. Elongated water-soaked lesions are visible on the stem and subsequently covered by a cottony mycelial growth of the fungus. Once the stem is entirely girdled by such lesions and the plant wilts and dries.[109] Appearance of the disease at an initial growth stage of the plant resulted in the death of whole plant.[14] Damping-off of *B. napus* seedlings was also reported.[110] Little sign of attack may be seen on foliage having round brown lesions converted soon in shot holes; however, at times may even start on leaves, which wilts and droops downward and then moves on to the stem. Occasionally, the infection is limited to a smaller area of pith, which results in slow stunting of the plant and premature ripening reasonably than the speedy collapse of the affected plants. The diseased stem tends to shred and abundant greyish-white-to-black and spherical sclerotia appear either on the surface or in the pith of the affected stem.[109] Root rot symptoms due to the disease are also reported.[111]

Management of the disease is very difficult and inconsistent and uneconomical because of wide host range and durable persistence of the sclerotia. It is anticipated to collect and burn all the infected stubbles to kill the sclerotia because the disease is carried-over through sclerotia with crop debris and refuse of the plants stimulate sclerotial formation.[14] Sclerotia-free clean seed should be sown to avoid entry of inoculum in the field. A wide host-range of about 408 species and the air-borne infection through ascospores restrict the scope of crop rotation for the control of the disease.[112] Though, deep summer ploughing, crop rotation with nonsusceptible hosts (rice and maize), application of recommended dose of nitrogenous fertilizer, adequate irrigation scheduling, restricted plant stand perform to minimize the sclerotial count in the soil, which then might demonstrate beneficial in control of the disease ensuing from soil-borne inoculum.[113] Eradication of broad-leaf weeds like *Chenopodium* spp. is important in minimizing the disease. Late sowing might be supportive under the conditions of Canada as well as India by shortening the overlap between phenological susceptibility and exposure to maximum ascospore load.[114] Soil application of compost inhibited carpogenic germination of apothesia and may reduce SR infection.[115]

Differences in common growth habit and morphological characters might be important characteristics for tolerance of the disease. *Brassica napus* cultivar "Omi nature" is resistant to the disease. It is of medium height, early maturing, and possesses a stiff stem with many branches. The greater extent of resistance in the "Isuzu" variety of *B. napus* has also been reported from Japan.[116] Response of Charlton genotype was observed comparatively stable irrespective of different isolates of the pathogen, whereas highly variable reactions were shown by other genotypes including Zhongyou-ang No. 4, and Purler against the similar isolates. There is an urgent need of highly resistant genotypes for utilization in the oilseed *Brassica* breeding programmes to enhance the level of field resistance in cultivated *B. napus* and *B. juncea.* Although complete resistance has not been identified in canola and partial field resistance has been identified in Chinese variety Zhongyou 821,[117] a cultivar Zhongshuang No. 9 and claimed to be better than Zhongyou 821 was reported resistant to Scleritinia rot.[118] Li et al.[42,53] reported resistant genotypes of *B. napus* namely, 06-6-3792 (China), ZY004 (China), and RT 108 (Australia) based on <3.0 cm mean stem lesion lengths.[119,120] In addition, the levels of resistance reported previously in *B. juncea* were and in particular far lower, for example, *B. juncea* JM06018 and JM 06006 with mean stem lesion lengths of 4.8 cm[120] as compared with *B. napus* genotypes. High levels of resistance reported against *S. sclerotiorum* in introgression lines derivative of *E. cardaminoides, D. tenuisiliqua* and *E. abyssinicum*.[121] They reported a unique genotype cv. Charlton, which show relatively consistent resistant reactions across different isolates may be best for commercial exploitation to breed for resistance in oilseed *Brassica* against SR.[121] *Brassica juncea* cv. Ringot I have been reported as resistant to SR.[122] Recently, it has been noted that defense to *S. sclerotiorum* in rapeseed-mustard is linked with the successive activations of salicylic acid signaling and jasmonic acid signaling.[123]

Since no single method can effectively control *S. sclerotiorum*, the best approach to control the pathogen is by integration of various eco-friendly measures. Certain fungicides including quintozene, fentin acetate, and calcium cyanamide have been found suitable to reduce the apothecial growth of the pathogen. Under field conditions, effectiveness of calcium cyanamide in disease management by 40–90% has been confirmed in Germany.[124,125] The infested seed-lot fumigated with methyl bromide with success to eradicate viable sclerotia.[126] However, this seems unlikely to be effective for bulk quantities of seeds. Foliar sprays of chemicals to minimize the secondary spread of the disease has been explored. Three foliar sprays of benomyl (0.025%) gave better results in controlling the disease with an increase in seed yield.[107] A single aerial spray of benomyl at the early bloom stage in the disease-prone regions of Canada is also advised.[114]

Nowadays, an increasing awareness about ecological pollution caused by higher consumption of fungicides in some extent and change of fungicide resistant pathogen strains has provoked the researchers to develop ecofriendly disease management. There are several reports on SR management of mustard by fungicides[127]; foliar spray of carbendazim at full bloom stage provided significant reduction in SR apart from highest seed-yield among different treatments over control.[128] Fungicide methyl benzimidazole has been reported to manage SR in oilseed Brassicas.[129] Seed treatment with *Trichoderma viride* and *Allium sativum* aqueous bulb extract (2% w/v) found effective against the SR disease combined with properties of growth promotion, better plant stand and yield improvement.[130,131] The integration of seed treatment with foliar sprays gained superior decrease of the disease.[132,133]

13.2.6 POWDERY MILDEW (PM)

Occurrence of PM on rapeseed and mustard is reported from France,[134] Germany,[135] India,[136,137] Japan,[138] Sweden, Turkey, UK,[135] and USA.[139] Recent reports on the existence of PM disease of rapeseed-mustard deal with *Erysiphe cruciferarum* Opiz ex. Junell..[140] Obviously, it is assumed that PM is unable to cause much loss to the oilseed *Brassica* crops except in sporadic severe epidemics, once the entire leaves and siliquae become covered with the powdery growth of the fungus at initial phonological stage. The PM disease has been found to appear rigorously in Gujarat, Rajasthan, and Haryana states of India, ensuing in substantial yield losses. But the exact data on yield losses are not available. Considering the differences in disease intensity from year to year, it appears that the loss is proportional to the disease intensity, which varies considerably depending on the stage at which it occurs. PM of oilseed Brassica was ranged from 5 to 100% incidence on various Brassica species, particularly in India.[141] Therefore, the situation is very upsetting with increased incidence and severity of the PM disease on oilseed Brassica crops broadly all over the world.

Initially, disease symptoms appear as dusty-white, round, and powdery patches on leaves of the diseased plants.[142] Under favorable weather situations (moderately high temperature), the powdery patches enlarged and merge to cover the whole plants. The plants infected severely may show stunted growth and produce a reduced amount of silique. The immature silique showed white patches in the early stage of disease occurrence. Subsequently such silique covered completely with a white mass of mycelia and conidia of the fungus and remains small in size with small shrivelled seeds.[143] Under favurable situations, resting spores of the fungus called cleistothecia are formed on affected leaves, stems, and silique, which are visible as black scattered and/or rigorous bodies.[14]

Suitable planting date seems to be an encouraging approach to control the pathogen. A limited form of resistance against the PM has been reported in *B. alba, B. alboglabra, B. rapa* var. *Brown sarson, B. chinensis, B. japonica,* and *Eruca sativa.*[144] There appears to be a range of level of resistance in rape, although immunity is not apparent. Foliar application of Karathane thrice at 10-days intervals is also found effective for the control of disease.[145] Seed treatment and foliar sprays with *T. viride* as bioagent and aqueous bulb extract of *A. sativum* found effective against the disease.[51]

13.2.7 CLUB ROOT

The club root disease of oilseed Brassica is caused by an obligate biotrophic fungus *Plasmodiophora brassicae* Woronin, normally in acidic and poorly drained soils. It occurs severely in the regions of high winters than spring type climates. The disease studied first time systematically and described the host pathogen interaction along with lice cycle of the fungus.[146] The disease has been reported on rapeseed-mustard from different countries of the world including East Germany, Malaya, New Zealand, Poland, Sweden, the UK, and the USA.[147] The disease is generally found on *Brassica* species in West Bengal, Orissa, and hills of Darjeeling[148] and Nilgiri[149] of India, with significant yield losses up to 50%.[150] Decline in crop acreage due to heavy losses caused by *Plasmodiophora brassicae* on *Brassica napus* in southern districts of New Zealand.[151] However, precise data on yield losses due to the disease on oilseed Brassica is still unknown.

Appearance of the affected plants looks common with healthy growth at the initial stage. With advancement of disease the plants become undersized displaying pale-green or yellowish leaves and subsequently died in a short span of time. After uprooting the plants and overgrowth or swollen of the main roots looks as small or rod or round-shaped knobs known as clubs. The club shape differs with the type of root of a species. The root system converted into different shaped malformations after the many infections comes together. Enormous resting spores are enclosed in the swollen root system. Although, the older and large size clubs of the root system collapse till end of the season.

Avoid further cultivation of the crop in club root-infested fields during preceding season. Appropriate drainage facilities and application of lime @ 1 kg/m^2 area to raise the pH upto 7.2 to be adopted to mitigate the disease incidence. It has been observed that the alkaline soils inhibited the germination of *P. brassicae* spores.[152] It is challenging to succeed the disease control due to prolonged existence of resting spores in soil. Breeding for resistance is complicated as the existences of numerous physiologic races of *P. brassicae* have varying aptitude to infect different species of the Brassica.

Resistant genotypes of *B. napus* (GSL-1, WBBN-1, -2, PCRS-80, WW-1507, ISN-700, MNS-3), *B. carinata* (HC-1, -4, -5, 9221, PC-3, PCC-2, PPSC-1, PC-5) and *B. nigra* (ACCBN-479) are described to the disease.[153] As sexual reproduction of the fungus is higher help in development of new pathotypes resulted in restrict durability of host resistance. Although, certain chemicals like benomyl, quintozene, and some soil fumigants reported as effective for management of club root.

13.2.8 FUSARIUM WILT

Fusarium wilt reported first on *B. juncea*[154] and *B. nigra*[155] in India is caused by *F. oxysporum* f. sp. *conglutinans* (Wr.) Snyder and Hansen. They further isolated group A and B isolates and observed that both the isolates are pathogenic to *B. Rapa* var. Toria, *B. rapa* var. Yellow Sarson, *B. oleracea* var. *botrytis, B. oleracea* var. *capitata, Eruca sativa, Mattiola incana,* and *Raphanus sativus.* The disease occurred on *B. rapa, B. nigra, Eruca sativa, Sinapis alba, Sysymb officinale, B. carinata, Crambe abyssinica,* and *C. hispanica.*[156] It seems there is a complex of *F. oxysporum* causing yellows in oilseeds Brassica and among which one that causes yellowing in *B. rapa* is proposed to be *F. oxysporum* f. sp. *rapae.*[157]

Appearance of the affected plants leaves is drooping, vein clearing and chlorosis, which may turned to wilting, drying, and ultimately death of the whole plant. The symptom expression may differ with the stage of the host plants.[154] Typical symptoms on affected plants do not appeared at the early developmental stage. Affected plants at pre- and early flowering phase displayed defoliation. Longitudinal edges and troughs develop externally on stem of the infected plants, which is never witnessed in the later stages. Prominent stunted growth of the infected plants was observed particularly in preflowering stages. No abnormality or decay of root tissues occurred till the infected plants completely dried. The mycelium and/or microconidia of the pathogen existed on vascular tissues of the stem and root. The typical characteristic symptom of vascular wilt is the browning of tissue walls and plugging with a dark gummy substance. Further the epidermis of roots sloughs off at later stage of the disease.

13.2.9 BACTERIAL STALK ROT

The stalk rot of oilseed Brassica is caused by *Erwinia carotovora* (Jones), Holland reported first time from India.[158] According to them, the disease appeared in epiphytotic proportion on the commonly cultivated *B. juncea* 1979 in the State of Rajasthan, India. Fast succulent growth of plants by excess nitrogen doses simultaneously inadequate drainage system in field may cause severe infection. Severe incidence of the disease at farmer's field in Rajasthan state of India was observed up to 60–80%.[159]

The water-soaked lesions appear at the collar area of plants is the typical symptom, which is ordinarily together with a white foaming. The enlarged lesions also cover the larger areas of tender branches and leaves express wilting and further become weak. The pith tissues of affected stem and branches become soft, pulpy, and create dull white ooze produces a foul smell. The diseased portion of collar turned into sunken and buff-white to pale-brown in color resulted collapse of plants in a short time. Other hosts of this bacterium are *Brassica oleracea* var. *botrytis, Daucus carota, Lycopersicon esculentum*, and *Nicotiana tabacum.*

Foliar spray of certain antibiotics like aureomycin (chlorotetracycline) 200 µg/ml is most effective in reducing the infection under field conditions. Use of clean seed and rouging of infected plants at initial stage of disease appearance may protect the crop from severe incidence.

13.3 CONCLUSION

During the last few years, pathogens in rapeseed-mustard have been documented as key factors responsible for financial losses with identification of the most important ones depending on their damage causing potential based on agro-ecological zones. Cultural practices for managing some diseases have been determined. Critical stages for the growth of some foliar diseases namely AB and WR have been recognized together with weather conditions that favor the disease development. Thus, now it is possible to recommend need-based sprays of botanical and chemical fungicides for effective, economic, and eco-friendly disease management. Loss-causing ability of oilseed Brassica diseases have been under-estimated resulting in uprise of relevant pathogens. Breeding for disease resistance has not received enough attention so far. Cultivars bred for agronomic benefits have been screened later for disease resistance. Rarely a systematic and sensible breeding effort was been made for disease resistance. Genetics of inheritance of disease resistance and its biochemical basis have been pinpointed in some cases. Use of modern biotechnological tools has also yielded success. But, sources of resistance to diseases have remained grossly under-utilized. Though research has provided some significant tools for managing the diseases, the adaptability of the same in the farmers' field has been lacking. An oilseed farmer is resource poor and farming under worst situations where disease management is having less attention. The need of the mustard grower is low monetary input production technology while research has devised mostly high input management strategies emulating cereals. There has been a total lack of extension wing of rapeseed-mustard pathology. Although assessing the effects of TMO and it is clearly reflected that transfer of technology has been a weak link in *Brassica* pathology. Even the simple practice of seed dressing is not followed by oilseed *Brassica* farmer regularly.

Perhaps a resource poor mustard grower would accept an ideal plant type resistant to many rather than to only one disease. Thus, the multiple-disease resistance concept needs to be advocated by overcoming the one of single disease resistance. The multiline cultivar concept and so successfully advocated in cereals, with resistance to major diseases, needs to be imported to rapeseed-mustard crops at least. Identification of sources of resistance to disease caused by *S. sclerotiorum* is required. Horizontal resistance to disease may be preferable for sustainability. There is also a need to transfer resistance, available in wild species, namely, resistance to rapeseed-mustard AB from wild related species. Studies on biochemical and physiological basis of disease resistance or pathogenic stress physiology need to be taken up. Generation of adequate information on bio-ecology, population dynamics, mass multiplication and artificial maintenance techniques, race/biotype pattern of pathogens, functional genomics of host–pathogen interaction, and rapid reliable screening and evaluation techniques against specific diseases as per priority needs to be addressed. For this purpose, there is a need for creation of national facility for reference (type) cultures to speed up work on genetic enhancement. Immediate confirmation and utilization of sources of resistance/tolerance to diseases to design well-thought-out crossing programmes targeted at resistance to diseases is required. Study of mechanisms and genetics of resistance to key diseases, wherever unknown, needs attention. Exploitation of modern tools for screening of large number of germplasm lines and for transfer of resistance in background of cultivated and popular varieties should be taken up. Studies on induced resistance, molecular, and biochemical basis of resistance could be alternative approach for immunization of plants. National crop laboratories may be given specific responsibilities and be commensurately equipped to cater to the assigned task.

However, a system approach in IDM needs to be deliberated for effective crop management. Through adaptive research at respective locations and in frontline demonstrations, component-wise and in-whole package, following measures may be combined for designing location-specific packages depending on the demand of the farmer's:

- Sowing time is directly affecting the disease incidence in a big way.
- Deep summer plowing.
- Use of disease-resistant varieties.
- Use of superior seeds for sowing.
- Use of seed protectants as bio-agents namely, *Trichoderma viride, Gliocladium virens,* or botanicals like *Allium sativum* bulb extract @ 1% w/v or mixture of carbendazim with metalaxyl 35 SD @ 6 g/ kg and 250 ppm streptomycin should be made mandatory.
- Application of recommended doses of N, P, K, and S fertilizers.
- Have only optimum plant population with recommended spacing 15 × 45 cm.
- Well-drained plots are very essential. If water accumulates on one side of a plot due to slope in the land, a channel may be dug at the lower end of the slope to drain out the access water.
- Clean cultivation, avoidance of weeds and stubbles. Roguing out diseased plant to avoid secondary spread of disease to reduce inoculum.
- Identify the right crop for rotation and also for noncrop season as sequence crop in order to discourage the pathogen build up. Choose a suitable intercrop as per the pathogen to be countered.
- Judicious use of irrigation, preferably two, depending on the stage of crop growth, soil type, and rainfall availability by avoiding irrigation at 60–70 days after sowing. In high rainfall areas like West Bengal and Orissa, it is better to avoid even second irrigation so that clubroot disease remains under check.

- Prophylactic spray of botanical/bioagent/fungicide/antibiotic when conditions favorable for disease occurrence prevail. Timing the application of such prophylactic sprays based on databank, expert systems, and epidemiological research findings.[160–162] Pest and disease forewarning/forecasting models ensure a cost-effective IDM for rapeseed-mustard. There is a need to ensure that the pesticidal sprays are avoided during peak activity period of the biocontrol agents.
- To make disease management easier and successful, there is a need for integrated seed treatment with well-formulated biocontrol agent(s) having tested delivery systems. Thus, *T. harzianum* tolerant to fungicide(s) and effective fungicide(s) like carbendazim and/or metalaxyl will be good for an effective biocontrol based IDM. Further evaluation of more effective biocontrol agents and botanicals is needed. An effective ecofriendly tool for soil-borne disease management like soil solarization needs to be made cost-effective.
- In view of the new liberal seed policy, quarantine should be strengthened to avoid introduction of pathogens or some of their biotype/races not existing in India. Major oilseed Brassica growing countries encounter the menace of Phoma stem canker (*Leptosphaeria maculans*, *L. biglobosa*) and light leaf spot (*Pyrenopeziza brassicae*), which fortunately do not occur in India. In fact, these countries (China, Australia, UK, France, Canada, etc.) spend a lot of their resources to tackle these problems.[163,164] Holistic understanding of the different agro-ecosystems and bioecology of the pathogens is essential. Biodiversity, preservation, and search for resistance in wild types will have to be focused to pool available genes in host and pathogen. Tools for gene mapping and pooling have to be augmented.[165]
- Judicious use of biotechnology and genetic strategies is now recognized as very rewarding. Standardization of methods for rapid identification of pathogens for diagnosis of and resistance to diseases will enable screening of large germplasm collection in a short time. The modern techniques may be very useful to develop novel biocontrol agents and even disease resistant transgenics namely, by use of R-gene mediated resistance.[164] These will necessitate rapeseed-mustard pathologists of this era to be geared for taking up such challenges to build patents.
- Risks related to incidence of a disease should be accounted for while programming models for whole-farm planning. Advances in crop production strategies have resulted in a spurt of high yields in major crops like cereals. To revolutionize rapeseed-mustard production,

skilful integration of various managerial tactics including disease risk analysis, as related to its incidence, and risk management should take into consideration the experience, beliefs, and preferences of the farmers' need as a participatory mode of technology development.

Today the threats to our environment are much greater in scope and number. There is need to monitor the race pattern of different pathogens and not just diseases. Expert systems and database for information retrieval need to be prepared for developing effective disease forecasting models possibly through simulation of epidemics for all major rapeseed-mustard diseases to enable the growers to take up need-based prophylactic measures. Keeping in view the options available for designing an effective and environment-compatible IDM strategy for the crops, conceptual models, and primary and secondary decision-making tools need to be used effectively. Huge gaps in accessible information on biocontrol, use of locally available cheap botanicals, etc. need to be bridged. There is also need to reduce gaps between lab and land or dissemination of information by converting IDM into a movement through programmes like National Network Projects and development of a community approach in recognition of pathogens. However, there is need to analyze the reasons for disappointment of IDM in oilseed rape still not reaching in farmers' fields. Technologies developed have been more a result of curiosity-driven research with lack of effective delivery systems. Pathologists have to be more pragmatic by rationalizing the targets for IDM with its implementation being remarkably visible. There is a need to view the disease problems from the farmers' point of view by interaction of the policy maker, scientist, extension worker, and the grower in an organized and harmonious way. Therefore, crucially required to formulate need-based IDM strategies or IDM with low input for sustainable agriculture and transfer them to farmers through proper adaptive research. Plant pathologists need to advocate eco-sustainable technology for disease management to protect the sphere. Bio-management of plant diseases is known to be economical and avoid ground-water pollution, death of nontarget beneficial microflora, and risk of new fungicide-resistant pathogen strains. Other advantages of promoting biocontrol of rapeseed-mustard diseases include low capital investment, high employment potential, use of local raw material and manpower resources, adaptable levels of sophistication, etc. This could be in tune with the provision of job-led growth that may be economically, environmentally, and socially sustainable with livelihood opportunities linked with agricultural competitiveness.

Recently, organic farming has speedily been renovated from a farmers' drive to an innovative part of farming policy in several European countries. In certification, agreement with available organic criteria is confirmed through annual inspections on farms.[166] The Indian Council of Agricultural Research has published a compilation of technologies available for bio-formulation. These technologies, if effectively utilized and industrialized, than merely planning and talking on the subject in various fora of seminars and symposia, can lead to early substitution of use of chemical fungicides for an ecofriendly and sustainable disease management program in oilseed crops. The problem of shelf life of bio-fungicides could be overcome by setting up small-scale bio-fungicide preparation systems at district level in bio-village frame with trained people and strict monitoring of quality norms, which may also provide employment while taking care of the availability aspect. Package of practices on different oilseed crops need to emphasize more on bio-fungicides. Proper extension coupled with appropriate trickle-down of information and feedback, timely procurement of the bio-formulations, conviction of farmers through frontline demonstration regarding cost-effectiveness and benefits of bio-fungicides through proper and timely application, intervention at dealers level to effect shift from chemicals to bio-fungicides, etc. are also essential for better coverage by bio-fungicides. There is also need for a mass movement for promotion of bio-fungicides and secure trust of end-users about their efficacy to enable them replace chemical fungicides to the extent possible similar to that witnessed in India for usage of liquid petroleum gas for cooking or the pulse polio immunization program. For the success of IDM, educating the oilseed planter with updated technology or oilseed grower awareness about diseases and IDM through trained extension personnel is very essential. Personnel involved in growing, handling and storing of oilseeds need to be equipped to diagnose and manage diseases in the field and store based on available options.

Intensive farming to produce more is now emerging in developing countries. The agricultural development strategies are prioritized on increasing yield, crop protection, human health, environmental hazards and social aspects. To improve the present situation there is need to determine the potential of a zone, the biophysical environment, etc. and improve on planned scale in cluster approach. These factors should be analyzed to set priorities for oilseed research and development. In case of a negative trend in sustainability, necessary adjustments should be made in the IDM strategy. Oilseed crops need to be and can be made productive by providing the required relief to the plant from stress provided by pathogens and latently associated deleterious microorganisms, through meticulous integration of the disease management options. Viewing of IDM in isolation has been a major folly and needs to be looked in totality. From a myth, interdisciplinary research has to be made a reality in oilseed scenario through intensive teamwork with a mission mode approach. Close collaboration of different disciplines on the same plane shedding "big brotherly attitude" of any discipline by broadening of horizons is another necessity. Novel pathogen management options have to be designed through reorientation of strategies may be through a shift from IDM to Total Pest Management, on ecosystem basis for rapeseed-mustard crop health. Thus, combining increase in oilseed productivity with its improved ecofriendly protection is needed for total oil *Brassica* crop management programme.

KEYWORDS

- **alternaria blight**
- ***brassica* species**
- **clubroot**
- **disease**
- **downy mildew**
- **management**
- **powdery mildew**
- **sclerotinia rot**
- **white rust**

REFERENCES

1. Anonymous. Annual Reports of All India Coordinated Research Project Rapeseed-Mustard). Directorate of Rapeseed-Mustard Research (ICAR), Bharatpur 321303, Rajasthan, **2020**.
2. Anonymous. Agricultural Statistics. Directorate of Economics and Statistics, Department of Agriculture and Cooperation, Ministry of Agriculture, Government of India, New Delhi, **2020**.
3. Chattopadhyay, C.; Seguin-Swartz, G. Breeding for Disease Resistance in Oilseed Crops in India. *Annu. Rev. Plant Pathol.* **2005,** *3*, 101–142.
4. Petri, G. A. Diseases of *Brassica* Species in Saskatchewan, 1970–72. III. Stem and Root Rots. *Canadian Plant Dis Survey* **1973a,** *53*, 88–92.

5. Mahmud, K. A. Damping-off of *Brassica Juncea* Caused by *Pythium Aphanidermatum* Fitz. *Sci. Cult.* **1950,** 16, 208.
6. Aulakh, K. S. Damping-off of Toria Seedlings due to *Pythium butleri. Indian Phytopathol.* **1971,** *24*, 611–612.
7. Srivastava, O. P. Dry Root and Bottom rot of Mustard Caused by *Rhizoctonia Solani. Indian J. Microbiol.* **1968,** *8*, 277–278.
8. Upadhyay, R.; Pavgi, M. S. Some New Hosts for *Pellicularia rolfsii* (Sacc.) West. from India. *Sci. Cult.***1967,** *33*, 71–73.
9. Srivastava, S. K.; Dhawan, S. Screening of *Brassica Juncea* Cultivars for Resistance Against *Macrophomina* Stem and Root Rot Disease. *Geobios* **1979,** *6*, 333–334.
10. Khan, R. U.; Kolte, S. J. Some Seedling Diseases of Rapeseed-Mustard and their Control. *Indian Phytopathol.* **2002,** *55*, 102–103.
11. Galletti, S.; Burzi, P. L.; Sala, E.; Marinello, S.; Cerato, C. Combining Brassicaceae Green Manure with *Trichoderma* Seed Treatment Against Damping-off in Sugarbeet. Bulletin-OILB/SROP. ***2006,*** *29* (2), 71–75.
12. Kolte, S. J. *Diseases of Annual Edible Oilseed Crops, II, Rapeseed-Mustard and Sesame Diseases*; CRC Press Inc.: Boca Raton, **1985;** 135 pp.
13. Porter, R. H. A Preliminary Report of Surveys for Plant Diseases in East China. *Plant Dis. Reporter* (Suppl.). **1926,** *46*, 153–166.
14. Vasudeva, R. S. Diseases of Rape and Mustard, In: *Rape and Mustard*; Singh, D. P., Ed.; Indian Central Oilseed Committee: Hyderabad, **1958;** pp 77–86.
15. Bains, S. S.; Jhooty, J. S. Mixed Infections by *Albugo Candida* and *Peronospora parasitica* on *Brassica Juncea* Inflorescence and Their Control. *Indian Phytopathol.* **1979,** *32*, 268–271.
16. Kolte, S. J.; Tewari, A. N. Note on the Susceptibility of Certain oleiferous *Brassicae* to Downy Mildew and White Blister Diseases. *Indian J. Mycol. Plant Pathol.* **1979,** *10*, 191–192.
17. Saharan, G. S.; Mehta, N.; Meena, P. D. *Alternaria Blight of Crucifers: Biology, Ecology and Management;* Springer Verlag: Singapore, **2017;** p 326.
18. Gupta, K.; Saharan, G. S.; Singh, D. Sources of Resistance in Indian Mustard Against White Rust and Alternaria Blight. *Cruciferae Newslett.* **2001,** *23*, 59–60.
19. Nashaat, N. I.; Heran, A.; Awasthi, R. P.; Kolte, S. J. Differential Response and Genes for Resistance to *Peronospora parasitica* (Downy Mildew) in *Brassica Juncea* (mustard). *Plant Breeding*, **2004,** *123*, 512–515.
20. Kaur, P.; Jost, R.; Sivasithamparam, K.; Li, H.; Barbetti, M. J. Proteome Analysis of the *Albugo Candida-Brassica Juncea* Pathosystem Reveals That the Timing of the Expression of Defence-Related Genes Is a Crucial Determinant of Pathogenesis. *J. Exp. Bot.* **2011a,** *62*, 1285–1298.
21. Sharma, K. D. Symptomatology, Yield Losses and Control of Downy Mildew and White Rust of Rapeseed and Mustard, M.Sc. (Agric.) Thesis. G. B. Pant Univ. Agric. Tech.; Pantnagar, **1980**.
22. Choi, Y. J.; Shin, H. D.; Thines, M. The Host Range of *Albugo Candida* Extends from Brassicaceae Through Cleomaceae to Capparaceae. *Mycol. Progress* **2009,** *8*, 329–335.
23. Voglmayr, H.; Riethmuller, A. Phylogenetic Relationships of *Albugo* Species (White Blister Rusts) Based on LSU rDNA Sequence and Oospore Data. *Mycol. Res.* **2006,** *110*, 75–85.

24. Choi, Y. J.; Shin, H. D.; Ploch, S.; Thines, M. Evidence for Uncharted Biodiversity in the *Albugo Candida* Complex, with the Description of a New Species. *Mycol. Res.* **2008,** *112*, 1327–1334.
25. Choi, Y. J.; Shin, H.; Ploch, S.; Thines, M. Three New Phylogenetic Lineages are the Closest Relatives of the Widespread Species *Albugo Candida. Fungal Biol.* **2011,** *115*, 598–607.
26. Choi, Y. J.; Shin, H. D.; Hong, S. B.; Thines, M. Morphological and Molecular Discrimination Among *Albugo Candida* Materials Infecting *Capsella Bursa-Pastoris* World-Wide. *Fungal Divers.* **2007,** *27*, 11–34.
27. Meena, P. D.; Chattopadhyay, C.; Singh, F.; Singh, B.; Gupta, A. Yield Loss in Indian Mustard due to White Rust and Effect of Some Cultural Practices on Alternaria Blight and White Rust Severity. *Brassica.* **2002,** *4*, 18–24.
28. Biga, M. L. B. Riesaminazione Della Speciedel *Genere Albugo* in Base Alia Morfologia Dei Conidi. *Sydowia.* **1955,** *9*, 339–358.
29. Saharan, G. S.; Verma, P. R.; Meena, P. D.; Kumar, A. *White Rust of Crucifers: Biology, Ecology and Management*; Springer Verlag: Frankfurt, Germany, **2014;** pp 244.
30. Mundkur, B. B. *Fungi and Plant Diseases*; MacMillan: London, **1959,** 75 pp.
31. Petrie, G. A. Diseases of *Brassica* Species in Saskatchewan, 1970–72. I. *Staghead Aster Yellows Can. Plant Dis. Survey* **1973b,** *53*, 19–25.
32. Pal, S. S.; Gupta, T. R.; Kumar, V.; Dhaliwal, H. S.; Kumar, V. Transfer of White Rust Resistance from *Brassica Napus* to *B. Juncea* cv. RLM 198. *Crop Improv.* **1999,** *26*, 249–251.
33. Mukherjee, A. K.; Mohapatra, T.; Varshney, A.; Sharma, R.; Sharma, R. P. Molecular Mapping of a Locus Controlling Resistance to *Albugo Candida* in Indian Mustard. *Plant Breed.* **2001,** *120*, 483–487.
34. Meena, S. S.; Meena, P. D.; Singh, V. V.; Meena, H. S.; Singh, D.; Yadav, R.; Singh, K. H.; Sharma, P.; Nanjundan, J.; Singh, B. K.; Singh, Y. P.; Ram, B. DRMR-2019 (IC0598622; INGR17077), An Indian Mustard (*Brassica juncea*) Germplasm with White Rust Resistance. *Indian J. Plant Genet. Resour.* **2019a,** *32* (2), 281.
35. Meena, S. S.; Meena, P. D.; Singh, V. V.; Meena, H. S.; Singh, D.; Yadav, R.; Singh, K. H.; Sharma, P.; Nanjundan, J.; Singh, B. K.; Singh, Y. P.; Ram, B. DRMR-2035 (IC0598623; INGR17078), An Indian Mustard (*Brassica juncea*) Germplasm with White Rust Resistance. *Indian J. Plant Genet. Resour.* **2019b,** *32* (2), 281.
36. Parui, N. R.; Bandyopadhyay, D. C. A Note on Screening of Rai [*Brassica Junca* (L.)]. *Curr. Sci.* **1973,** *42*, 798–799.
37. Singh, D.; Singh, H.; Yadava, T. P. Performance of White Rust (*Albugo candida*) Resistant Genotypes Developed from Interspecific Crosses of *B. Juncea x B. Carinata. Cruciferae Newslett.* **1988,** *13*, 110–111.
38. Kumar, S.; Saharan, G. S.; Singh, D. Inheritance of Resistance in Inter and Intraspecific Crosses of *Brassica Juncea* and *Brassica Carinata* to *Albugo Candida* and *Erysiphe Cruciferarum. J. Mycol. Plant Pathol.* **2002,** *32*, 59–63.
39. Mishra, K. K.; Kolte, S. J.; Nashaat, N. I.; Awasthi, R. P. Pathological and Biochemical Changes in *Brassica Juncea* (Mustard) Infected with *Albugo Candida* (White Rust). *Plant Pathol.* **2009,** *58*, 80–86.
40. Kole, C.; Williams, P. H.; Rimmer, S. R.; Osborn, T. C. Linkage Mapping of Genes Controlling Fesistance to White Rust (*Albugo Candida*) in *Brassica Rapa* (syn. *Campestris*) and Comparative Mapping to *Brassica Napus* and *Arabdiopsis Thaliana. Genome* **2000,** *45*, 22–27

41. Fan, Z.; Rimmer, S. R.; Stefansson, B. R.; Inheritance of Resistance to *Albugo Candida* in Rape (*Brassica napus* L.). *Can. J. Genet. Cytol.* **1983,** *25*, 420–424.
42. Panjabi-Massand, P.; Yadava, S. K.; Sharma, P.; Kaur, A.; Kumar, A.; Arumugam, N.; Sodhi, Y. S.; Mukhopadhyay, A.; Gupta, V.; Pradhan, A. K.; Pental, D. Molecular Mapping Reveals Two Independent loci Conferring Resistance to *Albugo. Theor. Appl. Genet.* **2010,** *121*, 137–145.
43. Li, C. X.; Sivasithamparam, K.; Barbetti, M. J. Complete Resistance to Leaf and Staghead Disease in Australian *Brassica Juncea* Germplasm Exposed to Infection by *Albugo Candida* (White Rust). Austr. Plant Pathol. **2009,** *38*, 63–66.
44. Singh, U. S.; Nashaat, N. I.; Doughty, K. J.; Awasthi, R. P. Altered Phenotypic Response to *Peronospora Parasitica* in *Brassica Juncea* Seedlings Following Prior Inoculation with an Avirulent Isolate of *Albugo Candida. Eur. J. Plant Pathol.* **2002,** *108*, 555–564.
45. Varshney, A.; Mohapatra, T.; Sharma, R. P. Development and Validation of CAPS and AFLP Markers for White Rust Resistance Gene in *Brassica Juncea. Theor. Appl. Genet.* **2004,** *109*, 153–159.
46. Gupta, I. J.; Sharma, B. J. Chemical Control of White Rust of Mustard. *Pesticides* **1978,** *12*, 45–46.
47. Verma, P. R.; Petrie, G. A. Effect of Fungicides on Germination of *Albugo Candida* Oospores in *Vitro* and on the Foliar Phase of the White Rust Disease. *Can. Plant Dis. Survey* **1979,** *59*, 53–59.
48. Dueck, J.; Stone, J. R. Evaluation of Fungicides for Control of *Albugo Candida* in Turnip Rape. *Can. J. Plant Sci.* **1979,** 59, 423–427.
49. Mehta, N.; Saharan, G. S.; Kaushik, C. D. Efficacy and Economics of Fungicidal Management of White Rust and Downy Mildew Complex in Mustard. *Indian J. Mycol. Plant Pathol.* **1996,** *26* (3), 243–247.
50. Berkenkamp, B. Effect of Fungicides on Staghead of Rape. *Can. J. Plant Sci.* **1980,** *60*, 1039–1040.
51. Meena, R. L.; Meena, P. D.; Chattopadhyay, C. Potential for biocontrol of white rust of Indian mustard (*Brassica juncea*). *Indian J. Plant Protection* **2003,** *31*, 120–123.
52. Li, C. X.; Sivasithamparam, K.; Walton, G.; Salisbury, P.; Burton, W.; Banga, S. S.; Banga, S.; Chattopadhyay, C.; Kumar, A.; Singh, R.; Singh, D.; Agnohotri, A.; Liu, S. Y.; Li, Y. C.; Fu, T. D.; Wang, Y. F.; Barbetti, M. J. Expression and Relationships of Resistance to White Rust (*Albugo candida*) at Cotyledonary, Seedling, and Flowering Stages in *Brassica juncea* Germplasm from Australia, China, and India. *Austr. J. Agric. Res.* **2007a,** *58*, 259–264.
53. Li, C. X.; Sivasithamparam, K.; Walton, G.; Fels, P.; Barbetti, M. J. Both Incidence and Severity of White Rust Disease Reflect Host Resistance in *Brassica juncea* Germplasm from Australia, China and India. *Field Crops Res.* **2008a,** *106*, 1–8.
54. Kolte, S. J.; Awasthi, R. P.; Vishwanath. Assessment of Yield Losses Due to Alternaria Blight in Rapeseed and Mustard. *Indian Phytopathol.* **1987,** *40*, 209–211.
55. Meena, P. D.; Kumar, A.; Vinod Kumar, Sikarwar, M.; Sharma, P.; Chattopadhyay, C.; Rai, P. K. Prevalence, epidemiology and forecasting of Alternaria blight of Indian mustard (*Brassica juncea*) in India. *J. Oilseed Brassica* **2020a,** *11* (1): 21–28.
56. Meena, P. D.; Awasthi, R. P.; Chattopadhyay, C.; Kolte, S. J.; Kumar, A. Alternaria Blight: A Chronic Disease in Rapeseed-Mustard. *J. Oilseed Brassica* **2010a,** *1*, 1–11.
57. Meena, P. D.; Meena, R. L.; Chattopadhyay, C.; Kumar, A. Identification of Critical Stage for Disease Development and Biocontrol of Alternaria Blight of Indian Mustard (*Brassica juncea*). *J. Phytopathol.* **2004,** *152*: 204–209.

58. Humpherson-Jones, F. M.; Maude, R. B. Studies on the Epidemiology of *Alternaria brassicicola* in *Brassica oleracea* Seed Production Crops. *Ann. Appl.Biol.* **1982,** *100,* 61–71.
59. Humpherson-Jones, F. M. The Development of Weather-Related Disease Forecasts for Vegetable Crops in the UK: Problems and Prospects. *Eur. Plant Protection Org. Bull.* **1992,** *21,* 425–429.
60. Meena, P. D.; Gupta, R.; Rani, A.; Sharma, P.; Singh, D. *Effect of Summer Temperatures on Survival of Alternaria brassicae* in Infected Indian Mustard Debris and Thermal Death Point Variations Amongst *Geographical* Isolates. *J. Oilseed Brassica* **2016,** *7* (1), 45–51.
61. Kolte, S. J.; Awasthi, R. P.; Vishwanath. Divya Mustard: A Useful Source to Create Alternaria Black Spot Tolerant Dwarf Varieties of Oilseed Brassicas. *Plant Variety Seeds* **2000,** *13,* 107–111.
62. Skoropad, W. P.; Tewari, J. P. Field Evaluation of the Epicuticular Wax in Rapeseed and Mustard in Resistance to *Alternaria brassicae. Can. J. Plant Sci.* **1977,** *57,* 1001–1003.
63. Meena, P. D.; Sujith Kumar, M. S.; Meena, H. S.; Jambhulkar, S.; Gohartaj, Pathak, D.; Srivastava, S.; Gupta, R.; Singh, D.; Gurung, B.; Rai, P. K. Confirmation of Induced Tolerance to Alternaria Blight Disease in Indian Mustard (*Brassica juncea* L.). *Appl. Biochem. Biotechnol.* **2020b,** *192*(3), 965–978.
64. Kolte, S. J.; Nashaat, N. I.; Kumar, A.; Awasthi, R. P.; Chauhan, J. S. Indo-UK Collaboration on Oilseeds: Towards Improving the Genetic Base of Rapeseed-Mustard in India. *Indian J. Plant Genet. Resour.* **2006,** *19,* 346–351.
65. Dueck, J.; Degenhardt, K. Effect of Leaf Age and Inoculum Concentration on Reaction of Oilseed *Brassica* spp. to *Alternaria brassicae. Proc. Am. Phytopathol. Soc.* **1975,** *2,* 59.
66. Brun, H.; Plessis, J.; Renard, M. Resistance of Some Crucifers to *Alternaria brassicae* (Berk.) Sacc. In *Proc. 7th International Rapeseed Congress,* Poznan, Poland, **1987;** pp 1222–1227.
67. Hansen, L. H.; Earle, E. D. Somatic Hybrids Between *Brassica oleracea* (L.) and *Sinapis alba* (L.) with Resistance to *Alternaria brassicae* (Berk.) Sacc. *Theor. Appl. Genet.* **1997,** *94,* 1078–1085.
68. Tewari, J. P. Current Understanding of Resistance to *Alternaria brassicae* in Crucifers. In *Proc. GCIRC 8th Intl. Rapeseed Congress,* 9–11 July 1991, Saskatoon, Canada, **1991,** *2,* 471–476.
69. Tewari, J. P.; Conn, K. L. Reaction of some wild crucifers to *Alternaria brassicae. Bull. OILB SROP,* **1993,** *16,* 53–58.
70. Westman, A. L.; Dickson, M. H. Disease Reaction to *Alternaria brassicicola* and *Xanthomonas campestris* pv. *campestris* in *Brassica nigra* and Other Weedy Crucifers. *Cruciferae Newslett.* **1998,** *20,* 87–88.
71. Sharma G.; Kumar V. D.; Haque, A.; Bhat, S. R.; Prakash, S.; Chopra, V. L. *Brassica* Coenospecies: A Rich Reservoir for Genetic Resistance to Leaf Spot Caused by *Alternaria brassicae. Euphytica* **2002,** *125,* 411–419.
72. Gupta, S. K.; Gupta, P. P.; Yadava, T. P.; Kaushik, C. D. Metabolic Changes in Mustard Due to Alternaria Leaf Blight. *Indian Phytopathol,* **1990,** 43, 64–69.
73. Singh, D. N.; Singh, N. K.; Srivastava, S. Biochemical and Morphological Characters in Relation to Alternaria Blight Resistance in Rapeseed-Mustard. *Ann. Agric. Res.* **1999,** *20,* 472–477.

74. Saharan, G. S. Disease Resistance. In *Breeding Oilseed Brassicas*; Labana, K. S., Banga, S. S., Banga, S. K., Eds.; Narosa Publ. House: New Delhi, India, 1992; pp 181–205.
75. Conn, K. L.; Tewari, J. P.; Hadziyev, D. The role of epicuticular wax in canola in resistance to *Alternaria brassicae.* Phytopathology.**1984,** *74*: 851.
76. Tewari, J. P. Subcuticular Growth of *Alternaria brassicae* in Rapeseed. *Can. J. Bot.* **1986,** *64*, 1227–1231.
77. Saharan, G. S.; Kadian, A. K. Analysis of Components of Horizontal Resistance in Rapeseed and Mustard Cultivars Against *Alternaria brassicae. Indian Phytopatholol.* **1983,** *36*, 503–507.
78. Krishnia, S. K.; Saharan, G. S.; Singh, D. Genetic Variation for Multiple Disease Resistance in the Families of Interspecific Cross of *Brassica juncea x Brassica carinata. Cruciferae Newslett.* **2000,** *22*, 51–53.
79. Zhang, F. L.; Xu, J.; Yan, H.; Li, M. Y.; Zhang, F. L.; Xu, J. B.; Yan, H.; Li, M. Y. A Study on Inheritance of Resistance to Black Leaf Spot in Seedlings of Chinese Cabbage. *Acta Agriculturae Boreali Sinica* **1997,** *12*, 115–119.
80. Kumar, B.; Kolte, S. J. Progression of Alternaria Blight of Mustard in Relation to Components of resistance. *Indian Phytopathol.* **2001,** *54*, 329–331.
81. Kant, L.; Gulati, S. C. Inheritance of Components of Horizontal Resistance to *Alternaria brassicae* (Berk.) Sacc. in Indian Mustard, *Brassica juncea* (L.) Czern & Coss. *J. Oilseeds Res.* **2002,** *19*, 17–21.
82. Meena, P. D.; Chattopadhyay, C.; Meena, S. S.; Kumar, A. **2011a,** *44*, 684–693.
83. Conn, K. L.; Tewari, J. P.; Dahiya, J. S. Resistance to *Alternaria brassicae* and Phytoalexin-Elicitation in Rapeseed and Other Crucifers. *Plant Sci.* **1988,** *55*, 21–25.
84. Dzurilla, M.; Kutschy, P.; Tewari, J. P.; Ruzinsky, M.; Senvicky, S.; Kovacik, V. Synthesis and Antifungal Activity of New Indolypthiazinone Derivatives. *Collection Czechoslovakia Chemists Community* **1998,** *63*, 94–102.
85. Pedras, M. S.; Smith, K. C. Sinalexin, a Phytoalexin from White Mustard Elicited by Destruxin B and *Alternaria brassicae. Phytochemistry* **1997,** *46*, 833–837.
86. Soledade, M.; Pedras, C.; Khallaf, I. Molecular Interactions of the Phytotoxins Destruxin B and Sirodesmin PL with Crucifers and Cereals: Metabolism and Elicitation of Plant Defenses. *Phytochemistry* **2012,** *77*, 129–139.
87. Vishwanath, Kolte, S. J.; Singh, M. P.; Awasthi, R. P. Induction of Resistance in Mustard (*Brassica juncea*) Against Alternaria Black Spot with an Avirulent *Alternaria brassicae* Isolate-D. *Eur. J. Plant Pathol.* **1999,** *105*, 217–220.
88. Sigareva, M.; Ping R. J.; Earle, E. D. Introgression of Resistance to *Alternaria brassicicola* from *Sinapis alba* to *Brassica oleracea* Somatic Hybridization and Backcrosses. *Cruciferae Newslett.* **1999,** *21*, 135–136.
89. Sigareva, M. A.; Earle, E. D. Camalexin Induction in Intertribal Somatic Hybrids Between *Camelina sativa* and Rapid-Cycling *Brassica oleracea. Theor. Appl. Genet.* **1999,** *98*, 164–170.
90. Mora, A. A.; Earle, E. D. Resistance to *Alternaria brassicicola* in Transgenic Broccoli Expressing a *Trichoderma harzianum* Endochitinase Gene. *Mol. Breed.* **2001,** *8*, 1–9.
91. Shivanna, K. R.; Sawhney, V. K. Pollen Selection for *Alternaria* Resistance in Oilseed Brassicas: Responses of Pollen Grains and Leaves to a Toxin of *A. brassicae. Theor. Appl. Genet.* **1993,** *86*, 339–344.
92. Kanrar, S.; Venkateswari, J. C.; Kirti, P. B.; Chopra, V. L. Transgenic Expression of Hevein, the Rubber Tree Lectin, in Indian Mustard Confers Protection Against *Alternaria brassicae. Plant Sci.* **2002,** *162*, 441–448.

93. Cramer, R. A.; Lawrence, C. B. Identification of *Alternaria brassicicola* Genes Expressed in Planta During Pathogenesis of *Arabidopsis thaliana. Fungal Genet. Biol.* **2004,** *41*, 115–128.
94. Cox, T.; Souche, J. L.; Grapel, H. The Control of *Sclerotinia, Alternaria* and *Botrytis* on Oilseed Rape with Spray Treatments of Flowable Formulation of Iprodione. *Abstr. 6th Int. Rapeseed Congr*, Paris, France, 17–19 May **1983;** pp 928–933.
95. Meena, P. D.; Meena, R. L.; Chattopadhyay, C. Eco-Friendly Options for Management of Alternaria Blight in Indian Mustard (*Brassica juncea*). *Indian Phytopathol.* **2008,** *61*, 65–69.
96. Meena, P. D.; Chattopadhyay, C.; Kumar, A.; Awasthi, R. P.; Singh, R.; Kaur, S.; Thomas, L.; Goyal, P.; Chand, P. Comparative Study on the Effect of Chemicals on Alternaria Blight in Indian Mustard-a Multi-Location Study in India. *J. Environ. Biol.* **2011b,** *32*, 375–379.
97. Sharma, S. R.; Kolte, S. J. Effect of Soil-Applied NPK Fertilizers on Severity of Black Spot Disease (*Alternaria brassicae*) and Yield of Oilseed Rape. *Plant and Soil.* **1994,** *167*, 313–320.
98. Vasilescu, B.; Avenot, H.; Bataille-Simoneau, N.; Laurent, E.; Guenard, M.; Simoneau, P. In Vitro Fungicide Sensitivity of *Alternaria* Species Pathogenic to Crucifers and Identification of *Alternaria brassicicola* Field Isolates Highly Resistant to Both Dicarboximides and phenylpyrroles. *Crop Protection.* **2004,** *23*, 481–488.
99. McCartney, H. A.; Lacey, M. E. Timing and Infection of Sunflowers by *Sclerotinia sclerotiorum* and Disease Development. *Aspect Appl. Biol.* **1999,** *56*, 151–156.
100. Hind, T. L.; Ash, G. J.; Murray, G. M. Prevalence of Sclerotinia Stem Rot of Canola in New South Wales. *Austr. J. Exp. Agric.* **2003,** *43*, 163–168.
101. Koch, S.; Dunker, S.; Kleinhenz, B.; Rohrig, M.; Tiedemann, A. A Crop Loss-Related Forecasting Model for Sclerotinia Stem Rot in Winter Oilseed Rape. *Phytopathology* **2007,** *97*, 1186–1194.
102. Malvarez, M.; Carbone, I.; Grunwald, N. J.; Subbarao, K. V.; Schafer, M.; Kohn, L. M. New Populations of *Sclerotinia sclerotiorum* from Lettuce in California and Peas and Lentils in Washington. *Phytopathology* **2007,** *97*, 470–483.
103. Singh, R.; Singh, D.; Li, H.; Sivasithamparam, S.; Yadav, N. R.; Salisbury, P.; Barbetti, M. J. Management of Sclerotinia Rot of Oilseed Brassica—A Focus on India. *Brassica* **2008,** *10*, 1–27.
104. Shaw, F. J. F.; Ajrekar, S. L. The Genus *Rhizoctonia* in India. Mem. Dep. Agric. *India Bot. Scr.* **1915,** *7*, 177.
105. Chattopadhyay, C.; Meena, P. D.; Kalpana Sastry, R.; Meena, R. L. Relationship Among Pathological and Agronomic Attributes for Soilborne Diseases of Three Oilseed Crops. *Indian J. Plant Protection* **2003,** *31*, 127–128.
106. Boland, G. J.; Hall, R. Index of Plant Hosts of *Sclerotinia sclerotiorum. Can. J. Plant Pathol.* **1994,** *16*, 94–108.
107. Roy, A. K.; Saikia, U. N. White Blight of Mustard and Its Control. *Indian J. Agric. Sci.* **1976,** *46*, 274–277.
108. Rai, J. N.; Dhawan, S. Studies on Purification and Identification of Toxic Metabolite Produced by *Sclerotinia sclerotiorum* Causing White Rot Disease of Crucifers. *Indian Phytopathol.* **1976,** *29*, 407–411.
109. Saharan, G. S.; Mehta, N. *Sclerotinia Diseases of Crop Plants: Biology, Ecology and Disease Management*; Springer Verlag: Singapore, **2008;** p 485.

110. Hims, M. J. Damping-off of *Brassica napus* (Mustard and Cress) by *Sclerotinia sclerotiorum*. Plant, **1979a,** *28*, 201–202.
111. Berkenkamp, B.; Vaartnou, H. Fungi Associated with Rape Root Rot in Alberta. *Can. J. Plant Sci.* **1972,** *52*, 973–976.
112. Hims, M. J. Wild Plants as a Source of *Sclerotinia sclerotiorum* Infecting Oilseed Rape. *Plant* **1979b,** *28*, 197–198.
113. Williams, J. R.; Stelfox, D. Occurrence of Ascospores of *Sclerotinia sclerotiorum* in areas of Alberta. *Can. Plant Dis. Survey* **1980,** *60*, 51–53.
114. Morrall, R. A. A.; Dueck, J. Sclerotiniastem Rot of Spring Rapeseed in Western Canada. Abstr. *6th Int. Rapeseed Congr.* Paris, France, 17–19 May 1983, **1983;** p 181.
115. Couper, G.; Litterick, A.; Leifert, C. Control of *Sclerotinia* Within Carrot Crops in NE Scotland: the Effects of Irrigation and Compost Application on Sclerotia Germination. Retrieved from http://www.abdn.ac.uk/organic/pdfs/sclerotia.pdf, **2001**.
116. Iwata, I.; Igita, K. On the Growth Characteristics of Direct Sowing Rape on Upland Field. *Bull. Kyushu Agric. Exp. Station* **1972,** *16*, 207.
117. Li, Y. C.; Chen, J.; Bennett, R.; Kiddle, G.; Wallsgrove, R.; Huang, Y. J.; He, Y. H. Breeding, Inheritance and Biochemical Studies on *Brassica napus* cv. Zhougyou 821: Tolerance to *Sclerotinia sclerotiorum* (Stem Rot). In: *Proc. 10th Intl. Rapeseed Congr.; Canberra, Australia*; Wratten, M., Salisbury, P. A., Eds.; **1999;** 61 pp.
118. Wang, H. Z.; Liu, G. H.; Zheng, Y. B.; Wang, X. F.; Yang, Q. Breeding of *Brassica napus* Cultivar Zhongshuang No. 9 with Resistance to *Sclerotinia sclerotiorum* and Dynamics of Its Important Defence Enzyme Activity. In: *Proc. 11th Intl. Rapeseed Congr. Copenhagen, Denmark*, **2003,** 43 pp.
119. Li, C. X.; Li, H.; Siddique, A. B.; Sivasithamparam, K.; Salisbury, P.; Banga, S. S.; Banga, S.; Chattopadhyay, C.; Kumar, A.; Singh, R.; Singh, D.; Agnihotri, A.; Liu, S. Y.; Li, Y. C.; Tu, J.; Fu, T. D.; Wang, Y.; Barbetti, M. J. The importance of the Type and Time of Inoculation and Assessment in the Determination of Resistance in *Brassica napus* and *B. juncea* to *Sclerotinia sclerotiorum*. *Austr. J. Agric. Res.* **2007b,** *58*, 1198–1203.
120. Li, C. X.; Liu, S. Y.; Sivasithamparam, K.; Barbetti, M. J. New Sources of Resistance to Sclerotinia Stem Rot Caused by *Sclerotinia sclerotiorum* in Chinese and Australian *Brassica napus* and *Brassica juncea* Germplasm Screened Under Western Australian Conditions. *Austr. Plant Pathol.* **2008b,** *38*, 149–152.
121. Garg, H.; Atri, C.; Sandhu, P. S.; Kaur, B.; Renton, M.; Banga, S. K.; Singh, H.; Singh, C.; Barbetti, M. J.; Banga, S. S. High Level of Resistance to *Sclerotinia sclerotiorum* in Introgression Lines Derived from Hybridization Between Wild Crucifers and the Crop *Brassica* Species *B. napus* and *B. juncea*. *Field Crops Res.* **2010,** *117*, 51–58.
122. Goyal, P.; Chahar, M.; Barbetti, M.; Liu, S. Y.; Chattopadhyay, C. Resistance to Sclerotinia Rot Caused by *Sclerotinia sclerotiorum* in *Brassica juncea* and *B. napus* Germplasm. *Indian J. Plant Protection* **2011b,** *39*, 60–64.
123. Wang, Z.; Tan, X.; Zhang, Z.; Gu, S.; Li, G.; Shi, H. Defense to *Sclerotinia sclerotiorum* in Oilseed Rape Is Associated with the Sequential Activations of Salicylic Acid Signalling and Jasmonic Acid Signalling. *Plant Sci.* **2012,** *184*, 75–82.
124. Hara, K.; Yanagita, Y. Using Calcium Cyanamide to Control Rape Sclerotinia Rot (Japanese). *Proc. Kyushu Assoc. Plant* Protection **1967,** *13*, 13.
125. Kruger, W. Control Measures for *Sclerotinia sclerotiorum* in Rape. *J. Phytopathol.* **1973,** *77*, 125–137.

126. Richardson, L. T.; Bond, E. J. Methyl Bromide Fumigation of Rapeseed Infested with Sclerotia of *Sclerotinia sclerotiorum. Can. J. Plant Sci.* **1978,** *58*, 267–268.
127. Singh, R.; Tripathi, N. N.; Kaushik, C. D.; Singh, R. Management of Sclerotinia Rot of Indian Mustard [*Brassica juncea* (L.) Czern and Coss.] by fungicides. *Crop Res. Hisar*, **1994,** *7*, 275–281.
128. Sharma, P.; Rai, P. K.; Meena, P. D.; Kumar, S.; Siddiqui, S. A. Relation of Petal Infestation to Incidence of *Sclerotinia sclerotiorum* in *Brassica juncea*. In Abstracts, Natl. Conf. Recent Advances Integrated Dis. Management Enhancing Food Prodn.; S.K.R.A.U.; Bikaner, 27–28 October, **2010;** p 76.
129. Penaud, A.; Huguet, B.; Wilson, V.; Leroux, P. Fungicides resistance of *Sclerotinia sclerotiorum* in French Oilseed Rape Crops. In: Abstracts, 11th International Rapeseed Congress, 06–10 July **2003,** Copenhagen, Denmark, 2003; pp 156.
130. Chattopadhyay, C.; Meena, P. D.; Kumar, S. Management of Sclerotinia Rot of Indian Mustard Using Ecofriendly Strategies. *J. Mycol. Plant Pathol.* **2002,** *32*, 194–200.
131. Meena, P. D.; Sharma, A. K.; Jha, S. K.; Chattopadhyay, C. Impact of Fungal Diseases on Yield of Indian mustard—Farmers' Perception and On-Farm Success of Garlic Clove Extract in Management of Sclerotinia Rot. *Indian J. Plant Protection* **2006,** *34*, 229–232.
132. Chattopadhyay, C.; Meena, P. D.; Meena, R. L. Integrated Management of Sclerotinia Rot of Indian Mustard. *Indian J. Plant Protection* **2004,** *32*, 88–92.
133. Chattopadhyay, C.; Vijay Kumar, R.; Meena, P. D. Biomanagement of Sclerotinia Rot of *Brassica juncea* in India—A Case Study. *Phytomorphology* **2007,** *57*, 71–83.
134. Darpoux, H. A Contribution to Study of the Diseases of Oleaginous Plants in France. *Annales Des Epiphyties* **1946,** *11*, 71–103.
135. Cook, R. J. Diseases of Oilseed Rape in Europe, Report of a Study Tour Undertaken in Germany and France. Ministry Agric.; Fisheries and Food, Agric. Develop. Service (mimeographed), **1975**.
136. Bhander, D. S.; Thakur, R. N.; Husain, A. A New Disease of Rapeseed and Mustard. *Plant Disease Reporter.* **1963,** *47*, 1039.
137. Sankhla, H. C.; Singh, H. G.; Dalela, G. G.; Mathur, R. L. Occurrence of Perithecial Stage of *Erysiphe polygoni* on *Brassica campestris* var. *sarson and B. juncea. Plant Disease Reporter*. **1967,** *51*, 800.
138. Hirata, K. *Host Range and Geographical Distribution of Powdery Mildews*; Niigata University Japan, **1966**.
139. Yarwood, C. F. Effect of Soil Moisture and Nutrient Concentrations on the Development Mildew. *Phytopathology* **1949,** *39*, 780–788.
140. Sharma, A. K. Powdery Mildew Diseases of Some Crucifers from Jammu and Kashmir State. *Indian J. Mycol. Plant Pathol.* **1979,** *9*, 29–32.
141. Meena, P. D.; Mehta, N.; Rai, P. K.; Saharan, G. S. Geographical Distribution of Rapeseed-Mustard Powdery Mildew Disease in India. *J. Mycol. Plant Pathol.* **2018,** *48*, 60–79.
142. Saharan, G. S.; Kaushik, J. C. Occurrence and Epidemiology of Powdery Mildew of *Brassica. Indian Phytopathol.* **1981,** *34*: 54–57.
143. Saharan, G. S.; Mehta, N.; Meena, P. D. *Powdery Mildew Disease of Crucifers: Biology, Ecology and Disease Management*; Springer Verlag: Singapore, **2019;** p 362.
144. Narain, A.; Siddiqui, J. A. Field Reaction of Species of *Brassica* to *Erysiphe polygoni. Oilseed J.* **1965,** 9, 153.
145. Singh, R. R.; Solanki, J. S. Fungicidal Control of Powdery Mildew of *Brassica juncea. Indian J. Mycol. Plant Pathol.* **1974,** *4*, 210.

146. Woronin, M. *Plasmodiophora brassicae.* Urheberder der Kohlpflanzen-Hemie. Jahrb. Wiss Bot.**1878,** *11*, 548–574.
147. Stout, G. L.; Altstatt, G. E.; Nakayama, R. M. Club root of *Crucifers.* Bull. Dept. Agriculture **1954,** *42*, 113.
148. Chattopadhyay, S. B.; Sengupta, S. K. Addition to Fungi of Bengal. *Bull. Bot. Soc. Bengal.* **1952,** *6*, 57–61.
149. Rajappan, K.; Ramaraj, B.; Natarajan, S. Knol-khol—A New Host for Club-Root Disease in the Nilgiris. *Indian Phytopatholol.* **1999,** *52*, 328.
150. Chattopadhyay, A. K. Studies on the Control of Clubroot Disease of Rapeseed-Mustard in West Bengal. *Indian Phytopathol.* **1991,** *44*, 397–398.
151. Lobb, W. R. Resistant Type of Rape for Areas with Club Root. *New Zealand J. Agric.* **1951,** *82*, 65–66.
152. Haenseler, C. M.; Moyer, T. R. Effect of Calcium Cyanamide on the Soil Microflora with Special Reference to Certain Plant Parasites. *Soil Sci.* **1937,** *43*, 133–152.
153. Chattopadhyay, A. K.; Moitra, A. K.; Bhunia, C. K. Evaluation of *Brassica* Species for Resistance to *Plasmodiophora brassicae* Causing Club Root of Rapeseed Mustard. *Indian Phytopathol.* **2001,** *54*, 131–132.
154. Rai, J. N.; Singh, R. P. Fusarial wilt of *Brassica juncea. Indian Phytopathol.* **1973,** *26*, 225–232.
155. Kanaujia, R. S.; Kishore, R. A New Wilt Disease of *Brassica nigra* Caused by *Fusarium* f. sp. *conglutinans. Indian Phytopathol.* **1981,** *34*, 84.
156. Armstrong, G. M.; Armstrong, J. K. Wilt of *Brassica carinata, Crambe abyssinica* and *C. hispanica* Caused by *Fusarium oxysporum* f. sp. *conglutinans* Race 1 or 2. *Plant Dis. Reporter* **1974,** *58*, 479–480.
157. Enya, J.; Togawa, M.; Takeuchi, T.; Yoshida, S.; Tsushima, S.; Arie, T.; Sakai, T. Biological and Phylogenetic Characterization of *Fusarium oxysporum* Complex, Which Causes Yellows on *Brassica* spp.; and Proposal of *F. oxysporum* f. sp. *rapae*, a Novel Forma Specialis Pathogenic on *B. rapa* in Japan. *Phytopathology* **2008,** *98*, 475–483.
158. Bhowmik, T. P.; Trivedi, B. M. A New Bacterial Stalk Rot of *Brassica. Curr. Sci.* **1980,** *49*, 674–675.
159. Meena, P. D.; Mondal, K.; Sharma, A. K.; Chattopadhyay, C.; Kumar, A. Bacterial Rot: A New Threat for Rapeseed-Mustard Production System in India. *J. Oilseed Brassica* **2010b,** *1* (1), 39–41.
160. Desai, A. G.; Chattopadhyay, C.; Agrawal, R.; Kumar, A.; Meena, R. L.; Meena, P. D.; Sharma, K. C.; Rao, M. S.; Prasad, Y. G.; Ramakrishna, Y. S. *Brassica juncea* Powdery Mildew Epidemiology and Weather Based Forecasting Models for India-a Case Study. *J. Plant Dis. Protection* (Zeitschrift für Pflanzenkrankheiten und Pflanzenschutz). **2004,** *111*, 429–438.
161. Chattopadhyay, C.; Agrawal, R.; Kumar, A.; Bhar, L. M.; Meena, P. D.; Meena, R. L.; Khan, S. A.; Chattopadhyay, A. K.; Awasthi, R. P.; Singh, S. N.; Chakravarthy, N. V. K.; Kumar, A.; Singh, R. B.; Bhunia, C. K. Epidemiology and Forecasting of Alternaria Blight of Oilseed *Brassica* in India—A Case Study. *Zeitschrift für Pflanzenkrankheiten und Pflanzenschutz* **2005,** *112*, 351–365.
162. Chattopadhyay, C.; Agrawal, R.; Kumar, A.; Meena, R. L.; Faujdar, K.; Chakravarthy, N. V. K.; Kumar, A.; Goyal, P.; Meena, P. D.; Shekhar, C. Epidemiology and Development of Forecasting Models for White Rust of *Brassica juncea* in India. *Arch. Phytopathol. Plant Protection* **2011,** *44*, 751–763.

163. Fitt, B. D. L.; Brun, H.; Barbetti, M. J.; Rimmer, S. R. World-Wide Importance of Phoma Stem Canker (*Leptosphaeria maculans* and *L. biglobosa*) on Oilseed Rape (*Brassica napus*). *Eur. J. Plant Pathol.* **2006,** *114*, 3–15.
164. Boys, E. F.; Roques, S. E.; West, J. S.; Werner, C. P.; King, G. J.; Dyer, P. S.; Fitt, B. D. L. Effects of R Gene-Mediated Resistance in *Brassica napus* (Oilseed Rape) on Asexual and Sexual Sporulation of *Pyrenopeziza brassicae* (Light Leaf Spot). *Plant Pathol.* **2011.** DOI: 10.1111/j.1365-3059.2011.02529.x
165. Pain, A.; Dhar, A. K.; Chattopadhyay, C. Journey of Phytopathogenic Fungi from Genetics to Genomics. In *Current Advances in Molecular Mycology*; Gherbawy, Y., Mach, R. L., Rai, M., Eds.; Nova Sci. Publ.: USA, 2008; pp 39–62.
166. Seppanen, L.; Juha Helenius. Do Inspection Practices in Organic Agriculture Serve Organic Values? A Case Study from Finland. *Agric. Human Values* **2004,** *21,* 1–13.

CHAPTER 14

Integrated Pest Management in Soybean-Based Cropping Systems

KUMARI SURBHI[1*], K. P. SINGH[1], T. ARAVIND[1], PARUL DOBHAL[2], POOJA BHATT[1], and HIMANI JEENA[1]

[1]*Department of Plant Pathology, College of Agriculture, Govind Ballabh Pant University of Agriculture and Technology, Pantnagar, Uttarakhand, India*

[2]*Department of Entomology, College of Agriculture, Govind Ballabh Pant University of Agriculture and Technology, Pantnagar, Uttarakhand, India*

[*]*Corresponding author. E-mail: surbhi.gbpuat2018@gmail.com*

ABSTRACT

Soybean is an important oilseed crop cultivated all over the world for its nutritional and nutraceutical benefit. Soybean being a leguminous crop is an integral part of different cropping systems. The full potential of production in soybean cropping systems is hindered by a wide range of insect pests and diseases. The challenge is to mitigate the damages and maximize the economic returns from the crop while minimizing the effect of plant protection chemicals and other products on the environment. Integrated pest management program not only help to manage pest problems in an eco-friendly, sustainable manner but it also increase the overall productivity of the cropping system by maintaining ecological balance between the microbial communities in rhizoshpere of the crop and natural enemy population in the field.

Integrated Pest Management in Diverse Cropping Systems. Devendra Pal Singh, PhD (Ed.)

14.1 INTRODUCTION

Food, nutritional security for the ever increasing population, is a major concern facing the world. For the purpose of fulfilling the nutritional demands, it is essential to grow healthy food to provide proper source of carbohydrate, protein, fat, vitamins, and minerals in sufficient amount. One such agricultural crop having immense nutritional value as an oilseed, source of protein is soybean (*Glycine max* L.). Seeds of soybean contain around 40% protein, making it an important source of protein in human diet. An oil content of about 18–22% in the seeds contributes an enriched source of unsaturated fatty acids.[1] The crop contributes about 25% to global vegetable oil production every year.[2] Apart from the nutritional qualities soybean also possess several nutraceutical benefits due to the presence of certain phytochemicals such as isoflavones. Soybean is cultivated throughout the world in areas of favorable agro-climatic conditions in different countries including USA, Brazil, Argentina, China, India, Paraguay, Canada, and Mexico.[3] During 2019–2020, an estimated 127.28 mha area was under soybean production in the world with an estimated production of 368.47 million metric tons and productivity of 2.89 t/ha.[4] In India, the total area under soybean production during 2020 is estimated around 11.3 mha with an estimated production of 13.2 million metric tons and productivity of 1.17 t/ha.[5] The major states in India involved in soybean cultivation include Madhya Pradesh, Maharashtra, Rajasthan, Uttar Pradesh, and Uttarakhand.

Soybean being a leguminous crop is an integral part of different cropping systems. Soybean-based cropping systems help in maintaining the ecological balance and fertility of soil. The major goal of sustainable crop production is to obtain maximum agricultural produce with minimum exploitation of available natural resources. Soybean aids in replenishing the nitrogen content of soil by fixing atmospheric nitrogen, making it available for the succeeding crop. Cultivation of cereal crops such as maize or wheat in rotation with soybean provides higher yield, greater economic returns, and also helps to rejuvenate the lost nutrients in the soil. Such cropping systems require lesser amount of fertilizer dose; thus cut down the cost of production. Growing nonleguminous crops in rotation or as inter-crop with soybean also help in reducing the attack of insect pests and diseases. However, the realization of full potential of production in soybean cropping systems is held back owing to the incidence of several important insect pests such as *Etiella zinkienella, Tetranychus urticae, Thrips tabaci, Spodoptera exigua, Helicoverpa armigera, Spodoptera litura, Spilarctia oblique*, and *Diachrysia orichalcea*.[6] Besides the insect pests, the crop is also infected by a number of fungal, bacterial, viral diseases including anthracnose, pod blight, aerial blight, root, stem rot, bacterial blight, bacterial pustules, yellow mosaic, and soybean mosaic.[7] Together these pests and diseases make the crop unproductive leading to huge yield losses every year.

The challenge is to mitigate the effect of these insect pests, diseases to maximize the economic returns from the crop while minimizing the effect of plant protection chemicals and other products on the environment. The chemical method of management of insect pests and diseases are most popular among farmers because of their faster results as compared to any other means of management. However, due to prolong use, dependence on chemical pesticides several issues regarding environmental pollution, pesticide residue in food products, pesticide resistance, and pest resurgence are arising. Therefore, for reducing the dependence on chemical methods, sustain improved yield from the crops, an integrated pest management (IPM) approach needs to be adopted. Integrated pest management includes all the suitable methods, namely, cultural, biological, chemical measures, in a compatible manner, keeping in view the environmental sustainability and population dynamics of the associated pest species. The concept of IPM is based on economic injury level (EIL) which is defined as the lowest population of insect pest that is capable of causing economic damage to the crop. However, management action is initiated at economic threshold level (ETL) so that the increasing pest population does not reach EIL. The application of insecticides is justified only when the ETL for the specific pest species has been reached. Thus, IPM furnishes a cost-effective, environmentally safe, and socially acceptable method of pest management in crop production.[8]

14.2 COMPONENTS OF INTEGRATED PEST MANAGEMENT

The IPM is "applied pest control which combines and integrates biological and chemical control." The concept of IPM was developed to deal with problems such as indiscriminate use of pesticides, imbalance of natural enemy population in agricultural fields, and pesticide resistance. IPM focuses on devising strategies for managing multiple pests of the crop simultaneously. It also includes keeping a record of population, time of emergence of natural enemies, and antagonists in the field. Decision for the application of pesticides is to be taken on the basis of ETL of the pest species. Ultimately, IPM provides an environmentally sound and economic method of pest management. The major principles comprising the IPM program are given in Annex III of Framework Directive (2009/128/EC) of European Union are discussed under the following heads.[9]

A. **Prevention:** The initial step in any management program is to prevent the entry of harmful organisms or infection by pathogenic organisms by means of cultural control practices. These cultural practices are started right from the time of sowing which includes manipulation in the date of sowing, planting density, tillage practices, zero tillage, destruction of crop residues, intercropping, and crop rotation with nonhost crops. Each one of these practices has beneficial effects in terms of preventing attack of pests and diseases. Besides these, selection of resistant varieties of the crop is also very important for preventing the pests. Certified seeds and planting material should be used for sowing. Proper agronomic practices such as recommended dose of fertilizers and timely irrigation should be supplied for proper growth of the crop which would otherwise serve to attract the pest population. Interculture operations should be performed with care and clean instruments so as to prevent dissemination of bacterial and viral diseases of the crop.
B. **Monitoring of Pest Population and Decision-Making:** There are several methods of monitoring the pest population, incidence of disease in the field which should be used to take an estimate, forecast the occurrence or prevalence of a pest problem in time to formulate management practices for it. The estimation of pest population gives an idea of the action threshold at which management practices should be started or pesticides should be applied. Here, use of expert systems such as forecasting models is essential in preventing losses. The ultimate decision is taken based on the ETL of the pest which is specific to the pest species.
C. **Focus on Alternative Methods of Pest Control:** The very core of IPM is to minimize the dependence on chemical means of management and devising alternatives which maintain biological equilibrium in the agricultural ecosystem. The use of biological control practices, conservation of natural enemy population in the field, physical, mechanical methods of attracting, and killing the pests, provide alternative and additional benefits in management of pest population with minimum harm to the environment.
D. **Selection of Suitable Pesticide and Reduction of Pesticide Use:** The chemical pesticides provide faster management and thus are

quite popular among the farmers. But for incorporation of a pesticide in IPM program, the effects of the chemical on environment and human health should be well thought through. The pesticide being applied should be target specific, least harmful to animal and human health as well as least toxic to the environment. Beside these, major focus should be given to reducing the dose, frequency of pesticide application through the use of pest monitoring, forecasting techniques, applying the chemicals only as and when required.

E. **Strategies for Dealing with Pesticide Resistance:** For reducing the risk of resistance development in pests against a class of chemical pesticides and combination products comprising of different chemicals with different mode of action should be used.
F. **Evaluation of Results:** Regular assessment of the results obtained from various management practices and techniques used is essential for taking continued benefits from pest management programs.

14.3 CULTURAL CONTROL OF DISEASES AND PESTS IN SOYBEAN-BASED CROPPING SYSTEMS

14.3.1 CROP ROTATION

Rotation of crops is an effective way of disrupting the life cycle of insect pests and pathogens. When different crops are planted in successive years and the resting stage of the organism is unable to survive in the absence of suitable plant host and thus reducing the inoculum of pathogen in the field. In case of organic agriculture and crop rotation is an essential strategy of managing the insect pests without application of chemicals. Moreover, some of the soil-borne pathogens are difficult to control and could only be economically managed with the help of crop rotation strategies. The rotation of leguminous crops such as soybean with cereal crops improves texture, structure of soil, improves water percolation, soil aeration, increases the water use efficiency, and nutrient uptake by the crop.[10] The rotation of maize crop with soybean has been found to be effective in reducing the infestation by maize stem borer and maize rootworm.

In an experiment conducted by Grabau and Chen[11] and found that the rotation of corn with soybean increased yield of both the crops. Besides this, a long term rotation was tested for nematode infestation with first 5 years of corn in monoculture then 5 years of soybean monoculture and then again planting corn in monoculture. To minimize the effect of nematodes under study and application of granular nematicides was also done to half of each plot. It was reported that population of *Heterodera glycines* and the soybean cyst nematode decreased due to monoculture of corn while Xiphinema, the dagger nematode was found in smaller densities. The population of *Pratylenchus* and *Helicotylenchus* was found to increase as a result of corn monoculture.

Rotation with nonhost crops such as wheat, maize, and alfalfa has been found effective in reducing the population of soybean cyst nematode. Soybean cyst nematode can survive on several alternate hosts including beans, cowpea, clover, and some weeds such as *Stellaria media* and *Cardamine parviflora.*[12] Thus, for proper management of nematode problem it is essential to manage weeds in the field. Due to long term survival of eggs of cyst nematode in the field and a corn-soybean crop rotation sequence of less than 2 years would be ineffective in managing the nematode.[13] Trap crops such as crimson clover (*Trifolium incarnatum*) has been found to attract and inhibit the reproduction of cyst nematode in greenhouse trials.[13,14]

The choice of crop to be taken in rotation with soybean varies with the nematode species to be managed, but other factors such as type of soil and cost of production are also important. For the management of root knot nematode (*Meloidogyne incognita*) in groundnut and *Sesame indicum* could be used in rotation with soybean. *Hoplolaimus galeatus* (lance nematode) could also be managed by rotating soybean with groundnut and tobacco or small grain crops and which serve as a poor host for the nematode.[13,15] There are few choices in case of ectoparasitic nematodes as they generally have a broad host range.

In case of seedling diseases and crop rotation has been found to be only of partial utility owing to a large host range of the seedling pathogens. Since, maize and soybean are affected by a number of similar soil-borne pathogens and their rotation is ineffective in eliminating these pathogens from the crop field.[16] However, corn-soybean-wheat long-term rotation has been found to increase diversity of less virulent *P. sojae* races in soil as compared to monoculture of soybean.[17] Similar is the case with charcoal rot pathogen and *Macrophomina phaseolina and* where there are few choices of nonhost crops for rotation with soybean. However, rotation with cotton has been reported to decrease the inoculums of *M. phaseolina* in the field. For pathogens such as *Sclerotinia sclerotiorum* which can survive by means of sclerotia for long durations of up to 8 years and it is not likely for crop rotation to prevent the incidence of disease effectively.[13] Rotation with small grain crops and tillage practices could help in prevention of sclerotial germination up to some extent. Root rot is another important disease affecting seed yield in soybean cropping systems. Maier et al. evaluated different cropping systems with respect to incidence of root rot and its effect on seed yield.[18] In all the four cropping systems evaluated, that is, soybean-oat and soybean-maize and soybean-wheat, soybean-pasture, greater than 50% incidence of major diseases such as root rot, charcoal rot, and sudden death syndrome were reported. The soybean-pasture cropping system was found to be least productive.

Sudden death syndrome caused by *Fusarium virguliforme* is one of the major yield reducing diseases of soybean. The disease could be prevented by following long-term rotation of corn and soybean. Crop rotation involving corn for 3–4 years has been reported to decrease the inoculum of the pathogen up to 5 times as compared to a 2 year corn-soybean rotation.[19] Proper weed and crop residue management is also essential in the field as the pathogen can survive in 15 different crops, weed species, and in corn residue for very long time.[13,20] Among the various cover crops, namely, cereal rye (*Secale cereal*), mustard (*Brassica juncea*), rapeseed (*Brassica napus*) tested for preventing the infection by *Rhizoctonia solani* on soybean crop, rye and rapeseed were found to improve the growth of soybean in pathogen inoculated plots. Cereal rye was also reported to enhance the suppressiveness of soil to *R. solani* and *F. virguliforme*.[21] The benefits of including rye as a cover crop in soybean cropping systems in terms of disease prevention and better crop production have also been emphasized by Da-Silva.[22]

Rotation of legume crops such as soybean with cereal crops with proper tillage operations with moldboard plough system has been reported to increase the population of beneficial arthropods in the field. Rotation of maize-soybean-wheat for 3 years has been reported to reduce the incidence of insect pests as a whole.[23,24]

14.3.2 INTERCROPPING

Intercropping refers to the practice of cultivating two or more agricultural crops on the same field simultaneously. The planting of crops without a proper row-to-row sequence is termed as mixed cropping while planting of one crop before the harvesting of other crop is known as relay intercropping. One of the most important strategies of IPM for sustaining agricultural production is maintaining a balance in biodiversity of the ecosystem. Intercropping provides an effective way of widening the diversity and productivity of an agricultural system. Along with the maintenance of ecological balance and intercropping also help in efficient utilization of natural resources and increasing farm production and safeguarding the crop from attack of insect pests, diseases, and weeds. An intercropping system is more dynamic in terms of activity of pests and their natural enemies as compared to a sole crop.

The effect of intercropping on the interspecific association of different pests of soybean including the *S. litura, Hedylepta indicate* and *D. Orichalcea* was evaluated.[6] Out of the nine intercropping systems of soybean tested, maize and pigeon pea were found to be most effective in preventing the infestation of defoliator pest (*S. litura*) and due to their phenological characteristics and insect repelling chemicals. These two intercropping systems recorded minimum damage to soybean leaves and highest seed yield. The other crops tested in intercropping system, namely, castor, chilli, foxtail millet, green gram, groundnut and pepper were found ineffective in reducing the infestation of defoliator pests in soybean.

The strip intercropping system of maize with soybean has been reported to give a substantial reduction in incidence of insect pests, weeds and diseases in both the crops. The infestation of stem borer of maize has been found to be effectively less in intercropped maize as compared to 21–43% incidence in the sole crop. The termite attack has also been reported to reduce under intercropping system.[25]

The rhizospheric interaction of different plant species in intercropping system also tend to reduce the occurrence of soil-borne diseases as seen in the case of maize-soybean intercropping. The root exudates of maize serve to attract and inhibit the motility of the zoospores of *Phytophthora sojae* and the causal organism of root rot of soybean. Maize roots were found to exude five different kinds of phenolic acids, namely, coumaric acid, cinnamic acid, ferulic acid, hydroxybenzoic acid and vanillic acid and which interfere with the infection of *Phytophthora sojae* in soybean.[26]

14.4 BIOLOGICAL CONTROL

Biological control refers to the method of suppressing the population of pests and incidence of plant diseases by using their natural enemies and antagonists. Biological control is an integral component of IPM programs. For sustaining ecological balance, the natural enemies of the pests are identified, encouraged in the field so as to maintain equilibrium between the population of pest and predator. Regular monitoring of the time of emergence of pest and the predator or parasitoid in the field is essential for proper planning of management strategies.

14.4.1 BIOLOGICAL CONTROL OF INSECT PESTS OF SOYBEAN

S. litura, S. oblique, Thysanoplusia orichalcea, Bemisia tabaci, and *Aphis gossypii* are among the major pests of soybean crop in India. These pests have been reported to emerge in the field during the period from middle of August to the first week of September. The emergence of natural enemies of the pests, namely, *Coccinella septumpunctata, Eocanthecona furcellata* coincide with the time of occurrence of pests and thus keep a check on the population of these insects in the field.[27] Antagonistic activity of three lady bird beetle species *Coccinella transversalis, C. septumpunctata,* and *Menochilus sexmaculata* on whiteflies and jassids have been reported in the soybean field. Other major predators of the insect pests of soybean include *E. furcellata* which feeds on lepidopteran larvae.[28,29] In North America, an invasive species of coccinellid beetle *Harmonia axyridis* has been reported as a predator of *Aphis glycines* (Soybean aphid) and an invasive pest of soybean in the country. *H. axyridis* has been found to be the most abundant species attacking the soybean aphid. The full potential of the coccinelid beetle as a predator in soybean agroecosystems around the world still needs to be explored.[30]

Some of the major parasitoids of soybean insect pests include *Trichogramma chilonis, Tetrastichus,* and *Telenomus,* which are egg parasitoids of *S. litura,* and *H. armigera.* Larval parasitoids include *Ichneumon promissorius, Carcelia* spp. *Diglyphus isaea* which parasitize the larvae of *Spodoptera* and *Helicoverpa.* The pupal parasitoids include *Encarsia Formosa* and *Eretmocerus* spp. which parasitize whitefly and *Lissopimpla excels* which is known to parasitize the pupa of *Helicoverpa.* Besides the predators, parasitoids and entomopathogenic organisms such as *Bacillus thuringiensis* var. *kurstaki* and SNPV have also been reported to provide efficient control of soybean pests.

14.4.2 BIOLOGICAL CONTROL OF DISEASES OF SOYBEAN

The use of antagonistic fungi and bacteria at the pre-sowing stage is important to prevent the incidence of major diseases of soybean. Application of *Trichoderma* and *Pseudomonas fluorescens*, PGPR (plant growth promoting rhizobacteria) is done in the form of seed, soil treatment before sowing and for managing the soil-borne diseases. Such antagonists not only provide protection from diseases but also improve the growth of the crop. Among the various biocontrol agents, Trichoderma is most widely used. There are several different species of *Trichoderma*, namely, *T. asperellum, T. hamatum, T. harzianum, T. viride*, and *T. virens*, which provide protection against a range of soil-borne pathogens including *Fusarium, Rhizoctonia, Phytophthora*, and *Sclerotinia.* Several essential properties of Trichoderma such as highly saprophytic ability and promotion of plant growth, production of secondary metabolites, antibiotics, active against broad range of pathogens, high ability for mass production, multiplication and makes it an excellent biocontrol agent. The plant growth promoting rhizobacteria on the other hand not only act as antagonists against plant pathogens but also help to induce systemic resistance in the plant.[31]

Two strains of *Trichoderma asperelloides* (T25, T42) have been reported to exhibit high potential for the biocontrol of white mold disease (*S. sclerotiorum*) on soybean. Methanolic and ethyl acetate extracts of these two strains of *T. asperelloides* strains were found to inhibit carpogenic germination of the fungus and also decreased the number of ascospores. The biocontrol efficiency of *T. asperelloides* srains was found to be greater than commercial strains of *Trichoderma harzianum*.[32] Some other microbial agents which reduce the incidence of sclerotinia stem rot include *Coniothyrium minitans* (Contans WG), *Sporidesmium sclerotivorum*, and *Streptomyces lydicus*. These biocontrol agents have been reported to cause up to 90% degradation of sclerotia of the fungus.[13,33]

Specific strains of several microbes have been reported to parasitize cyst nematode and inhibit the infestation of soybean root by nematodes by directly competing for space. Commercial formulation of these microbial strains may be used as seed treatment for protection against nematodes. The commercial products include *Bacillus firmus* strain 1-1582 (sold under trade name Votivo), *Pasteuria nishizawae* strain Pn1, *Bacillus amyloliquefaciens* strain PTA-4838, heat-killed *Burkholderia* spp. strain A396, and *Bacillus amyloliquefaciens* strain MBI 600. Few fungal species including *Hirsutella rhossiliensis* have been found to parasitize the cyst nematode and release certain toxic chemicals which inhibit the nematode infestation in soybean rhizosphere.[13,34]

Endophytic bacteria such as *Paenibacillus* HKA-15 and *Bacillus* HKA-121 have been reported to show antagonistic activity against charcoal rot pathogen (*M. phaseolina*). Seed treatment of soybean with *Rhizobacteria* has also been found to reduce the occurrence of disease in the field. In case of seedling diseases, some actinomycete bacteria and binucleate *Rhizoctonia* strains have found to be effective in limiting the growth of pathogen. Other antagonistic fungi include *Verticillium biguttatum* which parasitizes the sclerotial development of *R. solani* and *T. harzianum* which is known to reduce the incidence of both *R. solani, F. oxysporum. T. viride* has been reported to suppress the severity of *F. oxysporum* and *Pythium arrhenomanes*.[13]

The colonization of soybean rhizosphere by mycorrhizal fungi *Rhizophagus irregularis* and *Rhizophagus intraradices* has been reported to reduce the incidence of sudden death syndrome in soybean (Marquez et al. 2019; Pawlowski and Hartman 2020). Some isolates of *Trichoderma harzanium* are also known to suppress the growth of *F. virguliforme*.[35]

14.5 CHEMICAL CONTROL AS A COMPONENT OF IPM PROGRAM

The major focus of IPM programs being environmental safety, sustainability, chemical method of insect pest, and disease control is considered as the last option when no other method of management is able to suppress the pest population below damaging levels. Despite the availability of new advanced technologies of pest management, chemical control is an inseparable part of management strategies owing to complex nature of pest problems. Thus, need based and judicious use of pesticides continue to be an important part of IPM. Selection of pesticide should be done based on the minimum effect on human health, nontarget organisms, and environment

Seed treatment with azoxystrobin or carboxin (2 g/kg) has been reported to reduce the incidence of soil-borne diseases including *Rhizoctonia* root rot of soybean.[36] Foliar spray of fungicides including Pyraclostrobin (0.1%), Pencycuron (0.1%), and Fluopyram (0.1%) were found effective in managing aerial blight disease in the field.[37] Application of pyraclostrobin 20% WG or fluxapyroxad 167 g/l + Pyraclostrobin 333 g/l SC, have been recommended for the management of frog eye spot of soybean. While, foliar spray of tebuconazole 25.9% EC is reported to reduce the incidence of pod blight of soybean. Hexaconazole 5% EC is reported to reduce the incidence of soybean rust.

For the management of insect pests and Chlorantraniliprole 18.50% SC @ 150 mL/ha is reported to suppress the activity of defoliators (*S. litura* and *H. armigera*). Thiamethoxam 30% FS has been recommended for the management of white fly (*B. tabaci*) and stem fly.

14.6 CROP STAGE WISE IPM PRACTICES FOR SOYBEAN PESTS AND DISEASES

The list of major insect pests and diseases affecting soybean crop has been mentioned in Tables 14.1 and 14.2.

TABLE 14.1 Major Insect Pests of Soybean.

Insect	Scientific name
Stem fly	*Melanagromyza sojae*
Tobacco caterpillar	*S. litura*
Green semilooper	*Chrysodeixis acuta*
Girdle beetle	*Obereopsis brevis*
Pod borer	*Helicoverpa armigera*
White fly	*B. tabaci*
Bihar hairy caterpillar	*S. oblique*
Leaf defoliator	*Spodoptera exigua*

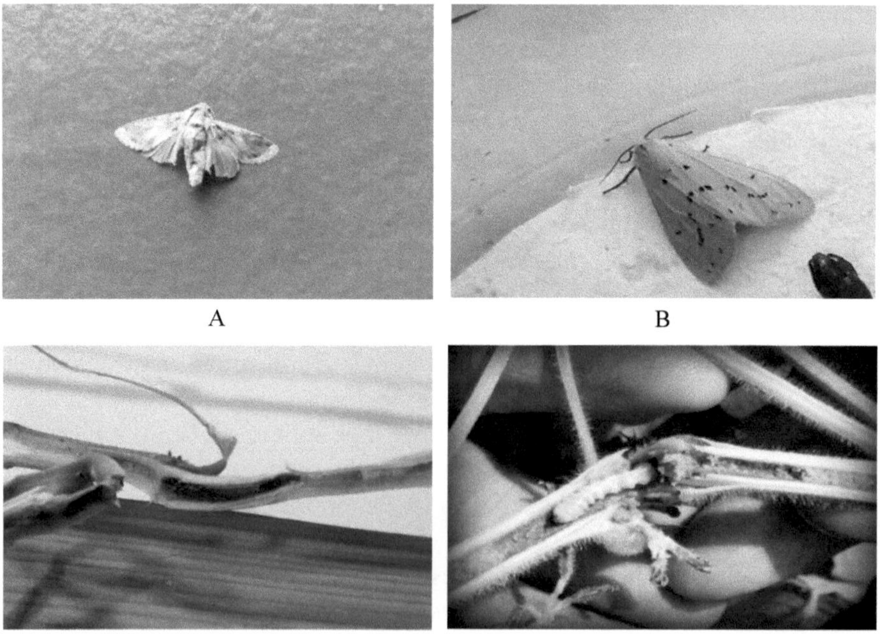

FIGURE 14.1 Major insect pests of soybean (A) *S. litura*. (B) *S. oblique*. (C) Maggot of *M. sojae*. (D) Maggot of *O. brevis*.

TABLE 14.2 Major Diseases of Soybean

Disease	Causal organism
Collar rot	*Sclerotium rolfsii*
Charcoal rot	*Macrophomina phaseolina*
Rhizoctonia root rot	*Rhizoctonia solani*
Rhizoctonia aerial blight	*Rhizoctonia solani*
Rust	*Phakopsora pachyrhizi*
Anthracnose	*Colletotrichum truncatum*
Myrothecium leaf spot	*Myrothecium roridum*
Alternaria leaf spot	*Alternaria alternate* and *A. tenuissima*
Bud blight	Groundnut bud necrosis virus
Purple seed stain	*Cercopora kikuchii*
Frog eye leaf spot	*Cercospora sojina*
Brown spot	*Septoria glycines*
Target leaf spot	*Corynespora cassiicola*
Bacterial blight	*Pseudomons syringae* pv. *glycinea*
Bacterial pustule	*Xanthomonas axonopodis* pv. *glycines*
Yellow mosaic	Mungbean yellow mosaic virus
Soybean mosaic	Soybean mosaic virus
Nematodes	*Rotylenchus reniformis, Meloidogyne* spp. *Pratylenchus* sp.

A

B

FIGURE 14.2 Major diseases of soybean (A) Anthracnose. (B) Pod blight. (C) Yellow mosaic. (D) Bacterial blight. (E) Bacterial pustule. (F) Aerial blight. (G) Root rot.

14.6.1 IN-BETWEEN CROP SEASON

Rotation of soybean with nonhost crops such as cereals is an important strategy to break the life cycle of soil-borne pathogens and pests. The planting of cover crops also serve to attract beneficial insect species, natural enemies of insect pests as well as improve soil structure and fertility. Addition of farm yard manure, crop residue, and vermi-composting also help in improving the nutrient availability of soil and enhance the diversity of microflora in rhizosphere of the crop. Destruction of crop residue by burning or flooding the field is important to eliminate the inoculum of pathogen and pests in the field. Tillage practices such as deep summer ploughing are also important to expose and eliminate the resting stages of pests.

14.6.2 PRE-SOWING STAGE

Proper tillage operations should be performed before sowing of the crop which help in desiccation of the larvae of insects and expose them to predation by natural enemies. Recommended dose of fertilizers should be applied so as to ensure proper growth and development of the crop. Excessive nitrogen dose could lead to more vegetative growth which attracts insect pests and diseases. Thus balanced application of fertilizers is essential. For management of soil-borne diseases and soil application of *T. harzianum* and *P. fluorescens* is recommended. Clean and disease free and certified seeds should be used for sowing. Seed treatment with *Trichoderma viride* at the rate of 5 g/kg seed or Carboxin 37.5% + Thiram 37.5% WS @ 3 g/kg seed is recommended to keep a check on the seedling blight and seed-borne diseases. Application of mycorrhiza and PGPR is recommended for promotion of plant growth and induction of systemic resistance in the crop. Treatment of seeds with Thiamethoxam 30% FS is recommended at the rate of 10 mL per hectare to prevent damage due to stem fly and reduce incidence of white flies.

Selection of resistant crop varieties is a key component of IPM program. Use of resistant varieties is the most efficient and eco-friendly way of managing pest problems in agricultural systems. Owing to the lower incidence of diseases, pests in resistant cultivars, the frequency and dose of chemical sprays is reduced. The resistant varieties of soybean to major insect pests and diseases as recommended in integrated pest management package for soybean by Directorate of Plant Protection and Quarantine & Storage[38] have been mentioned in Table 14.3.

TABLE 14.3 Resistant Varieties of Soybean to Insect Pests and Diseases.

Insect pest/ Disease	Resistant varieties
Stem fly	JS-335, PK-262, NRC-12, NRC-37, MACS-124, MAUS-2, MAUS-47
Tobacco caterpillar	JS 81-21, PS-564, PK-472
Green semilooper	NRC-7, NRC-37, PUSA-16, PUSA-20, PUSA-24, JS 93-05, JS 97-52, MAUS-47, JS 80-21
Girdle beetle	JS 71-05, NRC-7, JS 97-52, MAUS-32, Indira Soya-9
Rust	PK-1024, PK-1029, JS-80-21, Indira soyabean 9
Sclerotium blight	NRC 37
Charcoal rot	NRC-2, NRC-37, LSb-1, MACS-13, JS 97-52
Rhizoctonia aerial blight	PK-472, PK-1042, PK-564, SL-295
Anthracnose, pod blight	Bragg, Himso-1563, PK-472, VLS-21, NRC-12
Bacterial pustule	PK-1029, PK-1042, Bragg, Himso-1563, Indira soya 9
Yellow mosaic virus	PK-416, PK-472, PS-564, PK-1024, PK-1029, PK-1042, Pusa-37, JS- 97- 52
Soybean mosaic virus	JS-71–05, KHSb -2, LSb-1, MACS-58, MACS-124, VLS 2
Myrothecium leaf spot	JS-71-05, JS-335, MACS-13, MACS-124, MAUS-47, NRC-7
Frog eye leaf spot	Bragg, JS-80-21, KHSb-2,VLS 21
Alternaria leaf spot	KHSb-2, NRC-2, PK-327, PK-1042, Himso-1563, JS-80-21, Pusa-37, VLS-21

14.6.3 SOWING AND SEEDLING STAGE

Sowing of trap crops such as castor, dhaincha along with the main crop of soybean could help reduce the attack of insect pests on the main crop. Sowing should be done at the stage when sufficient soil moisture is present. Recommended seed rates and planting density should be followed to avoid dense canopy of crop which would otherwise attract pests and diseases. Intercropping with maize and sorghum help in management of pests and increasing the biodiversity of natural enemies and predators of pests in crop field and improve the rhizosphere microflora for avoidance of soil-borne diseases.

At the seedling stage and proper rouging should be done to remove any plants affected by collar rot. The field should be kept weed free by hand removal of weeds and use of pre-emergence herbicides. The plants infested with *S. litura* and *Spilarctia obliqua* should be removed.

14.6.4 VEGETATIVE AND FLOWERING STAGE

The insects that appear during vegetative stage of soybean crop include tobacco caterpillar, girdle beetle, hairy caterpillar, and green semiloopers. Collection and destruction of plants infested with girdle beetle and those with gregarious feeding by the caterpillars help in keeping the field clean. Application of Chlorantraniliprole 18.50% SC @ 150 mL/ha helps in managing green semilooper and stem fly and girdle beetle infestation in soybean. For the management of soybean rust application of hexaconazole 5% EC @ 100mL/ 100 L is recommended at vegetative stage of crop.

14.6.5 PODDING STAGE

During podding stage, several important diseases including aerial blight, anthracnose, pod blight, purple seed stain, frog-eye leaf spot, bacterial leaf blight, bacterial pustules, yellow mosaic, and soybean mosaic infect the soybean crop. For managing frog eye leaf spot and application of Pyraclostrobin 20% WG or Fluxapyroxad 167 g/l + Pyraclostrobin 333 g/l SC has been recommended. Pod blight could be managed by application of tebuconazole 25.9% EC.

14.7 CONCLUSION

Agricultural systems in today's scenario need to be focused on safer techniques of crop production with a thoughtful concern toward managing the ecological balance of nature. The intelligent use of natural resources is essential for getting continued benefits from the available resources in future. Integrated pest management by serving a way to minimize the use of chemical pesticides in crop production is an effective answer for the concerns related to pesticide residue in food products and development of resistance in pests and environmental sustainability. The combination of several management techniques from the sowing to harvesting stage of the crop ultimately reduces the frequency and dosage of pesticides used in production of soybean. Thus, adoption of IPM program not only assists in managing the pest problems but also increases the overall productivity of soybean cropping systems.

KEYWORDS

- **soybean**
- **IPM**
- **cropping system**
- **biocontrol**
- **pesticide**
- **economic threshold level**

REFERENCES

1. Wang, J.; Zhou, P.; Shi, X.; Yang, N.; Yan, L.; Zhao, Q.; Guan, Y. Primary Metabolite Contents are Correlated with Seed Protein, Oil Traits in Near-Isogenic Lines of Soybean. *Crop J.* **2019,** *7* (5), 651–659.
2. Singh, K. P.; Aravind, T.; Bhatt, P.; Surbhi K. Identification of Resistant Sources in Field Evaluation of Soybean (*Glycine max* (L.) Merr.) Genotypes Against Pod Blight, Rhizoctonia Aerial Blight. *Int. J. Curr. Microbiol. Appl. Sci.* **2019,** *8* (03), 1823–1827. https://doi.org/10.20546/ijcmas.2019.803.215
3. Surbhi, K.; Singh, K. P. Influence of Weather Factors on Severity of Aerial Blight of Soybean. *Indian Phytopathol.* **2020,** *73*, 493–497. https://doi.org/10.1007/s42360-020-00235-w
4. USDA Global Market Analysis, Food, Agricultural Service/ USDA. 2020. https://apps.fas.usda.gov/psdonline/circulars/production.pdf
5. ICAR-IISR. Soybean Monitor/ Market Watch. ICAR- Indian Institute of Soybean research: Indore, 2020. https://iisrindore.icar.gov.in/pdfdoc/soybeanmonitorsept2020.pdf
6. Bapatla, K. G.; Patil, R. H.; Yeddula, S. Impact of Leaf Damage by Defoliators on Yield of Soybean as a Sole Crop, as a Main Crop in Intercropping Systems. *Int. J. Pest Manage.* **2018,** *64* (1), 51–58.
7. Surbhi. K.; Singh. K. P.; Singh. N. K.; Aravind, T. Assessment of Genetic Diversity Among Soybean Genotypes Differing in Response to Aerial Blight (*Rhizoctonia solani* Kuhn) Using SSR Markers. *J. Phytopathol.* **2020**. https://doi.org/10.1111/jph.12956
8. Bortolotto, O. C.; Pomari-Fern,es, A.; Bueno, R. D. F.; Bueno, A. D. F.; da Kruz, Y. K. S.; Queiroz, A. P.; Ferreira, R. B. The Use of Soybean Integrated Pest Management in Brazil: A Review. *Agron. Sci. Biotechnol.* **2015,** *1* (1), 25 – 32.

9. Barzman, M.; Bàrberi, P.; Birch, A. N. E.; Boonekamp, P.; Dachbrodt-Saaydeh, S.; Graf, B.; & Lamichhane, J. R. Eight Principles of Integrated Pest Management. *Agron. Sustain. Dev.* **2015,** *35* (4), 1199–1215.
10. Khan, S. M.; Ali, S.; Nawaz, A.; Bukhari, S. A. H.; Ejaz, S.; Ahmad, S. Integrated Pest, Disease Management for Better Agronomic Crop Production. In *Agronomic Crops*; Springer: Singapore, 2019; pp 385–428.
11. Grabau, Z. J.; Chen, S. Determining the Role of Plant-Parasitic Nematodes in the Corn–Soybean Crop Rotation Yield Effect Using Nematicide Application: II. Soybean. *Agron. J.* **2016,** *108* (3), 1168–1179.
12. Bond, J.; G. Bird, F. Warner, S. Chen, D. Malvick, A. Dorrance, K. Harrison, P. Esker, A. MacGuidwin, J. Faghihi. *Soybean Cyst Nematode Management Guide*, 5th ed., 2019. https://soybeanresearchinfo.com/pdf_docs/SCNGuide_5thEd.pdf.
13. Roth, M. G.; Webster, R. W.; Mueller, D. S.; Chilvers, M. I.; Faske, T. R.; Mathew, F. M.; Smith, D. L. Integrated Management of Important Soybean Pathogens of the United States in Changing Climate. *J. Integr. Pest Manage.* **2020,** *11* (1), 17.
14. Harbach, C. J. Investigating the Interactions Between Cover Crops, the Soybean Cyst Nematode Through Lab, Greenhouse and Field Studies. 2019. Ph.D. dissertation, Iowa State University, Ames, IA.
15. Subbotin, S. A.; Chitambar, J. J. *Plant Parasitic Nematodes in Sustainable Agriculture of North America: 2—Northeastern, Midwestern and Southern USA*; Springer: New York, 2018; 453p.
16. Rojas, J. A.; A. Witte, J. L. Jacobs, Z. A. Noel, M. I. Chilvers. Diversity, Characterization of Oomycetes Associated with Corn Seedlings in Michigan. *Phytobiomes* **2019,** *3*, 224–234.
17. Li-ming, Z.; L. Shuang, S. Zhe, H. Jing, C. Qiu-ming, S. Bing, D. Jun-jie, L. Wei-ting, W. Jing-zhi. Impact of Crop Rotation on Pathotype and Genetic Structure of *Phytophthora sojae* in Fields. *J. Northeast Agr. Uni.* **2016,** *23*, 1–12.
18. Maier, M.; Souza, C. A.; Casa, R. T. Cropping Systems on Root Rot, Soybean Seed Yield. *Ciência Rural.* **2018,** *48*, 7.
19. Leandro, L. F. S.; Eggenberger, S.; Chen, C.; Williams, J.; Beattie, G.A.; Liebman, M. Cropping System Diversification Reduces Severity and Incidence of Soybean Sudden Death Syndrome Caused by Fusarium virguliforme. *Plant Dis.* **2018,** *102*, 1748–1758.
20. Navi, S.; Yang X. B. Impact of Crop Residue, Corn-Soybean Rotation on the Survival of *Fusarium virguliforme* a Causal Agent of Sudden Death Syndrome of Soybean. *J. Plant Pathol. Microbiol.* **2016,** *7*, 1–7.
21. Wen, L.; Lee-Marzano, S.; Ortiz-Ribbing, L. M.; Gruver, J.; Hartman, G. L.; Eastburn, D. M. Suppression of Soilborne Diseases of Soybean with Cover Crops. *Plant Dis.* **2017,** *101* (11), 1918–1928.
22. Da Silva, A. G. D. Studies on the Impact of Cover Crops on Soybean Productivity, Root Diseases. Ph.D. Thesis. Iowa State University, 2019; 134p.
23. Dunbar, M. W.; Gassmann, A. J.; O'Neal, M. E. Impacts of Rotation Schemes on Ground Dwelling Beneficial Arthropods. *Environ. Entomol.* **2016,** *45* (5), 1154.
24. Jabbour, R.; Pisani-Gareau, T.; Smith, R. G.; Mullen, C.; Barbercheck, M. Cover Crop, Tillage Intensities Alter Ground-Dwelling Arthropod Communities During the Transition to Organic Production. *Renew. Agric. Food Syst.* **2016,** *31*, 361–374.
25. Iqbal, N.; Hussain, S.; Ahmed, Z.; Yang, F.; Wang, X.; Liu, W.; Liu, J. Comparative Analysis of Maize–Soybean Strip Intercropping Systems: A Review. *Plant Prod. Sci.* **2019,** *22*(2), 131–142.

26. Zhang, H.; Yang, Y.; Mei, X.; Li, Y.; Wu, J.; Li, Y.; Wang, H.; Huang, H.; Yang, M.; He, X.; Zhu, S.; Liu, Y. Phenolic Acids Released in Maize Rhizosphere During Maize-Soybean Intercropping Inhibit Phytophthora Blight of Soybean. *Front. Plant Sci.* **2020,** *11*, 886. DOI: 10.3389/fpls.2020.00886.
27. Suyal, P.; Gaur, N.; KN, R. P.; Devrani, A. Seasonal Incidence of Insect Pests, Their Natural Enemies on Soybean Crop. *J. Entomol. Zool. Stud.* **2018,** *6* (4), 1237–1240.
28. Ahirwar, R.; Devi, P.; Gupta, R. Seasonal Incidence of Major Insect-Pests, Their Biocontrol Agents of Soybean Crop (Glycine max L. Merrill). *Sci. Res. Essays* **2015,** *10* (12), 402–406.
29. Kumari, C.; Yadu, Y. K.; Jha, S. K.; Jaiswal, S. K. Seasonal Incidence of Major Insect-Pests of Soybean, Their Bio-Control Agents. *J. Exp. Zool. India*. **2020,** *23* (1), 525–529.
30. Koch, R. L.; Costamagna, A. C. Reaping Benefits from an Invasive Species: Role of Harmonia Axyridis in Natural Biological Control of Aphis Glycines in North America. *Biocontrol* **2017,** *62* (3), 331–340.
31. Heinrichs, E. A.; Muniappan, R. Integrated Pest Management for Tropical Crops: Soyabeans. *CAB Rev*. **2018,** *13* (055), 1–44.
32. Sumida, C. H.; Daniel, J. F.; Araujod, A. P. C.; Peitl, D. C.; Abreu, L. M.; Dekker, R. F.; & Canteri, M. G. *Trichoderma asperelloides* Antagonism to Nine *Sclerotinia sclerotiorum* Strains and Biological Control of White Mold Disease in Soybean Plants. *Biocontrol Sci. Technol*. **2018,** *28* (2), 142–156.
33. Smolińska, U.; Kowalska, B. Biological Control of the Soil-Borne Fungal Pathogen *Sclerotinia sclerotiorum*. *J. Plant Pathol*. **2018,** *100*, 1–12.
34. Strom, N.; Hu, W.; Haarith, D.; Chen, S.; Bushley, K. E. Corn, Soybean Host Root Endophytic Fungi with Toxicity Towards the Soybean Cyst Nematode. *Phytopathology* **2020,** *110*, 603–614.
35. Pimentel, M. F.; Arnão, E.; Warner, A. J.; Subedi, A.; Rocha, L. F.; Srour, A.; Bond, J. P.; Fakhoury, A. M. *Trichoderma* Isolates Inhibit *Fusarium virguliforme* Growth, Reduce Root Rot and Induce Defense-Related Genes on Soybean Seedlings. *Plant Dis.* **2020,** *104*, 1949–1959.
36. Kumar, A.; Zacharia, S.; Maurya, A. K.; John, V. Effect of Fungicides, Neem Oil on the *Rhizoctonia* Root Rot of Soybean (*Glycine max* L.). *Int. J. Curr. Microbiol. Appl. Sci.* **2019,** *8* (1), 368–372. https://doi.org/10.20546/ijcmas.2019.801.037
37. Mathpal, M. Epidemiology, Management of *Rhizoctonia* Aerial Blight of Soybean (*Glycine max* L. Merrill). Ph.D. Thesis. G.B. Pant University of Agriculture & Technology, Pantnagar. 2016; 221p.
38. Sharma, A. N.; Gupta, G. K.; Verma, R. K.; Sharma, O. P.; Bhagat, S.; Amaresan, N.; Kapoor, K. S. Integrated Pest Management Package for Soybean. Faridabad (India): Directorate of Plant Protection, Quarantine & Storage, 2014; 55 p.

CHAPTER 15

Integrated Pest Management in Indigenous Oilseed Crops

P. DURAIMURUGAN* and M. SUJATHA

ICAR-Indian Institute of Oilseeds Research, Rajendranagar, Hyderabad 500030, Telangana, India

*Corresponding author. E-mail: duraimuruganp@rediffmail.com

ABSTRACT

India is one of the major producers of oilseeds contributing around 8% of the world's production. In spite of increase in production of oilseeds, the import of edible oil is increased due to increase in consumption levels creating huge demand supply gap. The country needs a threefold increase in the oilseeds production in the next 30 years. This gap can be decreased by increasing the production and productivity of oilseed crops. Though several high yielding varieties and hybrids are being cultivated and biotic problems hamper realizing the actual potential in these rainfed oilseed crops. One of the obvious reasons for low productivity is damage due to insect pests with yield losses of up to 25%. Hence, there is an earnest need to avoid the insect pests' damage following integrated pest management strategies to increase productivity in oilseed crops. It emphasizes on careful monitoring of insect pests and conservation of beneficial organisms like pollinators and biocontrol agents of insect pests and need-based critical intervention with appropriate tactics of pest management. The major insect pests of annual oilseed crops, namely, castor, sesame, safflower, sunflower, and their integrated management strategies are discussed in this chapter.

Integrated Pest Management in Diverse Cropping Systems. Devendra Pal Singh, PhD (Ed.)

15.1 INTRODUCTION

Vegetable oils are essential for the country's nutrition, energy, and economy, as well as worldwide commodities supplies. After the United States, China, and Brazil, India's vegetable oil economy is the world's fourth largest. With 12–15% of the global oilseeds area, 6–7% of vegetable oil output, 9–10% of total edible oil consumption, and 13.6% of vegetable oil imports, the country has a major position in the global oilseeds scenario. The different agro-ecological conditions in the country encourages cultivating all the nine annual oilseed crops, which comprise seven edible oilseeds, namely, soybean [*Glycine max* (L.) Merr.], rapeseed-mustard (*Brassica* spp.), groundnut (*Arachis hypogaea* L.), sesame (*Sesamum indicum* L.), sunflower (*Helianthus annuus* L.), safflower (*Carthamus tinctorius* L.), niger (*Guizotia abyssinica* Cass.) and two nonedible oilseeds, namely, castor (*Ricinus communis* L.) and linseed (*Linum usitatissimum* L.). India grows the annual oilseeds on an area of 25.02 million hectares and producing 32.38 million tonnes with productivity of 1295 kg/ha. Despite the largest cultivator of oilseeds in the world and the proportion of Indian imports in total availability of edible oils has increased from a meager 3% in 1970–1971 to almost 60% in 2019–2020 (14.72 million tonnes at an immense expense of Rs. 68,558 crores) due to increase in demand of vegetable oil with increase in population and increase in standard of living (income) and enhanced use for industrial and pharmaceutical and nutraceutical and cosmetics purposes.[1] The oilseed crops productivity is low in India when compared with the world average and the susceptibility of these oilseed crops to insect pests is one of the principal factors. Chemical insecticides have become indispensable in oilseeds crop protection due to the immediate and spectacular effects. In many occasions, they have been used as the principal method of control. However, the excess and unwise use of a wide spectrum of insecticides has resulted in many undesirable side effects, namely, destruction of natural enemies of pests and other nontargeted organisms of insect pests, resistance to insecticides, environmental, health hazards, and besides risking export embargo especially export potential oilseed crops due to contamination of seed with insecticide residues. With increased awareness of the adverse impact of insecticides on the eco-system and the concept of sustainable agriculture has gained recognition with integrated pest management (IPM) as one such major step toward sustainability of agriculture. Hence, the information on IPM of key insect pests of indigenous oilseed crops, namely, castor, sesame, safflower, and sunflower is discussed in this chapter.

15.2 CASTOR

15.2.1 INSECT PEST SCENARIO

Among biological constraints limiting castor production and undoubtedly insect pests dominate the scenario. More than 100 species of insects and six species of mites are known to infest castor at various phenological phases of the crop.[2] However, about a dozen of them are of economic importance in India (Table 15.1). The yield losses caused by insect pests on castor ranged from 17 to 63%.[3] The extent of the insect pests problem is relatively high in Southern India where castor is grown principally as rainfed crop and whereas it is low in Gujarat and Rajasthan under irrigated conditions.

15.3 INTEGRATED PEST MANAGEMENT STRATEGIES

15.3.1 CULTURAL PRACTICES

Deep summer ploughing is recommended to expose and eliminate the hibernating pupae of insect pests to predating birds or hot sun rays. Select triple or double waxy bloom castor hybrids, namely, GCH-4, GCH-5, GCH-7, DCH-519, YRCH-1, and ICH-66 which are resistant to leafhopper. Avoid growing of triple bloom cultivars in whitefly endemic areas.[4] Castor varieties/hybrids with nonspiny capsules (Jwala) or semicompact spikes (GCH-4 and GCH-7) are less harmed by capsule borer.[5] Intercropping with cluster bean or black gram or groundnut or cowpea (1:2 ratio) reduces semilooper, leafhopper, shoot and capsule borer infestation and builds up natural enemies (*Microplitis* parasitoids, coccinellids, spiders) population.[6] Sowing cucumber as trap crop along field borders ideally before sowing of castor attracts the migrating red hairy caterpillars and facilitates mechanical killing. Utilize vegetative twig traps like *Jatropha* or *Ipomoea* or *Calotropis* for assortment and killing of migrating larvae of red hairy caterpillars in endemic areas.[7]

15.3.2 PHYSICAL CONTROL

In regions where red hairy caterpillar is an issue, install light trap (using 200 W mercury lamp @ one trap/10 ha) coinciding with the first shower

TABLE 15.1 Insect and Mite Pests of Castor in India

Common name	Scientific name	Family	Order	Nature of damage
Pests of national significance				
Castor semilooper	*Achaea janata* (Linnaeus)	Noctuidae	Lepidoptera	Defoliation
Tobacco caterpillar	*Spodoptera litura* (Fabricius)	Noctuidae	Lepidoptera	Defoliation
Shoot and capsule borer	*Conogethes* (*Dichocrocis*) *punctiferalis* (Guenée)	Crambidae	Lepidoptera	Shoot, capsule boring, feeding
Leafhopper	*Empoasca flavescens* Fabricius	Cicadellidae	Hemiptera	Sap sucking
Whitefly	*Trialeurodes ricini* Misra	Aleyrodidae	Hemiptera	Sap sucking
Thrips	*Scirtothrips dorsalis* Hood	Thripidae	Thysanoptera	Sap sucking
Pests of regional significance				
Bihar hairy caterpillar	*Spilosoma obliqua* (Walker)	Arctiidae	Lepidoptera	Defoliation
Red hairy caterpillars	*Amsacta albistriga* (Walker), *A. mooreii* Butler	Arctiidae	Lepidoptera	Defoliation
Hairy caterpillars	*Euproctis fraterna* Moore	Lymantriidae	Lepidoptera	Defoliation
Tussock caterpillar	*Orgyia postica* (Walker)	Lymantriidae	Lepidoptera	Defoliation
Castor spiny caterpillar	*Ergolis (Ariadne) merione* (Cramer)	Nymphalidae	Lepidoptera	Defoliation
Castor slug	*Parasa lepida* (Cramer)	Limacodidae	Lepidoptera	Defoliation
Serpentine leaf miner	*Liriomyza trifolii* (Burgess)	Agromyziidae	Diptera	Leaf mining
Red spider mite	*Tetranychus telarius* (Linnaeus)	Tetranychidae	Acarina	Sap sucking

of monsoon to attract and kill the emerging moths and continue for 30–40 days on community basis. In circumstances, where operating electric light trap is not possible, a petromax light (using 200 candle power @ one trap/4 ha) is also effective in attracting and killing of moths. The peak period of emergence of red hairy caterpillar moths needs to be established before using of light traps. The planning of light trap ought to be so that majority moths being attracted must be trapped.[2]

15.3.3 MECHANICAL CONTROL

Handpicking and annihilation of egg-masses and gregarious phases of *S. litura* and hairy caterpillars alongside damaged leaves are successful for the management of defoliator pests in castor, which keep the defoliation level generally under 25%. Collect and destroy the shoots and capsules infested by capsule borer, wherever possible.[8] Keep 50 bird perches per hectare to attract predatory birds, which readily predates larvae of semilooper, *S. litura* and capsule borer. Install sex pheromone traps for *S. litura* (@ 10 traps/ha) for monitoring and applying timely management methods.[9,10]

15.3.4 BIOLOGICAL CONTROL

Wide array of biocontrol agents like parasitoids, predators, and microbial agents have been documented on castor semilooper. Among egg parasitoids, *Trichogramma chilonis* is dominant resulting in more than 40% semilooper egg parasitization in the field. In Gujarat, total natural parasitization by *Trichogramma australicum*, *Telenomus* sp., *Trissolcus* sp. on semilooper eggs ranged from 25 to 100%. An exotic scelionid egg parasitoid, *Telenomus proditor* imported from New Guinea parasitized 11–94% of castor semilooper. Among larval parasitoids on semilooper, the braconid, *Snellenius* (*Microplitis*) *maculipennis* is quite possibly the most potential parasitoid in the castor environment. About 60–90% larval parasitization by this parasitoid is reported.[11] During rainy season, a gregarious, ecto-parasitoid, *Euplectrus maternus* is found to parasitize semilooper to a lesser extent. General insect predators like ants, mantids, chrysoperla, reduviid bugs, pentatomid bugs, spiders have been found to predate on castor semilooper. Apart from insect predators, several species of insectivorous birds have been reported to predate on castor semilooper. Among them, Indian myna (*Acridotheres tristis*), house crow (*Corvus splendens*), black drongo (*Dicrurus adsinilis*), rosy paster (*Sternus roseus*), green bee eater (*Merops orientalis*), large gray babbler (*Trudoides malcolm*), and hoopoe (*Upupa epops*) were more efficient in the predation of late instar larvae of semilooper.[2] Among the microbial biocontrol agents, *Bacillus thuringiensis* var. *kurstaki* @ 1 g/L, suspension concentrate (SC) formulations of *B. thuringiensis* var. *kurstaki* (3 mL/L) developed singly and in combination with *Nomuraea rileyi*, *Beauveria bassiana* (3 mL/L) were found effective in reducing semilooper population.[11–13] Green muscardine fungi, *Nomuraea rileyi* was reported to be effective against semilooper.[14]

Among natural enemies of *S. litura,* egg parasitoid, *Trichogramma evanescens minutum*; larval parasitoids, *Apanteles prodeniae, Cotesia* sp.; pupal parasitoids, *Tetrastichus ayyari, Trichospilus pupivora* and predatory stink bug, *Cantheconidia furcellata* are important. The microbial agents like *Spodoptera* NPV (SlNPV), entomopathogenic fungi, namely, *Metarhizium* (*Nomuraea*) *rileyi* and entomopathogenic bacteria, *B. thuringiensis* can also be utilized for the management of *S. litura.*[15] SC formulations of *B. thuringiensis* var. *kurstaki* (DOR Bt-127) @ 3 mL/L, *Bt* var. *kurstaki* DOR Bt-127 + *Nomuraea rileyi* were found highly effective against *S. litura* on castor.[13] The ichneumonid larval parasitoids, namely, *Diadegma ricini*, *Theronia* sp., the braconid parasitoids, namely, *Apanteles* sp., *Bracon hebetor* were recorded on shoot and capsule borer. Insectivorous birds like black drongo, Indian myna, and bee eater were also found predating on capsule borer larvae present in the frass. *Chrysonotomyia* sp. has been observed to parasitize both eggs and maggots of leaf miner, *L. trifolii.*[16,17]

15.3.5 BOTANICAL CONTROL

Neem seed kernel extract (5%) and other neem-based formulations were found to be effective in reducing the population of lepidopteran pests, sucking pests, and leaf miner. Spraying of neem seed kernal extract (4% NSKE) synchronizing with egg oviposition and early larval stages of semi-looper helps in lessening the infestation. The leaf extracts of neem, *Nerium indicum*, neem seed kernel extract (4–5%), neem oil (1–3%), and commercial neem-based formulations were found effective against *S. litura.* Application of neem oil @ 5 mL/L on the under surface of the leaves was found effective against whitefly.[2,18]

15.3.6 CHEMICAL CONTROL

Castor crop can withstand at least 25% defoliation without any economic yield reduction.[17] Hence, economic threshold levels (ETLs)-based application of insecticides needs to be followed for economic management of insect pests. The ETL is 25% defoliation for lepidopteran defoliators (semilooper/spodoptera/hairy caterpillars) and 10% capsule damage for capsule borer and 10% of leaf curling symptom in a plant for leafhopper. Conserve larval parasitoid of semilooper (*S. maculipennis*) by evading spraying of chemical insecticides when 1–2 parasitoid cocoons are observed per plant. Pheromone trap catches based action threshold of 81 moths/trap/week has been reported for *S. litura* in castor.[19] If the injury by the insect pests surpasses ETL, spray insecticides, namely, flubendiamide 39.35SC @ 100 mL/ha, chlorantraniliprole 18.5SC @ 150 mL/ha, thiodicarb 75WP @ 500 g/ha, profenofos 50EC @ 500 mL/ha, quinalphos 25EC @ 1000 mL/ha for the effective management of semilooper, tobacco caterpillar, and hairy caterpillars.[20,22] Spinosad 45SC @ 75 mL/ha, indoxacarb 15.8EC @ 500 mL/ha, thiodicarb 75WP @ 500 g/ha, profenofos 50EC @ 500 mL/ha, flubendiamide 39.35SC @ 100 mL/ha, chlorantraniliprole 18.5SC @ 150 mL/ha were found effective against shoot and capsule borer.[21,22] Profenophos 50EC @ 1000 mL/ha, buprofezin 25SC @ 750 mL/ha were found efficient in lessening the whitefly infestation. Clothianidin 50WDG @ 50 g/ha, acetamiprid 20SP @ 100 g/ha, dimethoate 30EC @ 825 mL/ha were found effective against leafhopper and thrips.[20]

15.4 SESAME

Sesame, an important indigenous oilseed crop and is grown as rainfed *kharif*, semi-*rabi*, also as summer crop. The average productivity of 493 kg/ha in India is quite low with wide fluctuations from year to year. One of the key limitations in sesame production is the colossal damage caused by insect pests. A total of 201 species representing 55 genera were observed on sesame worldwide.[23] About 81 species of insect pests belonging to nine orders are known to harm this crop at one or different stages of plant growth in different agroecological regions in India.[24,25] Of these, about 10 insect pests are of economic importance and the pests attack tolls a heavy loss of 25–90% in seed yield (Table 15.2).[26,27]

TABLE 15.2 Insect Pests of Sesame in India

Common name	Scientific name	Family	Order	Nature of damage
Pests of national significance				
Leaf webber or roller, capsule borer	*Antigastra catalaunalis* Duponchel	Crambidae	Lepidoptera	Webbing, feeding of tender leaves, flowers and boring of capsules
Gall fly	*Asphondylia sesami* Felt	Cecidomyiidae	Diptera	Feeding floral buds results in the creation of a gall-like structure that does not mature into a flower/capsule
Leafhopper	*Orosius albicinctus* Distant	Cicadellidae	Hemiptera	Sap sucking, vector for phyllody
Pests of regional significance				
Bud fly	*Dasineura sesame* (Grover & Prasad)	Cecidomyiidae	Diptera	Feeding floral buds results in the creation of a gall-like structure that does not mature into a flower/capsule
Sphinx moth or Hawk moth	*Acherontia styx* Westwood	Sphingidae	Lepidoptera	Defoliation
Bihar hairy caterpillar	*S. obliqua* Walker	Arctiidae	Lepidoptera	Defoliation
Red hairy caterpillar	*A. albistriga* Walker	Arctiidae	Lepidoptera	Defoliation
Pod bug	*Elasmolomus sordidus* Fabricius	Lygaeidae	Hemiptera	Sap sucking
Pentatomid bug	*Nezara viridula* (Linnaeus)	Pentatomidae	Hemiptera	Sap sucking
Cotton aphid	*Aphis gossypii* Glover	Aphididae	Hemiptera	Sap sucking

15.5 INTEGRATED PEST MANAGEMENT STRATEGIES

15.5.1 CULTURAL PRACTICES

The practice of crop rotation with nonhost crops is effective in reducing pest populations, maintaining productivity and as significant yield reduction has been reported in growing sesame in the same field consistently for over 2 years. Field should be ploughed to expose pupae and larvae of pests for predation. In *kharif*, early sowing, use of early varieties, proper use of balanced doses of fertilizer make the crop to pass the vulnerable stage before pests build-up the population above ETL.[28] Intercropping with mungbean, urdbean, cowpea, groundnut, pigeonpea, pearl millet, moth bean proved to be more effective than sole crop for the management of leaf webber and capsule borer, and other insect pests. Sesame + green gram/black gram (3:3), sesame + cluster bean (3:3) intercropping minimized the incidence of leaf webber and capsule borer and bud fly.[29] Remove alternate weed hosts of phyllody, namely, *Parthenium hysterophorus, Crotalaria* sp., *Trifolium* sp. and disease infected plants. Keep the boundary and bunds of the field free from weeds.[27,30]

15.5.2 MECHANICAL CONTROL

Collection and destruction of caterpillars of sphinx moth, egg masses, and early instars of hairy caterpillars when in congregation effectively reduce the pest population. Collection and destruction of leaf rolls, shoots, capsules, buds infested by leaf webber and capsule borer help to check their population. Clipping of the galls and burning the shed buds are good prophylactic measures against gall fly. Rogue out disease infected plants periodically and keep the field free of infected plant debris.[31] Use of light trap @ 1/ha for monitoring activity of pest moths. Yellow sticky traps @ 10 traps/ha can be used for monitoring of sucking pests. Predatory birds readily eat the caterpillars of leaf roller and capsule borer and help to check when they are numerous. Hence, erecting 40–50 bird perches per hectare facilitates predation of larvae.[32,33]

15.5.3 BOTANICAL CONTROL

Application of neem cake @ 200 kg/ha was found effective in reducing diseases, nematodes, and insect pests. Spray of botanical pesticides, namely, neem seed kernel extract (5%), neem oil (2 to 5%), neem gold (0.3%), azadirachitin (0.03%), neem seed oil emulsion (3%), and mahua oil (2%) were found effective to control leaf webber and capsule borer, bud fly, and leafhoppers.[33]

15.5.4 BIOLOGICAL CONTROL

Several bio-control agents have been reported to control pests of sesame. *Bracon gelechi*, *Trathala flavoorbitalis* are the major parasitoids, whereas *C. furcellata* is the efficient predator of leaf webber and capsule borer. Parasitoids, namely, *Bracon hebetor*, *B. brevicornis*, *Trichogramma* spp., *Phanerotoma handecasisella*, *Campoplex* sp., *Erioborus* sp., *Apanteles* sp., and predators, namely, *Cicindella* sp., lacewing, ladybird beetle, spiders, and red ant were recorded as natural enemies of leaf webber and capsule borer. Parasitoid, *Pteromalus fasciatus* and predators, namely, spider, ladybird beetle, lacewing were reported on gall fly. Conserve parasitoids, namely, *Trichogramma evanescens* or *T. minutum* or *Apanteles obliquae* for management of Bihar hairy caterpillar. The spray of *B. thuringiensis* var. *kurstaki* @ 1 kg/ha has been found good for the management of lepidopteran pests. Entomopathogenic fungi, namely, *B. bassiana* (DOR strain) 30SC @ 0.3 mL/L and *Metarrhizum ansiopliae* (NBAII strain – 1 × 10^8 CFUs/g) @ 10 g/L were found effective against leaf roller and capsule borer.[15,27,34]

15.5.5 CHEMICAL CONTROL

Monitor insect pests through customary survey or by means of light traps or pheromone traps or sticky traps and make proper decisions for chemical insecticide sprays. ETL of 1–2 larvae/plant or two webbed leaves/m^2 or 10% damage for leaf webber and capsule borer and 2–5 galls/plant or 10% damage for gall fly has been reported. Seed treatment with thiamethoxam 25 WG @ 5 g/kg or imidacloprid 70 WS @ 7.5 g/kg found effective for management of sucking pests.[29] Spraying of spinosad 45SC @ 75 mL/ha or profenophos 50 EC @ 1000 mL/ha or flubendiamide 39.35 SC @ 150 mL/ha or chlorantraniliprole 18.5SC @ 200 mL/ha or novaluron 10 EC @ 1000 mL/ha or quinalphos 25 EC @ 1000 mL/ha was reported to be effective against leaf roller and capsule borer, sphinx, and hairy caterpillars. Application of dimethoate 30 EC @ 750 mL/ha or imidacloprid 17.8 SL @ 125 mL/ha, quinalphos 25EC @ 750 mL/L, carbaryl 50 WP @ 1000 g/ha at bud initiation stage was found to be effective against gall/bud fly. Spraying of acetamiprid 20 SP @ 150 g/ha or imidacloprid 17.8 SL @ 125 mL/L or dimethoate 30 EC @ 750 mL/L or thiacloprid 240 SC @ 500 mL/ha is reported to be effective for management of leafhopper.[27]

15.6 SAFFLOWER

15.6.1 INSECT PEST SCENARIO

Safflower crop is damaged by over 80 species of insect, mite, and nematode pests appearing throughout the world on this crop.[35] In India, 36 insect pests have been reported on this crop.[34,36] Among them, about 10 insect pest species are economically important which may cause heavy yield loss if these pests are not properly managed[33,35] (Table 15.3).

TABLE 15.3 Insect and Mite Pests of Safflower in India

Common name	Scientific name	Family	Order	Nature of damage
Pests of national significance				
Safflower aphid	*Uroleucon carthami* (Theo)	Aphididae	Hemiptera	Sap sucking
Safflower caterpillar	*Perigaea capensis* (Thunberg)	Noctuidae	Lepidoptera	Defoliation
Gram pod borer	*Helicoverpa armigera* (Hübner)	Noctuidae	Lepidoptera	Defoliation, capsule boring and feeding
Pests of regional significance				
Gujhia weevil	*Tanymecus indicus* Faust	Curculionidae	Coleoptera	Root boring, feeding
Capsule fly	*Acanthiophilus helianthi* (Rossi)	Tephritidae	Diptera	Capsule feeding
Semilooper	*Plusia orichalcea*	Noctuidae	Lepidoptera	Defoliation
Tobacco caterpillar	*S. litura* (Fabricius)	Noctuidae	Lepidoptera	Defoliation
Armyworm	*Spodoptera exigua* Hubner	Noctuidae	Lepidoptera	Defoliation
Flower head borer	*Eublemma parva* Hubner	Noctuidae	Lepidoptera	Capsule boring, feeding on developing seeds
Spider mite	*Tetranychus* spp.	Tetranychidae	Acarina	Sap sucking

15.7 INTEGRATED PEST MANAGEMENT STRATEGIES

15.7.1 CULTURAL PRACTICES

Plough the fields deeply before sowing to expose the resting stages of various insect pests to sun and predatory birds. Optimum time of sowing as suggested for different regions is the most important nonmonetary input for aphid and capsule fly management in safflower crop. Sowing the crop normally in the first week of October is desirable as it escapes from aphid attack. Late sown crop always succumbs to aphid and capsule fly attack. Avoid late sowing in November. Grow varieties like Bhima, A-1, NARI-H-15 which are resistant or moderately resistant to aphid.[37] Varieties, namely, A-300, JSF-1, Tara, EB-7, Giza. NS-75, 116-4-5 have been reported resistant to capsule fly. Generally, spineless varieties with succulent branches suffer more aphid infestation than spiny and woody branched varieties. Apply recommended dose of fertilizer and as application of more nitrogen aggravates aphid infestation. Follow crop sanitation and destroy alternate hosts (*Euphorbia* sp. and *Sonchus arvensis*). Intercropping with nonhost crops, namely, sorghum, wheat, coriander reduces aphid and safflower caterpillar infestation. Intercropping with niger should be avoided.[37]

15.7.2 MECHANICAL CONTROL

Hand collection and destruction of egg masses and early stage larvae of *S. exigua* and *S. litura* which will be in gregarious condition. The removal and destruction of aphid, capsule fly infested buds, late instar larvae of safflower caterpillar and gram pod borer are reported to be effective in reducing the pest infestation.[31] Install yellow sticky traps or pan water trap @ 25 traps/ha 15 cm above the canopy has been reported for monitoring of aphids. Locally accessible empty tins painted yellow and coated with castor oil or grease or vaseline on outer surface may additionally be used.[19,20,33]

15.7.3 BIOLOGICAL CONTROL

A high degree of parasitization (66.6%) of safflower aphid by dipteran endoparasitoid, *Pseudendaphis* sp. has been reported during the first week of March. Parasitization of *Aphidencyrtus aphidivorus* and *Aphelinus* sp. is also reported from safflower aphids. A number of coccinellid predators (*Brumus suturalis, Coccinella septempunctata, Menochilus sexmaculatus, Coccinella arcuate, Harmonia octomaculata, Coccinella repanda*), chrysopid predators (*Chrysopa formosa, Chrysoperla sinica, Chrysopa* sp.), syrphid predators (*Sphaerophoria indiana, Ischiodon scutellaris*) were reported.[35] Exploit the activity of predators and parasitoids if observed in the field and by spraying the crop with NSKE (5%) and avoiding usage of chemical pesticides as far as possible. As soon as aphid infestation is noticed, spray NSKE (5%) or neem oil (300 ppm) on the borders of 1.8 m length to prevent the spread of aphids.[32]

Three braconid parasitoids, namely, *Apanteles ruficrus, Rhogas percurrens, Rhogas* sp. and one eulophid parasitoid, *Euplectrus euplexiae* have been reported to parasitize safflower caterpillar. Parasitoids, namely, *Eriborus argenteopif, Pteromalus* sp., entomopathogenic nematode, namely, *Mermis* sp., green muscardine fungus, namely, *Metarhizium anisopliae* and a number of polyhedrosis viruses were also reported on safflower caterpillar.[36]

A number of hymenopterous parasitoids, namely, *Campoletis chlorideae, Banchospis ruficornis, Ecphoropsis perdistictus, Enicospilus* sp., *Eriborus argenteopilosus, Microchelonus curvimaculatus* and tachnids, namely, *Eucarcelia illota, Sturmiopsis inferens, Goniophthalmus halli, Drino imberbis* and *Palexorista solennis* have been reported parasitizing gram pod borer, *H. armigera*. The HaNPV, entomogenous fungal and bacterial diseases have been found to keep its population under check.[38] *Orymurus* sp. an important parasitoid is recorded on capsule fly, while *Chrysopa virgestes* predates on the maggots.[36]

15.7.4 CHEMICAL CONTROL

Monitoring of insect pests using pheromone traps or sticky traps and assessment of ETL can be used to make decisions for chemical pest management. The ETLs for aphid is reported to be 50–70 nymphs/adults/plant (5 cm apical twig). The seed treatment with thiamethoxam 30FS @ 10 mL/kg seed or imidacloprid 600FS @ 8 mL/kg of seed has been reported to protect safflower crop from aphids up to 30 days.[29] Spraying of insecticides, namely, pymetrozine 50WG @ 300 g/ha, thiamethoxam 25WG @ 125 g/ha, imidacloprid 17.8SL @ 125 mL/ha, clothianidin 50WDG @ 50 g/ha or a combination insecticide [Chlorpyriphos 50% + Cypermethrin 5%] 55EC @ 1000 mL/ha, dimethoate 30EC @ 1000 mL/ha or acephate 75SP @ 750 g/ha was found effective against safflower aphid. Dusting with quinalphos 4% D @ 25 kg/ha, phenthoate 2% DP @ 20 kg/ha, quinalphos 1.5% DP @ 20 kg/ha in water scarce areas were also found effective against aphid. Insecticides, namely, indoxacarb 15EC @ 0.3 mL/L, spinosad 45SC @ 0.15 mL/L were reported to be effective against safflower caterpillar. Spraying with quinalphos 25 EC @ 800–1200 mL/ha or dusting with quinalphos 1.5 D @ 25 kg/ha was found effective against *H. armigera* and *S. exigua.* Protection at germination stage against gujhia weevil (in Vidharbha area of Maharashtra) is of great economic importance. Soil application of phorate 10G granules @ 10 kg/ha into soil at the time of sowing and foliar spray with chlorpyriphos 20 EC@ 1250 mL/ha was reported to be effective for management of gujhia weevil.[1,38]

15.8 SUNFLOWER

Insect pests are the major production constraints in traditional sunflower areas. A total of 251 insect and mite pests have been documented on sunflower across the world. Insect pests of sunflower are diverse in tropical countries as compared to temperate countries. In India, over fifty insect pest species have been found to ravage the crop at different phenological phases. Among them and about a dozen of insect pests are of major economic importance and having high damage potential in India[39,40] (Table 15.4).

15.8.1 CULTURAL PRACTICES

In cutworm endemic areas, deep ploughing for exposing the over wintering larvae and pupae to predators and cultivating the crop on slopes of ridges (6–8 cm height) are some of the eco-friendly and effective methods in reducing damage to seedlings. Field cleanliness and use of completely decomposed FYM mixed with wood ash has been found effective against termites. Termites can also be effectively managed by eliminating queen by removing it out from the termetorium. Crop rotation of 3–4 years, use of healthy and quality seeds, removal and burning of crop residues of previous crop, removal of volunteer sunflower plants/*Parthenium* weed around the field and clean cultivation ensures destruction of alternate hosts of major pests and vectors of diseases.[41] Intercropping of sunflower with groundnut, soybean, pigeonpea, and finger millet were found effective in minimizing insect pest infestation. Intercropping with pigeonpea (short duration and determinate types) at 2:1 ratio is effective for the control of foliage feeders and borer population. Whitefly and leafhopper population was low in sunflower + groundnut (1:4) intercropping system. Intercropping of sunflower with cotton affected lesser thrips and leafhopper incidence on sunflower. Border

TABLE 15.4 Insect Pests of Sunflower in India.

Common name	Scientific name	Family	Order	Nature of damage
Pests of national significance				
Capitulum borer / Head borer	*H. armigera* Hübner	Noctuidae	Lepidoptera	Capitulum (head) boring, feeding on developing seeds
Tobacco caterpillar	*Spodoptera litura* Fabricius	Noctuidae	Lepidoptera	Defoliation
Green semilooper	*Thysanoplusia orichalcea* Fabricius	Noctuidae	Lepidoptera	Defoliation
Bihar hairy caterpillar	*S. obliqua* Walker	Arctiidae	Lepidoptera	Defoliation
Leafhopper	*Amrasca biguttula* Ishida	Cicadellidae	Homoptera	Sap sucking
Thrips	*S. dorsalis* Hood; *Thrips palmi* Karny	Thripidae	Thysanoptera	Sap sucking, vector of viral diseases
Pests of regional significance				
Cutworm	*Agrotis ipsilon* Rott	Noctuidae	Lepidoptera	Defoliation
Cabbage semilooper	*Trichoplusia ni* (Hubner)	Noctuidae	Lepidoptera	Defoliation
Stem borer	*Nupserha* sp.	Cerambycidae	Coleoptera	Stem boring
Whitefly	*Bemisia tabaci* Gennadius	Aleyrodidae	Hemiptera	Sap sucking
Armyworm	*S. exigua* (Hubner)	Noctuidae	Lepidoptera	Defoliation
Termite	*Odontotermes obesus* (Rambur) & *Microtermes obesi* Hol	Termitidae	Isoptera	Root, stem feeding

crops like sorghum, pearl millet, and maize in 5–7 rows has to be grown around sunflower crop that attracts the thrips population and additionally hampers landing of wind borne thrips or *Parthenium* pollen grains carrying inoculum on sunflower crops.[38,39]

15.8.2 MECHANICAL CONTROL

Monitoring using light traps or pheromone traps is essential for the polyphagous insect pests, namely, *S. litura and H. armigera*, so that initial sources of pests could be identified and destroyed. Pheromone traps of *S. litura* (10 per ha) and *H. armigera* (10 per ha) can be used to monitor the moth population in order to take up timely measures to contain the damage. Set up yellow sticky traps @ 10–12 traps/ha coated with grease/sticky oily materials is recommended for monitoring of sucking pests. Both tobacco caterpillar and Bihar hairy caterpillar lay eggs in masses preferably on lower surface of leaves and early instars larvae are gregarious. Defoliation of gregarious early instar larvae on lower side of leaves results a "mesh" like appearance to the leaf which can be effortlessly located, collected and demolished by reconnaissance through the field. Hand collection and destruction of young larvae of cutworm feeding gregariously on foliage and late instar larvae of capitulum borer and other defoliators was also found effective in reducing the pests population.[38,40]

15.8.3 BOTANICAL MANAGEMENT

Spraying of NSKE (5%) or commercial neem formulations synchronizing with oviposition, early larval stages of the lepidopteran pests will not only lessen injury but also benefits in preserving the activity of beneficial organisms like natural enemies of insect pests and pollinators like honey bees. NSKE (5%) was also found effective against leafhopper. *Cleodendrum inerme* dust (25%) and plant extracts (10%) not only proved to be effective against *H. armigera* larvae but also brought abnormality in pupal and adult stage.[42]

15.8.4 BIOLOGICAL CONTROL

Predators, namely, *Chrysoperla carnea*, *Menochilus sexmaculata* and parasitoids, namely, *Trichogramma* spp., *Campoletis* spp., and *Bracon* spp. have been observed as major key mortality factors of *H. armigera*. Release of the coccinellid/chrysopid predators @ 1 larva/head or parasitoids @ 50,000/ha was reported to be effective for the management of *H. armigera.* Egg parasitoid, *Trichogramma evanescens minutum;* larval parasitoids, *A. prodeniae, Cotesia* sp.; predatory stink bug, *C. furcellata* and pupal parasitoids, *T. ayyari, T. pupivora* are important natural enemies of *S. litura*. Five parasitoids, namely, *A. ruficrus, A. africanus, Euplecturs ceylonensis, Campoletis chloridae,* and *Exorista xanthaspis* were found parasitizing *T. orichalcea* larvae.[42] Microbial agents such as NPV and muscardine (white, green) have often been noticed in the field populations of lepidopteran pests. Use of SlNPV or HaNPV @ 250 LE/ha with UV protectants is effective for the management of *S. litura* and *H. armigera.* Spray *N. rileyi* @ 2×10^{11} conidia/l spray solution during low temperature, high humidity periods for management of *S. litura.* Spraying of *B. thuringiensis* var. *kurstaki* @ 1 g/L of water was found effective against early larval stages of lepidopteran defoliators. Suspension concentrate (SC) formulations of *Bt* var. *kurstaki* DOR Bt-127 @ 3 mL/L were found highly effective against semilooper, *S. litura* and *H. armigera* on sunflower.[15]

Coccinellid predators like *Coccinella septumpunctata, M. sexmaculata, Chilochorus nigritus, Scemnus nubillus*, and *Brumus suturalis* were found effective in reducing the sucking pests population like thrips and leafhopper. In addition, predatory lygaeid, *Geocoris tricolor, Anthocoris* sp. and several mantids such as *Eumantissa giglio, Cariagrion coromandeliarum, Ichneura* sp., *Dolichopus* sp., and *Therevia* sp. play vital role in suppression of sucking and other pests.[43]

15.8.5 CHEMICAL CONTROL

The usage of insecticides ought to be necessity based and their application ought to be avoided after anthesis as this may poison the pollinators. Honeybees play a very vital role as pollinators that are essential for cross-pollination in sunflower. Spraying of insecticides at bud stage of the crop coinciding with *H. armigera* eggs appear on them is the most perfect stage to manage the pest. It is recommended to use insecticides when the *H. armigera* crosses its ETL of 4.25 larvae per 10 capitulum or 15–20 leafhoppers/plant. Seed treatment with imidacloprid 70WS @ 5 g/kg or imidacloprid 48FS @ 5–9 mL/kg or thiamethoxam 70WS @ 4 g/kg or thiamethoxam 30FS @ 10 mL/kg seed were found effective against sucking pests, namely, leafhoppers (jassids), whitefly and thrips.[1,29] Spraying of spinosad 45 SC @ 150 mL/ha, profenofos 50 EC @ 1000 mL/ha, quinalphos 25EC @ 1000 mL/ha, chlorantraniliprole

150 mL/ha, cypermethrin 10 EC @ 650 mL/ha were found to be effective against lepidopteran pests, namely, capitulum borer, tobacco caterpillar, semiloopers and hairy caterpillars. Spraying of dimethoate 30 EC @ 850 mL/ha or imidacloprid 70 WS @ 500 mL/ha or oxydemeton methyl @ 1000 mL/ha has been recommended for sucking pests management.[1,4,39]

15.9 CONCLUSION

The dilemma of more production of vegetable oils for continuously expanding population on one hand and desire for clean, safe, ecologically sound environment on the other, demands careful planning for management of insect pests. IPM is the practice of optimizing pest treatment in a cost-effective and environmentally friendly manner. This is performed by combining numerous techniques in a coordinated manner to keep pest damage below the level of economic detriment while also protecting animals, plants, and the environment. Though several promising environmentally friendly IPM components have been evaluated, there is a need to develop location-specific IPM modules that include promising biocontrol agents, botanicals, resistant varieties, and cultural practises, as well as relatively safer insecticides for biocontrol agents and pollinators. Farmers must be educated in pest population monitoring and simple expert methods for pest management of oilseed crops. With the present knowledge and proper management of insect pests, the national productivity of oilseed crops can be considerably enhanced in India. It will make oilseeds cultivation more sustainable, profitable, and globally competitive.

KEYWORDS

- **indigenous oilseed crops**
- **castor**
- **insect pest scenario**
- **integrated pest management strategies**
- **cultural practices**
- **physical control**
- **mechanical control**
- **biological control**
- **chemical control**
- **sesame**
- **safflower**
- **sunflower**

REFERENCES

1. Reddy, V. A.; Kadirvel, P.; Sujatha, M.; Lavanya, C.; Mukta, N.; Sudhakara Babu, S. N.; Suresh, G.; Padmavathi, P.; Prasad, R. D.; Santha Lakshmi, P. M.; Srinivas, P. S.; Duraimurugan, P.; Prayaga, L. *Handbook on ICAR-IIOR Technologies (2009–2019)*; ICAR-Indian Institute of Oilseeds Research: Rajendranagar, Hyderabad, India, 2020; pp 42.
2. Lakshminarayana, M.; Raoof, M. A. *Insect Pests and Diseases of Castor and Their Management*; Directorate of Oilseeds Research, Hyderabad, India, 2005; pp 78.
3. Lakshminarayana, M.; Duraimurugan, P. Assessment of Avoidable Yield Losses Due to Insect Pests in Castor (*Ricinus communis* L.). *J. Oilseed Res.* **2014,** *31*, 140–144.
4. Varaprasad, K. S.; Duraimurugan, P. Integrated Pest Management—Minor Oilseeds Crops. In *Hand book of Integrated Pest Management*; Chattopadhyay, C.; Tanwar, R. K.; Sehgal, M.; Birah, A.; Kanwar, V.; Ahmad, N.; Mehta, N., Eds.; Indian Council of Agricultural Research: New Delhi, **2018**; pp 557–577.
5. Lakshminarayana, M. Studies on Antixenosis in Castor, *Ricinus communis* L. Against Major Insect Pests. *Indian J. Plant Protection* **2005,** *33*, 216–219.
6. Rao, M. S.; Rao, C. A. R.; Srinivas, K.; Pratibha, G.; Vidya Sekhar, S. M.; Sreevani, G.; Venkatswarlu, B. Intercropping for Management of Insect Pests of Castor, *Ricinus communis* in the Semi-Arid Tropics of India. *J. Insect Sci.* **2012,** *12*, 1–10.
7. Basappa, H.; Duraimurugan, P. Management of Pests of Oilseed Crops. In *Pests and Their Management*; Omkar, Ed.; Springer Nature: Singapore, 2018; pp 223–240.
8. Suresh, G.; Lavanya, C.; Duraimurugan, P.; Santha Lakshmi, P. M.; Manjunatha, T.; Prayaga, L.; Jawaharlal, J.; Reddy, V. A. *Castor Production Technology*, 7th Rev ed.; ICAR-Indian Institute of Oilseeds Research, Hyderabad, India, 2020; p 26.
9. Duraimurugan, P.; Srinivas, P. S.; Sampath Kumar, M. Evaluation of Pheromone Trapping for Monitoring of Seasonal Activity of *Spodoptera litura* in Castor. *Agrotechnology*, **2014,** *2*, 249.
10. Duraimurugan, P.; Alivelu, K. Determination of an Action Threshold for Tobacco Caterpillar, *Spodoptera litura* (F.) Based on Pheromone Trap Catches in Castor (*Ricinus communis* L.). *J. Entomol. Res.* **2018,** *42* (2), 189–194.
11. Duraimurugan, P.; Lakshminarayana, M.; Vimala Devi, P. S. Comparative Efficacy of Microbial, Botanical, Chemical Insecticides Against Lepidopteran Pests in Castor. *Ecoscan* **2015,** *9*, 7–10.
12. Vineela, V.; Vimala Devi, P. S.; Duraimurugan, P. Effect of Novel Formulation of *Bacillus thuringiensis* var. *kurstaki* (DOR Bt-127) Against Semilooper (*Achaea janata*) in Castor. *J. Oilseeds Res.* **2020,** *37* (Special Issue), 188–189.
13. Vimala Devi, P. S.; Duraimurugan, P.; Poorna Chandrika, K. S. V.; Vineela, V.; Hari, P. P. Novel formulations of *Bacillus thuringiensis* var. *kurstaki*: An Eco-Friendly Approach for Management of Lepidopteran Pests. *World J. Microbiol. Biotechnol.* **2020.** DOI: 10.1007/s11274-020-02849-8.
14. Lingappa, S.; Patil, R. K. *Nomuraea rileyi, a Potential Mycoinsecticide*; University of Agricultural Sciences: Dharwad, Karnataka, India, 2002; p 30.
15. Vimala Devi, P. S.; Duraimurugan, P. Exploitation of *Nomuraea rileyi* and *Beauveria bassiana* for the management of lepidopteran pests. *Kavaka* **2013,** *41*, 43–67.
16. Basappa, H.; Singh, H.; Patil, B. R. Non-Conventional Oilseeds as Source of Biopesticides for the Management of Sucking Pests in Sunflower. *J. Oilseeds Res.* **2009,** *26*, 431–433.

17. Lakshminarayana, M. Eco-Friendly Management of Insect Pests of Castor. In *Research and Development in Castor: Present and Future Strategies*; Hegde, D. M., Ed.; Indian Society of Oilseeds Research: Hyderabad, India, 2010; pp 141–158.
18. Satyagopal, K.; Sushil, S. N.; Jeyakumar, P.; Shankar, G.; Sharma, O. P.; Boina, D. R.; Sain, S. K.; Ram, A.; Kapoor, K. S.; Sanjay, A.; Subhash, K.; Patni, C. S.; Suresh, D.; Ekabote, K. Rajashekarappa, Lakshminarayana, M.; Narayanaswamy, H.; Shivanna, B. K. *AESA Based IPM Package for Castor*; National Institute of Plant Health Management: Hyderabad, Telangana, India, 2014; p 38.
19. Duraimurugan, P.; Alivelu, K. Development and Evaluation of Integrated Pest Management Module for Castor. *J. Exp. Zool., India* **2017a,** *20* (Supplement 1), 1389–1393.
20. Duraimurugan, P.; Alivelu, K. Field Efficacy of Newer Insecticides Against Sucking Insect Pests in Castor. *Indian J. Plant Protection* **2017b,** *45* (3), 246–250.
21. Duraimurugan, P.; Lakshminarayana, M. Efficacy of Newer Insecticides Against Defoliators and Capsule Borer in Castor. *Indian J. Plant Protection* **2014,** *42*, 307–311.
22. Narayanamma, V. L.; Reddy, V. A.; Singh, T. V. K. Evaluation of Newer Insecticides for the Management of Defoliators and Capsule Borer in Castor. *Indian J. Plant Protection* **2010,** *38* (2), 144–146.
23. Dilipsundar, N.; Chitra, N.; Gowtham, V. Checklist of Insect Pests of Sesame. *Indian J. Entomol.* **2019,** *81*. DOI: 10.5958/0974-8172.2019.00141.X.
24. Ahuja, D. B. Seasonal Occurrence of Insect Pests in *Sesamum indicum* Linn. *J. Entomol. Res.* **1989,** *13* (1–2), 116–120.
25. David, B. V.; Anathakrishnan, T. N. *General and Applied Entomology*; Tata Mc Graw-Hill Publishing Company Ltd.; New Delhi, 2004; p 1020.
26. Ahuja, D. B.; Kalyan, R. K. Evaluation of Various Plant Products Against *Antigastra catalaunalis* Dup.; *Polyphagotarsonemus latus* (Banks) on Sesame, *Sesamum indicum* L. *J. Insect Sci.* **2003,** *16* (2), 92–94.
27. Ranganatha, A. R. G.; Alok, J.; Deshmukh, M. R.; Rajani, B.; Panday, A. K.; Gupta, K. N.; Surabhi, J.; Seema, P. *Improved Technology for Maximizing Production of Sesame. All India Coordinated Research Project on Sesame and Niger*; Indian Council of Agricultural Research, JNKVV Campus: Jabalpur, Madhya Pradesh, India, 2013; p 28.
28. Kumaraswamy, H. H.; Jawaharlal, J.; Ranganatha, A. R. G.; Chander Rao, S. Safe Sesame (*Sesamum indicum* L.) Production: Perspectives, Practices and Challenges. *J. Oilseeds Res.* **2015,** *32*, 1–24.
29. Sujatha, M.; Chander Rao, S.; Vishnuvardhan Reddy, A. *Significant Achievements of 50 Years of AICRP on Oilseeds*; ICAR-Indian Institute of Oilseeds Research: Rajendranagar, Hyderabad, India, 2017; p 46.
30. Gopinath, K. A.; Venkateswarlu, B.; Venkateswarlu, S.; Yadav, S. K.; Balloli, S. S.; Srinivasa Rao, Ch.; Prasad, Y. G.; Maheswari, M. *Organic Sesame Production*; Central Research Institute for Dryland Agriculture: Santoshnagar, Hyderabad, India, 2011; 34 pp.
31. Singh, H.; Basappa, H.; Chander Rao, S.; Prasad, R. D. *IPM in Oilseed Crops*; Directorate of Oilseeds Research: Rajendranagar, Hyderabad, India, 2002; p 25.
32. Satyagopal, K.; Sushil, S. N.; Jeyakumar, P.; Shankar, G.; Sharma, O. P.; Boina, D. R.; Sain, S. K.; Lavanya, N.; Sunanda, B. S.; Ram, A.; Kapoor, K. S.; Sanjay, A.; Subhash, K.; Patni, C. S.; Suresh, D. E.; Rajashekarappa, K.; Lakshminarayana, M.; Narayanaswamy, H.; Shivanna, B. K.; Sathyanarayana, N.; Latha, S. *AESA Based IPM Package for Sesame*; National Institute of Plant Health Management: Hyderabad, Telangana, India, 2014; p 30.

33. Satyagopal, K.; Sushil, S. N.; Jeyakumar, P.; Shankar, G.; Sharma, O. P.; Boina, D. R.; Sain, S. K.; Ram, A.; Kapoor, K. S.; Sanjay, A.; Subhash, K.; Patni, C. S.; Ekabote, S. D.; Rajashekarappa, K.; Lakshminarayana, M.; Narayanaswamy, H.; Shivanna, B. K. *AESA based IPM Package for Safflower*; National Institute of Plant Health Management: Hyderabad, Telangana, India, 2014; p 34.
34. Singh, V.; Basappa, H.; Hegde, D. M. Safflower Pests and Their Integrated Management in India. *Sesame Safflower Newslett.* **1999,** *14,* 117–124.
35. Weiss, E. A. *Oilseed Crops*; Safflower Pests Publication: Longman, 1983; p 660.
36. Singh, V.; Prasad, R. D. Insect Pests and Diseases of Safflower and Their Management; Directorate of Oilseeds Research: Hyderabad, India, 2005; p 48.
37. Basappa, H.; Singh, H.; Srinivas, P. S. Multiplication of Safflower Aphid (*Uroleucon carthemi* H.R.L) on Different Varieties of Safflower and Alternate Hosts. *J. Oilseeds Res.* **2012,** *29,* 417–419.
38. Duraimurugan, P. Effect of Weather Parameters on the Seasonal Dynamics of Tobacco Caterpillar, *Spodoptera litura* (Lepidoptera: Noctuidae) in Castor in Telangana State. *J. Agrometeorol.* **2018,** *20* (2), 139–143.
39. Basappa, H.; Santhalakshmi, P. M. *Insect Pests and Diseases of Sunflower and Their Management*; Directorate of Oilseeds Research: Hyderabad, India, 2005; p 70.
40. Anonymous. *The AESA Based IPM—Sunflower (Helianthus annuus L.)*; National Institute of Plant Health Management: Hyderabad, Telangana, India, 2014; p 72.
41. Basappa, H. Integrated Pest Management in Sunflower: An Indian Scenario. *Proc. 16th Int. Sunflower Conf.*, Fargo North Dakota, USA, **2004,** *2,* 853–859.
42. Basappa, H. Biodiversity of Biocontrol Agents in Sunflower Ecosystem. *J. Biol. Control* **2011,** *25,* 182–192.
43. Basappa, H. Population of Predators Associated with Leafhopper in Sunflower. *J. Biol Control* **2007,** *21,* 85–87.

CHAPTER 16

Integrated Disease and Pest Management under Potato Based Cropping System

AMITAVA BASU* and AMITAVA KONAR

Department of Plant Pathology, Bidhan Chandra Krishi Viswavidyalaya, Mohanpur, Nadia, West Bengal, India

Corresponding author. E-mail: basuamitava1961@gmail.com

ABSTRACT

Indian agriculture has been continuously contributing to meet the rising demand of food to different stakeholders in the entire food chain. Agricultural production in India has been increasing remarkably leading to surplus food. This progress is achieved through adoption of high-yielding varieties of potato and other crops and proper use of fertilizers and pesticides. With the increase in global free trade, the chances of introduction of new invasive alien pests and pathogens have also escalated. Besides, there are possibilities of improper application of chemical pesticides and its ill effects on the human health, environment, and biodiversity have enhanced. These may be reduced through promotion of alternative technologies such as biopesticides, bioagents, cultural practices, etc., and less dependence on chemical fertilizers and pesticides used in crops. The recent advancements with respect to newer more eco-friendly pesticide molecules, biological tools, and host resistance, there is enough scope for refinement of this strategy towards effective application. The potato pest management techniques are therefore are the need of hour to manage potato health and getting healthy crop produce.

Integrated Pest Management in Diverse Cropping Systems. Devendra Pal Singh, PhD (Ed.)

16.1 INTRODUCTION

Potato is cultivated in 1.83 m ha in India. The annual production is around 48605000 tonnes and India was the second largest potato producer in world production during 2016–2017.[1] Potato offers possibilities for increasing food production and agricultural economy for our country. This is possible through the use of high-yielding cultivars and effective agro-techniques and plant protection measures and availability of quality seed. Insect pest is the major constraint that destabilizes potato crop productivity. The insect pests damage weaken the plant, inhibiting growth of potato tubers, and by feeding on the tubers.[2] The bacteria are very small in size, invisible to the naked eye, and usually grades plants to wilt and affect growth severely. The major potato diseases caused by bacteria are bacterial wilt, soft rot, ring rot, and scab. Viruses are smaller than bacteria and depend on other organisms for survival and transmission.[3] The symptoms may be stunted plants, wrinkled or curled leaves, and yellowish or purple leaves. The potato leaf roll virus (PLRV), potato viruses Y, A (PVY, PVA), and potato mop-top virus are major viral diseases. Nematodes are extremely small worms that live in the plant's root system causing nodules on roots and damage to tuber skins, etc. The most important nematodes causing damage to potato crops are root-knot nematodes and potato cyst nematodes (PCNs). The fungal diseases develop rot, spots, blight, and mildew covering with spores. The important fungal diseases on potatoes include late blight, powdery scab, black scurf, bacterial soft rot, blackleg, *Fusarium* wilt, and dry rot.[3] The diseases are transmitted by wind, insects, or water. Sometimes a disease may survive in a seed potato or leftover residues from an earlier crop. The survival and spread of disease organisms are dependent due to levels of humidity, temperature, and sunlight.[4]

Pests of National Importance in India

Diseases

Late blight (*Phytophthora infestans*), Black scurf *(Rhizoctonia solani*), Leaf spot complex (*Alternaria* spp, *Phoma* spp.), Early blight (*Alterneria solani* Sorauer), *Septoria* leaf spot (*Septoria lycopersici*), Common scab (*Strepto-myces scabies*), Bacterial wilt (*Ralstonia solanacearum.*), and Viral diseases (Potato virus X, S, Y, PLRV)

Nematode

Root-knot nematode (*Meloidogyne* sp.).[5]

Insect pests Aphids (*Myzus persicae* Sulzer, *Aphis gossypii* Glover), Potato tuber moth (PTM) (*Phthorimaea operculella* Zeller), Jassids/leaf hoppers (*Amrasca biguttula* Ishida., *Empoasca fabae* Harris), and Whitefly (*Bemisia tabaci* Gennadius).[5]

Pests of Regional Importance

Diseases

Virus (potato spotted wilt virus), charcoal rot *(Macrophomina phaseolina* (Tassi) Goid.), Viroid (potato spindle tuber viroid), pink rot (*Phytophthora erythroseptica* Pethybr.), soft rot (*Erwinia carotovora* Winslow et al.), and Black heart (Abiotic disorder due to Acute oxygen deficiency).

Nematodes

Golden nematode (*Globodera rostochinensis* Wollenweber), PCN (*Globodera pallida* Stone).

Insect Pests

Cutworm (*Agrotis segetum* Denis & Schiffermüller), White grub (*Holotrichia* spp. Hope), Serpentine Leaf miner (*Liriomyza huidobrensi* Blanchard), Spider mites (*Polyphago tarsonemus latus* Banks), Thrips (*Thrips* sp.}, Tobacco caterpillar (*Spodoptera litura*), Leaf eating caterpillar *(Helicoverpa armigera* Hübner), and Semi looper (*Trichoplusiani* Hübner).[5]

16.2 PRINCIPLES BEHIND IPM IN POTATO BASED CROPPING SYSTEM

Selection of tolerant cultivars, healthy seed/seedling/planting material tubers, treatment of seedling/planting material, tubers with especially biopesticides, adoption of proper spacing, and green manure to be applied.[6,7] The organic manures and biofertilizers are applied based on soil test results. The phosphate fertilizers may not be used in every season.[6,7] Proper irrigation is to be done and crop rotation is recommended.

16.3 REGULAR MONITORING IN THE FIELD BY POTATO GROWERS

The weekly monitoring of conditions of soil water, pests, natural enemies, and weather factors is essential. The evaluation of the field condition, Harmful: Beneficial pest ratio, Collection of egg masses, and removal of infested plants are all important tasks.[6,7] The results of sweep net and visual counts, etc. can be used to quantify these insects. The ration of harmful and beneficial insects varies with feeding potential of natural enemy and pest types.[6,7] The modern IPM technology is based on the economic threshhold level (ETL)[8] emphasis is always given to the agro-ecosystem. The plant health is dependent on its environment including physical factors and biological factors.[6,7] The previous experience of the farmers should be taken into consideration for decision making. They need to develop their own skills and knowledge from time to time.[7] They are also capable enough to improve farming practices by experimentation.[8,9]

16.4 OBSERVATIONS UNDER IPM PROGRAM

Weekly observations of plant health (pests, diseases, weeds; bioagents), conditions of soil, input costs, harvest cost, status of the beneficial insects, ratio of pests, beneficial insects, identification of all pests, diseases, crop health condition, and management practices, real-time for adoption of practices, and epidemiological condition for severe outbreak to be recorded.[7,8]

16.5 FARMER FIELD SCHOOL

Potato growers can learn about the pests, their nature of damage, the natural enemies, management of pests, management of water, nutrient, influence of weather factors on pest buildup, role of natural enemies in pest management, and decision-making process.[9] Farmers can monitor the major pest in their own fields at regular intervals through field scouting. Surveillance on pest occurrence should commence soon after crop establishment. Five spots are to be selected randomly in the field. Five random plants were selected at each spot for recording counts of insects.[10] In case of aphid, jassid, whitefly, and mite: Both nymphs, adults on five randomly selected leaves per plant to be counted, recorded.[6] In case of Thrips, the number of nymphs and adults of thrips present on five terminal leaves per plant to be counted and recorded.[6] For, leaf miner, only the number of live mines on five randomly selected leaves per plant should be taken into consideration. For potato tuber moth, *Spodoptera, Helicoverpa,* cutworm, total number of fruits, damaged fruits due to potato tuber moth, *Helicoverpa, Spodoptera,* cutworm, and number of larvae on individual plants should be counted and recorded.[6]

The disease scouting is based on diagnosing symptoms caused by fungal, bacterial, viral pathogens or abiotic factors such as weather, fertilizers, nutrient deficiencies, pesticides, and abiotic soil problems. Generally, fungal diseases produce symptoms with irregular growth, pattern, and color (except viruses), however, a biotic problems cause regular and uniform symptoms.[6] In case of infected root and cut roots are examined for internal infections. The total number of roots damaged/infested/infected due to rot should be evaluated and incidence should be recorded.[5] Assessment of disease severity for leaf infection is important. Leaf diseases cause most damage during the seedling and flowering stages. The number of leaves/plant infected due to disease and incidence should be recorded.[6] The stems and tubers should be split or taken apart and examined for discoloration caused by fungi and bacteria. The number of stems and tubers infected due to diseases and percent disease incidence should be recorded.[6]

The pest surveillance is an important aspect. The pheromone traps for monitoring potato tuber moth, *Agrotis*, *Helicoverpa, Spodoptera,* pheromone traps for insects, potato tuber moth, *Spodoptera, Helicoverpa, and* cutworm @ 4–5/acre are essential.[11] Yellow/blue pan water/sticky traps for monitoring aphid and white fly. Establishment of yellow pan water/sticky traps 15 cm above the canopy for monitoring whitefly and blue pan water/sticky trap for thrips @ 4–5 traps/acre to be installed. Setting up of light traps @1 trap/acre 15 cm above the crop height for monitoring and mass trapping insects is beneficial.[11] Nematode detection is done by the Baermann funnel method. The Baermann technique is based on the active migration or movement of larvae. In brief, faces are suspended in water and the larvae move into the water. Since they are unable to swim and they sink to the bottom and can be collected for identification.[12]

16.6 ECOLOGICAL PEST MANAGEMENT

The practices like growing of flowering plants or compatible cash crops along the field border by arranging shorter plants toward main crop and taller plants towards the border to attract natural enemies as well as to avoid immigrating pest population. The flowering plants are to be grown on the internal bunds. Growing of weed plants naturally like, *Tridax procumbens,*

Ageratum sp., *Alternanthera* sp., etc., which act as a nectar source for maintaining natural enemies. Broad spectrum chemical pesticides should not be applied when the P: D ratio is favorable.[11,13] Crop rotations to be done with leguminous plants which enhance nitrogen content. The soils should be covered year-round with living vegetation/or crop residue. The addition of organic matter in the form of FYM, vermicompost, and crop residue is practiced to enhance below ground biodiversity and reduce tillage for saving of hibernating natural enemies. Application of balanced dose of fertilizers, nutrients, application of mycorrhiza, plant growth promoting Rhizobacteria, *Trichoderma viride, Trichoderma harzianum, Pseudomonas fluorescens* to seed/seedling/planting material, nursery, and soil application (If commercial products are used and cross check for label claim).[11,13]

Disease resistant/tolerant varieties of potato

i) **Moderate resistant/ resistant to late blight:** Kufri Anand, Kufri Arun, Kufri Himalini, Kufri Badshah, Kufri Chamatkar, Kufri Chipsona-1, Kufri Chipsona-2, Kufri Frysona, Kufri Chipsona-3, Kufri Chipsona-4, Kufri Frysona, Kufri Gaurav, Kufri Garima, Kufri Girdhari, Kufri Giriraj, Kufri Himsona, Kufri Jawahar, Kufri Jeevan, Kufri Jyoti, Kufri Kanchan, Kufri Kashigaro, Kufri Khyati, Kufri Kumar, Kufri Kundan
ii) **Moderate resistant/ resistant to early blight:** Kufri Jeevan, Kufri Jyoti, Kufri Kanchan Kufri Kashigaro, Kufri Khyati
iii) **Resistant to PVX:** Kufri Chipsona-1
iv) **Resistance to PLRV:** Kufri Kuber
v) **Tolerant to Gemini virus:** Kufri Anand
vi) **Immune to wart disease:** Kufri Anand, Kufri Chamatkar, Kufri Chipsona-2, Kufri Frysona, Kufri Giriraj, Kufri Himsona Kufri Jawahar, Kufri Kumar
vii) **Resistant to charcoal rot:** Kufri Chamatkar, Kufri Kumar, Kufri Kundan
viii) **High tolerance to nutrient stress**: Kufri Gaurav.[14]

16.7 STANDARD CULTURAL PRACTICES TO BE ADOPTED DURING IPM

Summer deep plowing, soil solarization during summer, field sanitation, rouging, avoiding waterlogged conditions in the field, following crop rotation, application of manures, fertilizers as per soil test recommendations, sowing plant four rows of maize, sorghum, and Bajra (pearl millet) around the potato crop field as barrier crop.[13]

- FYM @ 8 t/acre or vermin compost @ 4–6 t/acre to be applied during field preparation.
- 2 kg each of *Azospirillum* and *Phosphobacterium* with 10 kg FYM / acre to be applied during field preparation in soil.[13]
- Weeds are to be eradicated by shallow plowing before sowing.[13]
- Light irrigation and covering the beds with polythene sheet of 45 gauge (0.45 mm) thickness to be done for 3 weeks before sowing.
- Marigold to be grown in the nursery 15 days prior to sowing against cyst nematode.
- Preparing seed beds of more than 35 cm height (for better water drainage).[14]
- Application of Neem cake@ 80 kg/acre is effective.[14]
- Selection of resistant/tolerant varieties.
- Selection of certified and pest and disease free seed tubers for planting.
- Application of 200 kg nitrogen (N), 150 kg phosphorus (P_2O_5), and 150 kg potassium (K_2O)/ha.[13,14] N to be applied half as basal and half during earthing up at 30 days after sowing.
- Recommended agronomic practices to be followed to achieve the healthy crop stand.[13,14]
- Pre-emergence herbicide Oxyflourfen 23.5% EC@170–340 mL in 200–300 L water/acre to be applied within 3–4 days after sowing.[13]
- Postemergence herbicide Paraquat dichloride 24% SL @ 200 g a i./ acre can be applied to control broadleaf, sedges, and grassy weeds.[15,16]
- Tubers stored in oxygen-deficient structures should be avoided.

16.8 STANDARD CHEMICAL AND BIOLOGICAL MANAGEMENT PRACTICES

Soaking of potato seed tubers in Boric acid (3%) for 20–30 min[14] before planting or in Streptocycline 40–100 ppm solution for half an hour prior to planting, or soaking potato seed tubers in combined formulation of Carbendazim 25%+ Mancozeb 50% WS @ 3 g/kg of seed. Application of *T. viride, T. harzianum,* and *P. fluorescens* for the treatment of seed tubers and soil application (if commercial products are used and label claim to be ensured. However, the biopesticides produced by farmers for own consumption in their fields and registration is not required).[13,15]

16.9 INTEGRATED MANAGEMENT OF BACTERIAL WILT OF POTATO THROUGH CULTURAL APPROACH

- Planting bacterial wilt-free seed tubers.
- Disinfecting the cutting knife using 1% sodium hypochlorite solution.
- Applying lime (dolomite) in the soil as acidic or alkaline soil is not conducive to the pathogen.
- Planting potatoes in soils free from bacterial wilt.
- Crop rotation with nonsolanaceous crops like corn, soybean, onion, and garlic for 2–3 years.
- Elimination of infected plant debris before planting and clearing weeds away
- Using composted organic fertilizer not contaminated with bacterial wilt.
- Biofumigation with Brassica sp crop residue in heavily infected soil.
- Managing irrigation in the field by digging channels that allow water to flow freely from the field. Bacterial wilt will spread rapidly in flooded fields.
- Using water not contaminated with bacterial wilt to irrigate the crop.
- Summer plowing during April–May.[17,18]

16.9.1 BIOLOGICAL MANAGEMENT

- Apply neem cake @ 80 kg/acre
- Treat the seed tubers with *P. fluorescens* @ 10 g/kg of seed.

16.9.2 CHEMICAL MANAGEMENT

Prophylactic field preparation with bleaching powder @12 kg/ha can be applied at 2–3 years interval. Two to three sprays of (streptomycin sulphate 9% + Tetracycline hydrochloride 1%) SP @ 40–50 ppm solution at an interval of 20 days. First, spray 30 days after planting.[17,18]

16.10 BLACK SCURF (*R. SOLANI*)

- The seed tubers should be treated before sowing with 3% boric acid for 20–30 min.
- seed treatment with *T. viride* @5 g/kg

- Soil can be amended with oilcakes (Neem Cake/Mustard Cake) at the rate of 20–30 kg per hectare. Combination of seed and soil treatment gives the best control of the disease.[19]

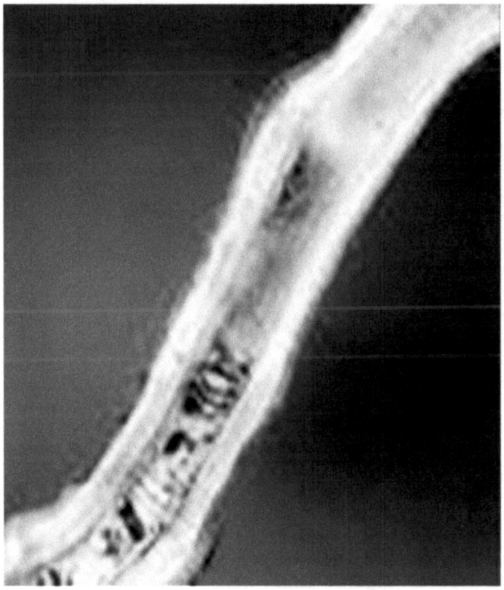

FIGURE 16.1 Infected stem due to *Ralstonia soalanacearum*

FIGURE 16.2 Infected tubers showing brown ring rot symptom

FIGURE 16.3 Infected plant showing Bacterial wilt symptom

FIGURE 16.4 Stem canker symptom produced by *Rhizoctonia solani*

FIGURE 16.5 Sclerotia on tuber infected by *Rhizoctonia solani*

FIGURE 16.6 Aerial tuber formation infected by *Rhizoctonia solani*

16.11 SOFT ROT AND BLACK LEG (*E. CAROTOVORA*)

- Avoiding planting in wet or flooded fields.
- Improving irrigation systems to allow water to flow in, out of the field more easily.
- Using healthy seed not contaminated with the disease.
- Destroying sources of the disease, particularly infected plants.
- Avoiding damage to tubers when weeding, hilling up, harvesting, and transporting harvest produce.
- Harvesting in dry weather conditions.
- Sanitation of the field for the whole season.[20]

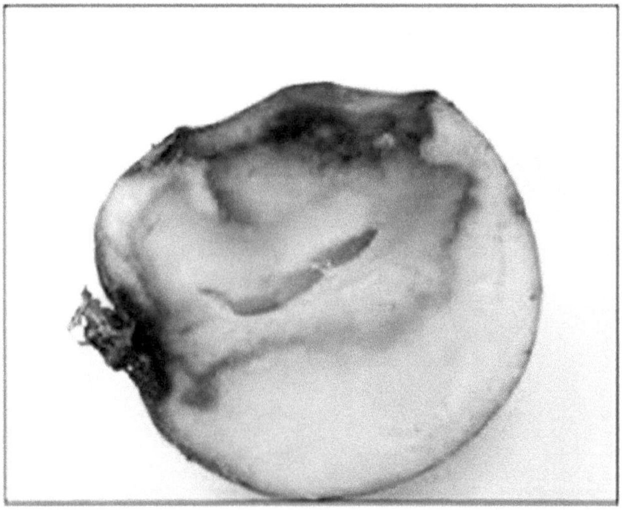

FIGURE 16.7 Soft rot infected tuber due to *Erwinia carotovora*

16.12 POTATO SCAB (*S. SCABIES*)

- Use of disease-free planting materials.
- Application of green manure/oilcakes before planting—effectively reduce disease incidence.
- Treating the tubers with 3% boric acid for 20–30 min before and after cold storage.
- Irrigating the land at regular intervals during tuberization stage.
- Crop rotation with nonhost crops such as cabbage or maize.

FIGURE 16.8 Soft rot infected stem due to *Erwinia carotovora*

- Summer plow of potato field to expose the potato field during April–May.
- Treating the seed tubers with *Bacillus subtilis* @10 g/kg of seed
- Avoiding planting in fields with low soil pH and or increasing soil pH through the application of lime.

1. Plant certified, disease-free seed potatoes, and resistant varieties are recommended. Red-skinned varieties have more resistance to the disease as compared with white skin.
2. Potato scab is more severe alkaline soils as compared to soil pH levels of 5.2 or lower.
3. Tilling in a cover crop—mustard, canola, and alfalfa—prior to planting potatoes.
4. Use of gypsum before planting @25 lbs per 2000 sq ft to enhance the calcium content of soil.
5. Adequate irrigation during early tuber development should be avoided.[2,21]

16.13 EAF SPOT COMPLEX

1. Septoria leaf spot 2. Phoma leaf spot 3. Stemphylium leaf spot 4. Cercospora leaf spot[2]

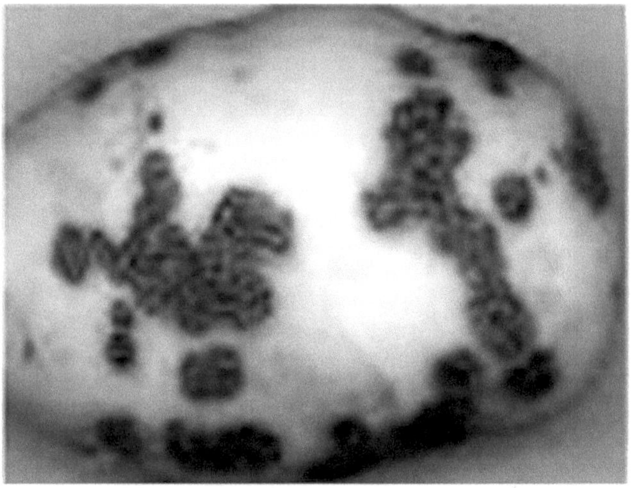

FIGURE 16.9 Scab infection on tuber

FIGURE 16.10 Septoria leaf spot symptom

16.13.1 MANAGEMENT OF LEAF SPOT COMPLEX

- Diseased plant parts should be destroyed properly.
- Avoiding irrigation in cool cloudy weather.

- *T. harzianum, T. viride,* and *P. fluorescens* are effective against the disease.
- Protectant fungicides (e.g., maneb, mancozeb, chlorothalonil) are also effective.[2]

FIGURE 16.11 Phoma leaf spot symptom

FIGURE 16.12 Stemphylium leaf spot symptom

FIGURE 16.13 Cercospora leaf spot symptom

16.14 MANAGEMENT ACTIVITY OF ROOT-KNOT NEMATODE

- Planting crops during the second week of October in autumn and in early January to limit root- knot nematode infestation on tubers.[14,22]
- Avoiding seed tubers from an infested area and the field should be kept free from weeds.
- Deep plowing and drying of soil in summer months facilitate drying of infected larvae. Burning of trash and following 2 years rotational sequence maize-wheat-potato-wheat reduces the root-knot damage significantly.
- Growing of allelopathic plants like *Tagetes patula and Tagetes erecta* (African marigold) in between two and three rows of potatoes.[14,23]
- Application of Carbofuran (Furadon 3G) @3 kg ai/ha is effective.
- Appling NSKE 4% and neem cake @ 80 kg/acre during soil preparation as preventive.

Root-knot nematode in North India is managed using nematode-free seed tubers and crop rotation with maize or wheat and application of 1–2 kg ai/ha Carbofuran 3% G at planting. Cyst nematode is managed by crop rotation with vegetables and particularly cabbage and carrot and intercropping potato with beans or wheat and alternating nematode-resistant potato variety "Kufri

Swarna" and application of 2 kg ai/ha Carbofuran 3% G at planting. A 2-year adoption of INM for root-knot and a 3-year.[23]

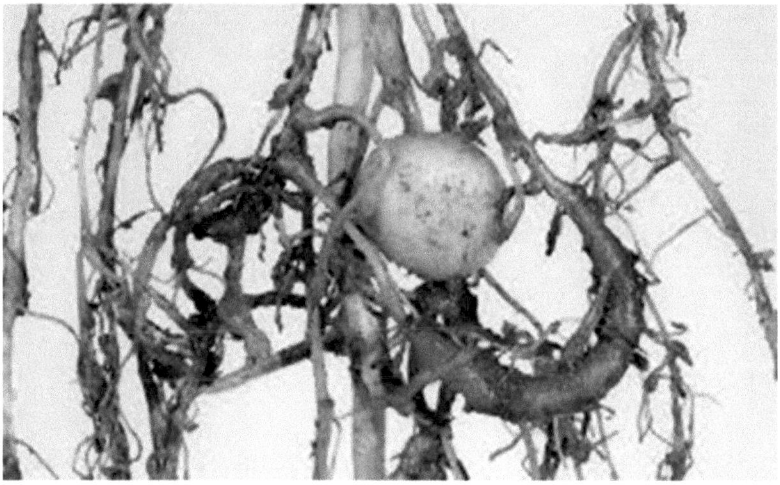

FIGURE 16.14 Root knot Nematode infection

16.15 DRY ROT

- Planting only clean and healthy seed tubers.
- Tuber damage and injury must be avoided during harvest and storage.
- Ensuring freedom from contaminated soil adhering to tubers.
- Washing of tubers followed by drying under shade substantially reduces the infection.
- Harvested tubers should be stored in cold stores.
- Tubers should be examined periodically and rotten tubers must be examined sorted out.
- Avoiding cut tubers for planting. Whole tuber planting to be encouraged.[2,14]

16.16 CULTURAL MANAGEMENT OF EARLY BLIGHT

- Removing sources of infection and destroying contaminated plants and crop rotation.
- Using good irrigation systems to prevent the disease from spreading through water.

FIGURE 16.15 Dry rot infected tubers

- Balanced use of fertilizer to produce healthy plants more resistant to infection.
- Cutting back stems at 80 DAP, to prevent the disease spreading to tubers.
- Use of tolerant varieties like Kufri Sindhuri, Kufri Naveen, Kufri Jeevan, etc.[2,14]

16.16.1 CHEMICAL MANAGEMENT OF EARLY BLIGHT

- Spray of Mancozeb 35% SC @ 500 g/100 L of water @ 0.5%. Or
- Chlorothalonil 75%WP @350–500 g 240–320 L of water/acre (second spray after14 days). Or
- Propineb 70% WP @ 300 g in 100 L of water or 0.30% (second spray after 15 days interval). Or
- Hexaconazole 2% SC @1.2 L in 200 L of water/acre (second spray after 21 days interval).[2,14]

16.17 CULTURAL APPROACH FOR LATE BLIGHT MANAGEMENT

- Use short-duration early maturing varieties to escape the disease.
- Sowing of late blight-resistant cultivars, Kufri Badshah, Kufri Chipsona, Kufri Chipsona-2, Kufri Sindhuri, Kufri Jawhar, and Kufri Sutlej Kufri Giriraj protect the crop.[14,24]

FIGURE 16.16 Early blight infected leaves

- High ridges to cover all daughter tubers properly for reducing chances of infection.
- During favorable temp. 10–20°C and RH>80%, split irrigation lightly can be continued.
- During heavy rains, a sticker like triton @ (0.1%) may be mixed with fungicide to prevent washing.
- Seven days moving sum of RH > 85% for 90 h along with a 7-day temperature range between 7.2 and 26.6°C for 115 h predicts the late blight appearance within 10 days of these conditions.[14,24]
- If 75% of crop foliage is killed by the disease, the haulms should be cut and buried deep in the field.

16.17.1 MANAGEMENT OF LATE BLIGHT THROUGH CHEMICAL APPROACH

- Prophylactic spray (Copper oxychloride 50% WP @ 4–5 g/L or Mancozeb @ 0.2% or Propineb 70% WP @ 2–3 g/L of water).[14,24]
- Use of curative fungicides like Metalaxyl 8% + Mancozeb 64% WP @ 0.25% @ 1 kg/ acre using 400 L water (second spray not <7 weeks) or Cymoxanil 8% + Mancozeb 64% WP @ 600–800 g using 200–300 L of water per acre (second spray after 10 days interval) or

Dimethomorph 50% WP + Mancozeb 75% WG @ 400 g in 300 L water per acre (second spray after 16 days interval.[14,24]

- The seed dressing with *T. viride* + *P. fluorescens* not only reduces the disease incidence and intensity but also increases the plant stand and yield of tubers.[26–28]

FIGURE 16.17(a–c) Late blight infected leaves, twigs and tubers.

16.18 WART OF POTATO (*SYNCHYTRIUM ENDOBIOTICUM*)

- Some of the immune varieties are detected—Kufri Jyoti, Kufri Sherpa, Kufri Kanchan, Pimpernel, Ackersegan, Voran, Sangdorje, and Ultimus.
- Avoid planting of susceptible varieties, Darjeeling Red Round, Kufri Naveen, Kufri Chandramukhi, Kufri Ashoka, Kufri Sindhuri, Kufri Lauvkar, Kufri Lalima, Kufri Dewa, Kufri Safed, and Up to Date.
- Practice of long crop rotation for 15 years or more.
- Rouging out of susceptible plants by burning, destroying warty tubers.
- Use of healthy tubers.
- Diseased tubers may be consumed or fed to the cattle only after boiling.
- Wart lumps of potato peelings should not be thrown in the field or in the manure pit but destroyed by burning.[25]

16.19 POTATO SPINDLE TUBER VIROIDS

Viroids are the smallest-known infectious agents causing diseases in higher plants. The warm weather results in aerial symptoms, which are masked in cooler conditions. Primary haulm symptoms are seldom evident in potato plants. The symptoms on leaflets appear as slightly reduced size with fluted margins with tendency to curve inward and overlap the terminal leaflet. The affected plants remain stunted and are harder. Tubers may be reduced in size and may be deformed with spindling and conspicuous eyes.[3]

FIGURE 16.18 Wart symptom produced on tuber

FIGURE 16.19 Stunted plant due to PSTVd

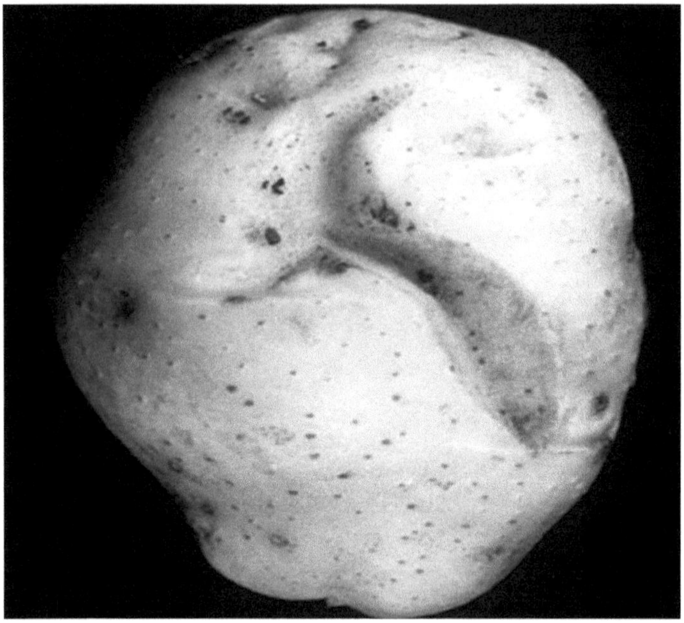

FIGURE 16.20 Surface cracking due to PSTVd

16.20 POTATO PINK ROT DISEASE

1. Pink rot is storage disease and caused by *Phytopthora erythroseptica,* which is more common in potato-growing regions throughout India, nowadays.
2. When the tuber is cut, exposed to air and the infected tissues turns from cream to salmon-pink color and then in to black.
3. The infected tubers have rubbery texture and give off an odor-like ammonia.
4. For managing pink rot, mefenoxam or metalaxyl are most commonly used.
5. Postharvest treatment to tubers before storage with phosphorus acid or fungicides like cyazofamid, oxathiapiprolin, and fluopicolide.
6. Do not select fields having a history of pink rot.
7. Follow crop rotation.
8. Maintain proper soil moisture.
9. The pink rot tubers should not be stored with potato bulk.[29]

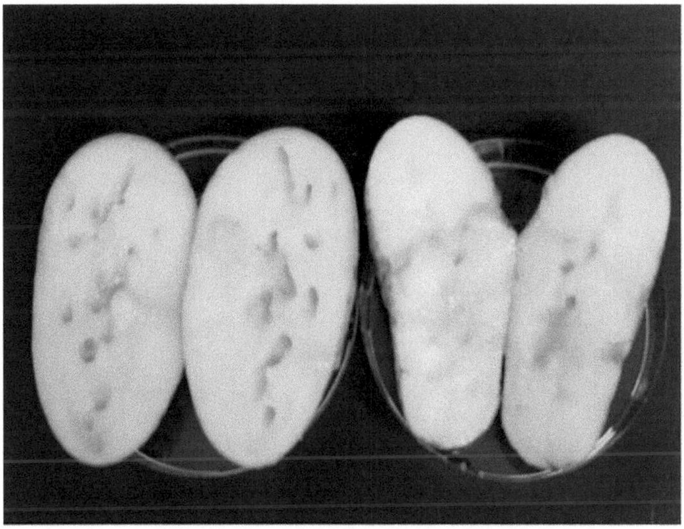

FIGURE 16.21 Pink rot symptom inside the tuber

19.21 POTATO VIRUS DISEASES (PVX, PVY, RUGOSE MOSAIC, PLRV ETC.)

- In case of whitefly, use of peppermint as repellant plant and planting of attractant plant like French bean is recommended, which attracts predatory thrips.
- Application of certified seeds from a reliable source and planting early in the season.[3]
- Rouging out infected plants as soon as detected in the field.
- Dehaulming should be done when aphid population crosses the critical level.[3]
- Control of aphid vectors at planting with Dimethoate @ 1 mL/L or use of Furadan @ 10 kg per hectare in furrows at planting.[3]
- Spray of metasystox @ 0.1% or Imidachlorprid @ 4 mL/16 L for controlling aphid population.[3]

16.22 WHITE GRUBS AND ROOT GRUBS

- Repeated plowing before monsoon (April–May) and exposing the grubs and pupae for predation by natural enemies such as crows, mynahs, etc. or maybe hand collected and destroyed.

FIGURE 16.22 PVX infected plant

FIGURE 16.23 PVY infected plant

FIGURE 17.1 Gummosis of kinnow

FIGURE 17.2 Greening of kinnow

FIGURE 16.24 PLRV infected plant

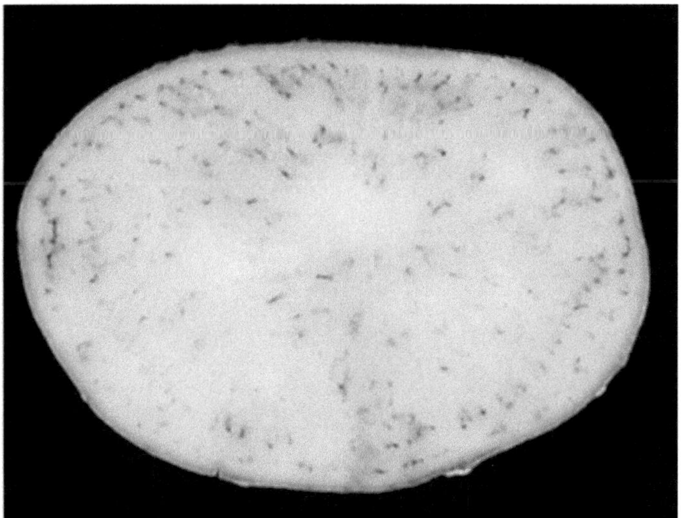

FIGURE 16.25 PLRV infected tubers

- Flooding of the fields wherever possible for 7–10 days.
- Intercropping of *T. patula* and *T. erecta* between two and three rows of potatoes.
- Adoption of suitable crop rotations.
- Application of well rotten FYM/Compost in the field.[30]

16.22.1 BIOLOGICAL CONTROL

- Use of entomopathogenic nematodes as spray @ 1 billion nematodes per acre in infested fields by white grub and root grub.
- The entomopathogenic nematode infected *Corcyra* larvae containing live infective juveniles are released in soil near the base of plant @ four cadavers per plant during May–June or in September in case of white grub and root grub.[30]

FIGURE 16.26 White Grub larvae

16.23 POTATO TUBER MOTH

- The regular monitoring of PTM adult males with sex pheromone baited water trap in the field.

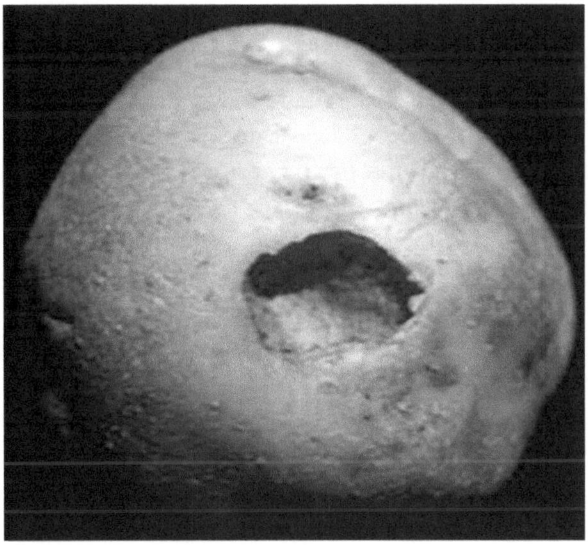

FIGURE 16.27 Tubers infested by White Grub

- Removal of leftover tubers and volunteer potato plants, deep planting, frequent irrigation, and covering the exposed tubers in the field with soil.
- Spraying of Dimethoate @ 1 mL/L. or metasystox at 0.1% gives effective control
- Application of 2–2.5 cm thick layer of dried lantana/Eucalyptus leaves below and on the top of the potato heaps checks PTM infection.
- Application of biocontrol agents namely *Bacillus thuringiensis*/ Granulosis virus against PTM control.[31,32]

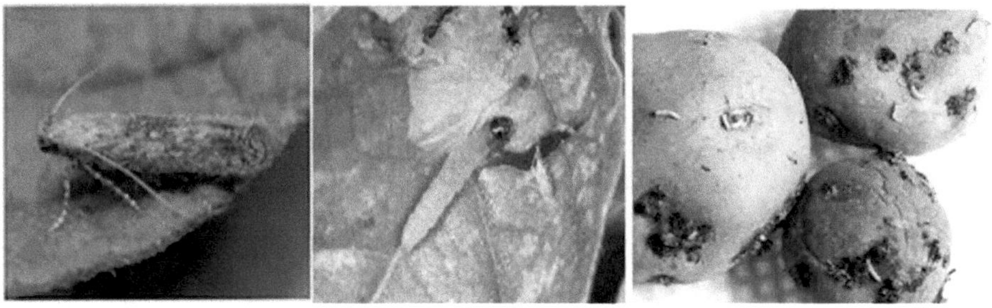

FIGURE 16.28(a–c) Potato tuber moth, larvae and infested tuber

Important potato pests and their time of appearance

Cutworm (*Agrotis ipsilon)* appears from February to March under Indian conditions and damages the seedlings at ground level. Both tomato fruit borer (*Helicoverpa* sp.), and tobacco caterpillar (*S. litura*) feeds on foliage and causes damage during the period from March to April and September to October, respectively. The white fly (*B. tabaci*) causes curling of leaves and sucks the sap during October to March Months. The adults of Epilachna beetle (*Epilachna viginti octopunctata*) also damages foliage. The Green Leaf hopper (*A. biguttula*) sucks the sap of leaves and causes hopper burn during September to March, whereas PTM (*P. operculella)* causes losses on leaves as well as tubers during same period. The aphids (*M. persicae, Aphis fabae, A. gossypii*) are active during the months of November to April. The adults and nymphs of aphids also suck the sap and spread the viruses.[31–33]

ETL of different potato pests

Following are ETL for different insect pests of potato:

- Aphid (20 aphids/100 compound leaves), Whiteflies (9–11 adults/plant), Leafhopper (5–10 adults/plant),
- Lepidoptera Defoliators (1–2 larvae/plant), Cutworms (1–2 larvae/10 plants), White grub (2 grubs per square feet), PTM (10–20 adults/trap/day, 2 or more holes/tuber in stores), Root knot nematode (>20 larva/200 mL of soil).[31,32]

Insect wise IPM schedule in Organic Crop Production

A. Sucking pests (Aphids, white flies and thrips)

1. Use of virus-free seed material from a reliable source.
2. Pest monitoring with the help of yellow sticky traps in case of aphids and white flies.
3. Rouging of infected plants with virus and volunteer plants.
4. Spraying of mixture of cattle urine (1 L) and neem mixture (CUNM) in water in the ratio of 1:9.

B. Defoliators (Grubs, beetles, and moth)

1. Deep plowing during summer.
2. Clipping of leaves having egg masses of insects.
3. Picking of insects by hand and destruction of larvae and grubs.
4. Installation of light or pheromone traps to attract/kill the adult beetles and moths.

C. Borers

1. Deep planting ~8 cm deep (against white grubs).
2. Use of rotten FYM.
3. Installation of light traps @ 1 trap/ha
4. Soil application of neem + 5% cattle urine at seedling and tuber formation stages.
5. Use of *Beauvaria bassiana* (Larvocel), or *Metarrhizium anisoplae* @5% as soil application
6. Application of cattle urine extracts or botanicals like *Melia* sp./ *Eucalyptus* sp. (5%) at the time of sowing, earthing up, and tuber formation.
7. Foliar spray of *Bacillus thuringiensis* (*Bt*) @ 3 gm/L or NSKE@ 2 mL/L of water at weekly intervals.

D. Potato Tuber Moth (PTM)

1. Sanitizing the store before tuber storage
2. Tuber treatment with *Bt* @ 0.05% or GV (mixed and dried 1 kg talc + 20 GV infected PTM larvae + 2 mL sticker + 1 L. water) @ 500 gm/bag
3. Use of pheromone trap @ 4 traps/100 m^3 in storage or in field @20 traps/ha.
4. Covering the stored potatoes with powdered *Lantana/Eucalyptus/*Neem leaves
5. Tuber treatment before storing using CIPC @ 40 mL/t of potatoes.

E. Nematodes

1. Deep ploughing
2. Planting of *T. erecta* as ntercrop in between 2 and 3 lines
3. Amendment with neem, groundnut, sesamum, castor, mahua, karanja, etc. de-oiled cakes.
4. Planting resistant variety Kufri Swarna/Kufri Neelima (Cyst nematode), Hybrid HC-294 (Root Knot Nematode).
5. Use of biotic agents, *Paecilomyces liliacinus*, *Aspergillius* spp., and *Bacillus* namely NB-I, B-68.[31–33]

For Seed Potato Production

Follow above measures and add spraying of Imidacloprid @ 2 g/5 L water and haulm cutting in case of sucking pests, spraying Carbaryl @2 g/L of water for management of defoliators, Carbofuran 3G (80–100 kg/ha) at the time of earthing up or use of Chlorpyriphos @ 2.5 mL/L of water at 25 days as drench on ridges in case of borers, use of malathion dusting @250 g/quintal tubers in control of viruses and foliar spray with Carbofuran @ 3 kg a.i./ha in case of nematodes.[31–33]

IPM packages in potato based on crop growth stages

A. Planting stage

1. Clean cultivation and deep plowing.
2. Elimination of collateral hosts.
3. Soil application of Phorate 10G or Furadan 3G @ 2–3 kg a. i. per ha in case of white grub and nematodes in Nigiris hills. Crop rotation with nonhost crops in case of Root Knot Nematode (RKN).
4. Deep plowing and use of resistant varieties (Kufri Swarna, Kufri Neelima) for management of PCN and PTM in Nilgiris plateau.
5. Application of Phorate 10G @ 1.0 kg/ha in soil in case of seed crop in mid hills for aphids.

B. Emergence

1. Spraying 5% cattle urine solvent extracts of Melia or Lantana plants in case of aphids, whitefly, leaf hoppers, mites, and cut worms in early crop in Northwestern plains. The spray of Dicofil @ 1–2 mL/L of water or Chlorpyriphos 20 EC @ 2.5 kg/ha at 35 days of sowing.

C. Tuber initiation

1. Application of Furadan 3G @ 2–3 kg a.i. /ha for control of white grubs, nematodes and aphids on Nilgiris hills and plains.

D. Bulking

1. Use of foliar spray of Methyl Demeton 25 EC @ 0.03% and removal of collateral hosts in case of aphids and white grubs in seed and commercial crop on hills.
2. Fixing light traps in the fields along with foliar sprays of Bt @ 0.05% or GV @ 4 LE/L water for control of PTM on mid hills and on plateau.

E. Leaf senescence

1. Haulm cutting and pheromone traps are used for mass trapping of PTM.

F. Harvesting

1. Early harvest along with the destruction of pest-damaged tubers.

G. Storage

1. Storing potatoes in cold storage, cleaning the store before storing, treating tubers with Bt @ 0.05% or GV (1 kg talc + 20 GV infected larvae of PTM + 2 mL of sticker + 1 L of water mixed, dried & apply @ 500 g/bag, 4 traps of pheromone per 100 m^3 of storage space, covering the stored potatoes with *Lantana/Eucalyptus* or Neem leaf, treatment of potato tubers with CIPC, and talc-based formulations of potential botanicals/entomopathogen for control of vector and PTM.[31–33]

16.24 PREPARATION OF CATLE URINE OR SOLVENT PLANT EXTRACTS

The extracts of plants were made either in cattle urine or in chemical solvent like methanol, acetone, ethyl alcohol, etc. The test plant materials were dried in shade and grinded to make powder and later mixed in urine in (1:1) for allowing fermentation. It was fermented for 15 days and later filtered through Whatsmann filter paper to get extract. It was further diluted using water according to concentrations for sprays. The powdered and dried plant material was subjected to Soxhlet's extraction for 8 h at 60°C using solvent to get chemical extract.[30–32]

16.25 METHOD OF MAKING TALC-BASED FORMULATIONS OF ENTOMOPATHOGENS/BOTANICALS

The cultures of bacterial and fungal pathogens were mixed in talk in the 40:60 ratio. It was then dried aseptically for 7 days and later under vacuum overnight. The material was then turned in to powder and the spores were counted using spread plate technique and were adjusted to 1×10^{10} spores per gram. For other ingredients like botanicals or urine of cattles and the solvent extracts were mixed in it in 1:2 ratio (1 part talc: 2 parts plant extract) and dried for 7 days before using it.[31–33]

16.26 CONCLUSION

The increasing expenditure due to plant protection chemicals and the agriculture nowadays become less profitable enterprise in India. The toxic agrochemicals pollute the environment and harm consumers' health. A more eco-friendly and cost-effective alternative option for Indian agriculture would be Integrated Pest Management. Under IPM, the monitoring-based measures of crop protection are practiced to reduce the crop losses. The success of the IPM program requires efficient management and continuous implementation of best practices. In both long and short-term program, the IPM has made an example toward pesticide-free environment in many countries in the world. These systems are dynamic in nature and new technologies are integrated in it as they become available with time and space.

KEYWORDS

- **integrated pest management**
- **potato**
- **pest**
- **disease**
- **agroecosystem**
- **crop growth stage**

REFERENCES

1. Monthly Report of Potato, Horticulture Division, Department of Agriculture Cooperation & Farmers Welfare, Ministry of Agriculture& Farmers Welfare, Govt. of India, New Delhi, 2021.
2. Paul Khurana, S. M. Diseases & Pests of Potato - A manual; Central Potato Research Institute, ICAR, Shimla, H.P. India, 2000; pp 1–66.
3. Paul Khurana, S. M. Potato Viruses & Viral Diseases, A Technical Bulletin No. 35 (Revised), CPRI, ICAR, Shimla, H.P, India, 1999; pp 1–73.
4. Singh, P. H.; Chandla, V. K.; Paul Khurana, S. M.; Garg, I. D.; Thakur, K. C.; Singh, S.; Lal, S. S. Extension Bulletin No 34 (E) Potato Cultivated in North Eastern India CPRI, ICAR, Shimla, H.P, India, 2003; pp 1–38.
5. www.newworldencyclopedia.org.entry/potato
6. Tantowijayo, W.; Van de Fliert, E. All About Potatoes: An Ecological Guide to Potato Integrated Crop Management, International. Potato Center, Food, Agriculture Organization Regional Vegetable IPM Program in Asia Bangkok, Thailand, 2006.
7. www. project ipm centers.org
8. www 2 ipm, UC anr.edu>agriculture>introduction
9. Sardana, H. R.; Trivedi, T. P.; Bambawale, O. M.; Naved, S.; Singh, R, V.; Arora, S.; Krishnamoorthy, P. N.; Pandey, P. K.; Pandey, K. K.; Sengupta, A. Vegetable Crops. In *Validated IPM Technologies for Selected Crops*; Singh, A., Sardana, H. R., Sabir, N., Eds.; National Centre for Integrated Pest Management: New Delhi, 2004; pp 119–147.

10. Van den Berg, H. *Facilitating Scientific Method as Follow-Up for FFS Graduate*; Food, Agriculture Organization: Rome, Italy, 2001.
11. Plant Pest Surveillance. *International Plant Protection Convention*; Food, Agriculture Organization of the United Nations, 2016; pp 1–55
12. plantpath.iaslate.edu/tylkalab/content/extracting-nematode
13. Altieri, M. A.; Nichollas, C. I. N. D. Ecologically Based Pest Management: A Key Pathway to Achieving Agro Ecosystem Health, 2014. http://www.unicamp.br/fea/ortega/agroceol/ecpestma.html
14. http://www.cpri.icar.gov.in
15. http://www.ipmdata.ipmcenters.org
16. Mohler, C. Ecology of Weed Management in Organic System. http://www.css.cornell.edu/weed Eco/Mohler.EcologyofWeedManagementPresentation.pdf, 2010.
17. Kumar, S.; Singh, P. H.; Garg, I. D.; Paul Khurana, S. M. Integrated Management of Potato Diseases. *Indian Hortic.* **2003,** *48*, 14.
18. http://www.theorganicfarmer.org/know-about-bacterial-wilt-,-control-it/
19. http://pnwh,books.org/plantdisease/potato-solanum-tuberosum-rhizoctonia-canker-black-scurf
20. http://pnwh,books.org/plantdisease/potato-solanum-tuberosum-bacterial-soft-rot-blackleg-,lenticel-rot
21. Basu, A.; Das, S. Integrated Management of Potato *(Solanum tuberosum)* Diseases in Hooghly Area of West Bengal. *Indian J. Agric. Sci.* **2003,** *73* (12), 649–651.
22. Basu, A. Screening of Potato Genotypes Against Common Scab Disease Caused by *Streptomyces scabies* in the Plains of West Bengal. *Indian J. Agric. Sci.* **2007,** *77* (3), 199–200.
23. Krishna Prasad, K. S. Management of Potato Nematodes: An Overview. *J. Hortl. Sci.* **2008,** *3* (2), 89–106.
24. Arora, R. K.; Sharma, S.; Singh, B. P. Late Blight Disease of Potato and Its Management. *Potato J.* **2014,** *41* (1), 16–40.
25. Singh, P. H. Shekhawat, G. S. Wart Disease of Potato in Darjeeling Hills. *Tech. Bull.* **2000,** *19*, 73.
26. Basu, A.; Konar, A; Mukhopadhyay, S. K.; Chettri, M. Biological Management of Late Blight of Potato Using Talc Based Formulations of Antagonists. *JIPA*. **2001,** *28* (1), 80–81.
27. Basu, A.; Hossain, M. M.; Konar, A.; Chettri, M. Integrated Management Of Late Blight Disease Of Potato. *Ann. Plant Protection Sci.* **2003,** *11* (1), 64–66.
28. Basu, A. Employing Eco Friendly Potato Disease Management Allows Organic Tropical Indian Production Systems To Prosper. *As. J. Food Ag-Ind. Spl. Iss.* **2009,** S80–S87.
29. http://www. Potatodiseases.org/pinkroot.html
30. Mehta, P. K.; Chandla, V. K.; Mathur, Y. S. Status of White Grub in North Western Himalaya. *J. Insect Sci.* **2010,** *23*, 1–14.
31. Chandel, R. S.; Chandla, V. K.; Garg, I. D. Integrated Pest Management of Potato Tuber Moth in India. *Trop. Agric.* **2008b,** *20*, 129–140.
32. Sharma, S.; Thakur, M.; Chandla, V. K.; Singh, B. P.; Chakrabarti, S. K.; Rawat, S. Integrated Management of Potato Pest: A Field Manual, Central Potato Research Station, Shimla, 2015.
33. Shankar, U.; Priya, S.; Kumar, D. *Vegetable Pest Management: Guide for Farmers*; International Book Distributing Company: Lucknow, India, 2008.

CHAPTER 17

Integrated Disease Management in Kinnow

P. N. MEENA, M. K. KHOKHAR, and D. RAGHAVENDRA

ICAR-National Research Centre for Integrated Pest Management, Pusa Campus, New Delhi 110012, India

**Corresponding author. E-mail: pnshera@yahoo.co.in*

ABSTRACT

Citrus is one of major tropical fruit of India, which have vital role in the fruit economy of the country next to mango and banana. It is affected by many viral, bacterial, and fungal diseases from nursery to fruit ripening stages. Among the citrus fruit groups and kinnow crop affected by many important diseases like, *Phytophthora* (gummosis), greening, ring spot virus, citrus canker, citrus scab, dieback, anthracnose, *Alternaria* fruit rot, and sooty mold causes economic losses from nursery to harvesting and information regarding disease management are scattered and scanty. Moreover, information on detection, diagnosis of pathogens, and its transmitting vectors are also scanty. Nowadays, strategies for management of the pests are mainly focused around the synthetic pesticides, which is overloading in the food chain, causing human health hazards, chemical residue in soil, lowering the natural enemies population, disturbing the natural beneficial microorganisms, and resurgence of pests and resistance problem. Therefore, it is immense need to curtail the chemical pesticide load from the food chain that could be achieved suitably by developing and implementing the latest IPM technology. This book chapter provides the latest information on economic impact, distribution, disease symptoms, epidemiology, disease cycle, cultural control, biological control, and chemical control that help in controlling these menacing, which are still challenges even after several years of their

Integrated Pest Management in Diverse Cropping Systems. Devendra Pal Singh, PhD (Ed.)

discovery. All of this information should lead to more efficient in Integrated Disease Management of kinnow pathogens.

17.1 INTRODUCTION

Citrus fruit is grown across the globe and has an enormous social, cultural, and economical influence on our society.[1] It is considered an important tropical fruit crop in India, which plays a vital role in the fruit economy of the country, next to mango and banana. In India, citrus fruits cultivated area is spread >100.3 million hectares, whereas production is increased by 12.54 million tonnes.[2] In India, citrus is primarily grown in Assam, Andhra Pradesh, Maharashtra, Punjab, Haryana, Karnataka, Uttaranchal, Bihar, Orissa, and Gujarat. Among the citrus fruits, kinnow is most valuable fruit crop that mostly grown in Haryana, Punjab, and Rajasthan states. The cultivated area under kinnow fruit is being stretched from arid and semiarid regions due to its growing demand in domestic and international consumer markets.[1] The cultivated area under citrus fruits was 19.4 thousand hectares with the production of 235.4 thousand MT in Haryana in which Kinnow occupied 80% in both area and production.[2] In Haryana, it is extensively grown in the districts of Hisar, Bhiwani, Sirsa, Mahendergarh, and Ambala.

The national average productivity of citrus is 9.69 tonnes/ha. Punjab state grows >50,000 hectares of kinnow and contributes more than 90% of this area and production.[1] Productivity and quality are main concerns in citrus that are severely affected by insect pests and diseases. Citrus globally hampered by different *Phytophthora* spp., causing tree decline and yield losses. Several *Phytophthora* species are reported in different regions[3] among which *Phytophthora nicotianae*, *Phytophthora palmivora,* and *Phytophthora citrophthora* are the most predominant species[4] in India (Assam, Nagpur, and Punjab).[5] Citrus greening disease (CGD) caused by *Candidatus Liberibacter asiaticus* is the most predominant species in citrus crop, which is vectored by Asian psyllid (*Diaphorina citri* Kuwayama).[6,7] Huanglongbing (HLB) was recorded in sixteen states of India using the RT-PCR.[4] Blackfly (*Aleurocanthus woglumi* Ashby) and whitefly (*Aleurolobus marlatti* (Quaintance) and aphids (*Aphis* spp. *Toxoptera* spp.) secreted honeydew and spread sooty mold disease from one plant to another plant.

Most of the farmers rely on pesticides for pest and diseases management. They apply huge quantity of pesticides injudicious form, which leads to pest resistance. Moreover, it also increases chemical residue in soil, high input agriculture on farmers and consumers and environment. Therefore, IPM needs to be augmented to reduce pesticide load from food chain. At present, kinnow crop is affected by many important pests, and information regarding pest management is scattered and scanty. This book chapter provides the latest information on IPM interventions against important pests and diseases of Kinnow.

17.2 GUMMOSIS OF CITRUS *(PHYTOPHTHORA* SPP.)

Citrus gummosis is most emerging and devastating disease now a day. It is caused by soil-borne *Phytophthora* spp., belongs to the kingdom Chromista, class Oomycetes and Peronosporales order. Among the pathogen orders, Peronosporales and Saprolegniales are major order.[3] Most of species of this pathogen comes in genus Phytophthora and are responsible for tremendous agricultural losses for every year.

17.2.1 HISTORY AND ECONOMIC IMPACT

In world, Gummosis is the longest-known diseases of citrus. First, *Phytophthora* epidemic was observed during 1832–1836 in Azore Ireland. Later in 1841 in France, then in 1845 in Portugal. Till 1914, it was confirmed all over the world.[8] Gummosis causes 10–30% yield losses each year.[9] It is extremely destructive and declining disease of orchards. A total of 90% orchards with an average disease incidence of 45% were observed in various surveys.[10] Pathogen is spread very fast.[11]

17.2.2 DISEASE SYMPTOMS

Gummosis is a major destructive disease, which produces gummy material on the trunk and affected branches of citrus plants. Each blister contains a gum pocket from where it secretes out. Affected tree bark changes in dark brown color and develops longitudinal bark cracks. In severe cases, bark rots, and the trees dry due to girdling effect. The disease is more conducive during heavy rains and excessive soil moisture. Pathogen is spread plant to plant through flood method of irrigation, and deep planting, Due to the increase severity of the pathogen, the necrotic lesions expand and cause the death of tree.[12] The gummy material is dissolved in water during the rain and appearance of discolored cortex.

FIGURE 17.3 Die-back of kinnow

FIGURE 17.4 Canker

FIGURE 17.5 Anthracnose disease

FIGURE 17.6 Fruit rot of kinnow

17.2.3 EPIDEMIOLOGY

Relative humidity, temperature, and rainfall are major climatic factors that affect *P. nicotianae* pathogen. Increased soil moisture and RH and rainfall influenced disease incidence and disease severity.[13] *P. nicotianae* produces chlamydospores when soil temperature fall down below 12°C and went into dormant till comes of favorable conditions. When temperature rise above 12°C, population of pathogen suddenly starts to increase.

17.2.4 PATHOGEN DISSEMINATION

Use of infected nursery stocks is a primary source of spread of *Phytophthora* spp. in citrus orchards. This pathogen is soil borne in nature and persists in soil or infected root in hidden manners. Pathogen is also spread in citrus orchards through vehicles equipments and flood irrigation method. Surface water following heavy rains may carry the pathogens and spread between the plants.

17.2.5 DISEASE MANAGEMENT

- *TRUNK PAINT:* Affected bark portion, branches, and green portion of trunk of infected plants may be scrap with the help of knife. The affected portion of plants such as barks and branches carefully collect and immediately destroyed. Apply the fungicidal paint of Ridomil Gold on affected portion of bark, trunk, and branches twice in a year whenever symptoms appear.
- *SOIL DRENCH:* Drenching of Ridomil Gold (25 g/tree in 10 L of water) give twice in the months of February–March and July–August.
- *FOLIAR SPRAYS:* Foliar spray of fosetyl aluminium (Aliette) (2.5 g/L) can be made during September-October months. Spray of Aliette and drenching of Ridomil Gold can be made with combination in root zone area of citrus plants.
- *IRRIGATION:* Double ring basin method is used instead of over flooding irrigation.
- *BIOAGENTS:* Application of *T. viride* (2×10^7 cfu/g) at 100 g/ plant + *P. fluorescens* (2×10^8 cfu/g) at 100 g/ plant with carrier material of FYM (10 kg) followed by spraying of tree trunk with Ridomil MZ-72 @ 0.2% for the management of *Phytophthora* in February, March, August, and September.
- *INTERCROPPING:* Intercropping of cowpea/garlic/chickpea between the rows of trees, which was useful for moisture conservation and resulting in to less frequency of irrigation that in turn prevented spread of *Phytophthora* disease and also suppression of weeds.

- *INTERCULTURE:* Pruning and thinning of lower branches reduce the infection risk of gummosis.

17.3 CITRUS GREENING

CGD is most destructive disease that infects almost all cultivars of citrus groups and causes huge economic losses.

17.3.1 HISTORY AND ECONOMIC IMPACT

Globally >60 million trees had been knockdown due to this disease. The name of the disease has been recently changed from "greening" to the Chinese word "HLB" (since the disease was first discovered in China).[14] Huanglongbing means "yellow dragon disease," the infected plants turn in yellow in color and start only one branch before they die. The term "greening" means green color of infected fruits after maturity. The pathogen is transfer by grafting and vector. This disease is caused by CGD fastidious bacteria, which is member of the phylum Proteobacteria.[15] The bacterium is systemic in nature in vectors and present in the sieves tubes of phloem tissues of infected trees. The CGD persist in the nature and in plants in three different forms due to combinations of different climatic factors. The causal organism of CGD disease is *Candidatus idatus, Liberibacter asiaticus*, a heat-tolerant form of bacteria persist in Asia, is transfer through *D. citrii* vector and the Asian citrus psyllid.[6] Incidence of the disease ranged from 8% to 43% in Mosambi sweet orange, 30–40% in Malta sweet orange, 9–46% in Sathgudi sweet orange, 15–47% in Coorg mandarin, 1–6% in Nagpur mandarin, 16–30% in Sikkim mandarin, 10–20% in Darjeeling mandarin, 10–53% in Jampui Hills mandarin, 3–15% in Kinnow mandarin, 8–38% in Assam lemon, and 2–13% in acid lime.

17.3.2 DISEASE SYMPTOMS

Greening infected citrus trees leaves are small, upright, and showed green veins along with chlorotic interveinal areas. The disease severity and range of infected plant symptoms varies according to extent of infection, nutritional status, age, and season. Infected leaves show chlorotic mottling and mainly greening is identified by Leaf mottle symptoms. In severity of disease chlorotic may be converting in scattered dark green islands. Infected plants show symptoms on leaf canopy, branches, twinges, and on fruits. Infected plants always bearing small green color fruits that fall prematurely, while those that remain on the tree do not color properly and remain green on the stylar (lower) end.

17.3.3 EPIDEMIOLOGY

C. Liberibacter spp. is a fastidious gram-negative proteobacterium, which is present in phloem tissues[15] and transmitted through psyllid vectors in Asia[6] and America, *Trioza erytreae* in (Africa) and also transferred by grafting in nursery. *Ca. Liberibacter africanus* is mainly present in African, "*Ca. Liberibacter americanus*" in Brazil, and "*Ca. Liberibacter asiaticus* (Las)" in Asia, North and South America.[6] Symptoms of Asian HLB mainly occur at the temperatures of 32°C–35°C.[17] When new flush is started, natural transmission of disease is increased due to build-up population and migration of psyllid. Symptom expression usually occurs during autumn and winter season and sometime may be continuous. Severe occurrence of the disease is noticed in sweet oranges, acid lime, and lemon. Heavy soil, high water table, high soil moisture, soil pH of 5.4–7.5, and temperature of 25°C–28°C are conducive for disease development. Low grafting and deep planting increase the chances for soil borne infection. The fungi survive on fallen fruits, twigs, leaves, cracks of the tree, spread by irrigation water, rain splashes, wind, insects to stems, leaves, and fruits.

17.3.4 PATHOGEN DISSEMINATION

In nursery, CGD is transmitted by infected budwoods, while in orchards through citrus psylla. Pathogen is acquired by Psylla at nymphal stage and transmits it when adult.

Through Insect

The greening pathogen can be transmitted by two species, *D. citri* Kuwayama (Asiatic psylla) and *T. erytreae* Del Guerico, (African psylla). Bacteria multiply in vector by persistent manner and present in the haemolymph and transmission occurs through salivary glands. Transmission rates up to 100% have been reported.[17] The acquisition latent period is 24 h for African psyllid.

Serial transfer experiments suggest that young nymphs of *T. erytreae* retain bacteria without transmit them.[17]

Through Graft Transmission

In China, Lin[18] confirmed that yellow shoot was graft transmissible. Graft transmissibility of African greening was confirmed in 1965 by Oberholzer.[19] Variability in graft transmission of *Candidatus liberibacter* spp. depends upon the plant part used for grafting, amount of tissue and pathogen isolated from single bud. Graft transmission of African CGD varied from 0% to 50%.[20]

Dodder Transmission

Cuscuta campestris transmitted the CGD from periwinkle to Sweet orange and within three months yellowing symptoms appeared on sweet orange.[19] Transmission of greening organism from affected citrus to periwinkle was achieved by graft inoculation.[21]

17.3.5 DISEASE MANAGEMENT

- Removal of affected trees, and their replacement by healthy plants developed on improved rootstock.
- Regulatory measures should be imposed on sale, movement, and infected nursery stock.
- Application of Bactronol @1000 ppm/Tetracycline @ 500 ppm for management of citrus greening.
- Spraying of 150 g Thiamethoxam 25% WG or 500 mL Monocrotophos 36 WSC in 500 L of water
- Petroleum oil sprays (2%).

17.4 TWIG BLIGHT/ DIEBACK *(BOTRYODIPLODIA THEOBROMAE)*

17.4.1 HISTORY, ECONOMIC IMPACT

Citrus decline (complex problem) gradually decline of production and productivity of orchards and finally, orchards become unproductive. Seriousness of this complex problem observed in the 19th century. It was first observed by Roghoji Bhonsale in the 18th century in the central province and in Assam first observed in the year 1888. The decline of grapefruit, mandarin, and sweet orange varieties grafted on sour orange has led to the loss of about 100 million. This disease was first observed in Punjab in 1940 and Department of Agriculture, Bombay in 1912. It became very severe in early 1950s. There are several biotic and abiotic factors responsible for this disease. Abiotic factors included nutritional imbalance and soil-related constraint, while biotic factors included pathogens and Phanerogamic parasites.

17.4.2 DISEASE SYMPTOMS

In India, twig blight is commonly found in mandarin plants. Among the citrus plants, Nagpur mandarin trees in central India commonly affected by twig blight and show drying of twigs. The disease progresses due to involving of complex stress factors, such as infestation of insect pests, incidence of diseases, drought stress, nutritional deficiencies, and *Phytophthora* root rot infection. The affected plants show drying of twig from tip to die back, which indirectly and directly reduced the quality and quantity of fruits yield. Affected plants prematurely drop fruits and develop stem end rot in fruits during storage.

17.4.3 EPIDEMIOLOGY

Colletotrichum gloeosporioides remains persists in dead twigs, and branches during dormant stage, which later leads to twig blight. The infection hyphae produced by appresoria remains latent even after the fruit matures. The conidium of the fungus mainly persists on affected dead twigs of the mother plants and dispersed by rain splashes to developing fruits.

17.4.4 PATHOGEN DISSEMINATION

Twig blight causing pathogen invaded the tree tissues after wounding caused by pruning and mechanical damage. It can also invade the citrus trees in extreme cold, heat, and drought conditions. Fungal pycnidia also spread the diseases from infected bark to healthy bark and twigs. Primarily, declining of twigs from tip to down ward mainly occurs due to some physiological deficiency that renders to more colonization of healthy tissues. Similarly, disease also spread from infected plants to healthy plants due to rain splash.

17.4.5 DISEASE MANAGEMENT

- Dried twigs carefully removed from infected trees, destroy them, and always apply Bordeaux paste on cut ends. Take spray of Copper Oxychloride@3 g/L of water in March, July, and September month.
- Spray of Streptocycline 100 ppm or COC 0.3%.
- The fungus is controlled by spraying of Carbendazim @ 0.1%, twice at monthly interval after pruning.

17.5 CITRUS CANKER (*XANTHOMONAS AXONOPODIS* PV. *CITRI*)

17.5.1 HISTORY AND ECONOMIC IMPACT

Canker is most feared disease of citrus trees, which causes different level of severity and losses in different citrus varieties. Canker is prevalent in endemic form in India, Japan, and South-East Asian countries except Europe. It was first reported in Punjab[22,23] and its occurrence was recorded in Andhra Pradesh, Karnataka, Rajasthan,[25] Tamil Nadu,[24] Madhya Pradesh, Assam, and Uttar Pradesh.[26] Canker is serious problem in acid lime (*C. aurantifolia*) growing region of central India, Nellore, southern India, it has also observed in Kinnow mandarin nursery in the state of Punjab. A total of 8–9% of fruit infection is common in susceptible citrus trees that have heavy foliage infection, defoliation, and leaving only bare twigs.[27] In Argentina, 83–97% of grapefruit trees were diseased during 1979–1980.[28]

17.5.2 DISEASE SYMPTOMS

Canker infection mainly occurs in the early growing stage; the lesions start as pinpoint spots and attain a diameter of 2–10 mm as infection spread. Initially, lesions are circular but later may aggregated at the leaf margin. Lesions become visible underside of leaves and on upper surface after 7 days of infection. Yellow halo of leaves is main characteristic symptom of the disease. Young lesions are pustule and later became corky and sunken center. The lesions later spread on fruits and causes fruits crack, which fall prematurely. Canker-infected fruits are not acceptable in market.

17.5.3 EPIDEMIOLOGY

The disease spread from plant to plant through canker legions present on twigs and branches, and constitute the primary source of inoculums infections. The infected leaves of affected tress drop early and retain bacteria up to 6 months,[29] whereas pathogen persist and survive up to 76 months in diseased twigs.[30] Vasudeva[31] reported that bacteria survive in sterilized soil for 52 days, and in infected leaves >6 months, while 11 days under desiccation at 30°C. Similarly, Paracer[32] observed that organism survive up to 120 days in ordinary temperature. The bacterium epiphytically survives without developing of symptoms on citrus host, associated weed, grass, and in soil.[33] The saprophytes are present in soil in absence of plants.[34] Graham[35] observed that infected lesions oozing out the bacterial cell during congenial weather conditions and serve to source of inoculums for disease development. This pathogen mainly enters in plant tissue through wounding, stomata creating by insect and wind. Bacterial multiplication and disease development mainly depends on host plant resistance.[36] Successful bacterial infection requires moisture on the host surface for 20 min.[24]

17.5.4 PATHOGEN DISSEMINATION

Xac is moved from infected plants to healthy plants by rain droplets. Xac has mucilaginous coat that suspend in water and is dispersed in droplets with in plants. While at short distance, it may be dispersed through wind and rain. Infected nursery plants serve source of movement or disperse for long distance. Disease development mainly progresses between temperature of 20°C–30°C.[24] *Phyllocnistis citrella* Staint (leaf miner) make small serpentine mines between leaves and infest stems and fruit and quickly spread the bacteria with creating of number of individual lesions, which later coalesce and turn into large irregular shaped lesions. When leaf miner feed in leaves, it makes feeding galleries, which later contaminated with bacteria resulting huge inoculums production and bacterial infection.[26,37]

17.5.5 DISEASE MANAGEMENT

- Eradication of all infected and suspected plants as for as possible and should be burnt.
- Used workers tools and cloths properly disinfected.
- Use only healthy nursery stock for planting and affected twigs of old orchards should be pruned than spraying with Bordeaux mixture (1.0%), COC (0.3%) at periodical intervals.
- Proper quarantine procedure should be followed to prevent the spread of canker in uninfected citrus growing areas.
- Use mandarin-resistant varieties and compared with lemon groups.
- Spray (Streptomycin Sulphate 9% + Tetracycline Hydrocloride 1%) 100 ppm solution at an interval of 15–20 days.

17.6 SCAB (*ELSINOE FAWCETTI*)

17.6.1 HISTORY AND ECONOMIC IMPACT

Elsinoë fawcettii Bitancourt and Jenkins causing citrus scab disease is widely occurring in citrus growing region, where weather conditions are congenial for infection. Incidence of *E. fawcettii* is mostly seen on mandarins, grapefruits, and sour oranges. Nursery rootstocks of rough lemons, oranges, *P. trifoliate,* and *C. limonia* are more susceptible to citrus scab disease. The disease is present in Assam, Andhra Pradesh, Uttar Pradesh, Karnataka, and West Bengal. A total of 55% spoilage of citrus fruits was observed in Assam by citrus scab.

17.6.2 DISEASE SYMPTOMS

Scab infection in seed beds and nurseries causes stunting of seedlings, wrinkled, and distortion of leaves. Scab mainly infects the citrus leaves, fruits, twigs, and produced wart-like outgrowth. Disease primarily pinkish in color on fruits and leaves but later turn into dark color. Spores of the fungus are produced in infected lesions present on fruits and leaves.

17.6.3 EPIDEMIOLOGY

Young leaves are highly susceptible. Fruits remain susceptible up to three months after fruit set. Wet weather for 2.0–2.5 h is essential for germination of conidia and fresh infection. The conidia are formed at temperature ranging from 7°C to 30°C with 66–100% RH. The pathogen infects wet tissues at optimum temperature of 16°C–23°C. The disease is more severe at low temperature and high relative humidity. The pathogen perpetuates and survives in off-season in scab pustules as ascospores. The fungus spread through conidia.

17.6.4 PATHOGEN DISSEMINATION

Overhead irrigation and rain splash are responsible for spread of the fungus spore in mandarin orchards. Dew and congenial humidity force to spore to be release from infected fruit and leaves lesions. But due to the limited splashing action, only localized dispersal may occur.

17.6.5 DISEASE MANAGEMENT

Cultural Methods

- Infected nurseries become primary source of inoculums to any area. Disease-free plants should be selected to new plantation to avoid introduction of inoculums. Always avoid overhead sprinkler. The diseased fruits, leaves, twigs, should be destroyed to minimize the inoculums load and re-infection.
- Prophylactic sprays of captafol and benomyl at new flush minimize the incidence.
- Spraying the tree with COC 0.3% before the new spring growth starts and a second spray could be applied after the petals have dropped.
- Water logging condition should be avoided in close citrus plants that favor development of pest population.

17.7 ANTHRACNOSE (*COLLETOTRICHUM GLOEOSPORIOIDES*)

17.7.1 HISTORY AND ECONOMIC IMPACT

Anthracnose is also known as wither tip. In recent years it is responsible for decline in lime production in North India. The diseases distributed in Punjab, Maharashtra, Madhya Pradesh, Uttar Pradesh, Assam, Rajasthan, and south India. A total of 32% disease incidence has been reported in South Africa, while 64.6% in Costa Rica during 1990.[38]

17.7.2 DISEASE SYMPTOMS

Anthracnose infected trees showing various disease symptoms such as dieback of twigs, leaf spot shedding of infected leaves, and fruit rot. Infected leaves initially show green spot but later changes in brown color. Infected twigs show black dot like acervuli in concentring ring. When infection rate is more severe, immature fruits prematurely dropped, while infected branches and tree dies due to die back in few years. Malta oranges and grapefruits suffer much more damage by this pathogen. In wither-tip disease, the branches and twigs looked to have been burned by fire. Drying of twigs from top to downward is witnessed. Dry twigs are ash color. Anthracnose-infected fruits show tear-stain symptoms.[39] *C. gloeosporoides* infected fruits show round, brown-colored water-soaked sunken lesions.[40]

17.7.3 EPIDEMIOLOGY

C. gloeosporioides persist on branches and twigs in a dormant form. This leads to the twig blight. The infection hyphae produced by appresoria remains latent even after the fruit matures. The conidia of the fungus are produced on dead twigs of the mother plants and dispersed by rain splashes to developing fruits. Extreme moisture and temperature around 25°C make more favorable for appearance of disease.

17.7.4 PATHOGEN DISSEMINATION

Sporulation in *C. gloeosporoides* follows two ways, saprophytic and pathogenic. Pathogenic germination occurs on plants characterized through mitosis followed by growth of single germ tube. This process started immediately and results in the development of appresoria. Saprophytic germination takes place in rich medium and it takes longer period of time with the development of two germ tubes that develop opposite sides of spore. But it do not form appresoria ultimately plants are not infected. Although pycnidia are commonly present in dead bark, wood, and the pathogen is source of primary infection after tissues are wounded by pruning, mechanical damage, or unfavorable environmental conditions, such as extreme heat and cold, flooding, or drought. Rain splash also help pathogen spread from one tree to another. Initially, *C. gloeosporoides* was thought to be responsible for postharvest anthracnose and but later on it was also reported as preharvest pathogen having symptoms such as stem end rot of fruit and wither-tip and tear stain on citrus fruits.[41]

17.7.5 DISEASE MANAGEMENT

- The diseases can be minimized by proper use of manuring, cultural practices, and pesticides measures.
- Properly pruned the affected dead twigs, destroyed during winter, and sprayed with 1% Bordeaux mixture in February, March, and September to reduce fruit drop. Cut end should be protected with Bordeaux mixture paste.
- Tree should be sprayed with 0.1% bavistin or 0.2% captafol 3 times after pruning. Prochloraz and benomyl gave the best results in controlling this disease during storage in coorg mandarin.

17.8 FRUIT DROP

Fruit drop in mandarin orchards mainly occur due to complex infection of pathogens and deficiency of minerals, which cause reduction in yield and quality of harvested fruits. In this fruit drop and the premature fruits fall down, give pain to the citrus farmers. Such fruits become unfit for marketing. This drop is of much economic importance and if left uncontrolled it can seriously reduce yield and returns. Therefore, adoptions of integrated approaches are more crucial for the sustainable management of fruit drops. Fruit drops due to complex pathogen infection mainly starts in the month of August, mid-September, and mid-October. September–October months are most detrimental in case of fruit drop, because in this month's fruits are on harvesting and they drawn more nutrition from plants. Pathological fruit drop is caused by different plant pathogens and most common of these are *C. gloeosporioides, Diplodia natalensis,* and *Alternaria citri* symptoms. Above mentioned pathogens and besides causing fruit drop are also responsible for die-back of twigs. The disease initiates on tree as drying of twigs from tip downward and black dot fruiting bodies usually seen in infected twigs. Pathological fruit drop is identified by the appearance of small, circular, light brown lesions around the stalk end of the fruit. As the area enlarges and a soft dark brown pliable rot develops. The dropped fruits are deep yellow to brown color and show dark brown coin-like rotting around the stalk end. In some fruits, lesions are present on both stalk-ends of fruits. *A. citri* causing stalk end rot on fruit at early stages and become weak parasite and it remains quiescent and express symptoms later on. At the maturity of fruits, fungus become active and exerts a stimulus resulting in abscission. Sometimes diseased fruits shrink due to late infection and become black, light in weight and mummified, and remain hanging to the stalks for a longer period.

17.9 SOOTY MOLD

17.9.1 HISTORY AND ECONOMIC IMPACT

The sooty mold is caused by fungi *Capnodium citri.* The disease is prevalent whenever honey dew substances secreting by insects, such as aphids, whitefly and psylla that are frequently found during lusting growth of Kinnow or citrus plants.

17.9.2 DISEASE SYMPTOMS

The disease is characterized by black velvety thin membrane covering on the leaf lamina. In severe cases, the tree twigs and leaves surface completely turns black with mold. Sooty mold infected leaves become curl and shrivel during dry conditions. The fungus multiplies on the honey dew secreted by the insects and spreads on the plant surface making it black by producing of black spores. During flowering times and its attack results in reduced fruit set and premature flowering and fruit drop. The fungus is non-pathogenic because their myceliums fail to absorb nutrients from host tissue but draws nutrients from the sweet honey dew. Thus, interferes of normal functioning of plants and photosynthesis is not affected by fungus. The mold deposit may delay in developing of fruit color and difficult to remove from the fruit surface, which down the market price and causes loss in income of farmers.

17.9.3 EPIDEMIOLOGY

The disease is more severe in dense orchards where light intensity is low. Citrus plants exposed to sunlight had less incidence, while tress in centers of the orchards had 95% incidence. High incidence of insect, the sugary substances secreted by them favor development of the sooty mold. Regular heavy rainfall washes down this substances but high humidity proved congenial for the growth of the fungus.

17.9.4 PATHOGEN DISSEMINATION

This fungus is spread in citrus plants during heavy infestation of sugary-producing insects. When insect feeds on the leaves of plant they also transmit the fungus on other plants. Similarly, heavy rainfall also spread this fungus on whole plants.

17.9.5 DISEASE MANAGEMENT

- Give three sprays of Ziram 27 SC (1250 mL) + Ethion 50 EC (1000 mL) in 500 L of water at monthly interval in August, September, and October.
- Fruits should be dipped in a bleaching solution and subsequently wash them.

17.10 ALTERNARIA FRUIT ROT

17.10.1 HISTORY AND ECONOMIC IMPACT

This pathogen is to spread after starting of fruits maturity. Citrus fruit rot is caused by *A. citri*. *Alternaria* fruit rot has been noticed in Punjab, West Bengal, Rajasthan, and Haryana states.

17.10.2 DISEASE SYMPTOMS

On naturally infected fruits, small discolored spots appear on fruits, which turn brown covering an area of 2–3 cm. Sometimes, the entire fruits may be girdled and fruit may drop. The temperature ranges for the development of the disease is between 15°C and 35°C with a peak between 25°C and 35°C.[42] External injury to rind of Malta orange was necessary for infection and disease development, which increased with the extent of injury.[43] The fungus reduced total soluble solids and reduced sugars and increased acidity in the affected fruits.

17.10.3 EPIDEMIOLOGY

The temperature ranges for development between 15°C and 35°C, high humidity favor the development of disease. In congenial conditions, pathogen spread very fast on all fruits and rotted the fruits and also reduced the market quality.

17.10.4 DISEASE MANAGEMENT

Phaltan at 5 ppm and Bordeaux mixture (4-4-50) were found effective against the fruit rot. Karathane at 0.1% as preharvest spray can also check the disease development.[43]

17.11 GENERAL MANAGEMENT PRACTICES

- First spray should be given, when new flush emerged.
- Always destroy the ant colonies because they are carried out insects on plants feeding site.
- Close spacing, water logging should be avoided in orchards to prevent pest population.
- Avoid pruning during active growth periods, to avoid the perpetuation of pest.
- Apply only need-based nitrogenous fertilizers because it promotes new flushes, which favor pest infestation.
- Always insure maximum light interception in orchards.
- Preparations of spray solution, spraying operations, insecticide residue, and compatibility in mixtures are important aspects to keep in mind before undertaking the sprays.
- Insecticide spray should be done on young and vulnerable life stages of the pest.
- Spray should be covered on canopy of plant and lower surface new flush.
- Avoid repeat application of a particular chemical pesticide and do not use pesticide beyond date of expiry.
- Do not spray in strong winds, cloudy days, and drizzling.

17.12 CONCLUSION

At present, strategies for management of the pests are mainly focused around the synthetic pesticides, which is overloading in the food chain and causing human health hazards and chemical residue in soil and lowering the natural enemies population and disturbing the natural beneficial microorganisms and resurgence of pests and resistance problem. Therefore, it is necessary to reduce the pesticide load from the food chain that could be achieved suitably by developing and implementing the latest IPM technology. To overcome these problems and to achieve our goals, it is essential to put all IPM strategies/ options available together and practiced in farmers' participatory mode to evolve a comprehensive IPM technology for the management of kinnow or citrus diseases. As future thrust, the survey and surveillance of the diseases should be a regular feature covering wider areas under Kinnow cultivation. Studies on population biology of various pathogens over time, space, to evaluate known resistance rootstock may be strengthened. Development of comprehensive system for race characterization of different pathogens is the need of the hour. Determining the way of different transmission mode of different pathogen and finger printing of resistant rootstock will be standardized, ITS and molecular markers for resistance to the disease can be undertaken, detection and diagnosis techniques to identify the pathogen, its variability using serological and molecular tools will be developed. Evaluate and develop resistance gene management strategies for durable resistance. Integrated management approaches need to be refined and output-oriented research should be focused. Develop location-specific integrated management modules by combining HPR, cultural practices and biological control for specific disease. Develop techniques and strategies to monitor virulence shifts in different pathogen through molecular techniques. Construction of refined molecular linkage maps using DNA analysis and understanding species relationship through the analysis of mitochondria and genetic transformation. The greatest gains from biotechnology in near future can possibly come from work of defensive traits especially for complex diseases like die-back.

KEYWORDS

- **kinnow**
- **diseases**
- **yield losses**
- **epidemiology**
- **integrated disease management**

REFERENCES

1. Vijaya, H. M.; Godara, R. K.; Shashank, S.; Sharma, N. Effect of Exogenous Application of Micronutrients on Growth, Yield of Kinnow Mandarin Under Semi-Arid Zone of Haryana. *J. Pharmacogn. Phytochem.* **2017,** *6*, 733–735.
2. Anonymous. National Horticultural Board, 2017–18 Indian Horticulture Database. 2017–18. http://nhb.gov.in/area-pro.
3. Erwin, D. C.; Ribeiro, O. K. *Phytophthora* Diseases Worldwide; APS Press: St. Paul, Minnesota, 1996.
4. Das, A. K.; Nerkar, S.; Kumar, A.; Bawage, S. Detection, Identification, Characterization of *Phytophthora* spp. Infecting Citrus in India. *J. Plant Pathol.* **2016,** *98*, 55–69.
5. Paracer, C. S.; Chahal, D. S. Diseases of Root, Crown, Trunk of Citrus. *Punjab Hort. J.* **1962,** *2*, 92–5.
6. Capoor, S. P.; Rao, D. G.; Viswanath, S. M. *Diaphorina citri* Kuway, a Vector of the Greening Disease of Citrus in India. *Indian J. Agric. Sci.* **1967,** *37*, 572–576.
7. Garnier, M.; Jagoueix, E. S.; Cronje, P. R.; Le Roux, H. F.; Bove, J. M. Genomic Characterization of a *Liberibacter* Present in an Ornamental Rutaceous Tree, *Calodendrum capense*, in the Western Cape Province of South Africa. Proposal of '*Candidatus liberibacter africanus* subsp. *capensis*'. *Int. J. System. Evol. Microbiol.* **2000,** *50*, 2119–2125.
8. Fawcett, H. S. Transmission of Psorosis of Citrus. *Phytopathology* **1938,** *28*, 669.
9. Mounde, L. G.; Ateka, E. M.; Kihurani, A. W.; Wasilwa, L.; Thuranira, E. G. Occurrence, distribution of citrus gummosis (*Phytophthora* spp.) in Kenya. *Afr. J. Hort. Sci.* **2009,** *2*, 56–68.
10. Mekonen, M.; Ayalew, A.; Weldetsadik, K.; Seidm, A. Assessing, Measuring of Citrus Gummosis (*Phytophthora* spp.) in Major Citrus Growing Areas of Ethiopia. *J. Hort.* **2015,** *2*, 154. DOI:10.4172/2376-0354.1000154.
11. Naqvi, S.; Singh S. Fungal Diseases of Citrus: Diagnosis and Management. *Techn. Bull. Natl. Res. Centre Citrus* (NRCC), **2002,** *5*, 61–67.
12. Naqvi, S. Managing *Phytophthora* Diseases on Citrus. *Indian Hort.* **2000,** *44*, 5–9.
13. Benson, D. M.; Grand, L. F.; Vernia, C. S.; Gottwald, T. R. Temporal and Spatial Epidemiology of *Phytophthora* Root Rot in Fraser Fir Plantations. *Plant Dis.* **2006,** *90*, 1171–1180.
14. Halbert, S. E.; Nuñez, C. A. Distribution of the Asian Citrus Psyllid, *Diaphorina citri* Kuwayama (Rhynchota: Psyllidae) in the Caribbean Basin. *Florida Entomol.* **2004,** *87*, 401–402.
15. Jagoueix, S.; Bové, J. M.; Garnier, M. The Phloem-Limited Bacterium of Greening Disease of Citrus Is a Member of the Alpha Subdivision of the Proteobacteria. *Int. J. Syst. Bacteriol.* **1994,** *44*, 379–386.
16. Gottwald, T. R. Current Epidemiological Understanding of Citrus Huanglongbing. *Annu. Rev. Phytopathol.* **2010,** *48*, 119–139.
17. Batool, A.; Iftikhar, Y.; Mughal, S. M.; Khan, M. M.; Jaskani, M.; Abbas, J. M.; Khan, I. A. Citrus Greening Disease a Major Cause of Citrus Decline in the World Review. *Hortic. Sci.* **2007,** *34*, 159–166.
18. Lin, K. H. Further Studies on Citrus Yellow Shoot. *Acta Phytophylact. Sin.* **1963,** *2*, 243–251.
19. Da graça J. V. Citrus Greening Disease. *Annu. Rev. Phytopathol.* **1991,** *29*, 109–136.
20. Van Vuuren, S. P. Variable Transmission of African Greening to Sweet Orange. In *Proc. Conf. Int. Organ. Citrus Virol. 12th IOCV*, University of California, Riverside, CA, 1993; pp 264–268.

21. Garnier, M.; Bové, J. M. Transmission of the Organism Associated with Citrus Greening Disease from Sweet Orange to Periwinkle by Dodder. *Phytopathology* **1983,** *73*, 1358–1363.
22. Luthra, J. C.; Sattar, A. Citrus Canker, Its Control in Punjab. *Punjab Fruit J.* **1942,** *6*, 179–182.
23. Bedi, K. S. Some Important Observations on the Citrus Canker in Punjab. *Punjab Hort. J.* **1961,** *2*, 89–91.
24. Ramakrishnan, T. S. *Common Diseases of Citrus in Madras State*; Govt. of Madras Publication, 1954.
25. Prasad, N. *Citrus Canker. Proc. Seminar on Disease of Horticultural Plants;* Shimla, 1959; pp 87–88.
26. Nirvan, R. S. Effect of Antibiotic Sprays on Citrus Canker. *Hort. Adv.* **1960,** *4*, 155–160.
27. Goto, M. Citrus Canker. In *Plant diseases of international importance*, Vol. III; Kumar, J.; Chaube, H. S.; Singh, U. S.; Mukhopadhyay, A. N., Eds.; Prentice-Hall: Englewood Cliff, 1992; pp 170–208.
28. Stall, R. E. Seymour, C. P. Canker, a Threat to Citrus in the Gulf- Coast States. *Plant Dis.* **1983,** *67*, 581–585.
29. Rao, Y. P. Hingorani, M. K. Survival of *Xanthomonas citri* (Hasse) Dowson in Leaves, Soil. *Indian Phytopath* **1963,** *16*, 362–364.
30. Chakravarti, B. P.; Porwal S.; Rangarajan, M. Studies on Citrus Canker in Rajasthan. Disease Incidence, Survival of the Pathogen. *Lab Dev. J. Sci. Tech.* **1966,** *4*, 262–265.
31. Vasudeva, R. S. Sci. Rep. Indian Agric. Res. Inst.; New Delhi, 1956–57; *1958*, 93.
32. Paracer, C. S. Some Important Diseases of Fruit Trees. *Punjab Hort. J.* **1961,** *1*, 45–47.
33. Leite, R. P.; Mohan, S. K. Survival of *Xanthomonas campestris* pv. *citri* (Hasse) Dye in Soil, in Association with Some Gramineous Plants. *Proc. Intn. Soc. Citric.* **1984,** *2*, 365–368.
34. Goto, M. Studies on Citrus Canker III. Survival of *Xanthomonas citri* (Hasse) Dowson in Soils, on the Surface of Weeds. *Bull. Fac. Agric. Shizouka Univ.* **1970,** *20*, 21–29.
35. Graham, J. H. Population Dynamics, Survival of *Xanthomonas campestris* in Soil in Citrus Nurseries in Maryland and Argentina. *Plant Dis.* **1989,** *73*, 423–427.
36. Koizumi, M. Ultrastructural Changes in Susceptible and Resistant Plants of Citrus Following Artificial Isolation with *Xanthomonas citri* (Hasse) Dowson. *Ann. Phytopothol. Soc. Japan* **1979,** *45*, 635–644.
37. Cook, A. A. Association of Citrus Canker Pustules with Leaf Miner Tunnels in North Yemen. *Plant Dis.* **1988,** *72*, 546.
38. Arauz, L. F.; Wang, A.; Duran, J. A.; Monterrey, M. Causes of Post Harvest Losses of Mango at the Wholesale Market Level in Costa Rica. *Agronomia Costarricense.* **1994,** *18*, 47–51.
39. Agrios, G. N. *Plant Pathology*, 5th ed.; Academic Press: New York, 2005.
40. Barker, I.; Pitt, D. Detection of the Leaf Curl Pathogen of Anemones in Corms by Enzyme Linked Immunosorbent Assays (ELISA). *Plant Path.* **1988,** *37*, 417–422.
41. Huang, F.; Chen G. Q.; Hou, X. et al. *Colletotrichum* Species Associated with Cultivated Citrus in China. *Fungal Divers.* **2013,** *61*, 61–74.
42. Agarwal, G. P.; Hasija, S. K. *Alternaria* Rot of Citrus Fruits. *Indian Phytopath.* **1967,** *20*, 259–260.
43. Chand, J. N.; Rattan, B. K.; Suryanarayan, D. Epidemiology, Control of Fruit Rot of Citrus Caused by *Alternaria citri* Ellis, Piece. *J. Res. PAU.* **1967,** *4*, 217–222.

CHAPTER 18

Harnessing the Antifungal and Antinemic Potential of *Trichoderma* in Integrated Crop Management System for Mitigating Fungal Nematode Complex in Cucumber

S. NAKKEERAN*, B. SREENAYANA, R. SARAVANAN, P. RENUKADEVI, and S. VINODKUMAR

Department of Plant Pathology, Tamil Nadu Agricultural University, Coimbatore 641003, India

**Corresponding author. E-mail: nakkeeranayya@tnau.ac.in*

ABSTRACT

The fungal saprophyte *Trichoderma* is ubiquitous in nature. This micro fungus plays a macro role in the suppression of soilborne fungal pathogens and nematodes infecting crop plants. Being a potential producer of hydrolytic enzymes and cellulases it colonizes the crop plants both at rhizosphere and spermosphere. Colonization by this unique fungus helps in preventing the entry of inimical microbes and nematodes infecting crop plants. Increase in the area under protected cultivation of vegetable crops has increased the outbreak of fungal nematode complex. Among the different vegetable crops, cucumber is an important vegetable crop cultivated across the globe. The area under protected cultivation of cucumber in India is being increased year after year. Continuous cultivation of cucumber for more than 2–3 years predisposes the crop for the infection of *Fusarium* nematode complex and thus becomes as a challenging task in the cultivation of cucumber under open field and protected conditions. However, judicious use of *Trichoderma* under integrated crop management will pave the way for successful cultivation of cucumber.

Integrated Pest Management in Diverse Cropping Systems. Devendra Pal Singh, PhD (Ed.)

preparation, and delivery. Various formulations of biocontrol agents, viz., dry flowable formulations, wettable powders, granules, dusts, gels, liquids, pellets, oil-based formulations, spray-dried freeze-dried and vaccum-dried formulations amenable for consumers are to be formulated. There are more than 40 products of biocontrol agents sold worldwide. Commonly used biocontrol agent of *Trichoderma* spp. is available commercially in different names, such as Root shield (*T. harzianum* T22), Trichopel (*T. harzianum+T. viride*), Trichodowels (*T. harzianum+T. viride*), Trichodex (*T. harzianum*), Biofungus (*Trichoderma* sp), Bioderma (*T. viride*) to control a variety of plant pathogens, such as *Botrytis, Fusarium, Pythium, Rhizoctonia, Sclerotinia, Sclerotium, Verticillium, Gaumannomyces*, wood rot fungi. Apart from *Trichoderma, Gliocladium, Ampelomyces quisqualis,* nonpathogenic *Fusarium oxysporum*, *Candida oleophila* are some of the important fungal biocontrol agents available commercially.[141] Recently, many research and review papers have discussed about the different formulations of biocontrol agents.

Most of the commercially available formulations are carrier-based beneficial microorganisms intended for seed and soil application.[142] The common carriers used are talc and kaolinite clay, diatomaceous earth, grain flour, and vegetable oils.[143] Talc formulation of *Trichoderma* developed with the addition of glycerol (osmoticant in the production medium) at the rate of 3%, 6% had a prolonged shelf life of about 7, 12 months, respectively with viability greater than 2×10^6 cfu/g compared with normal talc formulation without the addition of glycerol that had only 4 months of shelf life. This particular formulation protected the tomato plants suffering from *Fusarium* wilt considerably.[144] *Trichoderma*–chitin mixture-based talc formulations were used for the management of head rot (*Sclerotinia sclerotiorum*)—root-knot nematode (*M. incognita*) complex disease of cabbage.[145] They have applied talc-based formulation of biocontrol agents with or without chitin amendment individually as well as in mixtures against head rot-root-knot disease complex under greenhouse conditions and found that combined application of bioformulation mixture (*T. viride*-Tvc1, Tvc2+*T. harzianum*-Thc) with chitin has reduced the head rot incidence to about 8.4%. The combined application also enhanced the activities of defense-related enzymes, such as peroxidase (PO-PO1, PO2, PO3, PO4), polyphenol oxidase (PPO-PPO1, PPO2, PPO3), phenylalanine aAmmonia lyase (PAL), chitinase in cabbage. They also concluded that the mechanism associated with disease reduction may be due to the increased activities of defense-related proteins (PO, PPO, PAL, Chitinase) in cabbage.

Trichoderma produces hydrolytic enzymes, such as glucanases and chitinases. These enzymes interact with *Fusarium* and *Meloidogyne incognita* and thereby suppress fungal nematode complex. This microbe not only stops with it, but also produces an array of volatile and nonvolatile organic compounds of antimicrobial in nature. In addition, it also secretes ceratoplatanin-like small molecules with antimicrobial nature. Further, *Trichoderma* complements the crop growth through the production of growth hormones that in turn alters the root and shoot architecture of cucumber and thus results in enhanced yield and suppression of fungal nematode complex. Delivering of *Trichoderma* through seed treatment and soil drenching at periodical intervals aids in the proliferation of the antagonist in rhizoplane and serves as a line of control to prevent the ingress of fungal nematode complex into plant system. *Fusarial* wilt of cucumber is effectively managed under protected cultivation through biopriming and soil application of *Trichoderma.* Combined application of talc formulation of *Trichoderma* spp. viz., *T. virens* TRI 37, *T. harzianum* TRI 35, *T. harzianum* TRI 36 through seed treatment @ 10 g/kg and soil application 2.5 kg/1000 m^2 reduced the disease incidence and promoted plant growth, increased the yield, and number of fruits in cucumber in comparison with control. Hence, the application of *Trichoderma* spp., not only suppresses the biotic stress but also indirectly complements for the production of residue free produces.

18.1 INTRODUCTION

Cucumber (*Cucumis sativus* L.) being a major vegetable crop, India ranks 27th position in the world cucumber production. It is cultivated over an area of 1.07 m ha with an annual production and productivity of 1658 MT and 15.5 t/ha, respectively.[1] In recent days, production and productivity of cucumber is drastically reduced due to the outbreak of *Fusarium* wilt–nematode complex incited by *Fusarium oxysporum* f. sp. *cucumerinum* and *Meloidogyne incognita.* Yield losses are associated with root-knot nematode (RKN)—*Meloidogyne* spp. together with the soilborne pathogen *F. oxysporum.*[2] *F. oxysporum* f. sp. *Cucumerinum.* They invade the vascular system and secrete mycotoxins resulting in the development of necrotic lesions, followed by wilting and death.[3] Monoculturing of cucumber for three to four seasons increases the risk of *Fusarium* wilt as high as 70% with 10–50% reduction in yield and results in complete crop failure.[4] A 3-year survey results from Shandong province and China revealed that the fungal nematode complex in cucumber was noticed up to 67.6% in greenhouses. After four or more crops, the occurrence of RKN increased up to 95%[5] with an average yield loss of 20–30%. However, in some cases, the incidence reached 100%.[6] Hence, management of *F. oxysporum.* f. sp. *cucumerinum* and *Meloidogyne* complex in vegetables is a challenging task. Application of nematicides and fungicides curtails the fungal nematode complex in cucumber. However, due to indiscriminate application of chemicals, the nontargeted and beneficial and microflora and fauna are also affected. Soil application of carbendazim reduced the population of *Azotobacter, Azospirillum*, phosphate-solubilizing bacteria, and other beneficial bacteria and fungi.[7] These have reoriented the scientists to lookup for the need for an alternate method of management. Biological control of fungal and plant–parasitic nematode complex improves the beneficial flora and fauna in the soil, and hence improves the soil health through the recruitment of desirable rhizo/endomicrobiome.

Trichoderma spp. were effective against plant pathogenic fungi and oomycetes and nematodes.[8] *Trichoderma* spp. employ multifaceted mode of action which include competition[9] and mycoparasitism besides the secretion of diverse hydrolytic enzymes[10] and antimicrobial metabolites contributing to plant growth promotion and induction of resistance in plants.[11,12] *Trichoderma* has been reported to be a successful antagonist against various important plant pathogens, such as *Phytophthora* sp.,[13] *Fusarium* sp.,[14] *Botrytis*,[15] *Macrophomina*,[16] *Rhizoctonia solani*,[17] *Sclerotinia sclerotiorum*,[18] *Sclerotium rolfsii*,[19] and *Pythium*.[20] Pretreatment of stevia plants with *Trichoderma asperellum* reduced the wilt incidence by 90%.[21] Similarly, *Trichoderma* species have been reported to suppress *Fusarium* wilt in banana,[22] tomato,[23] soybean,[24] and many other crop plants. Furthermore, *Trichoderma* spp. were also reported to inhibit the juvenile mobility, egg hatching, galling, gall index, and soil population of root-knot nematode.[25] *Trichoderma* species reduced the gall index in green gram.[26] *T. harzianum* strains possess egg-parasitic capability in the model nematode *C. elegans*.[27] *Trichoderma* spp. secrete numerous nematicidal compounds, viz., acetic acid,[28] peptide cyclosporin A,[29] gliotoxin,[30] and trichodermin.[31] Three metabolites of *Trichoderma* spp., were identified as 1-β-vinylcyclopentane-1α, 3α-diol, 6-pentyl 2H-pyran-2-one, 4-(2-hydroxyethyl) phenol which were found to be nematicidal.[32] Nematicidal activities of *Trichoderma*-derived pyrones, such as 6-(1'pentenyl)-2H-pyran-2-one, massoialactone, δ- decalactone, and viridepyronone were demonstrated.[33] Root exudates are complex and comprise phenolics, organic acids, amino acids, vitamins, purines, and enzymes, etc. Root exudates provide necessary nutrient for the ramification of rhizomicrobiome. Further, they also promote chemotaxis of microbes toward the rhizosphere by the exudation of specific chemoattractants. Organic acids could be utilized as a nutritional base for rhizosphere fungi (*Trichoderma* spp.) and supports roots colonization.[34] Application of *Trichoderma* spp. as seed treatment in wheat crop has resulted in the identification of 2900 compounds from root exudates.[35] Further, application of *Trichoderma* sp. to the soil altered the rhizomicrobiome in cucumber[34] and banana.[36]

18.2 SIGNIFICANCE OF *FUSARIUM* WILT

The pathogen *Fusarium* responsible for causing wilt on greenhouse cucumber is internally seed-borne.[37] In China, monoculture of cucumber resulted in the development of *F. o.* f. sp. *cucumerinum* causing *Fusarium* wilt with increased yield loss.[38] *Fusarium* wilt of cucumber is widespread in North China and very serious in green house cropping systems.[39] *F. o.* f. sp. *cucumerinum* is host-specific to cucumber and several cucurbit species, such as muskmelon (*Cucumis melo* L.), water melon (*Cirullus vulgaris*), and cantaloupe (*Cucumis melo* L.).[40,41] The pathogen *F. o.* f. sp. *cucumerinum* enter through roots and it is being restricted to vascular tissues leading to plant death. Subsequently, it colonizes the exterior parts of cucumber plants and vascular tissues.[42] The pathogen resides in the soil for a long period and invades at all stages of development. The pathogen enters into the roots of host plants by direct penetration or wounds and colonize the vascular vessels.[43] Among the causal agent of wilt diseases in cucumber, *F. o.* f. sp. *cucumerinum* is economically important which results drastic in yield losses in green house. Monoculturing of cucumber for three to four continuous seasons enhances *Fusarium* wilt up to 70% with 10–50% yield reduction and results in complete crop failure.[4]

18.3 SYMPTOMS OF *FUSARIUM* WILT

Cucumber infected with *Fusarium* wilt caused yellowing of leaves, browning of vascular tissues, stunting wilting, and death of the plant.[40] Besides, wilt-infected cucumber plants or parts of infected plants lose turgidity. Loss of turgidity is due to clogging of xylem vessels by mycelium, spores, and polysaccharides produced by the fungus. It blocks water flow from roots to the upper part of the plants. Moreover, toxins secreted by the fungus may be transported to leaves and reduce chlorophyll synthesis, thereby decreasing photosynthesis. During advanced stages of infection, white mycelial growth, orange spore masses were observed on diseased tissues during high-humid conditions. Seedlings may show symptoms occasionally but the most distinctive symptoms generally appear after fruit set or at the time of drought stress leading to sudden wilt.[44] Cucumber plantlets develop root rot or stem canker or crown rot, leading to wilting and plant death. Symptoms of crown rot are associated with brown discoloration at crown and vertical lesions on the stem extending up to 30 cm from the crown.[42]

18.4 SYMPTOMS OF ROOT-KNOT NEMATODE INFECTION

Root-knot nematodes are plant parasitic nematodes of the genus *Meloidogyne*, Family: Meloidogynidae. Their common name refers to the characteristic galls or root-knots, associated with infestation. This genus comprises more than 95 species and with some species having several races.[45] Overall, these root-knot nematodes parasitize more than 5500 plant species, including annual and perennial crops.[46] Out of 95 species, only 18 species were reported from different states of India. Among them, *M. incognita, M. javanica, M. graminicola*, and *M. arenaria* were the most common devastating species. *Meloidogyne* spp. are difficult to manage because of their wide host range, short life cycle and high reproductive rates.[47] Feeding by nematode induces large galls throughout the root system of infected plants, alters the uptake of water, nutrients, and interfere with the translocation of photosynthates[48] and increases the incidence and severity of *F. o.* f. sp. *cucumerinum* complex in cucumber.[42]

18.5 *TRICHODERMA* AS ANTIFUNGAL AND ANTINEMIC BIOCONTROL AGENT

Trichoderma was first described more than 200 years ago[49] and was later envisaged into four genera. A Monograph on *Trichoderma* has recognized this genus into nine groups and several biological species under each group.[50,51] Rifai species group was elevated to special level and has recognized several species within each of five sections of the genus *Trichoderma*.[52–55] *Trichoderma* are free-living fungi, ubiquitous, nonpathogenic, highly interactive with roots and soil and reside inside the plants.[56] Various compounds released by the *Trichoderma* spp. stimulate localized and systemic resistance response in plants and are recognized as effective biological agents for disease control and plant growth promotion.[57] *Trichoderma* spp. are the most widely studied biocontrol agents (BCAs) against plant pathogens because of their ability to suppress the population of soilborne plant pathogens.[58] Soilborne *Trichoderma* spp. are effective against wide range of phytopathogenic fungi.[9] They produce small green or white conidia from phialides on profusely branched conidiophores, which facilitate successful colonization of cellulosic materials, decaying plant material through the production of antibiotics (secondary metabolites) and enzymes.[59]

18.6 MODE OF ACTION OF *TRICHODERMA*

18.6.1 MYCOPARASITISM

Mycoparasitism is a complex process which involves recognition, attack, penetration, and lysis of pathogen. Hyphal coiling and mycoparasitic activity by *Trichoderma* on *R. solani* has been well-documented.[60,63] It penetrates and colonize both sclerotia and mycelium of *S. rolfsii*[61] and thus effectively reduced sclerotial viability.[62] *T. harzianum* hyperparasitized *Pythium* and resulted in cell wall lysis, followed by vacuolation and coagulation of protoplasm.[64] Mycoparasitism by the isolates of *T. harzianum* reduced the radial growth of *P. aphanidermatum* and lysed the damping-off of tomato and cauliflower.[65] Three *Trichoderma* spp. were effective against *F. o.* f. sp. *carthami* and inhibited the growth of pathogen by an interaction that took place in close contact with host hypha and caused swelling, coiling, and lysis of mycelia and reduced mycelial growth of the pathogen.[66,67]

18.6.2 ANTIBIOSIS

Trichoderma produces trichodermin, sesquiterpene, gliotoxin, viridin volatile, and nonvolatile metabolites with antimicrobial action.[68] The volatile component from *Trichoderma* spp. inhibited the mycelial growth of *S. rolfsii*[64] and *M. phaseolina.*[69] Among these metabolites, production of harzianic acid, alamenthicins, tricholin, peptaibols antibiotics, 6-pethyl-a–pyron, viridian, gliovirin, gliosoprenins, and heptelidic acid were produced by *Trichoderma* spp. *T. virens* produced gliovirin,[70] *T. harzianum* produced pyrone antibiotics against *Gaeumannomyces graminis var. tritici.* The synergistic action of antibiotics and hydrolytic enzymes increased the level of antagonism than the individual compounds.[8,70] Synergistic interaction between endochitinase, gliotoxin, hydrolytic enzymes, and peptaibols from *T. harzianum,* effectively reduced the conidial germination of *Botrytis cinerea.*[11] Low molecular weight diffusible compounds or antibiotics produced during the interaction by *Trichoderma* spp. suppressed the growth of other harmful microorganisms.[71] The volatile compounds of *Trichoderma* inhibited the growth of different pathogens, viz., *F. oxysporum, Alternaria zinniae, R. solani*.[72] The nonvolatile metabolites secreted by *T. viride* in culture medium significantly restricted the growth of all pathogens. The volatile antifungal compounds extracted from the *Trichoderma* strain SQR-T037 inhibited the growth of *F. o.* f. sp. *niveum* causing watermelon wilt.[73]

18.6.3 COMPETITION AND RHIZOSPHERE COMPETENCE

Rhizosphere competence of *Trichoderma* spp. favors its rapid colonization in the developing root system.[74] *Trichoderma polysporum* and *T. viride* were bestowed with the ability to tolerate biotic stress coupled with enhanced competitive saprophytic ability.[75] The concept associated with competition and rhizosphere competence was responsible for endogenous fungi replacement on the root surface. *T. virens*-treated root bits collected from the soil heavily infested with *M. phaseolina* yielded only *T. virens*. However, when these same cultures were incubated at 40°C, pathogen growth was observed and not the growth of *Trichoderma*.[76] CSA of mutant strains of *Trichoderma* were comparatively higher than the parent leading to effective disease control.[77]

18.6.4 PRODUCTION OF HYDROLYTIC ENZYMES

The genus *Trichoderma* produces hydrolytic enzymes as chemical weapons responsible for effective mycoparasitism and biocontrol of plant pathogens. Further, *Trichoderma* plays a major role in the tritrophic interaction with plant and pathogen. During the interaction, *Trichoderma* spp. produce lytic enzymes, proteolytic enzymes, diffusible volatile metabolites, other secondary metabolites active against their hosts, that is, they succeed by impairing growth conditions of pathogens.[59] The presence of *ech42* (endochitinase) gene, β-1, 3-glucanase (*bgn*13.1) gene in *Trichoderma* was effective in increasing the antifungal activity.[78,79] Coculturing of *Trichoderma* with pathogens increased the activity of cellulase, chitinase, pectinase over monoculture, which confirmed positive induction of enzyme secretion by *Pythium* spp., *Alternaria* spp. and *Fusarium* spp. Presence of pathogens induced the total enzyme secretion of *T. viride,* which are directly associated with increased antagonistic activity of *T. viride* against pathogenic fungi.[80] The genes encoding cellobiohydrolase and endochitinase in *T. asperellum,* were effective in suppressing the growth and development of *S. sclerotiorum*.[81]

18.6.5 ANTIMICROBIAL GENES OF TRICHODERMA SPP.

Several plant pathogens that cause major diseases in crop plants were controlled by *Trichoderma*.[82] Chitinase, tubulins, protease, xylanase, galacturonase, glucanase, stress tolerant, and cell adhesion proteins are the major genes characterized in *Trichoderma*. Cell wall degradation, parasitic activity, and stress tolerance are the major biocontrol mechanisms. From *T. harzianum*, genes encoding for proteinase *prb1,* endochitinase *ech42* were isolated and characterized. These genes were involved in mycoparasitic activity against *R. solani and S. rolfsii*. For the production of these enzymes, genes were induced by lectin–carbohydrate interaction with a diffusible factor. This factor regulated the production of proteinase and endochitinase which helps in the mycoparasitism.[83] In *T. harzianum*, disruption *ech42* gene affected the biocontrol activity.[84] The genes encoding for N-acetyl-β-D-glucosaminidase *(exc1, exc2),* chitinase (*chit*42, *chit*33), protease (*prb*1), β-glucanase (*bgn* 13.1) had a major role in the mycoparasitic activity against *F. oxysporum*.[85] The gene, *qid74* isolated from *T. harzianum* CECT 2413 had a significant role in cell protection and adherence to hydrophobic surfaces that helps the fungus in mycoparasitic activity against *R. solani*. The *qid74* gene was responsible for adhesion to the hydrophobic surface of the pathogenic fungi and helped in the antagonistic activity.[86] In *T. virens*, an adenalyte-cyclase encoding gene *tac1* was isolated and cloned. It improved the mycoparasitic activity against *R. solani and P. ultimum*.[87] Antifungal activity of *T. brevicompactum* was attributed to the presence of *tri5* gene coding for trichodermin production.[88] Expression of genes coding for β-1, 3-endoglucanase, chitinase, and endochitinase was induced after contact with pathogen. It resulted in cell wall degradation of *F. solani*.[89] Furthermore, the pathogens, viz., *F. oxysporum, R. solani*, and *S. sclerotiorum* were mycoparasitized by *T. harzianum* due to upregulation of β-1, 3-endoglucanase, chitinase 33, and endochitinase 42 genes.[90]

18.6.6 TRICHODERMA AND PLANT GROWTH HORMONES

Application of beneficial microorganisms improves the plant growth and yield. Many *Trichoderma* spp. are well-known for their ability to promote plant growth and health. Rhizospheric *Trichoderma* isolates of *Calophyllum brasiliense* were involved in the synthesis of phytohormones IAA, GA, and cytokinins.[91] Siderophore-producing *T. asperellum* Q1 increased the fresh and dry weight of cucumber seedlings compared with the control. Plant growth promotion was attributed to the synthesis of IAA.[92] The plant growth-promoting activity by *T. harzianum* strain T-22 and the involvement of different classes of plant growth regulator in improving root and shoot growth have been reported.[94] The root growth and shoot growth increased to an extent of 76% and 61%, respectively. The level of GA3 and IAA increased significantly in leaves to 71% and 49%, respectively, while in roots to about 40% and 143%, respectively. Production capacity of IAA varied between the species. *T. koningiopsis* TRI 41 produced 15.25 μg/mL, *T. asperellum* TRI 15 produced 12.25 μg/mL, and *T. virens* TRI 37 produced 8.25 μg/mL.[95] Similarly, all the effective isolates of *Trichoderma* spp. produced gibberellic acid. However, maximum GA production was observed in *T. koningiopsis* TRI 41 (10.4 μg/mL), followed by *T. asperellum* TRI 15 (9.6 μg/mL) and *T. virens* TRI 37 (9.1 μg/mL). *T. koningiopsis* and *T. harzianum* produced more of IAA, GA, and cytokinin.[91,96] Similarly, *T. virens* expressed characteristic auxin-related phenotypes, such as enhanced biomass production, lateral root development in *Arabidobpsis thaliana* seedlings. When plants were grown under axenic conditions, the auxin-related compounds indole-3acetic acid, indole-3-acetaldehyde, indole-3-ethanol were produced by *T. virens*. An analysis of those three indolic compounds furnished detailed information about their efficacy at modulating root system architecture and activation of auxin-regulated gene expression in Arabidopsis mutant which indicated the major role of *T. virens* in auxin signalling for plant growth promotion.[93] These results suggest the change in phytohormone level produced by *Trichoderma* spp. that plays a key role to improve the root and shoot architecture of cucumber grown under protected cultivation.

18.6.7 TRICHODERMA AS A BIOCONTROL AGENT AGAINST ROOT-KNOT NEMATODE

Trichoderma spp., are ubiquitous, soil-dwelling, and self-establishing saprophytic fungus. They are common in the root systems of higher plants. Colonization by *Trichoderma* spp. promoted plant growth and yield.[76,97] *T. atroviride, T. harzianum, T. asperellum,* and *T. viride* have resulted in significant control of root-knot nematodes.[98,99] Egg mass production in root-knot nematode *M. arenaria* was decreased by treating soil with *T. harzianum* and

T. koningii.[100] *T. harzianum* was an effective egg parasite of *M. incognita* and promoted the growth of tomato.[101] Formulations of *Burkholderia cepacia* (Bc-2, Bc-F) and *T. virens* (Gl-3) significantly suppressed the number of root-knot nematode eggs and J2 production on the roots of pepper plants.[97] In India, laboratory tests revealed the biocontrol efficacy of *T. harzianum* in field conditions, which was accounted for up to 55.30% reduction over control.[102] Application of *T. harzianum* and *T. hamatum* or *T. virens* suppressed the root-knot nematodes.[103] Application of *T. virens* or *T. harzianum* 1 week prior to inoculation of nematode @ 4, 8g/kg soil drastically reduced the number of galls and J2s penetrated plant root system and thereby suppressed the population of *M. graminicola* in the soil.[104] Galling, gall index, final soil population of *M. graminicola* in aerobic rice were reduced in the fields applied with *T. viride.*[25] Different isolates were reported to exhibit varying degrees of virulence on egg parasitism by *T. asperellum*-203, *T. atroviride*, *T. atroviride* IMI 206040, and *T. harzianum*-248 which had the ability to parasitize all stages of *M. javanica* in vitro.[105] Different concentrations (10^6 spores/mL) of *T. harzianum* BI are capable of reducing the nematode infection, egg mass production, number of eggs per egg mass as compared with control.[106] *T. harzianum* induced 49% mortality of *M. javanica* J2s after 48 h.[107]

Trichoderma spp., caused more than 50% mortality of *M. javanica* after 48 h in vitro.[108] They possessed the strongest egg-parasitic activity.[27] Isolates of *T. longibrachiatum* had lethal and parasitic effect on *M. incognita* juveniles (J2s) at a conidial concentration of 1.5×10^7 conidia/mL and effectively inhibited and parasitized 88% J2s in cucumber seedlings after 14 days.[109] An isolate of *T. harzianum* from U.P. exhibited 69.11% parasitization of *M. incognita* eggs and compared with an isolate of Hisar which had 46.75% parasitization.[110]

18.6.8 ANTINEMIC BIOMOLECULES OF TRICHODERMA SPP. AGAINST ROOT-KNOT NEMATODE

The gelatinous matrix is an important barrier for the entry of microorganisms. After the gelatinous matrix, the egg forms an important barrier mainly because of the middle chitinous layer. The egg shell contains different amino acid compositions, including collagen-like proteins, which make the eggshell tough, and the combination of chitinolytic and proteolytic enzymes disrupt the eggshell.[112] *Trichoderma* spp. produce various enzymes or metabolites to digest the nematode body contents or egg contents.[111] Biomolecules produced by *Trichoderma* spp. are 2,4-Dihydroxy-2,5-dimethyl-3(2H)-furan-3-one, formic acid, 2-propenyl ester, 4H-pyran-4-one, 2,3-dihydro-3,5-dihydroxy-6-methyl,

5-hydroxymethylfurfural, benzeneacetic acid, 4-hydroxy-, methyl ester, glucitol, 6-O-octyl, n-hexadecanoic acid, oleic acid, octadecanoic acid, dl-alanine-beta-naphthylamide, pyrazino[1,2-a]indole-1,4-dione, 2,3-dihydro-2methyl-3-methylene, thiophene, 2,3-dihydro-, α-D-glucopyranose, 1,6-anhydro-, benzene ethanol, 4-hydroxy-, α-methyl mannofuranoside, pentanoic acid, 3-hydroxy-2-methyl-, methyl ester, methyl stearate, 6-(p-tolyl)-2-methyl-2-heptenol, trans-, heptadecane, benzeneacetic acid, and 4-hydroxy- methyl ester. These were detected from the effective isolates TRI 41 (*T. koningiopsis)*, TRI 37 (*T. virens)*, and TRI 15 (*T. asperellum)* with antifungal property. Since the fungus has chitin and glucan as its major cell wall component, the nematode cell wall morphogenesis also reveals the same compounds to be present. It might be also responsible for the suppression of RKN.[95] Nematicidal compounds have been identified from *Trichoderma* spp., including acetic acid,[28] peptide cyclosporin A,[29] gliotoxin,[30] and trichodermin.[31] Trypsin-like serine proteases from *T. viride, T. harzianum* (PRA1) were purified.[113] The nematicidal activity in PRA1 protease of *T. harzianum* was purified by chromatofocusing followed by gel filtration. It significantly reduced the number of hatched eggs of *M. incognita* egg mass.[113] Peptide cyclosporin has nematicidal activity against *M. incognita* which was isolated from *T. polysporum*.[29] A drastic reduction has been observed up to 30% in the number of galls and J2s of *M. incognita* in okra when treated with 1000 ppm concentrations of ethyl acetate extracts of *T. harzianum* and *T. viride* and 1000 ppm of methanol extract of *T. harzianum*.[114] Evaluation of the nematicidal activity with the culture filtrates of 329 *Trichoderma* strains indicated that 29 strains exhibited nematicidal activity.[31] They also extracted trichodermin from *Trichoderma* YMF on ethyl acetate by bioassay-guided fractionations.

18.7 DIVERSITY OF ANTIMICROBIAL BIOMOLECULES PRODUCED BY *TRICHODERMA* AGAINST *FUSARIUM* DURING DITROPHIC INTERACTION

Ditrophic interaction between *Trichoderma* and *Fusarium* induced the expression of diverse nonvolatile organic compounds responsible for the suppression of *Fusarium* wilt of cucumber. Differentially expressed biomolecules from the zone of interaction could be capitalized for the synthesis of smart formulations with antifungal and antinemic biomolecules. The differential expression of antimicrobial nonvolatile organic compounds from the PDA medium was observed in the hyperparasitized zone. The antimicrobial compounds identified from the zone of hyperparasitization were several folds greater than from the NVOC detected from nonhyperparasitized region of the *T. koningiopsis* TRI 41. The compounds, such as dl-alanine-beta-naphthylamide, heptaoxa-2-cycloheneicosanone, formic acid, 2-propenyl ester, n-hexadecanoic acid, baicalein trimethyl ether, pyrazino[1,2-a] indole-1,4-dione, 2,3-dihydro-2methyl-3-methylene-, propanamide, N-(1-naphthyl)-2-methyl- were identified from the hyperparasitized region. Multifold increase in the expression of these compounds compared with the nonhyperparasitized region of *T. koningiopsis* TRI 41 might have favored the maximum antifungal activity by the isolate TRI 41.[95] The antifungal activity of the dl-alanine-beta-naphthylamide-based compounds was also confirmed.[115,116] Similarly, 1,4,7,10,13,16,19-Heptaoxa-2-cycloheneicosanone-based compound was reported to possess antimicrobial activity against gram-negative, gram-positive bacteria.[117,118] The NVOC compound might also play a pivotal role in the suppression of bacterial pathogens as well as the fungal pathogens. Antifungal activity of baicalein trimethyl ether, formic acid, 2-propenyl ester, and n-hexadecanoic acid was also reported.[119–121] Likewise, the compounds pyrazino[1,2-a]indole-1,4-dione, 2,3-dihydro-2methyl-3-methylene, propanamide, N-(1-naphthyl)-2-methyl- were reported to have antibacterial, herbicidal activity, respectively.[122,123]These findings indicated that *T. koningiopsis* TRI 41 isolate has both antifungal and antibacterial activity. Thus, the isolate TRI 41 might have broad spectrum of action against both bacterial and fungal pathogens.

18.8 INFLUENCE OF *TRICHODERMA* ON ROOT ARCHITECTURE

Root system is the most important for plant health since it provides support to withstand in soil and contributes to water-use efficiency and acquisition of mineral and nutrients from the soil.[124] Auxin plays an important role in the root system architecture regulation. Application of auxins promotes lateral root, root hair development and whereas auxin transport inhibitors decrease root branching.[125,126] In maize plants, *Trichoderma* inoculation influenced the root system architecture that was related to increased yield, stimulated root biomass production, and root hair development.[127,128] Root growth of bitter-gourd and cucumber was promoted by species of *Trichoderma* as compared with untreated control.[129] Greenhouse experiment showed that *Trichoderma* spp. significantly increased the seedling height and root exploration, leaf area, root dry weight 15 days after sowing. *Trichoderma harzianum* 1295-22 increased the root growth as well as suppressed the disease. Mechanism of growth promotion is that the fungus counteracts auxin inhibition of root cell elongation.[127] *T. harzianum qid74* gene encodes a cysteine-rich cell wall protein that has an important role in adherence to hydrophobic surfaces and cellular protection.[130] They observed more and longer secondary root hairs in cucumber plants treated with *qid74*-overexpressing mutants. Fewer, shorter hairs were observed in roots treated with *qid74*-disrupted transformants compared with control. *qid 74* has induced modifications in root architecture and increased the total absorptive surface and facilitated the nutrient uptake and translocation of nutrients in the shoots and increased the plant biomass through an efficient use of NPK and micronutrients. A fungal invertase is essential for the initiation of the several mechanisms associated with root colonization..[131] In maize, *T. hamatum* T614 increased the leaf area, fresh root, and shoot weight of seedling.[132] Whereas, *T. hamatum* T447, *T. harzianum* T969 significantly reduced the seedling fresh root weight and shoot weight as compared with the control.

18.9 IN VITRO EFFICACY OF TRICHODERMA AGAINST FUSARIUM WILT AND RKN COMPLEX

T. asperellum CCTCC-RW0014 was highly effective against *F. oxysporum* f. sp. *cucumerinum*. In vitro studies resulted in the maximum antifungal activity of *T. virens* TRI 37 against cucumber vascular wilt pathogen.[133] Screening of different *Trichoderma* spp. against juvenile mortality/immobility of root-knot nematode (*M. incognita*) increased the juvenile mortality with the conidial suspension of *T. koningiopsis* TRI 41 (73.0%). It was followed by *T. virens* TRI 37 and *T. asperellum* TRI 15 which induced the mortality to an extent of 69.3 and 66.3%, respectively.[95] Conidial suspension of *T. harzianum* suppressed egg hatching and juvenile mortality of root-knot nematode, prevented penetration, colonization of conidia on egg mass of root-knot nematode (*M. javanica*) in tomato.[98] Isolated indigenous fungi, such as *T. viride and T. harzianum* from egg masses of root-knot nematodes collected from vegetable fields were tested for their efficacy against juvenile mortality and egg hatching of root-knot nematode in tomato caused by *M. incognita* in vitro.[134] Conidial suspension of *T. longibrachiatum* at 10^7 effectively infested the egg mass up to 89.3%, whereas the cultural filtrate of *T. longibrachiatum* at 100% concentration caused 64.5% juvenile mortality.[135] Similarly, conidial suspension, culture filtrate of *T. harzianum*, *T. virens*, *T. hamatum*, *T. koningii*, *T. viride* inhibited egg hatching, induced juvenile mortality of root-knot nematode at different concentrations of 25, 50, 75, 100%.[136] Among them, the conidial suspension of *T. harzianum* inhibited the hatching of eggs up to 80%, and 100% juvenile mortality was observed at 100% concentrations, whereas the culture filtrates of *T. harzianum* at 75% and 100% concentration induced 100% inhibition in egg hatching and juvenile mortality. These results confirmed the effect of *Trichoderma* spp. conidial suspension on the egg hatching and juvenile mortality of root-knot nematode in cucumber.[136]

18.10 ULTRAMICROSCOPIC EVIDENCES ON THE INTERACTION OF *TRICHODERMA* WITH *FUSARIUM*

Field emission scanning electron microscopy analysis reflected that the hyphae of *F. o.* f. sp. *cucumerinum* F1 excised from the hyperparasitized region resulted in lysis, loss of cell turgor, ramification, and coiling by *T. koningiopsis* TRI 41[95] (Figs. 18.1–18.5). Hyphal coiling and wall alterations were noticed along with the rapid collapse and loss of cell turgor of *F. o. radicis-lycopersici* under SEM investigation.[137] Hyphae of *T. harzianum* grew toward and hyperparasitized around the *S. sclerotiorum* hyphae.[138] Microscopic examinations of interaction zones revealed antagonistic interactions between *F. oxysporum* and *T. harzianum* isolate CEN287 which were evidenced by hyphal coiling.[139] Similarly, hyphae of *F. solani* were encoiled by *Trichoderma*.[140] In our study, the effective isolate TRI 41 (*T. koningiopsis*) significantly reduced the mycelial growth of *F. oxysporum* f. sp. *cucumerinum*. *T. asperellum*—NVTA2 colonized sclerotial bodies, and the entire surface of the sclerotia was covered with abundant sporulation of *T. asperellum* NVTA2.[81] The conidia of *T. asperellum* (NVTA2) lodged over the hyphae of *S. sclerotiorum* and lysed it. The healthy sclerotia had undulations in the surface and are accompanied with numerous cavities. Thus, the ultramicroscopic images of mycoparasitism confirmed the hyperparasitic behavior of *T. koningiopsis* TRI 41. It was equally responsible for reducing the inoculum potential of the pathogen.[81]

18.11 EFFICACY OF *TRICHODERMA* BIOFORMULATION ON PLANT GROWTH PROMOTION AND MANAGEMENT OF FUNGAL NEMATODE COMPLEX OF CUCUMBER

Success of any bioagent is highly dependent on the kind of formulation made. Formulation affects the product' shelf life, ability to multiply, survival in the environment, its potential to control disease, preparation cost, ease of

Invert emulsion of *T. harzianum* Rifai was effective in the management of postharvest diseases caused by *B. cinerea, Rhizopus stolonifer*, and *Pencillium expansum. T. harzianum* delivered in the form of invert emulsion suppressed the infection of *R. stolonifer* on apple, pear, peach, and strawberry. Interestingly, it was also effective in the suppression of *B. cinerea* infecting grape, pear, strawberry, and kiwi fruit. Furthermore, *P. expansum* infecting grape, pear, and kiwi fruit was also effectively controlled. Application of these invert emulsion on unwounded apple fruit was able to protect against infection of *R. stolonifera* till 59 days with 89% reduction.[146] Oil formulation of *T. asperellum* with 2.7×10^7 conidia/mL comprising soybean oil (74%), emulsifying agent (Tensiofix NTM-15%), structural agent (Tensiofix 869-5%), carbon source (Glucose-4%), and *T. asperellum* conidia (2%) was effective in the suppression of cocoa black pod rot.[147]

Dry flowables/water dispersible granules improved the safety for environment and end users. Dry flowable formulations avoid clump formations and make it easy for application. Dry flowable formulations of *Trichoderma* strains (*T. harzianum* Td50b, *T. asperellum* T36) having the conidial load of 5×10^9 cfu/g with the shelf life of 12 months were effective in the management of soilborne pathogens. Encapsulation of *Trichoderma* strains in sodium bicarbonate, alginic acid, and calcium chloride followed by drying of encapsulated fungi and mixing of granulated powder in sodium bicarbonate, magnesium stearate, and sodium ascorbate was also effective in the management of soilborne diseases. The resultant mixture was spray-dried and formulated using polyacrylic acid, citric acid, sodium bicarbonate, polyvinyl alchohol as adhesives and lecithin as wetting agent.[148] Encapsulated granules of *Trichoderma* spp. in sodium alginate matrix with 10^6 cfu/g had shelf life up to 14 months of storage at 28°C.[149]

Recently, preservation of microorganisms by drying has emerged as a preferable method for long-term storage of microbial cultures. Freeze-drying preserves the microorganisms for many years and is a desirable technology across the universe. Apart from this, incorporation of protective agents during different growth phases of microorganism prior to freezing or drying has increased the shelf life and maintained the viability for a longer period. The commonly used protectants are nonfat milk solids, serum, trehalose, glycerol, betaine, glucose, sucrose, lactose, polymers, such as dextran and PEG. Apart from freeze-drying, spray drying, vacuum drying, fluidized-bed drying were also followed for the preservation of microorganism.[150]

Hydrogel-based bioinoculants enhance the shelf life of biocontrol agents and made them suitable for application under drought conditions. Treatment of wheat seeds with hydrogel-based bioinoculant consortia consisting of nitrogen-fixing bacteria *Azotobacter chrococcum*, phytostimulator *T. viride*, and phosphate solubilizer *P. fluorescens* significantly enhanced the plant growth.[151] Storage of *T. harzianum* conidia in sterile deionized water extended the shelf life up to 480 days with a viable count of 3.2×10^6 cfu/mL and to about 720 days with a viable count of 2×10^5 at 28–32°C.[152] This ready-to-use liquid carrier formulation was effective against black pepper plants infected with *Phytophthora capsici*. It indicated the possibility of using sterile deionized water as a storage medium for conidial storage of *Trichoderma* spp. Thus, the same technique could be also extended for the management of soilborne diseases of cucumber.

The success of *Trichoderma* formulations is due to their ability to survive under unfavorable conditions, high reproductive capacity, capacity to modify the rhizosphere, efficiency in the utilization of nutrients, efficacy against pathogens, and efficiency in stimulating the plant growth and induction of plant defense mechanisms.[71] Biological control in protected cultivation is economical and durable. It suppressed the inoculum potential and maintained the soil health.[153] Combined applications of seed treatment and soil application with the consortia of *T. koningiopsis* TRI 41, *T. asperellum* TRI 15, *T. asperelloides* TNAU Tad 1 at 15, 30, and 45th days after sowing recorded 13.45% fungal nematode incidence. Many of the *Trichoderma* spp. were effective against several plant pathogenic fungi and nematodes.[8] Since *Trichoderma* strains are normally resistant to many toxic compounds and they grow rapidly upon inoculation in the soil. The capability of *Trichoderma* spp. as biocontrol agents of pathogens in several crops against *Fusarium* spp. and *Rhizoctonia* spp. are well-known.[154,155]

Bioformulation of *Trichoderma* spp. were effective in controlling *R. solani* and *F. o.* f. sp. *phaseoli*, the causal agent of bean damping-off and wilt diseases, respectively.[156] Among them, *T. harzianum* (Th1) was most effective in the suppression of damping-off and wilt of beans. The reduction in disease severity and increase in the yield of different crops after treatment with different formulations of *T. harzianum and T. viride* have been reported by several workers.[154,155,157] Several reports have revealed that addition of isolates of *Trichoderma* to rhizosphere enhanced plant growth.[158,159] The efficacy of *Trichoderma* spp. in suppressing *F. o.* f. sp. *lycopersici* under protected cultivation was demonstrated.[160] Under protected conditions, application of *T. harzianum* (N-8) decreased disease incidence (14.75%). Tomato plants treated with *T. harzianum* (N-8) isolate promoted plant growth. *T. harzianum* T35 controlled *F. oxysporum* by competing for rhizosphere space and nutrients.[65]

The root-knot nematode is an important pathogen on cucumber cultivated under protected cultivation. Several beneficial microorganisms can parasitize nematodes and their eggs. The combined applications of *Trichoderma* as seed treatment and soil application along with the consortia of *T. koningiopsis* TRI 41, *T. asperellum* TRI 15, and *T. asperelloides* TNAU Tad 1 at 15, 30, and 45 th days after sowing recorded 13.45% fungal nematode incidence.[95] But, in recent years' soil application of *Trichoderma* spp. has suppressed the occurrence of plant parasitic nematodes.[136] *Trichoderma* spp. antagonize nematodes by mycoparasitism, antibiosis, enzymatic hydrolysis.[161] Application of *Trichoderma* spp. especially *T. hamatum, T. harzianum,* and *T. virens* was effective in suppressing the root-knot nematode populations infesting vegetable crop.[103,136] Application of *Trichoderma* spp. reduced the galling and egg mass production. However, the degree of reduction varied with *Trichoderma* spp..[136] Application of conidial suspension of *T. harzianum* recorded 38% decrease in the number of galls/root system, whereas the culture filtrate reduced the galling by 29% over control. In greenhouse experiment, application of *Pseudomonas aeruginosa* alone or in combination with *T. viride* has resulted in remarkable biocontrol activity against *M. phaseolina, R. solani, F. solani, F. oxysporum, M. javanica* infesting okra roots as well as increased the plant height, fresh shoot weight, and root length.

Simultaneous infestation with root-knot nematodes and *F. o.* f. sp. *cucumerinum* leads to the formation of a disease complex that enhances crop losses than individual effect of either *F. o.* f. sp. *cucumerinum* or root-knot nematodes. The severity of *Fusarium* wilt, population of *M. javanica* J2s was reduced in soil due to *T. harzianum*. Delivering *T. harzianum* in protected cultivation was effective in reducing *Fusarium* wilt of tomato plants.[162] The reduction in root-knot nematode galls and reduced incidence of *Fusarium* wilt in tomato was enhanced due to the application of *T. harzianum*.[163] Application of *Trichoderma* conidial suspension (10^2–10^8 spores/mL) of *T. harzianum* reduced nematode infection.[106] *T. harzianum* penetrated nematode egg mass matrix and significantly decreased the nematode egg hatching level. Defense-related enzymes, viz., PO, PPO, and PAL increased in *T. harzianum*-inoculated plants.

However, more research work needs to be carried out to develop stable, cost-effective, and easy applicable formulations. Combined application of *T. virens* through seed treatment and soil application was effective in enhancing the growth and suppression of wilt incidence in tomato.[164] *T. harzianum* controlled *Fusarium* wilt of tomato by induction of resistance and through weakening of the hormonal disruption caused by the pathogen.[165] Synergetic action of different delivery systems, such as biopriming and soil application of *Trichoderma* spp. could be used well for efficient management of *Fusarium* wilt of cucumber under protected cultivation.[133] Combined application of talc formulation of *T. virens* TRI 37, *T. harzianum* TRI 35, *T. harzianum* TRI 36 through seed treatment @10g/kg, soil application 2.5 kg/1000 m^2 reduced the disease incidence, promoted plant growth, and increased the yield and number of fruits in cucumber in comparison with control. Application of *T. harzianum* and *T. viride* increased the growth parameters, viz., dry weight, root length, grain yield as well as reduced the chick pea wilt incidence. On the basis of present study, the biocontrol agents might be well exploited for sustainable disease management and to safeguard the environment.[166]

18.12 CONCLUSION

Parasitic nematodes of crop plants are the hidden enemies of the farmers since nematodes dwell in subterranean habitats. Because of this, the losses imposed by the nematodes are very high, and sometimes, they also affect the economic sustainability of the farming community. Continuous cultivation of vegetable crops, either in open field or under protected cultivation resulted in severe outbreak of nematode infection and it rendered very difficult to cultivate subsequent crop. Hitherto, the synergistic interaction of fungal pathogens and nematodes threatened the growers to switch over the crop for the ensuing season. A tricky issue beneath the diagnosis of nematode damage is often highly confusing because they often resemble the symptoms of several biotic and abiotic stress. A late identification of disease lowered the yield and economic returns of the crop. Besides, nonavailability of effective biomolecules for the management of nematodes and breakdown of resistance by nematodes and intensive cultivation rendered the management of plant parasitic nematodes as a challenging task. This necessitates identifying a suitable product that could effectively quench the proliferation and infection of nematodes.

Considering these issues, the research on biocontrol has witnessed the exploration of *Trichoderma* as a suitable alternative for the management of plant parasitic nematodes. Sustainability of bionematicides always remains as a challenging task and since they are always preventive and there is a large gap in the performance between lab and land. A focus on the delivery of bionematicide, mass production, increased shelf life, rhizosphere, and spermosphere engineering has to be addressed as researchable issue for harnessing the desirable attributes of a promising bionematicide to counter the attack of fungal nematode complex in cucumber and other vegetable crops.

FIGURE 18.1 *Trichoderma koningiopsis* hyperparasitizes *Fusarium oxysporum* f.sp. *cucumerinum.*

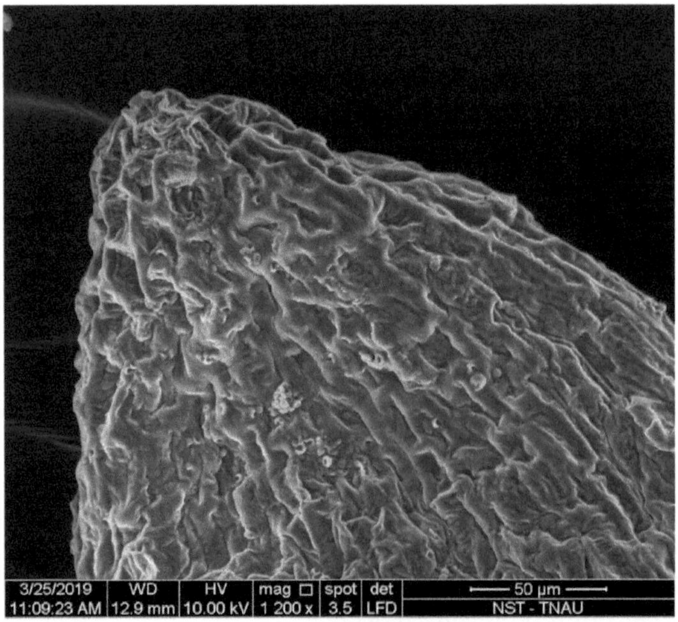

FIGURE 18.2 Rhizosphere colonization of cucumber by *Trichoderma koningiopsis.*

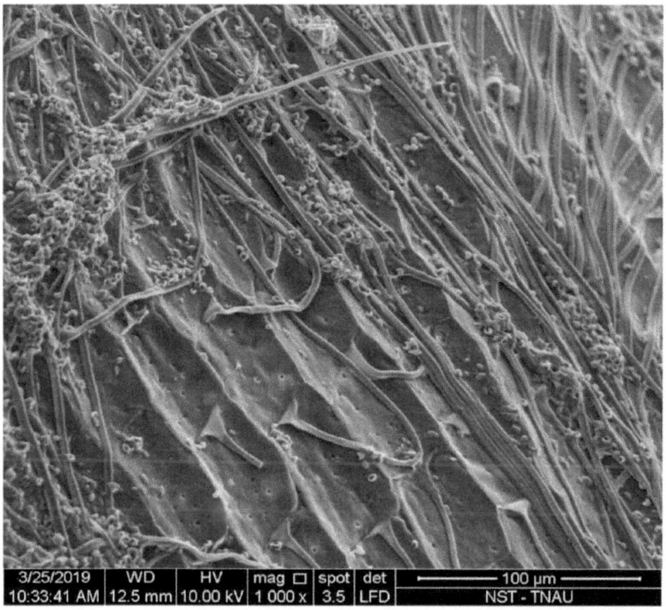

FIGURE 18.3 Colonization of spermosphere by *Trichoderma koningiopsis.*

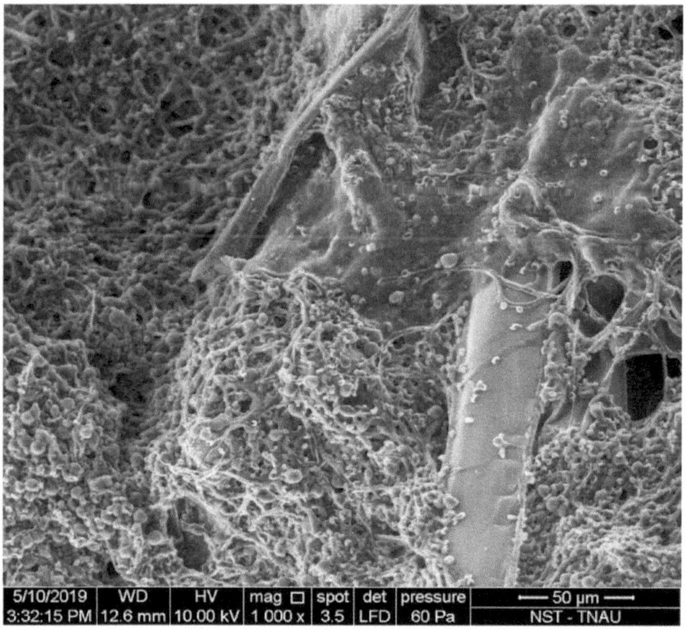

FIGURE 18.4 Colonization of egg mass of *Meloidogyne incognita* by *Trichoderma koningiopsis.*

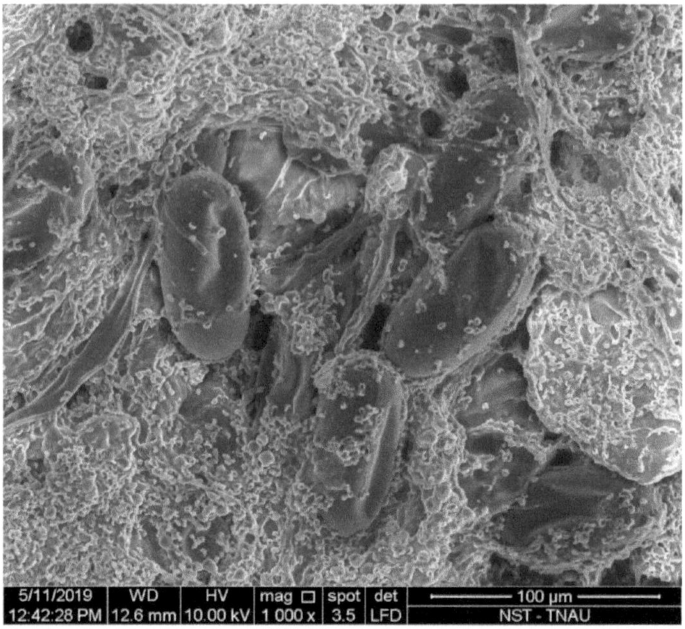

FIGURE 18.5 Colonization of Root-knot nematode eggs by *Trichoderma koningiopsis*.

KEYWORDS

- **trichoderma**
- **nematode**
- **antifungal**
- **antinemic**
- **nematicide**
- **formulations**
- **efficacy**
- **hyperparasitism**

REFERENCES

1. Indian Horticulture Database, Statistical Data, 2018. http://nhb.gov.in/area-pro/NHB_database_2019.pdf.

2. Qu, B.; Li, B. J.; Fan, H. Y. The New Development Tendency, Strategy of Non-Polluted Control of Cucumber Disease. *Northern Hortic.* **2003,** *5*, 7–8.
3. Zhang, S.; Raza, W.; Yang, X.; Hu, J.; Huang, Q.; Xu, Y.; Liu, X.; Ran, W.; Shen, Q. Control of Fusarium Wilt Disease of Cucumber Plants with the Application of a Bioorganic Fertilizer. *Biol. Fertil. Soils* **2008,** *44* (8), 1073–1080.
4. Chen, L.; Yang, X.; Raza, W.; Li, J.; Liu, Y.; Qiu, M.; Zhang, F.; Shen, Q. *Trichoderma harzianum* SQR-T037 Rapidly Degrades Allelochemicals in Rhizospheres of Continuously Cropped Cucumbers. *Appl. Microbiol. Biotechnol.* **2011,** *89* (5), 1653–1663.
5. Weibo, D.; Yanmao, S.; Rongguang, L.; Ruide, J.; Zhiqiang, Z.; Qing, L. Species Identification, Occurrence Investigation of Vegetable Root-Knot Nematodes Under Protected Cultivation in Shandong Province. *J. Laiyang Agric. Coll.* **2004,** *21* (2), 106–108.
6. Peng, D. L. Occurrence, Prevention of Vegetable Root Knot Nematode Disease. *China Veg.* **1998,** *4*, 57–58.
7. Mohiuddin. M.; Mohammed, M. K. Influence of Fungicide (Carbendazim), Herbicides (2, 4-D, Metribuzin) on Non-Target Beneficial Soil Microorganisms of Rhizospheric Soil of Tomato Crop. *IOSR J. Environ. Sci. Toxicol. Food Technol.* **2013,** *5*, 47–50.
8. Monte, E. Understanding *Trichoderma*: Between Biotechnology, Microbial Ecology. *Int. Microbiol.* **2001,** *4* (1), 1–4.
9. Elad, Y.; Chet, I.; Henis, Y. Degradation of Plant Pathogenic Fungi by *Trichoderma harzianum*. *Can. J. Microbiol.* **1982,** *28* (7), 719–725.
10. Troian, R. F.; Steindorff, A. S.; Ramada, M. H.; Arruda, W.; Ulhoa, C. J. Mycoparasitism Studies of *Trichoderma harzianum* against *Sclerotinia sclerotiorum*: Evaluation of Antagonism, Expression of Cell Wall-Degrading Enzymes Genes. *Biotechnol. Lett.* **2014,** *36* (10), 2095–2101.
11. Howell, C. R. Mechanisms Employed by *Trichoderma* Species in the *Biol. Control* of *Plant Diss*: The History, Evolution of Current Concepts. *Plant Dis.* **2003,** *87* (1), 4–10.
12. Martínez-Medina, A.; Alguacil, M. D.; Pascual, J. A.; Van Wees, S. C. Phytohormone Profiles Induced by *Trichoderma* Isolates Correspond with Their Biocontrol and Plant Growth-Promoting Activity on Melon Plants. *J. Chem. Ecol.* **2014,** *40* (7), 804–815.
13. Bae, S. J.; Mohanta, T. K.; Chung, J. Y.; Ryu, M.; Park, G.; Shim, S.; Hong, S. B.; Seo, H.; Bae, D. W.; Bae, I.; Kim, J. J. *Trichoderma* Metabolites As Biological Control Agents Against *Phytophthora* Pathogens. *Biol. Control.* **2016,** *92*, 128–138.
14. Saravanakumar, K.; Yu, C.; Dou, K.; Wang, M.; Li, Y.; Chen, J. Synergistic Effect of Trichoderma-Derived Antifungal Metabolites, Cell Wall Degrading Enzymes on Enhanced Biocontrol of *Fusarium oxysporum* f. sp. *cucumerinum*. *Biol. Control.* **2016,** *94*, 37–46.
15. You, J.; Zhang, J.; Wu, M.; Yang, L.; Chen, W.; Li, G. Multiple Criteria-Based Screening of *Trichoderma* Isolates for Biological Control of *Botrytis cinerea* on Tomato. *Biol. Control.* **2016,** *101*, 31–38.
16. Pastrana, A. M.; Basallote-Ureba, M. J.; Aguado, A.; Akdi, K.; Capote, N. Biological Control of Strawberry Soil-Borne Pathogens *Macrophomina phaseolina, Fusarium solani*, Using *Trichoderma asperellum*, *Bacillus* spp. *Phytopathologia Mediterranea* **2016,** *1*, 109–120.
17. Daryaei, A.; Jones, E. E.; Glare, T. R.; Falloon, R. E. pH, Water Activity in Culture Media Affect Biological Control Activity of *Trichoderma atroviride* Against *Rhizoctonia solani*. *Biol. Control.* **2016,** *92*, 24–30.
18. Sumida, C. H.; Daniel, J. F.; Araujod, A. P.; Peitl, D. C.; Abreu, L. M.; Dekker, R. F.; Canteri, M. G. *Trichoderma asperelloides* Antagonism to Nine *Sclerotinia sclerotiorum*

Strains, Biological Control of White Mold Disease in Soybean Plants. *Biocontr. Sci. Technol.* **2018,** *28* (2), 142–156.
19. Islam, M. M.; Hossain, D. M.; Nonaka, M.; Harada, N. Biological Control of Tomato Collar Rot Induced by *Sclerotium rolfsii* Using *Trichoderma* Species Isolated in Bangladesh. *Arch. Phytopathol. Plant Protection.* **2017,** *50* (3), 109–116.
20. Bedine-Boat, M. A.; Sameza, M. L.; Iacomi, B.; Tchameni, S. N.; Boyom, F. F. Screening, Identification, Evaluation of *Trichoderma* spp. for Biocontrol Potential of Common Bean Damping-Off Pathogens. *Biocontr. Sci. Technol.* **2020,** *30* (3), 228–242.
21. Díaz-Gutiérrez, C.; Arroyave, C.; Llugany, M.; Poschenrieder, C.; Martos, S.; Peláez, C. *Trichoderma asperellum* as a Preventive, Curative Agent to Control Fusarium Wilt in *Stevia rebaudiana. Biol. Control.* **2021,** *155*, 104–137.
22. Damodaran, T.; Rajan, S, Muthukumar, R. G.; Yadav, K.; Kumar, S.; Ahmad, I.; Kumari, N.; Mishra, V. K.; Jha, S. K. Biological Management of Banana Fusarium Wilt caused by *Fusarium oxysporum* f. sp. *cubense* Tropical Race 4 Using Antagonistic Fungal Isolate CSR-T-3 (*Trichoderma reesei*). *Front. Microbiol.* **2020,** *11.*
23. Cucu, M. A.; Gilardi, G.; Pugliese, M.; Gullino, M. L.; Garibaldi, A. An Assessment of the Modulation of the Population Dynamics of Pathogenic *Fusarium oxysporum* f. sp. *lycopersici* in the Tomato Rhizosphere by Means of the Application of *Bacillus subtilis* QST 713, *Trichoderma* sp. TW2, Two Composts. *Biol. Control.* **2020,** *142*, 104–158.
24. Pimentel, M. F.; Arnão, E.; Warner, A. J.; Subedi, A.; Rocha, L. F.; Srour, A.; Bond, J. P.; Fakhoury, A. M. *Trichoderma* Isolates Inhibit *Fusarium virguliforme* Growth, Reduce Root Rot and Induce Defense-Related Genes on Soybean Seedlings. *Plant Dis.* **2020,** *7*, 1949–1959.
25. Priya, M. S. Biomanagement of Rice Root Knot Nematode, *Meloidogyne graminicola* Golden, Brichfield in Aerobic Rice. *Int. J. Manage Soc. Sci.* **2015,** *3*, 591–608.
26. Mukhtar, T.; Tariq-Khan, M.; Aslam, M. N. Bioefficacy of *Trichoderma* Species Against Javanese Root-Knot Nematode, *Meloidogyne javanica*, in Green Gram. *Gesunde Pflanzen* **2021,** 1–8. https://doi.org/10.1007/s10343-021-00544-8
27. Szabó, M.; Csepregi, K.; Gálber, M.; Virányi, F.; Fekete, C. Control Plant-Parasitic Nematodes with *Trichoderma* Species, Nematode-Trapping Fungi: The Role of chi18-5, chi18-12 Genes in Nematode Egg-Parasitism. *Biol. Control.* **2012,** *63* (2), 121–128.
28. Djian, C.; Pijarowski, L.; Ponchet, M.; Arpin, N.; Favre-Bonvin, J. Acetic Acid: A Selective Nematicidal Metabolite from Culture Filtrates of *Paecilomyces lilacinus* (Thom) Samson, *Trichoderma longibrachiatum* Rifai. *Nematologica* **1991,** *37* (1–4), 101–112.
29. Li, G.; Zhang, K.; Xu, J.; Dong, J.; Liu, Y. Nematicidal Substances from Fungi. *Recent Patents Biotechnol.* **2007,** *1* (3), 212–233.
30. Watanabe, A.; Kamei, K.; Sekine, T.; Waku, M.; Nishimura, K.; Miyaji, M.; Tatsumi, K.; Kuriyama, T. Effect of Aeration on Gliotoxin Production by *Aspergillus fumigatus* in Its Culture Filtrate. Mycopathologia. **2004,** *157* (1), 245–254.
31. Yang, Z. S.; Li, G. H.; Zhao, P. J.; Zheng, X.; Luo, S. L.; Li, L.; Niu, X. M.; Zhang, K. Q. Nematicidal Activity of *Trichoderma* spp.; Isolation of an Active Compound. *World J. Microbiol. Biotechnol.* **2010,** *26* (12), 2297–2302.
32. Yang, Z.; Yu, Z.; Lei, L.; Xia, Z.; Shao, L.; Zhang, K.; Li, G. Nematicidal Effect of Volatiles Produced by *Trichoderma* sp. *J. Asia-Pacific Entomol.* **2012,** *15* (4), 647–650.
33. Degenkolb, T.; Vilcinskas, A. Metabolites from Nematophagous Fungi, Nematicidal Natural Products from Fungi as an Alternative for Biological Control. Part I: Metabolites from Nematophagous Ascomycetes. *Appl. Microbiol. Biotechnol..* **2016,** *100* (9), 3799–3812.

34. Zhang, F.; Yuan, J.; Yang, X.; Cui, Y.; Chen, L.; Ran, W.; Shen, Q. Putative *Trichoderma harzianum* Mutant Promotes Cucumber Growth by Enhanced Production of Indole Acetic Acid, Plant Colonization. *Plant Soil* **2013,** *368* (1), 433–444.
35. Lucini, L.; Colla, G.; Moreno, M. B.; Bernardo, L.; Cardarelli, M.; Terzi, V.; Bonini, P.; Rouphael, Y. Inoculation of *Rhizoglomus irregulare* or *Trichoderma atroviride* Differentially Modulates Metabolite Profiling of Wheat Root Exudates. *Phytochemistry* **2019,** *157*, 158–167.
36. Xue, C.; Penton, C.R.; Shen, Z.; Zhang, R.; Huang, Q.; Li, R.; Ruan, Y.; Shen, Q. Manipulating the Banana Rhizosphere Microbiome for Biological Control of Panama Disease. *Sci. Rep.* **2015,** *5* (1), 1–11.
37. Yildiz, M.; Delen, N. Studies on the Occurrence of Fusarium Wilt of Cucumber Ege Region of Turkey. *J. Turkish Phytopathology* **1977,** *6*, 111–117.
38. Ye, S.F.; Yu, J.Q.; Peng, Y.H.; Zheng, J.H.; Zou, L.Y. Incidence of Fusarium Wilt in *Cucumis sativus* L. Is Promoted by Cinnamic Acid, an Autotoxin in Root Exudates. *Plant Soil* **2004,** *263* (1), 143–150.
39. Huang, X.; Chen, L.; Ran, W.; Shen, Q.; Yang, X. *Trichoderma harzianum* Strain SQR-T37, Its Bio-Organic Fertilizer Could Control *Rhizoctonia solani* Damping-Off Disease in Cucumber Seedlings Mainly by the Mycoparasitism. *Appl. Microbiol. Biotechnol.* **2011,** *91* (3), 741–755.
40. Owen, J. H. Fusarium wilt of Cucumber. *Phytopathology* **1955,** *45* (8).
41. Gerlagh, M.; Blok, W. J. *Fusarium oxysporum* f. sp. *cucurbitacearum* Embracing all Formae Speciales of *F. oxysporum* Attacking Cucurbitaceous Crops. *Netherlands J. Plant Pathol.* **1988,** *94* (1), 17–31.
42. Vakalounakis, D. J.; Fragkiadakis, G. A. Genetic Diversity of *Fusarium oxysporum* Isolates from Cucumber: Differentiation by Pathogenicity, Vegetative Compatibility and RAPD Fingerprinting. *Phytopathology* **1999,** *89* (2), 161–168.
43. Pietro, A. D.; Madrid, M. P.; Caracuel, Z.; Delgado-Jarana, J.; Roncero, M. I. *Fusarium oxysporum*: Exploring the Molecular Arsenal of a Vascular Wilt Fungus. *Mol. Plant Pathol.* **2003,** *4* (5), 315–325.
44. Punja, Z. K.; Parker, M. Development of Fusarium Root, Stem Rot, a New Disease on Greenhouse Cucumber In British Columbia, Caused by *Fusarium oxysporum* f. sp. *radicis-cucumerinum*. *Can. J. Plant Pathol.* **2000,** *22* (4), 349–363.
45. Ghule, T. M.; Singh, A.; Khan, M. R. Root Knot Nematodes: Threat to Indian Agriculture. *Popular Kheti*. **2014,** *2* (3), 126–130.
46. Trudgill, D. L.; Blok, V. C. Apomictic, Polyphagous Root-Knot Nematodes: Exceptionally Successful, Damaging Biotrophic Root Pathogens. *Annu. Rev. Phytopathol.* **2001,** *39* (1), 53–77.
47. Jain, R. K.; Mathur, K. N.; Singh, R. V. Estimation of Losses Due to Plant Parasitic Nematodes on Different Crops in India. *Indian J. Nematol.* **2007,** *37* (2), 219–221.
48. Williamson, V. M.; Hussey, R. S. Nematode Pathogenesis, Resistance in Plants. *Plant Cell.* **1996,** *8* (10), 1735–1745.
49. Persoon, C. H. Disposita Methodical Fungorum. *Romers Neues Mag Bot.* **1794,** *1*, 81–128.
50. Rifai, M. A. A Revision of the Genus *Trichoderma*. *Mycol. Papers*. **1969,** *116*, 1–56.
51. Samuels, G. J.; Petrini, O.; Kuhls, K.; Lieckfeldt, E.; Kubicek, C. P. The *Hypocrea schweinitzii* Complex, *Trichoderma* Sect. *Longibrachiatum*. *Stud. Mycol.* **1998,** *41*, 1–54.
52. Bissett, J. A Revision of the Genus *Trichoderma*. II. Infrageneric Classification. *Can. J. Bot.* 1991a, *69* (11), 2357–2372.

53. Bissett, J. A Revision of the Genus *Trichoderma*. III. Section Pachybasium. *Can. J. Bot.* 1991b, *69* (11), 2373–2417.
54. Bissett, J. A Revision of the Genus *Trichoderma*. IV. Additional Notes on Section Longibrachiatum. *Can. J. Bot.* 1991c, *69* (11), 2418–2420.
55. Bissett, J. *Trichoderma atroviride*. *Can. J. Bot.* **1992,** *70* (3), 639–641.
56. Verma, M.; Brar, S. K.; Tyagi, R. D.; Surampalli, R. Y.; Valero, J. R. Antagonistic Fungi, *Trichoderma* spp.: Panoply of Biological Control. *Biochem. Eng. J.* **2007,** *37* (1), 1–20.
57. Ranasingh, N.; Saurabh, A.; Nedunchezhiyan, M. Use of *Trichoderma* in Disease Management. *Orissa Rev.* **2006,** *63* (2–3), 68–70.
58. Papavizas, G. C. *Trichoderma*, *Gliocladium*: Biology, Ecology and Potential for Biocontrol. *Annu. Rev. Phytopathol.* **1985,** *23* (1), 23–54.
59. Schuster, A.; Schmoll, M. Biology, Biotechnology of *Trichoderma*. *Appl. Microbiol. Biotechnol.* **2010,** *87* (3), 787–799.
60. Weindling, R. *Trichoderma lignorum* as a Parasite of Other Soil Fungi. *Phytopathology* **1932,** *22* (8), 837–845.
61. Henis, Y.; Adams, P. B.; Lewis, J.A.; Papavizas, G. C. Penetration of Sclerotia of *Sclerotium rolfsii* by *Trichoderma* spp. *Phytopathology* **1983,** *73* (7), 1043–1046.
62. Khattabi, N.; Oihabi, A.; Louali, L.; Ezzahiri, B. Effect of Fungicides, *Trichoderma harzianum* on Sclerotia of *Sclerotium rolfsii*. **2001,** 1000–1006.
63. Upadhyay, R. S.; Rai, B. Hyphal Interaction, Parasitism Amongst Some Rhizosphere Fungi of Pigeon-Pea. *Indian Phytopathology* **1983,** *36*, 302–306.
64. Upadhyay, J. P.; Mukhopadhyay, A. N. Biological Control of *Sclerotium rolfsii* by *Trichoderma harzianum* in Sugarbeet. *Int. J. Pest Manage.* **1986,** *32* (3), 215–220.
65. Kumar, A. M.; Pratibha, S. Confrontation Assay for *Trichoderma* as a Potential Biocontrol Agent Against *Pythium aphanidermatum*, *Sclerotinia sclerotiorum*. *Pest Manage. Horti. Ecosyst.* **2012,** *18* (1), 74–77.
66. Prasad, R. D.; Navaneetha, T.; Rao, L. V. Plant Growth Promotion, Induced Defense Response in Safflower (*Carthamus tinctorius* L.) by *Trichoderma*. *J. Biol. Control.* **2016,** *30* (1), 40–48.
67. Srinivasa, N.; Devi, T. P.; Sudhirkumar, S.; Kamil, D.; Borah, J. L.; Prabhakaran, N. Bioefficacy of *Trichoderma* Isolates Against Soil-Borne Pathogens. *Afr. J. Microbiol. Res.*. **2014,** *8* (28), 2710–2723.
68. Dennis, C.; Webster, J. Antagonistic Properties of Species-Groups of *Trichoderma*: I. Production of Non-Volatile Antibiotics. *Trans. Br. Mycol. Soc.* **1971,** *57* (1), 25–39.
69. Tamimi, K. M.; Hutchinson, S. A. Differences Between the Biological Effects of Culture Gases from Several Species of *Trichoderma*. *Trans. Br. Mycol. Soc.* **1975,** *64* (3), 455–463.
70. Howell, C.R. The Role of Antibiosis in Biocontrol. *Trichoderma*, *Gliocladium*. **1998,** *2*, 173–184.
71. Benitez, T.; Rincón, A. M.; Limón, M. C.; Codon, A. C. Biocontrol Mechanisms of *Trichoderma* Strains. *Int. Microbiol.* **2004,** *7* (4), 249–260.
72. Tapwal, A.; Singh, U.; Singh, G.; Garg, S.; Kumar, R. In Vitro Antagonism of *Trichoderma viride* Against Five Phytopathogens. *Pest Technol.* **2011,** *5* (1), 59–62.
73. Raza, W.; Faheem, M.; Yousaf, S.; Rajer, F. U.; Yamin, M. Volatile, Non-Volatile Antifungal Compounds Produced by *Trichoderma harzianum* SQR-T037 Suppressed the Growth of *Fusarium oxysporum* f. sp. *niveum*. *Sci. Lett.* **2013,** *1* (1), 21–24.
74. Howell, C. R.; Hanson, L. E.; Stipanovic, R. D.; Puckhaber, L. S. Induction of Terpenoid Synthesis in Cotton Roots, Control of *Rhizoctonia solani* by Seed Treatment with *Trichoderma virens*. *Phytopathology* **2000,** *90* (3), 248–252.

75. Widden, P.; Scattolin, V. Competitive Interactions, Ecological Strategies of *Trichoderma* Species Colonizing Spruce Litter. *Mycologia* **1988,** *80* (6), 795–803.
76. Harman, G. E.; Howell, C. R.; Viterbo, A.; Chet, I.; Lorito, M. *Trichoderma* Species—Opportunistic, Avirulent Plant Symbionts. *Nat. Rev. Microbiol.* **2004a,** *2* (1), 43–56.
77. Singh, R.; Maurya, S.; Upadhyay, R. S. The Improvement of Competitive Saprophytic Capabilities of *Trichoderma* Species Through the Use of Chemical Mutagens. *Braz. J. Microbiol.* **2016,** *47* (1), 10–17.
78. Singh, R. K.; Kumar, P.; Tiwari, N. N.; Singh, S. P.; Tiwari, A. K.; Vishwakarma, S. K.; Singh, A.; Kumar, A. Role of Endochitinase Gene, Efficacy of *Trichoderma* Against *Colletotrichum falcatum* Went. Causing Red Rot Disease in Sugarcane. *Sugar Technol.* **2014,** *16* (2), 180–188.
79. Kheiri, H. R.; Motallebi, M.; Zamani, M. R.; Deljo, A. Beta Glucanase (Bgn13. 1) Expressed in Transgenic *Brassica napus* Confers Antifungal Activity Against *Sclerotinia sclerotiorum*. *J. Crop Protection.* **2014,** *3* (1), 31–42.
80. Smitha, C.; Finosh, G. T.; Rajesh, R.; Abraham, P. K. Induction of Hydrolytic Enzymes of Phytopathogenic Fungi in Response to *Trichoderma viride* Influence Biocontrol Activity. *Int. J. Curr. Microbiol. Appl. Sci.* **2014,** *3* (9), 1207–1217.
81. Vinodkumar, S.; Indumathi, T.; Nakkeeran, S. *Trichoderma asperellum* (NVTA2) as a Potential Antagonist for the Management of Stem Rot in Carnation Under Protected Cultivation. *Biol. Control.* **2017,** *113*, 58–64.
82. Massart, S.; Jijakli, H. M. Use of Molecular Techniques to Elucidate the Mechanisms of Action of Fungal Biocontrol Agents: A Review. *J. Microbiol. Methods.* **2007,** *69* (2), 229–241.
83. Cortes, C.; Gutierrez, A.; Olmedo, V.; Inbar, J.; Chet, I.; Herrera-Estrella, A. The Expression of Genes Involved in Parasitism by *Trichoderma harzianum* Is Triggered by a Diffusible Factor. *Mol., Gen. Genet. MGG.* **1998,** *260* (2), 218–225.
84. Woo, S. L.; Donzelli, B.; Scala, F.; Mach, R.; Harman, G.E.; Kubicek, C. P.; Del Sorbo, G.; Lorito, M. Disruption of the ech42 (Endochitinase-Encoding) Gene Affects Biocontrol Activity in *Trichoderma harzianum* P1. *Mol. Plant-Microbe Interact.* **1999,** *12* (5), 419–429.
85. López-Mondéjar, R.; Ros, M.; Pascual, J.A. Mycoparasitism-Related Genes Expression of *Trichoderma harzianum* Isolates to Evaluate Their Efficacy as Biological Control Agent. *Biol. Control.* **2011,** *56* (1), 59–66.
86. Rosado, I. V.; Rey, M.; Codón, A.C.; Govantes, J.; Moreno-Mateos, M. A.; Benítez, T. QID74 Cell Wall Protein of *Trichoderma harzianum* Is Involved in Cell Protection, Adherence to Hydrophobic Surfaces. *Fungal Genet. Biol.* **2007,** *44* (10), 950–964.
87. Mukherjee, M.; Mukherjee, P. K.; Kale, S. P. cAMP Signalling Is Involved in Growth, Germination, Mycoparasitism, Secondary Metabolism in *Trichoderma virens*. *Microbiology* **2007,** *153* (6), 1734–1742.
88. Tijerino, A.; Cardoza, R. E.; Moraga, J.; Malmierca, M. G.; Vicente, F.; Aleu, J.; Collado, I.G.; Gutiérrez, S.; Monte, E.; Hermosa, R. Overexpression of the Trichodiene Synthase Gene tri5 Increases Trichodermin Production, Antimicrobial Activity in *Trichoderma brevicompactum*. *Fungal Genet. Biol.* **2011,** *48* (3), 285–296.
89. Steindorff, A. S.; do Nascimento Silva, R.; Coelho, A. S.; Nagata, T.; Noronha, E. F.; Ulhoa, C. J. *Trichoderma harzianum* Expressed Sequence Tags for Identification of Genes with Putative Roles in Mycoparasitism Against *Fusarium solani*. *Biol. Control.* **2012,** *61* (2), 134–140.

90. Sharma, P.; Kumar, V.; Ramesh, R.; Saravanan, K.; Deep, S.; Sharma, M.; Mahesh, S.; Dinesh, S. Biocontrol Genes from *Trichoderma* Species: A Review. *Af. J. Biotechnol.*. **2011,** *10* (86), 19898–19907.
91. Resende, M. I.; Jakoby, I. C.; dos Santos, L. C.; Soares, M. A.; Pereira, F. A.; Souchie, E. L.; Silva, F. G. Phosphate solubilization, phytohormone production by endophytic, rhizosphere Trichoderma isolates of guanandi (*Calophyllum brasiliense* Cambess). *Afr. J. Microbiol. Res.* **2014,** *8* (27), 2616–2623.
92. Qi, W.; Zhao, L. Study of the Siderophore-Producing *Trichoderma asperellum* Q1 on Cucumber Growth Promotion Under Salt Stress. *J. Basic Microbiol.* **2013,** *53* (4), 355–364.
93. Contreras-Cornejo, H. A.; Macías-Rodríguez, L.; Cortés-Penagos, C.; López-Bucio, J. *Trichoderma virens*, a Plant Beneficial Fungus, Enhances Biomass Production, Promotes Lateral Root Growth Through an Auxin-Dependent Mechanism in Arabidopsis. *Plant Physiol.* **2009,** *149* (3), 1579–1592.
94. Sofo, A.; Scopa, A.; Manfra, M.; De Nisco, M.; Tenore, G.; Troisi, J.; Di Fiori, R.; Novellino, E. *Trichoderma harzianum* strain T-22 Induces Changes in Phytohormone Levels in Cherry Rootstocks (*Prunus cerasus× P. canescens). Plant Growth Regulation* **2011,** *65* (2), 421–425.
95. Sreenayana, B.; Nakkeeran, S.; Muthulakshmi, P. Hyperparasitic Interaction of *Trichoderma virens* with *Fusarium oxysporum* f.sp. *cucumerinum* Induce Differential Display of NVOC Against Cucumber Vascular Wilt Pathogen. *J. Mycol. Plant Pathol.* **2019,** *49* (3), 237–251.
96. Saxena, A.; Raghuwanshi, R.; Singh, H.B. *Trichoderma* Species Mediated Differential Tolerance Against Biotic Stress of Phytopathogens in *Cicer arietinum* L. *J. Basic Microbiol.* **2015,** *55* (2), 195–206.
97. Meyer, S. L.; Massoud, S. I.; Chitwood, D. J.; Roberts, D. P. Evaluation of *Trichoderma virens*, *Burkholderia cepacia* for Antagonistic Activity Against Root-Knot Nematode, *Meloidogyne incognita. Nematology* **2000,** *2* (8), 871–879.
98. Sharon, E.; Bar-Eyal, M.; Chet, I.; Herrera-Estrella, A.; Kleifeld, O.; Spiegel, Y. Biological Control of the Root-Knot Nematode *Meloidogyne javanica* by *Trichoderma harzianum. Phytopathology* **2001,** *91* (7), 687–693.
99. Affokpon, A.; Coyne, D. L.; Htay, C. C.; Agbèdè, R. D.; Lawouin, L.; Coosemans, J. Biocontrol Potential of Native *Trichoderma* Isolates Against Root-Knot Nematodes in West African Vegetable Production Systems. *Soil Biol. Biochem.* **2011,** *43* (3), 600–608.
100. Windham, G. L.; Windham, M. T.; Williams, W. P. Effects of *Trichoderma* spp. on Maize Growth, *Meloidogyne arenaria* Reproduction. *Plant Dis.* **1989,** *73* (6), 493–495.
101. dos Santos, M. A.; Ferraz, S.; Muchovej, J. J. Evaluation of 20 Species of Fungi from Brazil for Biocontrol of *Meloidogyne incognita* Race 3. *Nematropica* **1992,** 183–192.
102. Rao, M. S.; Reddy, P. P.; Nagesh, M. Management of Root-Knot Nematode, *Meloidogyne incognita* on Tomato by Integration of *Trichoderma harzianum* with Neem Cake. *J. Plant Diss.* Protection **1997,** 423–425.
103. Siddiqui, I. A.; Shaukat, S, S. *Trichoderma harzianum* Enhances the Production of Biocontrol of *Meloidogyne javanica* by *Pseudomonas fluorescens* in Tomato. *Lett. Appl. Microbiol.* **2004,** *38*, 169–175.
104. Pathak, K. N.; Ranjan, R.; Kumar, M.; Kumar, B. Bio-Management of *Meloidogyne graminicola* by *Trichoderma harzianum*, *T. virens* in Rice. *Ann. Plant Protection Sci.* **2005,** *13* (2), 438–440.

105. Sharon, E.; Chet, I.; Viterbo, A.; Bar-Eyal, M.; Nagan, H.; Samuels, G. J.; Spiegel, Y. Parasitism of *Trichoderma* on *Meloidogyne javanica*, Role of the Gelatinous Matrix. *Eur. J. Plant Pathol.* **2007,** *118* (3), 247–258.
106. Sahebani, N.; Hadavi, N. *Biol. Control* of the Root-Knot Nematode *Meloidogyne javanica* by *Trichoderma harzianum*. *Soil Biol. Biochem.* **2008,** *40* (8), 2016–2020.
107. Shawky.; S. M.; Khalil, A. E.; Soliman, M. M.; Manal, M. Nonchemical Control Of Root-Knot Nematode, *Meloidogyne javanica* on Peanut in Egypt. *Zagazig J. Agric. Res.* **2010,** *37* (1), 185–206.
108. Qureshi, S.; Ruqqia, A.; Sultana. V.; Ara, J.; Ehteshamul-Haque, S. Nematicidal Potential of Culture Filtrates of Soil Fungi Associated with Rhizosphere, Rhizoplane of Cultivated and Wild Plants. *Pak. J. Bot.* **2012,** *44* (3), 1041–1046.
109. Zhang, S.; Gan, Y.; Xu, B.; Xue, Y. The Parasitic, Lethal Effects of *Trichoderma longibrachiatum* Against *Heterodera avenae*. *Biol. Control*. **2014,** *72*, 1–8.
110. Singh, N.; Ch, B.; Singh, A.; Sharma, M. C.; Goswami, B. K. Performance of a Highly Potent Strain of *Trichoderma harzianum* Possessing Egg Parasitic Character Against Root Knot Nematode in Addition to Mycoparasitism Isolated from Vegetable Growing Areas of Northern India. *Plant Arch.* **2015,** *15* (2), 667–670.
111. Sanz, L.; Montero, M.; Grondona, I.; Vizcaíno, J. A.; Llobell, A.; Hermosa, R.; Monte, E. Cell Wall-Degrading Isoenzyme Profiles of Trichoderma Biocontrol Strains Show Correlation with rDNA Taxonomic Species. *Curr. Genet.* **2004,** *46* (5), 277–286.
112. Morton, O.; Hirsch, P.; Kerry, B. Infection of Plant-Parasitic Nematodes by Nematophagous Fungi–a Review of the Application of Molecular Biology to Understand Infection Processes, to Improve Biological Control. *Nematology* **2004,** *6* (2), 161–170.
113. Uchikoba, T.; Mase, T.; Arima, K.; Yonezawa, H.; Kaneda, M. Isolation and Characterization of a Trypsin-Like Protease from *Trichoderma viride*. *Biol. Chem.* **2001,** *382* (10), 1509–1513.
114. Prasad, D.; Anes, K. M. Effect of Metabolites of *Trichoderma harzianum*, *T. viride* on Plant Growth, *Meloidogyne incognita* on Okra. *Ann. Plant Protection Sci.* **2008,** *16* (2), 461–465.
115. Yu, Y.; Zhang, G.; Li, Z.; Cheng, Y.; Gao, C.; Zeng, L.; Chen, J.; Yan, L.; Sun, X.; Guo, L.; Yan, Z. Molecular Cloning, Recombinant Expression, Antifungal Activity of BnCPI, a Cystatin in Ramie (*Boehmeria nivea* L.). *Genes* **2017,** *8* (10), 265.
116. Liao, J.; Xu, G.; Mevers, E. E.; Clardy, J.; Watnick, P. I. A High-Throughput, Whole Cell Assay to Identify Compounds Active Against Carbapenem-Resistant *Klebsiella pneumoniae*. *PloS One*. **2018,** *13* (12), e0209389.
117. Balla, O. Y.; Ali, M. M.; Garbi, M. I.; Kabbashi, A. S. Chemical Composition, Antimicrobial Activity of Essential Oil of *Mentha viridis*. *Biochem. Mol. Biol.* **2017,** *2* (5), 60–66.
118. Hameed, I. H.; Altameme, H. J.; Idan, S. A. *Artemisia annua*: Biochemical Products Analysis of Methanolic Aerial Parts Extract, Anti-Microbial Capacity. *Res. J. Pharm. Biol. Chem. Sci.* **2016,** *7* (2), 1843–1868.
119. Valerio, F.; Favilla, M.; De Bellis, P.; Sisto, A.; de Candia, S.; Lavermicocca, P. Antifungal Activity of Strains of Lactic Acid Bacteria Isolated from a Semolina Ecosystem Against *Penicillium roqueforti*, *Aspergillus niger* and *Endomyces fibuliger* Contaminating Bakery Products. *Syst. Appl. Microbiol.* **2009,** *32* (6), 438–448.
120. Kang, K.; Fong, W. P.; Tsang, P. W. Antifungal Activity of Baicalein Against *Candida krusei* Does Not Involve Apoptosis. *Mycopathologia* **2010,** *170* (6), 391–396.

121. Banaras, S.; Javaid, A.; Shoaib, A.; Ahmed, E. Antifungal Activity of *Cirsium arvense* Extracts Against Phytopathogenic Fungus *Macrophomina phaseolina*. *Planta Daninha* **2017,** 35.
122. Goetz, M. A.; Zhang, C.; Zink, D. L.; Arocho, M.; Vicente, F.; Bills, G. F.; Polishook, J.; Dorso, K.; Onishi, R.; Gill, C.; Hickey, E. Coelomycin, a Highly Substituted 2, 6-dioxo-Pyrazine Fungal Metabolite Antibacterial Agent Discovered by *Staphylococcus aureus* Fitness Test Profiling. *J. Antibiot.* **2010,** *63* (8), 512–518.
123. Priyadharsini, P.; Dhanasekaran, D.; Kanimozhi, B. Isolation, Structural Characterization of N-(naphthalene-1-yl) Propanamide, a Herbicidal Compound from *Streptomyces* sp. KA 1-3. In *Microbiological Research in Agroecosystem Management*, **2013,** 187–195.
124. López-Bucio, J.; Hernández-Abreu, E.; Sánchez-Calderón, L.; Pérez-Torres, A.; Rampey, R. A.; Bartel, B.; Herrera-Estrella, L. An Auxin Transport Independent Pathway Is Involved in Phosphate Stress-Induced Root Architectural Alterations in Arabidopsis. Identification of BIG as a Mediator of Auxin in Pericycle Cell Activation. *Plant Physiol.* **2005,** *137* (2), 681–691.
125. Reed, R. C.; Brady, S. R.; Muday, G. K. Inhibition of Auxin Movement from the Shoot Into the Root Inhibits Lateral Root Development in Arabidopsis. *Plant Physiol.* **1998,** *118* (4), 1369–1378.
126. Casimiro, I.; Marchant, A.; Bhalerao, R. P.; Beeckman, T.; Dhooge, S.; Swarup, R.; Graham, N.; Inzé, D.; Sandberg, G.; Casero, P. J.; Bennett, M. Auxin Transport Promotes Arabidopsis Lateral Root Initiation. *Plant Cell.* **2001,** *13* (4), 843–852.
127. Björkman, T. Effect of *Trichoderma* Colonization on Auxin-Mediated Regulation of Root Elongation. *Plant Growth Regulation* **2004,** *43* (1), 89–92.
128. Harman, G. E.; Petzoldt, R.; Comis, A.; Chen, J. Interactions between *Trichoderma harzianum* strain T22, Maize Inbred Line Mo17, Effects of These Interactions on Diseases Caused by *Pythium ultimum*, *Colletotrichum graminicola*. *Phytopathology* **2004b,** *94* (2), 147–153.
129. Lo, C. T.; Lin, C. Y. Screening Strains of *Trichoderma* spp. for Plant Growth Enhancement in Taiwan. *Plant Pathol. Bulletin.* **2002,** *4*, 215–220.
130. Samolski, I.; Rincon, A. M.; Pinzón, L. M.; Viterbo, A.; Monte, E. The qid74 Gene from *Trichoderma harzianum* Has a Role in Root Architecture, Plant Biofertilization. *Microbiology* **2012,** *158* (1), 129–138.
131. Vargas, W. A.; Mandawe, J. C.; Kenerley, C. M. Plant-Derived Sucrose Is a Key Element in the Symbiotic Association Between *Trichoderma virens* and Maize Plants. *Plant Physiol.* **2009,** *151* (2), 792–808.
132. Hajieghrari, B. Effects of Some Iranian *Trichoderma* Isolates on Maize Seed Germination and Seedling Vigour. *Af. J. Biotechnol.* **2010,** *9* (28), 4342–4347.
133. Vasumathi, S.; Aiyanthan, E. K. A.; Nakkeeran. S. Exploring the Diversity of *Trichoderma* spp.; the Synergistic Activity of Delivery Systems for the Management of *Fusarium* Wilt of Cucumber Under Protected Cultivation. *J. Mycol. Plant Pathol.* **2016,** *46* (4), 193–211.
134. Singh, S.; Mathur, N. In Vitro Studies of Antagonistic Fungi Against the Root-Knot Nematode, *Meloidogyne incognita*. *Biocontr. Sci. Technol.* **2010,** *20* (3), 275–282.
135. AL-Shammari, T. A.; Bahkali, A. H.; Elgorban, A. M.; El-Kahky, M. T, Al-Sum, B. A. The use of *Trichoderma longibrachiatum*, *Mortierella alpina* Against Root-Knot Nematode, *Meloidogyne javanica* on tomato. *J Pure Appl. Microbiol.* **2013,** *7*, 199–207.
136. Khan, M. R.; Ahmad, I.; Ahamad, F. Effect of Pure Culture, Culture Filtrates of *Trichoderma* Species on Root-Knot Nematode, *Meloidogyne incognita* Infesting Tomato. *Indian Phytopathology* **2018,** *71* (2), 265–274.

137. Chérif, M.; Benhamou, N. Cytochemical Aspects of Chitin Breakdown During the Parasitic Action of a *Trichoderma* sp. on *Fusarium oxysporum* f. sp. *radicis-lycopersici*. *Phytopathology* **1990,** *80* (12), 1406–1414.
138. Inbar, J.; Abramsky, M.; Cohen, D.; Chet, I. Plant Growth Enhancement, Disease Control by *Trichoderma harzianum* in Vegetable Seedlings Grown Under Commercial Conditions. *Eur. J. Plant Pathol.* **1994,** *100* (5), 337–346.
139. Carvalho, D. D.; Lobo Junior, M.; Martins, I.; Inglis, P. W.; Mello, S. Biological Control of *Fusarium oxysporum* f. sp. *phaseoli* by *Trichoderma harzianum* and Its Use for Common Bean Seed Treatment. *Trop. Plant Pathol.* 2014. *39* (5), 384–391.
140. Louzada, G. A.; Carvalho, D. D.; Mello, S. C.; Lobo Júnior, M.; Martins, I.; Braúna, L. M. Antagonist Potential of *Trichoderma* spp. from Distinct Agricultural Ecosystems Against *Sclerotinia sclerotiorum* and *Fusarium solani*. *Biota Neotropica*. **2009,** *9* (3), 145–149.
141. Fravel, D. R.; Connick, W. J.; Lewis, J. A. Formulation of Microorganisms to Control *Plant Diss*. In *Formulation of Microbial Biopesticides*, **1998,** 187–202.
142. Sivasakthivelan, P.; Saranraj, P. *Azospirillum*, Its Formulations: A Review. *Int. J. Microbiol. Res*. **2013,** *4* (3), 275–287.
143. Schisler, D. A.; Slininger, P. J.; Behle, R. W.; Jackson, M. A. Formulation of *Bacillus* spp. for Biological Control of *Plant Diss*. *Phytopathology* **2004,** *94* (11), 1267–1271.
144. Sriram, S.; Roopa, K. P.; Savitha, M. J. Extended Shelf-Life of Liquid Fermentation Derived Talc Formulations of *Trichoderma harzianum* with the Addition of Glycerol in the Production Medium. *Crop Protection* **2011,** *30* (10), 1334–1349.
145. Loganathan, M.; Sible, G. V.; Maruthasalam, S.; Saravanakumar, D.; Raguchander, T.; Sivakumar, M.; Samiyappan, R. *Trichoderma*, Chitin Mixture Based Bioformulation for the Management of Head Rot (*Sclerotinia sclerotiorum* (Lib.) deBary)–Root-Knot (*Meloidogyne incognita* Kofoid, White; Chitwood) Complex Diseases of Cabbage. *Arch. Phytopathol. Plant Protection* **2010,** *43* (10), 1011–1024.
146. Batta, Y. A. Control of Postharvest Diseases of Fruit with an Invert Emulsion Formulation of *Trichoderma harzianum* Rifai. *Postharvest Biol. Technol.* **2007,** *43* (1), 143–150.
147. Mbarga, J. B.; Begoude, B. A.; Ambang, Z.; Meboma, M.; Kuaté, J.; Schiffers, B.; Ewbank, W.; Dedieu, L.; Ten Hoopen, G.M. A New Oil-Based Formulation of *Trichoderma asperellum* for the Biological Control of Cacao Black Pod Disease Caused by *Phytophthora megakarya*. *Biol. Control*. **2014,** *77*, 15–22.
148. Oancea, F.; Raut, I.; Şesan, T. E.; Cornea, P. C. Dry Flowable Formulation of Biostimulants *Trichoderma* strains. *Agric. Agric. Sci. Procedia* **2016,** *10*, 494–502.
149. Locatelli, G. O.; dos Santos, G. F.; Botelho, P. S.; Finkler, C. L.; Bueno, L. A. Development of *Trichoderma* sp. Formulations in Encapsulated Granules (CG) and Evaluation of Conidia Shelf-Life. *Biol. Control*. **2018,** *117*, 21–29.
150. Morgan, C.A.; Herman, N.; White, P.A.; Vesey, G. Preservation of Micro-Organisms by Drying; a Review. *J. Microbiol. Methods* **2006,** *66* (2), 183–193.
151. Suman, A.; Verma, P.; Yadav, A.N.; Srinivasamurthy, R.; Singh, A.; Prasanna, R. Development of Hydrogel-Based Bio-Inoculant Formulations, Their Impact on Plant Biometric Parameters of Wheat (*Triticum aestivum* L.). *Int. J. Curr. Microbiol. Appl. Sci*. **2016,** *5* (3), 890–901.
152. Suseela Bhai, R.; Anandaraj, M. Enhancing Shelf Life of *Trichoderma harzianum* by Conidial Storage in Sterile Deionized Water. *J. Spices Aromatic Crops* **2014,** *23* (2), 243–249.

153. Abdullah, M. T.; Ali, N. Y.; Suleman, P. Biological Control of *Sclerotinia sclerotiorum* (Lib.) de Bary with *Trichoderma harzianum*, *Bacillus amyloliquefaciens*. *Crop Protection.* **2008,** *27* (10), 1354–1359.
154. Poddar, R. K.; Singh, D. V.; Dubey, S. C. Integrated Application of *Trichoderma harzianum* Mutants, Carbendazim to Manage Chickpea Wilt (*Fusarium oxysporum* f. sp. *ciceris*). *Indian J. Agric. Sci.* **2004,** *74* (6), 346–348.
155. Rojo, F. G.; Reynoso, M. M.; Ferez, M.; Chulze, S. N.; Torres, A.M. *Biol. Control* by *Trichoderma* species of *Fusarium solani* Causing Peanut Brown Root Rot Under Field Conditions. *Crop Protection* **2007,** *26* (4), 549–555.
156. Sallam Nashwa, M. A.; Abo-Elyousr, K. A.; Hassan, M. A. Evaluation of *Trichoderma* Species as Biocontrol Agent for Damping-Off, Wilt Diseases of *Phaseolus vulgaris* L. and Efficacy of Suggested Formula. *Egypt. J. Phytopathol.* **2008,** *36* (1–2), 81–93.
157. Singh, S.P.; Singh, H.B. Effect of Mixture of *Trichoderma* Isolates on Biochemical Parameters in Leaf of *Macrophomina phaseolina* Infected Brinjal. *J. Environ. Biol.* **2014,** *35* (5), 871–876.
158. Naseby, D. C.; Pascual, J. A.; Lynch, J. M. Effect of Biocontrol Strains of *Trichoderma* on Plant Growth, *Pythium ultimum* Populations, Soil Microbial Communities and Soil Enzyme Activities. *J. Appl. Microbiol.* **2000,** *88* (1), 161–169.
159. Alexandru, M.; Lazăr, D. A.; Ene, M.; Sesan, T. E. Influence of Some *Trichoderma* Species on Photosynthesis Intensity and Pigments in Tomatoes. *Roman Biotechmol. Lett.* **2013,** *18*, 8499–8510.
160. Barari, H. Biocontrol of tomato Fusarium Wilt by *Trichoderma* Species Under In Vitro, In Vivo Conditions. *Cercetari Agronomice Moldova* **2016,** *49* (1), 91–98.
161. Khan, M. R.; Haque, Z. Soil Application of *Pseudomonas fluorescens*, *Trichoderma harzianum* Reduces Root-Knot Nematode, *Meloidogyne incognita* on Tobacco. *Phytopathologia Mediterranea* **2011,** *50* (2), 257–266.
162. Mwangi, M. W.; Muiru, W. M.; Narla, R. D.; Kimenju, J. W.; Kariuki, G. M. Management of *Fusarium oxysporum* f. sp. *lycopersici*, Root-Knot Nematode Disease Complex in Tomato by Use of Antagonistic Fungi, Plant Resistance and Neem Oil. *Biocontr. Sci. Technol.* **2019,** *29* (3), 229–238.
163. Dababat, A.; Sikora, R. A. Importance of Application Time, Inoculum Density of *Fusarium oxysporum* 162 for Biological Control of *Meloidogyne incognita* on Tomato. *Nematropica* **2007,** *37* (2), 267–276.
164. Christopher, D. J.; Raj, T. S.; Rani, S. U.; Udhayakumar, R. Role of Defense Enzymes Activity in Tomato as Induced by *Trichoderma virens* Against Fusarium Wilt Caused by *Fusarium oxysporum* f. sp. *lycopersici*. *J. Biopest.* **2010,** *3* (1), 158–162.
165. Martínez-Medina, A.; Pascual, J. A.; Pérez-Alfocea, F.; Albacete, A.; Roldán, A. *Trichoderma harzianum*, *Glomus intraradices* Modify the Hormone Disruption Induced by *Fusarium oxysporum* Infection in Melon Plants. *Phytopathology* **2010,** *100* (7), 682–688.
166. Dar, W. A.; Ganie, S. A.; Bhat, J. A.; Mir, G. H.; Lawrence, R.; Narayan, S.; Singh, P. K. Comparative Efficacy of *Trichoderma viride*, *Trichoderma harzianum* Against *Fusarium oxysporum* f sp. *ciceris* Causing Wilt of Chickpea. *Afr. J. Microbiol. Res.* **2013,** *7* (50), 5731–5735.

CHAPTER 19

Integrated Pest and Disease Management in Medicinal and Aromatic Plants-Based Cropping System

K. T. SHIVAKUMARA, SATYAJIT ROY*, and R. P. MEENA

ICAR-Directorate of Medicinal and Aromatic Plants Research, Boriavi, Anand, Gujarat 387310, India

Corresponding author. E-mail: sroycob1959@gmail.com

ABSTRACT

Medicinal and aromatic plants (MAPs) care the public health in globally, because of their curative properties present in the form of complex chemical substances, such as alkaloids, glycosides, and essential oils. The cultivation of MAPs is increasing day by day because of their conservation, reducing the burden of forest source as well as meeting demand of global pharmaceutical industries. Besides, most of the plants have been facing problem of endangered status. The quantity and quality are negatively affected by various biotic and abiotic factors, such as insects, diseases, nematodes, and abiotic factors, such as climatic variation, lack of irrigation, and nutrient deficiencies. Among these factors, insects and disease are the most important factors for causing a significant amount of damage to plants. Pesticides are used in order to manage the harmful pest and diseases to enhance crop yields. Additionally, these synthetic chemicals might alter the therapeutic values of herbal products. Indiscriminate usage of pesticides causes serious ill effects to public health and environment. Research on pest and diseases is helpful in understanding the life history of pest species to formulate the suitable

Integrated Pest Management in Diverse Cropping Systems. Devendra Pal Singh, PhD (Ed.)

management strategies. Moreover, eco-friendly management practices are important. Monitoring of pest populations and disease epidemiology, identification, and choosing a combination of strategies to keep pest and disease at an acceptable level is necessary. The present chapter deals with overviews on diversity of pest and disease of MAPs and integrated pest management strategies for dealing in an eco-friendly manner.

19.1 INTRODUCTION

The medicinal and aromatic crops play an important role in public health care globally. As per the report of World Health Organization, 80% of world's population known about the traditional medicine. The diversity of 6000 species of indigenous flora from tropical regions has medicinal properties[1] and over 1000 plants are categorized as aromatic.[2] Medicinal and aromatic plants (MAPs) have in general natural habitat in the forest or cultivated round the year in developing and less developed countries.[3] Their use in rural areas is common for local treatments of various health issues and means of livelihood by collection of raw materials from the forest and their selling in local markets.[4] Over 90% of the herbal industries depend on raw materials coming from the wild leading to destruction and over exploitation of natural sources.[5] These are cultivated in about 6,17,000 ha have production of 11,56,000 Mt of biomass in India.[6] The medicinal plants (Indian gooseberry, sandalwood, asparagus, Brahmi, holy basil, and blond psyllium) are in high demand. The phytochemicals extracted from these are used for Ayurveda drugs, homoeopathic, cosmetics, dietary supplements, and functional foods.[7] In addition, MAPs have been used for biopesticides against various pest and disease due to their insecticidal property.[8,9] Synthetic drugs are in use in increased proportions having deleterious side effects. The people are inclined toward the natural herbal products of MAPs origin. Normally, anything from the nature is harmless compared with synthetic drugs. Improved cultivation practices are required to augment regular large-scale supply of biomass to various industries.[2] For preventing the forest resources, extinct and continued supply of raw materials for pharmaceutical industries, cultivation of MAPs is necessary. To meet out the present demand of the pharmaceutical industries, large number of MAPs are cultivated in the fields. Presently, raw material supplies are inadequate due to their less productivity and inferior quality.[10] The problems being currently faced by the farmers are availability of quality planting material, improved or recommended cultivars, overexploitation of natural resources, no systematic cultivation, long gestation period of some plant species, misidentification of herbs, indiscriminate collection from the forest, and standardized package of practices in different agroecological conditions.[11,12] MAPs are cultivated on a large-scale areas and also with climatic variation. The intensive cultivation and high-yielding verities are prone to attack of various insects, mites, pathogens, and nematodes causing significant losses and hampers the successful cultivation of the MAP. Sometimes, epidemic nature of the pest and diseases which leads to complete failure of the crops.[13] Additionally, some of the insects, pathogens, nematodes and mites are problem to one crop in an ecosystem.[14,15] So far IPM strategies are not properly exploited in case of medicinal and aromatic plants. Prior to adoption of any pest management technique, assessment of population dynamics, estimation of the damage intensity becomes prerequisite to determine their status. Pest control practices like use of synthetic chemicals may alter the quality and contents of the active ingredient and deteriorate their therapeutic values of the medicinal plants. The conventional pest management practices being adopted by farming community like use of synthetic pesticides for effective control of insect pests and diseases. However, the inappropriate chemicals, extra doses, spurious formulations, faulty spray equipment, more numbers of applications, expiry pesticides, and illiteracy of applicators is common problems in chemical control in developing countries.[16] Farmers are more interested to application of chemicals frequently to collect fresh and healthy raw herbage particularly for marketing. The repeated sprayings without needing to the interval period from last spaying to harvesting time leads to pesticide residues in plants parts.[17]

The chemical pesticides not only mean for pest and disease management as it leads to pesticide residue, pest resistance and resurgence problem. Resultantly, serious environmental pollution affects the life of nontarget organisms. Therefore, environmentally safe IPM strategy comes as best option, IPM thus is the best alternative for MAP-based crop ecosystem. It contains timely monitoring of pest populations, working out disease epidemiology, identification and choosing a combination of practices. The IPM includes physical, mechanical, cultural, biological, and chemical management practices in combination or alone to reduce the pest population. Methods are being used to manage the pest and diseases are integrated and eco-friendly. The current chapter deals with an attempt to provide a glimpse MAPs pest and disease and their IPM status (Table 19.1).

19.2 COMPONENTS OF IPM

19.2.1 PREVENTION

Prevention of pest and disease is one of the important steps in IPM and it can reduce the susceptibility of cultivars against the pest and disease. Before going to actual management, identification of pest species is necessary to formulate the critical management solution.

19.2.2 MONITORING AND PEST IDENTIFICATION

It involves pest monitoring and identification technologies to measure the infestation level of insect/disease to make a pragmatic decision about when and how to apply for management practices for pest and diseases.

19.2.3 ESTIMATION OF TOLERANCE AND ECONOMIC THRESHOLD LEVEL

Before going to adopt any management program there is need to arise at tolerance, injury level, and economic threshold.

19.2.4 DEVELOPING OF MANAGEMENT STRATEGIES

Based on the species identification and infestation level, there is need to formulate most effective and economic technologies like cultural, physical, mechanical, botanical, biological, and chemical methodologies to manage the pest and diseases.

19.2.5 EVALUATING THE PEST MANAGEMENT STRATEGY

On completion of adoption technology, there is need to evaluate the IPM program as mandatory step to know the how and why the methodology is successful/unsuccessful for future improvement. This requires the maintenance of systemic records.

19.2.5.1 PREVENTION

Prevention is cheapest and often most reliable method to avoid the many pest/disease problems. These types of methods include adjustments in cultural practices like altering as selection the planting time, Soil management, nutrient management, water management, and resistant varieties, etc.

19.2.5.2 MONITORING

Regular inspection of pest and disease on crop/plants is necessary to assertion real problems and economic threshold level. Therefore, understanding the how pests and diseases will significantly reduce the yield, there is need to monitor the crop on regular basis. When insect or pathogens are misidentified, there is all likelihood of the resulting management action to be ineffective. Many pests might appear similar to each other alike to beneficial or harmless organisms. Other abiotic factors, such as climatic condition, inappropriate irrigation, poor soil fertility, toxic contamination in soil, water, air, and genetic disorders are also known to cause disease-like symptoms.

19.2.5.3 TOLERANCE LEVELS AND ECONOMIC THRESHOLDS

An acceptable injury level to ascertain the extent of damage that can be tolerated has to be established in particular against the pest species. The Economic Threshold Level of particular crop on pest and disease is important in working out an economic problem.

19.2.5.4 DEVELOPING A PEST MANAGEMENT STRATEGY

After having identified the pest/disease a decision should follow about the extent of damage being acceptable or if unacceptable, a suitable pest/disease management strategy is needed.

19.2.5.5 EVALUATING RESULTS

After implementation of the IPM program, the evaluation of program is mandatory. maintaining records and reviewing in regular basis are necessary for establishing the achievable goals in IPM.

19.3 PESTS AND DISEASES

The present chapter deals with the respective pest and diseases of important MAPs in terms of its cultivation, biomass, economic importance, and export. The pests and diseases are summarized for study in Table 19.1.

19.4 INCIDENCE AND SEVERITY LEVEL

Aphis affinis appears to cause a significant damage to mint crop in Punjab during February–April month.[104] *Syngamia abruptalis* infects *Mentha citrate, M. piperita*, and *M. arvensis* in Punjab.[99] *Thysanoplusia orichalcea* Fab and *Heliothis armigera* Hubn are reported simultaneously from April till harvest in the case of mentha in Punjab.[105] Incidence of menth leaf webber (*S. abruptalis* Walk) from south India during the month of May-June[93] and in northern parts of India during August–October.[99,106,107] Incidence of hadda beetle, *E. vigintioctopunctata* and *H. armigera* on *W. somnifera* was reported from 10th to 12th and 12th to 14th standard week, respectively.[92]

Hadda beetle and mealybug, *Phenococcus solanopsis* cause significant damage on *W. somnifera* during the month of September–December and September–February, respectively.[108] Aerial apical parts of ashwagandha are affected by Cowbug, *Oxyrachis tarandus* on *W. somnifera* plants in Punjab.[109] The leaves of sarpagandha were destroyed by *Psilogramma menephron.* Similarly, *W. somnifera* roots and rootlets were also eaten by grubs of *Holotrichia serrata.*[110] The defoliation in Ashwagandha is caused by the larvae *Deilephila nerii* and *H. vigintioctopunctata.*[111] The grubs and adults of hadda beetle also cause defoliation in plants.[108] In tulsi, the sucking pests attack the tender parts of the plant and suck the sap. It secretes the honey dew-like substance which invites sooty moulds and affect the photosynthetic process and crop yield.[54] The lace bug (*Cochlochila bullita*) appears during 52nd standards week with 33.33% herbage yield losses in unprotected plots of tulsi.[112] New insect pest, *Psilogramma menephron* infected 40% on *Rauwolfia serpentine* and *H. serrata* with 20% incidence on *W. somnifera* from Madhya Pradesh.[110] The Jasmine bud borer, *Elasmopalpus jasminopalpus* is an important lepidopteran pest which causes a 38.85% incidence during the first fortnight of January.[92]

19.5 INTEGRATED PEST AND DISEASE MANAGEMENT PRACTICES

Based on the economic importance of pest and diseases, the management strategies were formulated by different workers which are summarized.

TABLE 19.1 Insect Pests and Diseases of Medicinal and Aromatic Plants-Based Cropping System.

Plants	Pest/disease	Pest species/pathogens	References
Aloe vera (*Aloe barbadensis*)	Collar & root rot	*Penicillium purpurogenum*	[18]
	Rot	*Fusarium oxysporum*	[19]
	Anthracnose	*Colletotrichum gloeosporioides*	[20]
	leaf spot	*Curvularia lunata*	[21]
		Alternaria brassicae	[19]
		Alternaria alternata, A. brassicae	[22]
	Basal stem rot	*Fusarium oxysporum*	[23]
	Rust	*Phakopsora pachyrhizi, Uromyces aloes*	[24]
Senna (*Cassia angustifolia*)	Aphids	*Aphis craccivora*	[25]
	Pierid Butterfly	*Catospsilia pyranthe and Eurema hecabe*	[25]
	Pod Borer	*Etiella zinckenella*	[25]
	Damping off	*Rhizoctonia bataticola*	[26]
	Leaf spot	*Alternaria alternata*	[26]
	Leaf blight	*Phyllosticta* spp.	[26]
	Dry rot	*Macrophomina phaseolina*	[26]
	Alternaria Blight	*Alternaria alternata*	[27]
	Cercospora Leaf Spot:	*Cercospora* sp.	[27]
	Leaf and Pod Spot	*Phomopsis* spp.	[27]
	Wilt	*Fusarium oxysporum*	[28]
Brahmi (*Bacopa monnieri)*	Armyworm	*Spodoptera litura*	[19]
	Leaf blight	*Alternaria alternata*	[19]
	Leaf shot hole	*Phomopsis*	[19]
Musli (*Chlorophytum borivilianum*)	Leaf roller	*Hyposidra successaria* Wlk.	[29]

TABLE 19.1 *(Continued)*

Plants	Pest/disease	Pest species/pathogens	References
	Blue butterfly	*Junonia orithya* L.	[29]
	Leaf blight	*Colletotrichum dematium, C. capsica*	[30]
	Leaf blight	*Alternaria alternata*	[31]
	Root rot	*Rhizoctonia bataticola*	[32]
Indian ginseng (*Withania somnifera*)	Ladybird beetle	*Henosepilachna vigintioctopunctata* (Fb.)	[33]
	Fruit borer	*Helicoverpa armigera*	[34]
	Sphingid,	*Deilephila nerii* (Linn.)	[34]
	Cowbug	*Oxyrachis tarandus* Fb.	[35]
	Mealybug	*Phenacoccus solenopsis*	[36]
	Damping off	*Rhizoctonia solani*	[37]
	Leaf spot	*Colletotrichum gloeosporioides*	[38]
	Leaf Spot Disease	*Alternaria alternata*	[39]
	Wilt	*Fusarium oxysporum* and *Fusarium solani*	[39]
	Tetranychid mite	*Tetranychus urticae*	[40]
Kalmegh [*Andrographis paniculata*]	Brown scale	*Parasaissettia nigra*	[41]
	Wart	*Synchytrium lepidagathidis*	[42]
	Leaf web blight	*Rhizoctonia solani*	[42]
	Leaf spot	*Cercospora* sp	[42]
	Root rot	*Macrophomina phaseolina*	[42]
	Yellow Vein Leaf Curl Disease	Eclipta yellow vein virus	[43]
	Witches' Broom of Kalmegh	*Phytoplasma*	[44]
Serpentine (*Rauwolfia serpentine L. Benth. Ex Kurz)*	Weevil	*Indomia cretaceous*	[45]

TABLE 19.1 *(Continued)*

Plants	Pest/disease	Pest species/pathogens	References
	Sphingid	*Deilephila nerii*	[45]
	Anthracnose	*Colletotrichum gloeosporioides*	[19]
	Leaf blight	*Alternaria tenuis* Nees	[19]
	Leaf blotch	*Cercospora serpentina* (L.) Benth ex Kurz.	[19]
	Leaf spot	*Cercospora rauvolfiae*	[46]
	Leaf blight and bud rot	*Alternaria tenuis, Alternaria alternata*	[46]
	Leaf blotch	*Cercospora serpitinae*	[47]
	Anthracnose	*Colletotrichum gloeosporioides*	[48]
	Die back	*Colletotrichum dematium*	[49]
	Inflorescence/ fruit rot	*Rhizopus stolonifera*	[50]
	Mosaic	*Cucumber mosaic virus*	[51]
Ocimum spp.	Tingid bug	*Cochlochila bullita* (Stal.)	[52]
	Perila leaf moth	*Pyrausta panopealis*	[53]
	Bihar hairy caterpillar	*Spilarctia oblique*	[29]
	Whiteflies	*Aleurodicus disperses* Russell, *Dialeurodes* sp.	[54]
	aphids	*Macrosiphum* sp.	[54]
	Downy mildew	*Peronospora belbahrii*	[55]
	Leaf spot	*Colleotrichum gloeosporioides, C. capsica*	[56]
	Leaf spot	*Cercospora ocimicola*	[170]
	Fusarium wilt	*Fusarium oxysporum* f. sp. *Basilicum*	[170]
	Gray mold	*Botrytis cinerea*	[170]
	Root rot	*Rhizoctonia solani, Pythium* spp.	[170]
Indian gooseberry, (*Phyllanthus emblica)*	Butterfly	*Deudorix isocrates* (Fb.)	[58]

TABLE 19.1 *(Continued)*

Plants	Pest/disease	Pest species/pathogens	References
	Aphid	*Schoutedenia emblica*	[59]
		Schoutedenia lutea van der Goot	[60]
	Semilooper	*Plusia signata* Fabricius	[61]
	tobacco cutworm	*Spodoptera litura* Fabricius	[61]
	thrips	*Thrips tabaci* Lind	[61]
	Root Rot:	*Macrophomina phaseolina*	[62]
	Leaf Blight	*Alternaria alternata*	[63]
Makoi (*Solanum nigrum*)	Melybugs	*Paracoccus marginatus, Phenococcus solenopsis*	[64]
	Aphids	*Aphis craccivora*	[64]
	Thrips	*Thrips tabaci*	[64]
	White Flies	*Bemisia tabaci*	[64]
	Red Spider Mite	*Tetranychus urticae*	[64]
	Yellow Mite	*Polyphagotarsonemus latus*	[64]
	Tobacco Caterpillar	*Spodoptera litura*	[64]
	Hadda Beetle	*Henosepilachna vigintioctopunctata*	[64]
	Powdery Mildew	*Leveillula taurica*	[65]
Coleus (*Coleus forskohlii* Briq.)	The root-knot nematode	*Meloidogyne incognita*	[66]
	Mealybug	*Coccidohystrix insolitus and Paracoccus marginatus*	[66]
	Leaf spot	*Corynespora cassiicola, Botryodiplodia theobromae*	[67]
	Arial blight	*Rhizoctonia solani*	[68]
	Bacterial Wilt	*Ralstonia solanacearum*	[69]
	Root Rot	*Macrophomina phaseolina*	[70]
	Root rot and wilt	*Fusarium chlamydosporum, Macrophomina phaseolina, Ralstonia solanacearum*	[71]

TABLE 19.1 *(Continued)*

Plants	Pest/disease	Pest species/pathogens	References
	Downy mildew	(*Peronospora belbahrii, P. lamii*)	[72]
Asparagus (*Asparagus officinalis* L.)	purple spot	*Stemphylium vesicarium*	[73]
	Fusarium crown	*Fusarium oxysporum* f. sp. *Asparagi*	[73]
	Phytophthora crown	*Phytophthora asparagi Phytophthora megasperma* var. *sojae*	[73]
	Rust	*Puccinia asparagi*	[57]
Amruthaballi *Tinospora cordifolia* (Thunb.) Miers	Leaf spot	*Xanthomonas campestris*	[74]
	Fruit sucking moth	*Othreis maternal*	[75]
	Grass hopper	*Neorthacris acuticeps*	[45]
	Phoma leaf spot	*Phoma putaminum*	[76]
	Leaf hopper	*Kolla ceylonica*	[45]
	Flat stem disease	Phytoplasma	[77]
Noni (*Morinda citrifolia*)	Green sting bug	*Nezara gramineae*	[78]
	White fly	*Bemisia tabaci*	[78]
	Mealy bug	*Maconellicoccus hirsutus*	[78]
	Fruit fly	*Bactrocera dorsalis*	[79]
	Anthracnose:	*Colletotrichum gloeosporioides*	[80]
	Soft rot	*Pantoea agglomerans*	[80]
	Root-knot nematode	*Meloidogyne incognita*	[81]
Madhunasini, *Gymnema sylvestre*	Aphid	*Aphis nerii*	[29]
	Mealybug	*Paracoccus marginatus*	[82]
Lemongrass (*Cymbopogon citratus*)	Spindle bug	*Clovia bipunctata*	[83]
	Stem boring caterpillar	*Chilotrea* sp.	[83]

TABLE 19.1 *(Continued)*

Plants	Pest/disease	Pest species/pathogens	References
	Nematodes	*Tylenchorhynchus vulgaris, Rotylenchulus reniformis*	[83]
Citronella [*Cymbopogon citratus* (DC.) Stapf]	Leaf blight	*Curvularia andropogonis*	[84]
	Lethal yellowing	*Pythium aphanidermatum*	[85]
	Rust	*Puccinia nakanishikii*	[86]
Palmarosa (*Cymbopogon martinii*)	Aphids	*Aphis gossypii*	[87]
	Thrips	*Haplothrips* sp.	[87]
	White grub	*Holotrichia consanguinea*	[87]
	Termite	*Microtermis* sp.	[87]
	Ellisella blight	*Ellisella caudate*	[87]
	Curvularia blotch	*Curvularia andrographis*	[87]
Jasmine sp.	Bud borer	*Elasmopalpus jasminophagus* (Hampson)	[88]
	Bud borer	*Hendecasis duplifascialis* (Hampson)	[89]
	Leaf web worm	*Nausinoe geometrialis (Guenee)*	[90]
	Red spider mite	*Tetranychus* sp.	[91]
	Jasmine whitefly	*Aleuroclava jasmini* (Takahashi)	[92]
	Tingid bug	*Corythanma ayyari* Drake	[92]
	Banana flower thrips	*Thrips florum* Schmutz	[90]
Mint sp.	Lepidoptera	*Syngamia abruptalis* Walk (Pyralidae)	[93]
	Lepidoptera	*Syngamia abruptalis* Walk (Pyralidae)	[94]
		Pyrausta fumalis Gn.	[95]
	Lepidoptera	*Thysanoplusia orichalcea*	[96]
	Lepidoptera	*Heliothis armigera*	[96]

TABLE 19.1 *(Continued)*

Plants	Pest/disease	Pest species/pathogens	References
	Aphids	*Aphis gossypii*	[97]
		Aphis affinis del Guercio	[98]
	Bug	*Nisia atrovenosa*	[99]
	cutworms	*Spilosoma obliqua*	[100]
	Mite	*Tetranychus uritcae* Koch	[101]
	Rust	(*Puccinia menthae*)	[102]
	Leaf spot	*Alternaria alternata*	[102]
	Powdery mildew	*Erysiphe cichoracearum*	[102]
	Aerial blight	*Rhizoctonia solani*	[102]
	Stolon rot	*Rhizoctonia solani, Rhizoctonia bataticola*	[102]
	Wilt	*Verticillium alba-atrum* var. *menthae*	[102]
	Sclerotinia blight	*Sclerotinia sclerotiorum*	[102]
Sandawood (*Santalum album*)	Beetle	*Sympiezomias cretaceous*	[103]
	Grasshopper	*Holochlora albida*	[103]
	Shield scale	*Pulvinaria psidii*	[103]
	Bug	*Letana inflate*	[103]
	Caterpillar	*Cryptothelea cramerii*	[103]
	Preemergence rot	*Fusarium* or *Phytophthora*	[103]
	Damping off	*Rhizopus* spp.	[103]
	Fusarium wilt	*Fusarium* or *Phytophthora*	[103]

Commonly, most of the pests and diseases are observed in various medicinal and aromatic crop-based ecosystems, such as leaf sucking insect, defoliators, fruit borers, flowers feeders, and other phytopathogens. Plant-based pesticides like neem are the major bio pesticides under IPM program to curb the pest and disease incidence. Various liquid formulations are available in commercial market.[79] Growing of two rows of maize as a barrier crops for avoiding the thrips and whiteflies population effectively. Rouging is an important agronomic practice for destroying the affected shoots and diseased plants from field. Monitoring of various sucking pests by installing the yellow sticky traps at 50/ha. Application of botanical pesticides, such as neem seed kernel @5%, neem oil (3%) can effectively reduce the incidence of sucking pests. Besides, various entomopathogens like *Beauveria bassiana, Metarhizium anisopliae, Verticillium lecanii, Bacillus thuringiensis*, and NPV and entomopathogenic nematodes are contributing for insect pest management. Application of insect growth regulators, biorational pesticides, and other green pesticides are used for the management of some of the pests and diseases.

19.5.1 AGRONOMIC PRACTICES

19.5.1.1 MEDICINAL PLANTS

Hand picking of larvae and poison baiting techniques are effectively controlled by cutworms in Indian ginseng.[113] Spotted leaf beetles were effectively controlled with farmyard manure @ 12.5 T/ha + Azophos @ 2 kg/ha + neem cake @1000 kg/ha in Ashwagandha. The soil digging, hand picking, and planting of nerium (*Nerium oleander*) as trap crop are few means for effective management of Hawk moth caterpillar, *Deilephila nerri.*[113] Larvae and pupae of black cutworm, *Agrotis ipsilon* were managed by summer deep ploughing technique.[113] *Meloidogyne incognita* has been effectively managed in Indian ginseng fields by the application of organic amendments.[114] In ginseng plants, for the management of *M. incognita* using a mixture of vermicompost+*T. harzianum* (1:1) @ 2 t/ha in green house condition is effective.[115] Deep ploughing is used for the management of stone borer (*Curculio* sp.) and fruit midge (*Clinodiplosis* sp.), while pruning effectively managed the incidence in Indian gooseberry.[58] The cv. Hatizola has proved tolerant against the vascular wilt (*Fusarium oxysporum*) compared with Indian gooseberry.[116] The planting from October to December in western India has resulted least infestation of *Aphis gossypii* and crop sowing in the third week of October helped in managing aphids in Blond psyllium.[117]

19.5.1.2 AROMATIC PLANTS

Planting of marigold (*Tagetes* sp.) as an intercrop, soil solarization and mulching of crop residue techniques were effectively managed to Reniform nematode (*Rotylenchulus reniformis*), spiral nematode *Helicotylenchus* sp., stunt nematode (*Tylenchorhynchus vulgans*), and lesion nematode (*Pratylenchus* spp.) in lemon grass ecosystem.[118] The plants of palmarosa are attacked by occasional pest like sugarcane white grub (*Holotrichia consanguinea*), which feeds on plant roots and rootlets. Flooding the infested field and application of chlorpyriphos 20 EC is recommended for its management.[52] The practices like incorporation of good cultural practices, such as proper drainage, collection and destruction of fallen flower buds, garden sanitation, pruning of bushes during the winter, raking of soil, avoiding alternative host crops, such as tomato, eggplant, and bitter gourd around the main crops are used for the management of Jasmine blossom midge (*Contarinia maculipennis*.[119,120]

In geranium, southern bacterial wilt caused by *Ralstonia solanacearum* is prevented by using prophylactic measures like the use of healthy material for planting and destruction of affected plant parts.[121] Fungal leaf blight (*Botrytis cinerea*) is the most common pathogen in leaf, stem, and flowers. Good management like wide spacing and sanitation practices as a prophylactic measure and use of *T. harzianum* at 5 kg/ha were effective.[122] Virus diseases, such as pelargonium flower break virus, cucumber mosaic virus, and tomato spotted wilt virus are common during the crop cycle. The planting of virus-free cuttings, weed and insect vectors management, and removal and destruction of diseased plants are few practices to manage viral diseases.[121,122] Physiological disorder (edema) is the major problem in geranium. It is managed with wide spacing and avoiding overhead irrigation and proper plant nutrients.[121,122] Patchouli plantations were found to have minimum number of pests and diseases like leaf roller, *Pachyzacia* sp. in the northeast region of India. As precautionary measures, planting should be avoided in shade condition.[123] In India, the preplant soil application of carbofuran 3G @ 2 kg/ha is recommended for managing the Southern root-knot nematode (*M. incognita*).[113,123] The soil application of compost can effectively suppress the population of lesion nematodes *Pratylenchus* sp. by 58.0–67.0% in root and 40.0–59.0% in soil in Indonesia.[124] Vetiver cultivation in sandy soils in the subterranean region predisposes it to the attack of termites, *Microtermes* spp. Application of 1% lindane dust in the soil before planting has been recommended.[52,125] In severe infestation, seedlings succumb to attacks by white grubs *Eupladia* sp. and *Phyllophaga serrata*. Flooding of termites infested fields with chlorpyriphos 20 EC mixed in irrigation water is recommended.[52] Vetiver plants irrespective of the growth stages are infested by *Aulacaspis madiunensis* and black scales (*Aclerda takahashii*) in the nursery and in the field.[126] The vetiver plants are affected due to root-knot nematodes, (*M. incognita* and *M. javanica*) and *Heterodera zeae*. The use of healthy stock planting and soil application of neem cake @ 5 t/ha is recommended to manage these nematodes.[52,113]

19.5.2 PHYSICAL METHODS

The light and pheromone traps are used to effectively manage the insect pest and this technology is economically viable for farmers. Pheromone traps for the management of aswagandha for *H. armigera* has been advocated which bores into fruits and tender leaves. Besides, the hawk moth caterpillar, *Deilephila nerri,* a minor defoliator, is managed by the installation of light traps (1 trap/ha).[113] Black cutworm, *Agrotis ipsilon* is the pest of serpentine, installation of one to two light traps/ha or 20 pheromone traps/ha is recommended.[113]

19.5.3 BOTANICAL PESTICIDES

Botanical biopesticide has received special attention for managing the pest and diseases. Success has been achieved by using neem oil (10,000 ppm of Azadiractin) 2–3 mL/L or tobacco decoction @ 20 mL/L or juice of garlic extract.[127] NSKE 5% and Neemazal 1% application at 900 mL/ha were effective on *Gloriosa superba*.[128] Same kind of management practice has been advised for *Solanum nigrum* pest.[129–131] *Bacillus thuringiensis* application @ 1% followed by neem oil @ 1% was found effective in the case of *P. gloriosae* on *G. superba*.[132] Efficacy of neem in controlling the aphids on isabgol (*Plantago ovata*) has been reported.[133] The repeated application of neem oil @ 0.2% has been proved to be effective against polyphagous grasshoppers.[134] The spraying of 0.5% oil extract of *Zingiber officinale*, *Allium sativum*, *Ocimum sanctum*, *Datura metel*, and *Mentha spicata* significantly reduces the leaf soft rot (*Alternaria alternata*) and *Fusarium* sp. As well as increased the rhizome yield.[135] Destruction of gooseberry aphid could effectively be controlled by the application of NSKE at 0.45% having 1500 ppm of Azadiractin..[136]

Application of 0.2% neem oil at 5 mL/L was effective in the management of nymphs and adult tinged bug (*Cochlochila bullita*) in holy basil.[52] Combination of *Trichoderma viride* with Neemato® (neem-based product) in soil at 500 g (5 m^2 area) resulted in 12.7% decrease in disease incidence under field condition.[137] Mealy bug, *Coccidohystrix insolitus* and epilachna beetle populations were significantly reduced with the application of farmyard manure (12.5 t/ha) and 3% neem oil as mixture combination and application of Azophos® biofertilizer at 2 kg/ha when incorporated into soil.[138] The six plant extracts were examined in both laboratory and field condition against the jasmine bud worm, *H. duplifascialis*. Out of six plant extracts, neem cake 5% showed superior result followed by the application of 5% NSKE and 1% neem oil.[139] The soil application of neem cake (100 g/plant) and neem oil (2%) spray or *pongamia* kernal 4% water extract significantly checked the damage and infestation of gallmite and *Aceria jasmini* in *Jasminum auriculatum*.[140] The soil amendments of neem cake @250 kg/ha combined with NSKE (10%) foliar application was found to be effective in the control of *A. jasmine*.[141] The Indian gooseberry aphid, *Schoutedenia emblica* populations can be managed by the application of NSKE (0.045%) @1500 ppm of azadirachtin.[136]

19.5.4 BIO AGENTS

The naturally occurring biocontrol agents like predators (ladybird beetle, rove beetle, damsel fly, spiders, mirid bugs, lace wings, *Gambusia* fish), parasitoids (*Trichogramma, Goniozus, Bracon,*) and microbial (bacteria, fungi, virus, protozoa, nematodes) play a pivotal role in IPM for the management of pest and diseases. Application of Gronim followed by *B. thuringiensis* has given the best result against the red bug and aphid incidence. Besides, reducing the larval defoliator population could be done by the application of *Chrysoperla cornea* @ 500 per 100 m^2 followed by *Trichogramma chilonis* at 1500 per 100 m^2.[132] The field release of *T. chilonis* at 100,000 eggs/ha and spraying of Bt @ 0.5 kg/ha or HaNPV at 250 larval equivalent ha^{-1} for the management of *Helicoverpa armigera* in *W. somnifera* field.[113] The Epilachna beetle was monitored by parasitic wasp (*Pediobius foveolatus*) in Ashwagandha..[142] Reduction of the aphid population was observed due to the activity of natural enemies from Western India. The peak activity of predatory Coccinellid beetles was observed during the fourth week of March.[143]

In seedling stage, the damping off is a major disease caused by *Fusarium* spp. under glasshouse condition. The mixture of *Azotobactor* and

Trochoderma sp. has increased the survival rate of seedlings from 78.5% to 97.4%.[144] The pomegranate butterfly (*Deudorix isocrates*) is a minor pest of gooseberry. Release of egg parasitoid, *T. chilonis* at 2,50,000 eggs/ha at an early stage of the insect incidence and repeating the release at 15 days interval reduce the pest population.[58] *Coleus* plants are affected by *Fusarium* wilt, and the combination of *T. viride* + *Pseudomonas fluorescens* is used for managing the disease complex.[145] Soil application of a mixture of bioagents *T. viride* + *P. fluorescens* @ 2.5 kg/ha significantly reduces the *Macrophomina phaseolina* incidence in *Coleus*.[146]

The seedlings of coleus were dipped in bacterial suspension of *P. fluorescens* to reduce the *Meloidogyne incognita* population and increasing the yield by 22.7–30.0%.[147,148] The mortality was 73% in the case of *Bacillus thturingiensis* var. kurstaki (Btk) (1 g/L) as compared with the control.[149] Three fungal pathogens, *Beauveria bassiana, Metarhizium anisopliae*, and *Nomuraea rileyi* are found effective against the gallery worm of jasmine. Among these bioagents *M. anisopliae* has shown higher mortality against the gallery worm. The application of *T. harzianum* a day before the planting @ 4 g/pot and cakes of *Mentha longifolia* for 15 days prior to planting at 100 g/pot, least incidence (21.8% and 25.0%) was observed at 60 and 120 days of planting (DAP), respectively, when compared with 83.0% and 92.0% in control.[150] The application of *T. viride, P. fluorescens*, and *Bacillus subtilis* reduced the wilt incidence up to 33.3%, 22.2%, and 22.3%, respectively compared with 94.4% in control. Seed treatment with mixture of *T. viride+P. fluorescens* at 4 g/kg and carbendazim + mancozeb @2.5 g/kg reduced the incidence of root rot in safed musli.[151] In geranium, the leaf blight caused by *Botrytis cinerea* is the most common disease on leaf, stem, and flowers parts. The disease can be managed by soil application of *T. harzianum* @5 kg/ha.[122]

19.5.5 CHEMICAL METHODS

19.5.5.1 MEDICINAL PLANTS

Gall midge infestation in Indian gooseberry could effectively be reduced with chlorpyriphos @ 0.05%.[58] For the management of stone borer and fruit midge, the application of carbaryl (0.2%) or quinalphos (0.05%) were recommended. Spraying with dimethoate 30 EC (0.06%) or spinosad (0.25 mL/L) was effectively used for reducing the population of mealybugs (*Nipaecoccus viridis, N. vestator*, and *Ferrisia virgata*).[58] In Indian gooseberry, the pomegranate butterfly, *Deudorix (Virachola) isocrates* is a minor pest and praying of carbaryl @ (0.2% or spinosad @ 0.25 mL/L reduced its populations. For the management of Bark eating caterpillar, swab of cotton wool soaked with 0.025% dichlorvos (76 EC) was put in shelter holes and holes were plugged.[58] The fungus *Colletotrichum gloeosporioides* causes anthracnose disease and spraying with 0.1% carbendazim (50 WP) before harvesting of fruits was recommended.[58] Blue mold, *Penicillium citrinum* is another fungal disease on Indian gooseberry. Before storing as a precautionary measure, fruits are treated with 8% sodium hypochloride or 2% borax. During the storage, the use of 0.1% carbendazim or 0.1% thiophanate methyl (70 WP) reduces the disease incidence.[58] Dimethoate 30 EC (0.03%) followed by Imidaclorpid 17.8 SL (0.005%) can significantly reduce the *Aphis gossypii* incidence in blond psyllium.[152] Use of two sprays of oxydemeton-methyle 25 EC @ 0.025% that can effectively manage the *A. gossypii* population and increase the crop yield by 40% in blond psyllium.[153]

Black cutworm (*Agrotis ipsilon*) is considered as a major pest of serpentine and may be controlled with drenching of chlorpyriphos @ 2 mL/L. It feeds on the collar region during the evening hours.[113] In the field, foliar spots caused by *A. alternata* were controlled by treating the seeds with mancozeb (75 WP) or propineb (75 WP) at (2 g/kg) (70 WP).[46] Spraying with mancozeb 0.25% or carbendazim 0.25% gave a yield of 11.7–13.6 g/plant and 8.9–10.1 g/plant, respectively against the blossom blight of serpentine.[154] In Safed musli, the foliar and root diseases caused by a complex of *R. bataticola, S. Rolfsii, Phoma* sp., and *C. dematium,* influence the quality of harvestable product, it can effectively be controlled by spraying of tebuconazol 250 EC @0.01%.[151] In *Aloe* plantations, the insect pests were managed by using acephate 75 SP (0.06%), Malathion 50 EC (0.05%) or pyrethrin (0.2%). The seed treatment or sprays with mancozeb M-45 (0.25%) or copper oxychloride @ 0.4% control the damping off.[37] The minimum infestation by the epilachna beetle was observed by spraying of Dimethoate 30 EC 1 mL/L in ashwagandha.[155] *T. urticae* on ashwagandha was effectively managed by dicofal.[156] To check the severe damage of epilachna beetle, *H. vigintioctopunctata* damage on Indian ginseng was sprayed with 2 mL either endosulfan (35 EC) or quinalphos (25 EC).[155]

Leaf blight, *Alternaria* sp. was managed by spraying with mancozeb (75 WP) @ 0.25% followed by propiconazole (25 EC) @ 0.2%, respectively.[29] A foliar application of mancozeb 0.25% was found effective against leaf blight and root yield (5 q/ha) and seed yield (75 kg/ha) in ashwagandgha.[29] Fungal leaf spots incited by *Colletotrichum dematium* were managed with 0.2% dithane M-45.[157] Leaf blight caused by *Alternaria* sp. was controlled with the application of Bordeaux mixture (1%) and fusarium wilt, *Fusarium oxysporum* was managed by the application of mercurial fungicides like agallol-3 @ 0.25%.[52]

19.5.5.2 AROMATIC PLANTS

The sugarcane white grub, *H. consanguinea* Blanch. (Coleoptera: Scarabaeidae) occasionally feeds on palmarosa plant roots. Flooding the infested field with irrigation water (chlorpyriphos 20 EC added to water) has been recommended.[52] Both leaf blight of palmarosa (*Ellisella caudata* Sacc.) (Ellisellidae) and leaf blotch due to *Curvularia trifolii* (Kauffman) Boed. (Pleosporaceae) were controlled by Bordeaux mixture (1%) as spray using 500 L volume ha^{-1} at 15 days interval starting from disease appearance[52].[52] Lemon grass and citronella grass are infected by *Sitobion miscanthi* Takahashi, *Sipha flava* (Forbes), *Trialeurodes* sp., whitefly and red spider mite, *Tetranychus telarius* L. for which sprays of dimethoate 30 EC at 1 mL/L reduce the incidence of pest population.[113] Regular diseases, such as gray mold, leaf rust, powdery mildew and downy mildew, stem canker, wilt of *Verticillium* are reported on lemon and citronella grass. Application of 0.25% mancozeb (80 WP) or zineb (80 WP) is recommended in lemon grass from South Africa and India.[113,158] The lethal yellowing disease of lemon grass (*Pythium aphanidermatum*) becomes severe in India during wet season. Spraying of metalaxyl 72 WP @ 2.5 kg/ha+mancozeb 80 WP @ 0.25% manages the disease well.[159] Spraying of mancozeb 80 WP or zineb 75 WP at 0.25% at 15–20 days interval reduced the leaf blight intensity and incidence in lemon grass in South Africa and India.[113,158] The geranium plants affected by anthracnose due to *C. gloeosporoides* were controlled by two sprays of chlorothalonil 75 WP at 0.2% at 15 days interval.[15] Leaf spots are caused by airborne fungus, *A. alternata* and this disease was controlled by spraying fungicide (mancozeb 80 WP at 0.25%, chlorothalonil 75 WP at 1%, triadimefon 25 WP at 0.01% or thiophanate methyl 70 WP at 0.2%). These are the effective control measures.[121,122] Mint plants have disease incidence with few soilborne diseases are very common, viz., stolon rot, *M. phaseolina.* Stolons dipping method with 0.25% solution of captan 50 WP for 2–3 min before planting is necessary for management. Further, the wilt caused by (*F. oxysporum, Verticillium* sp.) and spraying with carbendazim or benomyl at 0.2% is effective.[160] Vetiveria grass is frequently attacked by subterranean termites, *Microtermes* spp. and their effectively controlled by spreading with lindane dust 1% (5 g pit^{-1}) at preplanting in the soil.[52] Flooding of infested field with chlorpyriphos 20 EC mixed in irrigation water.[52] Vetiver leaf blight, *Xanthomonas* sp. causes a serious damage during the wet season of the crop. Drenching the soil with copper oxychloride (50 WP) @ 0.3% or Bordeaux mixture @ 1% followed by carbendazim or mancozeb has been found effective.[52]

19.6 BIOINTENSIVE MODULES

1. The use of biointensive module having vermicompost @ 2 t/ha + karanj cake 250 kg/ha + bio fertilizers 2 kg/ha + NPK (40:40:40 kg/ha); *Chrysoperla carnea* 50,000 eggs/ha (three times release at 15, 30 DAS, and 15 DAH); *T. chilonis* 6.25 cc/ha (two times release at 45 and 60 DAS); *Bacillus thuringiensis* 750 g/ha (foliar sprays on 50, 80 DAS and 50 DAH); neem oil 3% (foliar sprays at 20, 40, and 70 DAS); and application of NSKE 5% as foliar sprays at 30 and 70 DAH) had significantly reduced the incidence of aphids, defoliators, and pod borer in main and ratoon crop of Senna. It also had highest yield of 2028.0 kg dry leaves/ha and 834.8 kg pods/ha. The cost–benefit ratio was 1:1.55.[161]
2. Vermicompost 2 t/ha + karanj cake 250 Kg/ha + biofertilizers @ 2 kg/ha + NPK (20:60:50 kg/ha); *Chrysoperla carnea* 50,000 eggs/ha (three times releases at 15, 30, and 105 DAP); *T. chilonis* 6.25 cc/ha (two release on 45 and 60 DAP); *Bacillus thuringiensis* 750 g/ha (three sprays on 50, 80, and 140 DAP; and fish oil resin soap 25 g/L (five sprays on 35, 55, 95, and 115 DAP) were more effective in reducing the incidence of scale, thrips, and defoliators on *Coleus* plant ecosystem.[162]

19.7 PRESENT AND FUTURE CHALLENGES OF IPM IN MEDICINAL AND AROMATIC PLANTS

19.7.1 MONITORING OF PESTS AND DISEASES

In practical, In the medicinal and aromatic plants, the incidence, economic loss, and disease severity have not been reported.[163] In recent days, adverse climatic condition affects the growth and development of pest disease and fluctuating the incidence level and severity. Monitoring of pest and disease frequently is necessary to know the incidence and severity level of pest and disease. Apart from this, before going to management program, blindly we need to assess the economic threshold level or plant injury and disease incidence is compulsory of insect pest.[164] Further, monitoring and encouragement of beneficial organisms, such as predators, parasitoid, and beneficial microorganisms by reducing the usage of indiscriminate pesticides unnecessarily in crop ecosystem and growing pollen- and nectar-rich plants is necessary in nearby crop in order to increase the activity of parasitoid. This information is necessary to decide whether spraying is needed. By this way, there is scope for reducing the plant protection operation, number of spraying, and unnecessary cost burden to farmers.

19.7.2 PESTICIDE RESIDUE

Pesticides contaminated in raw or processed medical and aromatic plant products will create a risk of human health and reduce the quality of medicinal property as well as will reduce the export commodity in international market. So pesticide residues are an important issue. Organochlorine pesticides are contaminated with plant materials[165] and herbal products are common.[166,167] The pesticides, such as dicofol, carbaryl, ethion, chlorpyriphos, dimethoate, dichlorvos, Malathion triazophos, and quinalphos are commonly applied in medicinal plants ecosystem. These toxic chemicals stay in the soil till the final products, and subsequently come to final users. Ultimately, these pesticides cause the ill effects to human health. The residual effect of synthetic pesticides on these crop plants adversely affects the predator, parasitoid, and biocontrol agents of various pests and diseases.[16] Pesticide residual analyses in MAPs products are very much essential.[16] Toxicity of chemical pesticides may occur at any stage of consumption on medicinal products. Therefore, the government has to follow strict rules and regulations toward addressing the pesticides residue problems in medicinal and aromatic crop-based system. Moreover, Farmers need awareness about the organic and inorganic pesticides, dose, and dosage of their usage, banned pesticides, label and color indication of pesticides, and correct application methods for addressing the issue.

19.7.3 PHYTOTOXICITY

In general, the usage of chemical pesticides in MAPs ecosystem is minimum due to the use of synthetic pesticides which may alter the quality of biochemical constituents and residue problem during postharvest time. Apart, till time, there is no phytotoxic effects of chemical pesticides observed in medicinal and aromatic plants.[168,169] Besides, modern agriculture uses very much synthetic neonicotinoids formulations to MAPs ecosystem for the management of leaf sucking insect. These may be because of the phytotoxicity effect on plants pertaining to phytotoxic effect of medicinal plants.

19.8 CONCLUSION

MAPs are used for treating the different diseases in preventive and curative manner. The plants having great demand in national and international market due to consumption rate is increased day by day globally. The MAPs have more good quality MAPs to pharmaceutical industries which are catching attention besides their conservation and cultivation. Besides, MAPs have more threatened by biotic factors, such as leaf sap sucking insects, defoliators, mites, flower attacking insects, diseases, and root-knot nematodes. Management of these factors are still not getting much attention in MAPs ecosystem. Management of pest and disease by synthetic pesticides is not sustainable and is not environment-friendly. Various management strategies are required for the production of good quality raw materials, but before implementing any IPM practices assessment of insect population as well as knowing the disease severity is necessary. Once the level of severity is known we could find out suitable management practices to mange them. IPM method is the most effective and environment-friendly technology for minimizing the pest and diseases. The proper understanding of important insect pests and diseases makes the IPM in medicinal and aromatic crop-based ecosystem more effective.

KEYWORDS

- **medicinal plants**
- **pest population**
- **crop loses**
- **botanical**
- **emtomopathogens**
- **pesticides**
- **eco-friendly**

REFERENCES

1. Khare, C. P. *Indian Medicinal Plants: An Illustrated Dictionary*, 1st ed.; Springer-Verlag: Germany, 2008; 900 p.
2. Panda, H. *Aromatic Plants: Cultivation, Processing and Uses*; Asian Pacific Business Press Inc.: Delhi, India, 2015.
3. Kumar, S. P. Adulteration and Substitution in Endangered ASU Medicinal Plants of India: A Review. *Int. J. Med. Aromat. Plants* **2014,** *4* (1), 56–73.
4. Trivedi, P. C. *Medicinal Plants and Traditional Practices*; Indian Books and Periodicals: Delhi, India, 2015.
5. Mathivanan, N.; Sithanantham, S.; Marimuthu, T.; Peter, K. V.; Rethinam, P.; Brahma, S.; Peter, P. I.; Kirti, S. Therapeutic and Commercial Potential of Medicinal Plants with Special Focus on *Morinda Citrifolia* L. (Noni). Souvenir cum Abstracts, Second World Noni Congress, March, 2016, SRM University, Kattankulathur, Tamil Nadu, 2016.
6. DA. *Annual Report 2016–17. Department of Agriculture, Cooperation and Farmers Welfare*; Ministry of Agriculture, Government of India: New Delhi, India, 2017.
7. Yaniv, Z.; Bacharach, U., Eds. *Handbook of Medicinal Plants*; Food Products Press & Haworth Medical Press: Binghamton, New York, 2005.
8. Isman, M. B. Plant Essential Oils for Pest and Disease Management. *Crop Protection* **2000,** *19*, 603–608.
9. Bakkali, F.; Averbeek, S.; Averbeek, D.; Idaomar, M. Biological Effects of Essential Oils-a Review. *Food Chem. Technol.* **2008,** *46*, 446–475.
10. Nagpal, A.; Karki, M. A. Study on Marketing Opportunities for Medicinal, Aromatic and Dye Plants in South Asia. In *Medicinal and Aromatic Plants Program in Asia*; New Delhi, India, 2004.
11. Chatterjee, S. K. Cultivation of Medicinal and Aromatic Plants in India: A Commercial Approach. *Acta Hortic.* **2002,** *576*, 191–202.
12. Mittal, R.; Singh, S. P. Shifting from Agriculture to Agribusiness: The Case of Aromatic Plants. *Agric. Econ. Res. Rev.* **2007,** *20*, 541–550.
13. Verma, J. S. Insect Pest Problem in Medicinal Plants–A Review. *Agric. Rev.* **2006,** *27* (3), 130–136.
14. Shukla, R. S.; Chauvan, S. S.; Gupta, M. L.; Singh, V. P.; Naqvi, A. A.; Patra, N. K. Foliar Diseases of *Mentha Arvensis*, Their Impact on Yield and Major Constituents of Oil of *Mentha Arvensis. J. Med. Aromat. Plants Sci.* **2000,** *22*, 453–455.
15. Parameswaran, T. N.; Kalra, A.; Mehta, V. K.; Radhakrishnan, K. Chemical Control of Tip Burn and Blight of Scented Geranium (*Pelargonium Graveolens*) Caused by *Colletotrichum Gloeosporoides* Under South Indian Hill Condition. *J. Med. Aromat. Plants Sci.* **2000,** *22*, 666–668.
16. Gahukar, R. T. Potential and Utilization of Plant Products in Pest Control. In *Insect Pest Management: Current Concepts and Ecological Perspectives*; Abrol, D. P., Ed.; Elsevier Inc.: New York, 2014; pp 125–139.
17. Zuin, V. G.; Vilegas, J. H. Y. Pesticide Residues in Medicinal Plants and Phytomedicines. *Phytother. Res.* **2000,** *14*, 73–88.
18. Avasthi, S.; Gautum, A. K.; Bhadauria, R. First Report of *Penicillium Purpurogenum* Causing Collar and Root Rot Infection in Aloe Vera. *Plant Pathol. Quarantine* **2015b,** *5* (1), 20–24.

19. Banerjee, S.; Ghosh, S. K. The Study of Emerging Fungal Diseases of Some Important Medicinal Plants in West Bengal, a Threat. *NY Sci. J.* **2015,** *8* (3), 98–104.
20. Avasthi, S.; Gautum, A. K.; Bhadauria, R. First Report of Anthracnose Disease of Aloe Vera by *Colletotrichum Gloeosporioides*. *J. Res. Biol.* **2011,** *6*, 408–410.
21. Avasthi, S.; Gautum, A. K.; Bhadauria, R. Occurrence of Leaf Spot Disease on *Aloe Vera* (L.) Burm. F. Caused by *Curvularia* Species from Madhya Pradesh. *Biodiversitas* **2015a,** *16* (1), 79–83.
22. Pandey, M.; Trivedi, S.; Gurha, S. N.; Tripathi, U. K. Dry Rot and Leaf Spot of *Aloe Barbadensis* Caused by *Alternaria Alternata*. *Ann. Plant Protection Sci.* **2008,** *16*, 485–486.
23. Kawuri, R.; Suprapta, D. N.; Nitta, Y.; Homma, T. Destructive Leaf Rot Disease Caused by *Fusarium Oxysporum* on *Aloe Barbadensis* Miller in Bali. *Agric. Sci. Res. J.* **2012,** *2* (6), 295–301.
24. Soni, K. K.; Pyasi, A.; Tiwari, P.; Verma, R. K. Occurrence of *Aloe Vera* Rust (*Uromyces Aloës*): A New Record from Madhya Pradesh, India. *J. Mycol. Plant Pathol.* **2011,** *41* (4), 644–646.
25. Marimuthu, T.; Suganthy, M.; Nakkeeran, S. Common Pests and Diseases of Medicinal Plants and Strategies to Manage. In *The New Age Herbals*; Singh, B., Peter, K. V., Eds., 2018; pp 289–312.
26. Chandel, S.; Dubey, K.; Kaushal, P. Major Diseases of Medicinal and Aromatic Plants Recorded in Himachal Pradesh-India. *J. Pl. Dis. Sci.* **2014,** *9* (2), 145–153.
27. Lenne, J. M. Diseases of Cassia Species—A Review. *Trop. Grassl.* **1990,** *24*, 311–324.
28. Magar, D. B.; Barhate, B. G. Studies on Wilt of Senna and in Vitro Evaluation of Fungicides and Bioagents Against *Fusarium Oxysporum*. *J. Plant Dis. Sci.* **2013,** *8* (2), 182–186.
29. DMAPR. *Annual Report of Directorate of Medicinal & Aromatic Plants Research*; Anand: Gujarat, India, 2014–2019.
30. Gautam, A. K. The Genera *Colletotrichum*: An Incitant of Numerous New Plant Diseases in India. *J. New Biol. Rep.* **2014,** *3* (1), 9–21.
31. Muthukumar, A.; Venkatesh, A. A New Record of Leaf Blight of Ribben Plant Caused by *Alternaria Alternata* in India. *J. New Biol. Rep.* **2013,** *2* (3), 228–230.
32. Tekade, A.; Koche, M.; Raut, B. Influence of Weather Factor on Fungal Diseases of Musli. *Plant DisJ. Plant Dis. Protection.* **2009,** *4* (2), 173–175.
33. Sharma, A.; Patil, P. K. First Record of 28-Spotted Ladybird Beetle, *Henosepilachna Vigintioctopunctata* (F.) Infesting *Withania Somnifera* (L.) Dunal in Punjab Province of Northern India. *Pest Technol.* **2011a,** *5*, 91–92.
34. Kumar, H. R. Survey of Pests of Medicinal Plants with Special Reference to Biology and Management of Epilachna Beetle, *Henosepilachna Vigintioctopunctata* Fabricius (Coleoptera: Coccinellidae) on Ashwagandha. M.Sc. (Agri) Thesis, University of Agricultural Sciences, Dharwad, 2007; p 92.
35. Sharma, A.; Patil, P. K. First Record of *Withania somnifera* (L.) Dunal, as a New Host of Cowbug, *Oxyrachis Tarandus* Fab in Plains of Punjab, Northern India. *World Appl. Sci. J.* **2011b,** *14*, 1344–1346.
36. Sahu, C.; Ganguli, J. L.; Joshi, S. Seasonal Incidence of the Mealybug, *Phenacoccus Solenopsis* Tinsley (Hemiptera: Pseudococcidae) on Ashwagandha in Raipur (Chhattisgarh). *Int. J. Plant Protection* **2017,** *10* (2), 270–274.
37. Jetawat, R. P. S.; Mathur, K.; Singh, K.; Singh, A. Effect of Fungicides Against *Fusarium Solani* and *Rhizoctonia Solani* Infecting Ashwagandha. *Ecoscan* 2015 *7*, 75–78.

38. Sarkar, S.; Dasgupta, B. First Report of Leaf Spot of Ashwagandha (*Withania somnifera* Dunal) Caused by *Colletotrichum gloeosporioides* from West Bengal, India. *J. Mycopathol Res.* **2017,** *55* (3), 257–259.
39. Shivanna, M. B.; Parashurama, T. R.; Somashekhara Achar, K. G.; Vasanthakumari, M. M. Fungal Foliar Diseases in *Withania Somnifera* and its Effect on Secondary Metabolites. *Plant Biosyst.* **2014b,** *148* (5), 907–916.
40. Kapadiya, T. B.; Shivkumara, K. T.; Meena, R. P.; Manivel, P.; Roy, S. Occurrence of Two Spotted Spider Mite, *Tetranychus Urticae* Koch (Arachnida: Acari: Tetranychidae) on Ashwagandha in Gujarat. *J. Entomol. Zool. Stud.* **2019,** *7* (3), 77–79.
41. Rani, J. B.; Sridhar, Y. Screening of Polyhouse Grown Rose Varieties for Resistance to Thrips, *Scirtothrips Dorsalis* Hood. *J. Ecofriendly Agric.* **2005,** *10* (2), 145–149.
42. Mondal, G. Occurrence of *Synchytrium* Inducing Gall on Punarnava (*Boerhavia Diffusa*) from West Bengal, India. Presented in 6th International Conference on "Plant, Pathogens and People" with the Mission "Challenges in Plant Patholog to Benefit Humankind" Held at NASC Complex, New Delhi During February 23–27, 2016, Organized by IPS, Div. Pl. Pathol.; IARI, New Delhi 110012, India. Abstract, 2016; p 236.
43. Khan, A.; Saeed, S. T.; Samad, A. New Record of *Catharanthus Yellow Mosaic Virus* and a Betasatellite Associated with Lethal Leaf Yellowing of Kalmegh (*Andrographis paniculata*) in Northern India. *Plant Dis.* **2015,** *99* (2), 292.
44. Saeed, S. T.; Khan, A. and Samad, A. First Report on the Molecular Identification of Phytoplasma (16SrII-D) Associated with Witches' Broom of Kalmegh (*Andrographis paniculata*) in India. *Plant Dis.* **2015,** *99* (1), 155.
45. Rehaman, S. K.; Pradeep, S.; Dhanapal, R.; Chandrashekara, G. V. Survey Studies on Insect-Pests Associated with Important Medicinal Plants in Shivamogga. *Karnataka J. Entomol. Zool. Stud.* **2018,** *6* (1), 848–857.
46. Parashurama, T. R.; Shivanna, M. B. Fungal Foliar Disease of *Rauvolfia serpentina* in Wild, Its Seasonal Occurrence, Seed Transmission and Disease Management. *Arch. Phytopathol. Plant Protection.* **2013,** *46*, 1609–1621.
47. Chandra, V. Leaf Blotch Disease of *Rauvolfia serpentina. Sci. Cult.* **1957,** *23*, 99.
48. Varadarajan, P. D. Anthracnose disease of *Rauvolfia serpentina* caused by *Colletotrichum gloeosporioides*. *Curr. Sci.* **1964,** *33*, 564–565.
49. Lele, C. V.; Ashram. Die-Back Disease of *Rauvolfia serpentina* Caused by *Colletotrichum dematium*. *Indian Phytopathol.* **1968,** *20*, 349–350.
50. Shukla, R. S.; Alam, M.; Sattar, A.; Khaliq, A.; Singh, H. N. First Report of *Rhizopus stolonifer* Causing Inflorescence and Fruit Rot of *Rauvolfia serpentina* in India. Bull. OEPP/EPPO Bull. **2006,** *36*, 11–13.
51. Raj, S. K.; Kumar, S.; Pratap, D.; Vishnoi, R. and Snehi, S. K. Natural Occurrence of Cucumber Mosaic Virus on *Rauvolfia serpentina*, A New Record. *Plant Dis.* **2007,** *91* (3), 322.
52. Smitha, G. R.; Varghese, T. S.; Manivel, P. *Cultivation of Aromatic Plants*; Directorate of Medicinal and Aromatic Plants Research: Anand, Gujarat, India, 2014.
53. Shivakumara, K. T.; Venkatesan, T.; Keerthi, M. C.; Shashank, P. R.; Pradeeksha, N.; Polaiah, A. C.; Reddy, R. N.; Saran, P. L.; Manivel, P. Occurrence of *Pyrausta panopealis* on Sweet Basil *Ocimum basilicum* in India. *J. Environ. Biol.*. **2021,** *42*, 265–270.
54. Sathe, T. V.; Sathe, N. T.; Ghodake, D.; Sathe, A. Sucking Insect Pests and Medicinal Value of Tulsi *Ocimum sanctum* L. (Lamiaceae). *Indian J. Appl. Res.* **2014,** *4* (3), 31–33.
55. Hansford, C. G. Annual Report of the Mycologist. *Rev. Appl. Mycol.* **1933,** *12*, 421–422.

56. Nilamudeen, M.; Girijal, V. K.; Nandakumar, C. *Colletotrichum* Leaf Spot of Holy Basil (*Ocimum tenuiflorum L.)*—A New Report from Kerala. *Progr. Hortic*. **2013,** *45* (2), 383–384.
57. Morrison, W. R.; Linderman, S.; Hausbeck, M. K.; Werling, B. P.; Szendrei, Z. Disease and Insect Pests of Asparagus: *Extension Bulletin* E3219, Michigan State University, 2014. http://msue.anr.msu.edu/uploads/resources/pdfs/Disease and Insect Pests of *Asparagus* (E3219) pdf.
58. Prakash, O. IPM Schedule for Aonla Pests. Extention Bulletin. National Horticulture Mission Ministry of Agriculture and Cooperation, Government of India, Delhi, India, 2012.
59. Sridhar, Y. Field Evaluation of Different Insecticides Against Aonla Aphid, *Schoutedenia emblica. Indian J. Plant Protection* ***2011*****,** *39*, 35–37.
60. Garg, V. K.; Patel, Y. A new aphid, *Scoutedenia lutea* van der Goot infesting aonla, *Emblica officinalis* Gaertn. *Indian Forester* **2010,** *136*, 557–558.
61. Suganthy, M.; Sakthivel, P. Field Efficacy of Biopesticides Against *Plusia signata* (Fabricius) on *Gloriosa superba. Madras Agric. J.* **2012,** *99* (4–6), 368–370.
62. Meena, B.; Rajamani, K. Biological Management of Root-Rot Disease in *Gloriosa superb. Int. J. Noni Res.* **2016,** *11* (1 & 2), 82–85.
63. Maiti, C. K.; Sen, S.; Paul, A. K.; Acharya, K. First Report of Leaf Blight Disease of *Gloriosa superba* L.; Caused by *Alternaria alternata* (Fr.) Keissler in India. *J. Gen. Plant Pathol.* **2007,** *73* (5), 377–378.
64. Suganthy, M.; Rajamani, K.; Nalina, L.; Meena, B. *Insect Pests of Medicinal and Aromatic Crops*; Tamil Nadu Agricultural University Press, TNAU: Coimbatore, 2011; p 52.
65. Sudha, A.; Lakshmanan, P. *Solanum nigrum,* a New Host for Powdery Mildew Disease of *Capsicum annuum* in the Madurai District of Tamil Nadu, India. *Aust. Plant Dis. Notes* **2007,** *2*, 97. https://doi.org/10.1071/DN07040.
66. Suganthy, M.; Vijayakumar, R. M. *Insect Pests of Medicinal and Aromatic Crops and Their Management*; A. E. Publications: Coimbatore, 2013; p 95. (ISBN: 93-81972-20-6).
67. Ramprasad, S. Studies on Collar Rot Complex of *Coleus forskohlii* (Wild.) Briq. M.Sc. Thesis, UAS Dharwad, India. 2005.
68. Shukla, R. S.; Kumar, S.; Singh, H. N.; Singh, K. P. First Report of Aerial Blight of *Coleus forskohlii* Caused by *Rhizoctonia solani* in India. *Plant Dis* **1993,** *77*, 429.
69. Chandrashekara, K. N.; Prasannakumar, M. K. New Host Plants for *Ralstonia solanacearum* from India. *New Dis. Rep.* **2010,** *22*, 6.
70. Kamalakannan, A.; Mohan, L.; Valluvaparidasan, V.; Mareeswari, P.; Karuppiah, R. First Report of *Macrophomina* Root Rot (*Macrophomina phaseolina*) on Medicinal Coleus (*Coleus forskohlii*) in India. *Plant Pathol.* **2006,** *55*, 302.
71. Singh, R.; Gangwar, S. P.; Singh, D.; Singh, R.; Pandey, R. and Kalra, A. Medicinal Plant *Coleus forskohlii* Briq.: Disease and Management. *Med. Plants*. **2011,** *3* (1), 1–7.
72. Ito, Y.; Takeuchi, T.; Matsushita, Y.; Chikuo, Y.; Satou, M. Downy Mildew of Coleus Caused by *Peronospora belbahrii* in Japan. *J. Gen. Plant Pathol* **2015,** *81*, 328–330.
73. Falloon, P. G.; Tate, K. G. Major Diseases of Asparagus in New Zealand. *Proc. Agron. Soc. N.Z.* **1986,** *16*, 17–28p.
74. Achar, K. G. S.; Parashurama, T. R.; Shivanna, M. B. A New Record of Leaf Spot Caused by *Xanthomonas campestris* in *Tinospora cordifolia* in India. *Int. J. Curr. Microbiol. Appl. Sci.* **2014,** *3* (1), 269–273.
75. Hanumanthaswamy, B. C.; Rajgopal, D.; Farooqui, A. A.; Chakravarthy, A. K. Insect Pests of Costus *speciosus* Linn.; A Medicinal Plant. *My For.* **1993,** *29* (2), 158–160.

76. Shivanna, M. B.; Achar, K. G. S.; Vasanthakumari, M. M.; Mahishi, P. Phoma Leaf Spot Disease of *Tinospora cordifolia* and Its Effect on Secondary Metabolite Production. *J. Phytopathol.* **2014a,** *162*, 302–312. https://doi.org/10.1111/jph.12191.
77. Lee, I. M.; Davis, R. E.; Gundersen-Rindal, D. E. Phytoplasma: Phytopathogenic Mollicutes. *Annu. Rev. Microbiol.* **2000,** *54*, 221–255p.
78. Malarvannan, S. Surveillance of Insect Pests of *Morinda citrifolia* L. and *Morinda pubescencs* J.E. Sm. In *West Coast of Kerala and Karnataka*. WNRF Technical Bulletin-02, World Noni Research Foundation and M. S. Swaminathan Research Foundation, Chennai, 2010; 28p.
79. Sithanantham, S.; Suganthy, M.; Janarthanan, S.; Mathivanan, N.; Marimuthu, T.; Venkatachalam, A.; Judy, S.; Rethinam, P. Organic Pest Control in Cultivated Medicinal Plants: Current Indian Scenario and Future R & D Needs with Case Study of Noni. *Int. J. Noni Res.* **2016,** *11* (1 & 2), 60–73.
80. Nakkeeran, S.; Marimuthu, T.; Raguchander, T. Exploring DAPG and Phenazine Producing PGPR Strains and Fungal Antagonists for the Management of Diseases of Noni (*Morinda citrifolia* L.), WNRF Technical Bulletin-11, World Noni Research Foundation, 2013; p 329.
81. Kavitha, P. G.; Jonathan, E. I.; Sankari Meena, K. Host-Parasite Relationship and Pathogenicity of Root Knot Nematode, *Meloidogyne incognita* in Noni. *Madras Agric. J.* **2012,** *99* (10–12), 862–866.
82. Shivkumara, K. T.; Joshi, S.; Venkatesan, T.; Pradeeksha, N.; Polaiah, A. C.; Manivel, P. A Report on the Occurrence of Invasive Papaya Mealybug, *Paracoccus marginatus* Williams & Granara de Willink (Hemiptera: Pseudococcidae) on a Medicinal Herb, *Gymnema sylvestre* (R.Br) in Gujarat. *India Pest Manage. Hortic. Ecosyst.* **2019,** *24* (2), 279–280.
83. Joy, P. P.; Mathew, S.; Mathew, G.; Joseph, A.; Sreevidya, P. P. *Lemongrass*. Aromatic and Medicinal Plants Research Station Odakkali, Asamannoor, Kerala, India, 2006; p 32.
84. Sato, T.; Ohkubo, H. Newly Discovered Leaf Blight of Citronella Grass Caused by *Curvularia andropogonis* in the Philippines. *Japan Agric. Res. Quart.* **1990,** *23* (3), 170–175p.
85. Alam, M.; Sattar, A.; Janardhanan, K. K.; Husain, A. Lethal Yellowing of Java Citronella (*Cymbopogon winterianus*) Caused by *Pythium aphanidermatum*. *Plant Dis.* **1992,** *76*, 1074–1076.
86. Gardner, D. E. Lemongrass Rust Caused by *Puccinia nakanishikii* in Hawaii. *Plant Dis.* **1985,** *69*, 1100.
87. Gingade, S.; Manivel, P.; Varghese, T. S. Cultivation of Palmarosa, Technical Report, ICAR-DMAPR, Anand, Gujarat, India, 2015.
88. Roopini, G. A. Studies on Seasonal Incidence of Pests on *Jasminum* spp. Emphasizing on Bud Borer Complex and Their Management Strategies on *J. multiflorum* M. Sc. Thesis, University Horticultural Sciences, Bagalkot, 2016.
89. Shobitha, C. A. Studies on the Bud Borer Complex of Jasmine. M. Sc. Thesis, University Agricultural Sciences, Bengaluru, 2002.
90. Neelima, Y. Bio-Ecology and Management of Jasmine Pests, M. Sc. (Agri.) Thesis, Acharya N. G. Ranga Agricultural University Rajendra Nagar, Hyderabad, 2005.
91. Rajakumar, E.; Hugar, P. S.; Kattimani, K. N. Seasonal Incidence of Red Spider Mite, *Tetranychus urticae* Koch. (Acari: Tetranychidae) on Jasmine. *Karnataka J. Agric. Sci.* **2005,** *18*, 150–153.

92. Kiran, C. M.; Hegde, J. N.; Chakravarthy, A. K.; Thippesha, D.; Kalleshwarswamy, C. M. Seasonal Incidence of Major Insect and Mite Pests of Jasmine. *Int. J. Curr. Microbiol. Appl. Sci.*. **2017,** *6* (10), 5060–5070.
93. Kareem, A. A. *Bombay Nat. Hist. Soc.* **1960,** *57*, 681–683.
94. Mathur, A. C.; Srivastava, J. B. *RRL Bull.* **1962,** *1*, 49–50.
95. Berry, R. E. *Ann. Entomol. Soc. Am.* **1974,** *67*, 580–582.
96. Sagar, P. *Indian Perfumer* **1988,** *32*, 186–89.
97. Hameed, S. F. *Indian J. Entomol.* **1975,** *37*, 203–205.
98. Sagar, P. *Indian Perfumer* **1989,** *33*, 23–24.
99. Sagar, P. *Indian Perfumer* **1983,** *27*, 22–23.
100. Bhutani, D. K. *Pesticides* **1984,** *8*, 24–28.
101. El-Nabi, L. M. A.; Alexandria J. *Agril. Res.* **1996,** *41*, 61–75.
102. Kalra, A.; Singh, H. B.; Pandey, R.; Samad, A.; Patra, N. K. and Kumar, S. Diseases in Mint: Causal Organisms, Distribution, and Control Measures. *J. Herbs Spices Med. Plants* **2008,** *11*, 71–91.
103. Remadevi, O. K.; Nagaveni, H. C.; Muthukrishnan, R. Pests and Diseases of Sandalwood Plants in Nurseries and Their Management. Working Papers of the Finnish Forest Research Institute, 2005; pp 69–76.
104. Sagar, P.; Singh, D. P. *Indian Perfumer* **1981,** *25*, 105–109.
105. Sagar, P. *Indian J. Ent.* **1985a,** *47*, 183–185.
106. Mathur, A. C. Anand, M. In *Pests of Medicinal Plants: Entomology in India, Entomological Soc. if India*, New Delhi, 1964; pp 271–277.
107. Sagar, P. Symp. Ecology and Resource Management, Oct. 2–4, Muzzaffar Nagar, India, Abst. 1982; pp 33–34.
108. Chaudhary, V. Arthropods Associated with Ashwagandha (*Withania somnifera*) in Semi-Arid Region of Gujarat. *Karnataka J. Agric. Res.* **2013,** *26* (3), 433–435.
109. Sharma, P. C.; Kumar, A.; Mehta, P. K.; Singh, R. Survey Studies on Insect-Pests Associated with Important Medicinal Plants in Himachal Pradesh. *Indian J. Sci. Res. Technol.* **2014,** *2* (4), 2–7.
110. Meshram, P. B. New Reports of Defoliator, *Psilogramma menephron* on *Rauwolfia serpentine* and Whitegrub, *Holotrichia serrata* on *Withania somnifera. Indian Forester* **2005,** *13* (7), 969–970.
111. Nirmal, A.; Jayaram, C. S.; Ganguli, J. L.; Tirkey, A. Scenario of Insect Pests on Ashwagandha (*Withania somnifera*) in the Plains of Chhattisgarh. *Insect Environ.* **2015,** *20* (4), 142–143.
112. Kumari, S.; Kumar, N.; Kumar, A. Seasonal Incidence and Damage Intensity of Lace Bug, *Cochlochila bullita* (Stal) (Hemiptera: Tingidae) on Tulsi, *Ocimum basilicum* L. *Int. J. Sci. Environ. Technol.* **2016,** *5* (6), 4312–4319.
113. TNAU Agritech Portal. Tamil Nadu Agricultural University, Coimbatore, Tamil Nadu, India, 2013–2016.
114. Ramakrishanan, S.; Senthilkumar, T. Non-Chemical Management of Root-Knot Nematode, *Meloidogyne incognita* in Ashwagandha (*Withania somnifera* Dunal) and Senna (*Cassia angustifolia* Vahl.) *Indian Journal of Nematology* **2009,** *39* (2), 170–174.
115. Pandey, R.; Kalra, A. P. Root-Knot Disease of Ashwagandha, *Withania somnifera* and Its Eco-Friendly Cost Effective Management. *J. Mycol. Plant Pathol.* **2003,** *33*, 240–245.
116. Meshram, P. B.; Soni, P. K. Screening of Certain Varieties of *Emblica officinalis* Gaertn. for Resistance to Insect Pests and Diseases in Central India. *J. Hortic. Forest.* **2011,** *3* (6), 190–194.

117. Patel, H. M.; Borad, P. K. Natural Enemies of *Aphis gossypii* in Isabgol. *J. Med. Aromat. Plant Sci.* **2006a,** *27* (2), 287–290.
118. DAFF. Directorate of Communication Services. Department of Agriculture, Forest and Fisheries, Pretoria, South Africa, 2015.
119. David, P. M. M.; Hanifa, A. M.; Natarajan, S. Biology and Control of Blossom Midge, *Contarinia* sp. (Diptera: Cecidomyiidae) on *Jasminum sambac* in Tamil Nadu. *Entomon.* **1990,** *15*, 193–196.
120. Suganthy, M.; Kamala, I. M. Jasmine Pest Management. *Kisan World.* **2016,** *43* (8), 33–36.
121. Moorman, G. W. *Geranium Diseases*. Penn State College of Agriculture Science, University Park: Pennsylvania, 2017.
122. Douglas, S. M. *Geranium Diseases*. The Connecticut Agriculture Experiment Station: New Haven, 2003.
123. ASFAC. In Assam Small Farmers Agribusiness Consortium. Guwahati, Assam, India, 2015.
124. Subarjah, C.; Himawan, T.; Puspitarini, R. D. Effects of Compost on Nematode, Pratylenchus sp. (Tylenchida: Pratylenchidae) Population in Patchouli. *J. Trop. Life Sci.* **2016,** *6* (2), 101–106.
125. CCARI. Jasmine, Central Coastal Agriculture Research Institute. ICAR, Panaji, Goa, India, 2016.
126. Muniappan, R. Scale Insects on Vetiver in Guam. *Vetiver Newslett.* **2001,** *23*, 16.
127. Biswas, B. C. Cultivation of Medicinal Plants: Success Stories of Two Farmers Fertilizer Marketing News. **2010,** *41* (3), 1–4.
128. Rathikannu, S. Bio-Ecology, Management of Pests of *Gloriosa superba* (Linn) and *Phyllanthus amarus* (Schum and Thonn) and Influence of Insect Damage on the Medicinal Properties. M.Sc. (Ag.) Thesis, Tamil Nadu Agricultural University, Coimbatore, 2005; p 68.
129. Rathikannu, S.; Sivasubramanian, P.; Suganthy, M. Field Efficacy of Botanicals on Lepidopteran Pests of Glory Lily, *Gloriosa superba* (L.) In Insect Pest Management a Current Scenario; Dunston, P. A., Ed.; Entomology Research Unit, Tamil Nadu, 2011; pp 505–508.
130. Suganthy, M.; Sakthivel, P. Field Evaluation of Botanicals on Pest Complex of *Solanum nigrum* Linn. *Madras Agric. J.* **2013,** *100* (4–6), 592–596.
131. Suganthy, M.; Rajamani, K.; Nalina, L.; Meena, B. Development of IPM Module Against Major Pests of Black Nightshade, *Solanum nigrum*. In 3rd International Symposium on Underutilized Plant Species Held at Agricultural College & Research Institute, Madurai, Tamil Nadu, 2015, 5–8 August; p 385.
132. Meshram, P. B.; Mawai, S.; Malviya, R. Biological Control of Insect Pests of Medicinal Plants: *Abelmoschus moschatus*, *Floriosa superba* and *Withania somnifera* in Forest Nursery and Plantation in Madhya Pradesh, India. *Am. J. Agric. Forest.* **2015,** *3* (2), 47–51.
133. Upadhyay, S.; Mishra, R. C. Efficacy and Economics of Insecticides and Neem (Azadirachta Indica) Based Products on Incidence of Aphid (*Aphis gossypii*) on Isabgol (*Plantago ovata*). *Indian J. Agric. Sci.* **1999,** *69*, 161–162.
134. CIMAP. Organic Cultivation of and *Bacopa monnieri* and *Ocimum sanctum*. Central Institute of Medicinal and Aromatic Plants, Resource Centre, Bangalore India, 2007.
135. Sharma, J. R.; Amrate, P. K.; Gill, B. S. Leaf Soft Disease of *Aloe barbadensis* and Its Management. *J. Res. Punjab Agric. Univ.* **2010a,** *47* (1–2), 22–24.
136. Sridhar, Y. Field Evaluation of Different Insecticides Against Aonla Aphid, *Schoutedenia emblica. Indian J. Plant Protection* **2011,** *39*, 35–37.

137. Kulkarni, M. S.; Ramaprasad, S.; Hedgey, Y.; Laxminarayan, H.; Hedge, N. K. Management of Collar Rot Complex Disease of *Coleus forskohlii* (Willd.) Briq. Using Bioagents, Organic Amendments and Chemicals. *Biomed.* **2007,** *2*, 37–40.
138. Ravikumar, A.; Rajedran, R.; Chinniah, C.; Irulandi, S.; Pandi, R. J. Evaluation of Certain Organic Nutrient Sources Against Mealy Bug, *Coccidohystrix insolitus* (Green) and the Spotted Leaf Beetle, *Epilachna vigintioctopunctata* Fab. on Ashwagandha, *Withania somnifera. J. Biopest.* **2008b,** *1*, 28–31.
139. Nelson, S. J.; Venugopal, M. S.; Janarthanan, S, Natarajan, S. Efficacy of Certain Plant Products Against Jasmine Budworm, *Hendecasis duplifascialis* Hmp. (Pyraustidae: Lepidoptera). *Indian Perfumer* **1993,** *37*, 26–239.
140. Rani, B. J.; Mohan, N. J. Pest Management in Ornamental, Medicinal and Aromatic Crops. Annual Report, IIHR, Bangalore, 1994; p 57.
141. Umapathy, G.; Rajendra, B. Management of the Eriophyid Mite, *Aceria jasmine* Chan. (Eriophyidae: Acari) on Jasmine. *Madras Agri. J.* **1999,** *86*, 83–86.
142. Venkatesha, M. G. Seasonal Occurrence of (*Hemosepilachna vigintioctopunctata* (F.) (Coleoptera: Coccinellidae) and Its Parasitoid on Ashwagandha in India. *J. Asia Pacific Entomol.* **2006,** *9* (3), 1–4.
143. Patel, H. M.; Borad, P. K. Natural Enemies of *Aphis gossypii* in Isabgol. *J. Med. Aromat. Plant Sci.* **2006b,** *27* (2), 287–290.
144. Bhat, S. A.; Lone, S. A.; Ahmad, S. S. Biocontrol of Damping Off in *Withania somnifera* (L.) Dunal. *Int. Res. J. Microbiol.* **2014,** *5* (4), 73–79.
145. Paramasivan, N.; Mohan, S.; Muthukrishnan, V. Management of Coleus Dry Root Rot Pathogen, *Macrophomina phaseolina* by Fungal and Bacterial Antagonists. *Indian J. Plant Protection* **2007,** *35*, 133–135.
146. Senthamarai, K.; Poornima, K.; Subramanian, S.; Sudheer, J. Nematode-Fungal Disease Complex Involving *Meloidogyne incognita* and *Macrophomina phaseolina* on Medicinal Coleus, *Coleus forskohlii* Briq. *Indian J. Nematol.* **2008,** *38*, 30–33.
147. Seenivasan, N.; Devrajan, K. Integrated Approach for the Management of Root-Knot Nematode, *Meloidogyne incognita* in Medicinal Coleus. *Indian J. Nematol.* **2008,** *38*, 154–158.
148. Pillai, K. G.; Nethravathi, C. J.; Visalakshy, P. N. G.; Swathi, C. Bioefficacy of Fungal Pathogens Against Jasmine Gallery Worm, *Elasmopalpus jasminophagus* (Lepidoptera: Phycitidae). *Pest Manage. Hortic. Ecosyst.* **2016,** *22* (1), 67–70.
149. Gopalakrishna, K.; Pillai, C. J.; Nethravathi, P. N.; Visalakshy, G.; Swathi, C. Bio-Efficacy of Fungal Pathogens Against Jasmine Gallery Worm, *Elasmopalpus jasminophagus* (Lepidoptera: Phycitidae). *Pest Manage. in Hortic. Ecosyst.* **2016,** *22* (1), 67–70.
150. Ramamoorthy, V.; Alice, D.; Meena, B.; Muthusamy, M.; Seetharaman, K. Biological Management of *Sclerotium* Wilt of Jasmine. *Indian J. Plant Protection* **2000,** *28* (1), 102–104.
151. Ingle, R. W.; Shende, S.; Deshmukh, V.; Joshi, M. S. Efficacy of Chemicals and Bioagents Against Foliar and Root Rot Disease of Safed Musli (*Chlorophytum borivilianum*) in Vitro. *Agric. Sustain. Dev.* **2013,** *1* (1), 35–38.
152. Khinchi, S. K.; Kumawat, K. C. Effect of Dates of Sowing of Blond Psyllium, *Plantago ovata* on the Incidence of Aphid, *Rhopalosiphum maidis. J. Insect Sci.* **2014a,** *37* (2), 246–249.
153. Jat, R. S.; Reddy, R. N.; Bansal, R.; Manivel, P. (Eds.). Good Agricultural Practices: Ashwagandha, Isabgol. Extension Bulletin, Directorate of Medicinal and Aromatic Plants Research, Anand, Gujarat, India, 2015.

154. Shukla, R. S.; Khaliq, A.; Alam, M. Chemical Control of Blossom Blight Disease of Sarpagandha Caused by *Colletotrichum capsici*. *Afr. J. Biotechnol.* **2010,** *9* (38), 6397–6400.
155. Chandranath, H. T.; Katti, P. Management of Epilachna Beetle on Ashwagandha. *Karnataka J. Agric. Sci.* **2010,** *23* (1), 171.
156. Sharma, M.; Sharma, A.; Ram, N. Bioefficacy of Non-Conventional Pesticides Against *Tetranychus urticae* Koch Infesting Ashwagandha (*Withania somnifera* Duval). *Pest Manage. Hortic. Ecosyst.* **2013,** *19* (2), 254–255.
157. Mehrotra, A. Diseases of Some Medicinal Plants and Their Management. *J. Med. Aromat. Plants Sci.* **2000,** *22* (1), 71.
158. Linse, T.; Drost, D. *Lemon Grass in the Garden*; Utah State University, Logan, UT, 2009.
159. Alam, M.; Sata, A. Epidemiology and Management of Lethal Yellowing Disease of *Java citronella*. *J. Med. Aromat. Plants Sci.* **2000,** *22* (B), 499–503.
160. Szezeponek, A.; Mazur, S. Occurrence of Fungal Diseases on Lemon Balm (*Mentha officinalis* L.) and Peppermint (*Mentha piperita* L.) in the Region of Malopolska. *Commun. Agric. Appl. Biol. Sci.* **2006,** *71* (3), 1109–1118.
161. Senthilkumaran, S. Studies on Development of Organic Package for the Management of Major Insect Pests of Senna (*Cassia angustifolia* Vahl.). M.Sc. (Ag) Thesis, The Tamil Nadu Agricultural University, Coimbatore, 2008; p 144.
162. Thangavel, K. Studies on the Development of Bio-Intensive Module for the Management of Major Insect Pests of Coleus (*Coleus forskohlii* Briq.). M.Sc. (Ag) Thesis, Tamil Nadu Agricultural University, Coimbatore, 2010; p 136.
163. Ramana, D.; Kumar, P.; Basavanna Goud, K. Pest Complex of Medicinal Plants. *Karnataka J. Agric. Sci.* **2010,** *23*, 197–199.
164. Bout, A.; Boll, R.; Mailleret, L.; Poncet, C. Realistic Global Scouting for Pests and Diseases of Cut Rose Crops. *J. Econ. Entomol.* **2010,** *103* (6), 2242–2248.
165. Srivastava, L. P.; Kumar, N.; Gupta, K. P.; Raizada, R. B. Status of HCH Residues in Indian Medicinal Plant Materials. *Bull. Environ. Contamin. Toxicol.* **2006,** *76* (5), 782–790.
166. Srivastava, L. P.; Gupta, K. P.; Raizada, R. B. Organochlorine Pesticide Residues in Herbal Ayurvedic Preparations. *Bull. Environ. Contamin. Toxicol.* **2000,** *64* (4), 502–507.
167. Rai, V.; Katkar, P.; Misra, C.; Ojha, S. K.; Srivastava, N.; Mehrotra, S. Metals and Organochlorine Pesticide Residues in Some Herbal Ayurvedic Formulations. *Bull. Environ. Contamin. Toxicol.* **2007,** *79* (3), 269–272.
168. Kumar, A.; Simon, S.; Yogi, K. Studies on the Management of Important Pests of Roses with Botanical Pesticides. *Pestology* **2013,** *37* (4), 23–28.
169. Kamala, M.; Chinniah, C.; Kennedy, J. S.; Kelyanasundaram, M.; Suganthy, M. Pesticidal Effect of Indigenous Plant Extracts Against Jasmine Bud Worm, *Hendecasis duplifascialis* Hampson in Jasmin (*Jasminum sambac* L.). *Int. J. Trop. Agric.* *35* (2), 315–323.
170. Moreira, Z. P. M.; dos Santos, P. O.; de Oliveira, T. A. S.; de Souza, J. T. Occurrence of Basil Leaf Spot Caused by *Pseudomonas cichorii* in Bahia State, Brazil. *Summa Phytopathologica* **2015,** *41* (1), 73p.

Index

I

L

M

P

R

S